AF520459

Scientific Lentil Production

Scientific Lentil Production

Anil Kumar Singh
N. Bhakta
U.R. Sangle
Manibhushan
P.K. Sundaram
Sanjeev Kumar
J.K. Yasin

SATISH SERIAL PUBLISHING HOUSE
403, Express Tower, Commercial Complex
Azadpur, Delhi-110033 (India)
Phone : 011-27672852, Fax : 91-11-27672046
e-mail : info@satishserial.com, hkjain1975@yahoo.com
Website : www.satishserial.com

Published by :

SATISH SERIAL PUBLISHING HOUSE
403, Express Tower, Commercial Complex, Azadpur, Delhi-110033 (INDIA)
Phone : 011-27672852 Fax : 91-11-27672046
E-mail : info@satishserial.com, hkjain1975@yahoo.com

ISBN 978-93-86200-51-8

Composed, Designed & Printed in India

About the Editors

Dr Anil Kumar Singh is a Principal Scientist (Agronomy) at ICAR-Research Complex for Eastern Region, Patna (Bihar). With more than 17 years of professional experience in the field of Natural Resources Management (NRM) and Plant Genetic Resources Management (PGRM), served as Head-in-charge of Crop Research Programme, Pusa Bihar and Officer-in-charge NBPGR regional Station Srinagar (Jammu & Kashmir). Have worked extensively on resource conservation technologies, nutrient management in cropping system, cropping system, developed refined agro-techniques, integrated farming system, including rainfed rice based crop production system. Developed five faba bean varieties and successfully completed 30 research projects of which five are externally funded projects in collaboration with international organizations like the World Bank, USAID, and various GC Centres like IRRI and CIMMYT. Having more than 300 publications including 130 research articles in international / national peer reviewed journals and 6 books on various aspects of agricultural economy including one Encyclopedic Dictionary of Agriculture. Conducted several training as a course director designed for state agriculture/development officers. Contribution of Dr Singh has been recognized by the professional societies by appointing him as editorial board member, of Journal of Agricultural Sciences (ISSN: 1916-9752) and Journal of Plant Disease Sciences (ISSN: 0973-7456). Dr Singh has also been recognized by the professional societies by awarding him National Academy of Agricultural Sciences Associateship 2014, Rashtriya Gavrav Award 2013, Certificate of Recognition -2012, Young Scientist Award- 2011, Kunwar Sexana Bahadur SRDA Award -2010, Scientist of the Year Award -2009, Rural Scientist of Biennium -2008-09.

Dr. Narayan Bhakta is presently working as Senior Scientist, Plant Breeding at ICAR-Research Complex for Eastern Region, Patna (Bihar). As a member of the team, he established a research station of Central Rice Research Institute at Gerua in Assam in 1997 as Regional Rainfed Lowland Research Station, Gerua. He contributed in developing the infrastructure of the research station and acted as Officer in-charge of the research station for more than four years. With more than 18 years of professional experience in the field of rice breeding and rice genetic resources management for rainfed lowland flood prone ecosystem, he has developed and released three varieties of *boro* rice and one variety of aromatic short grain rice. He has also developed a number of elite lines of rice including genetic stocks of deep water rice for rainfed lowland flood prone ecosystem. He has received the best scientist award of central Rice Research Institute, Cuttack during the year 2008-09. He has 14 publications including seven research papers in peer reviewed journals.

Dr. Manibhushan is presently working as Scientist (SS), Computer Application in Agriculture at ICAR Research Complex for Eastern Region, Patna, Post- Bihar Veterinary College, Patna (Bihar). With more than 15 years of professional experience in the field of Natural Resources Management (NRM), Database Management System (DBMS), Decision Support System (DSS), website development, programming in C, C++, SQL, Visual Basic, HTML, JAVA, etc. and land suitability classification using different digital image processing software. He has developed different information systems, decision support systems, spatial data analysis and development of land suitability maps. Having more than 15 research papers in national and international journals and several popular articles, technical bulletins. Contribution of Dr Manibhushan has also been recognized by the professional societies by awarding young scientist award in 2013. He has been recognized by appointing as a member and editor of journal Biospectra- ISSN: 0973-7057 and Journal of AgriSearch journal.

Dr. Umedrao Ramrao Sangle did his Graduation in Agriculture and Master degree in Agriculture (Plant Pathology) and Ph.D. from CCS University, Meerut and part worked at Division of Plant Pathology, IARI, New Delhi. Presently he is working as senior scientist (Plant Pathology) at ICAR Research Complex for Eastern Region, Patna (Bihar). He has vast experience on different crops, Microbial pesticides, biofungicides molecular characterisation and PGPR. He has visited Thailand to be a part of Asian Countries on Network on cotton in Asia and North Africa (INCANA). Member of 3rd Board Meeting of US-India Agriculture Knowledge Initiative (AKI) US Department of Agriculture of Agriculture, Washington DC at New Delhi. He has published 30 research papers, many book chapters and popular articles. He has completed two projects as PI and running four externally funded projects of which three are International projects. Developed four varieties including Makhana variety. He has been also recognized by the professional societies and recipient of Young Scientist Award and Life

Fellowship-2013 by the FMSET & FICCB, Ranchi. He is member of many national and international professional bodies, societies and committees. Dr Sangle has developed strains of bioagents and their mass production technology and established mushroom and bioagents production unit in different States.

Dr. Prem K Sundaram has born at Nalanda district of Bihar state and completed school education in Bokaro Steel City, Jharkhand. He has completed his graduation (B. Tech, Agricultural Engg.) from CCS Haryana Agricultural University, Hissar, Master's (M. Tech, Farm Machinery and Power Engg.) from G.B. Pant University of Agriculture and Technology, Pantnagar. He has more than 6 years of professional experience in the field of natural resources management (NRM), Biofuel, design and development of agriculture machines. He has significantly contributed in the field of Farm Machinery and Power Engineering, He has been a part of project funded by IFAD and IRRI. He has one patent, more than 14 national and international research articles and various popular articles to his credit. He has been editor and reviewer of *Journal of AgriSearch* and *Hortflora Research Spectrum*. He is life member of Indian Society of Agricultural Engineering (ISAE).

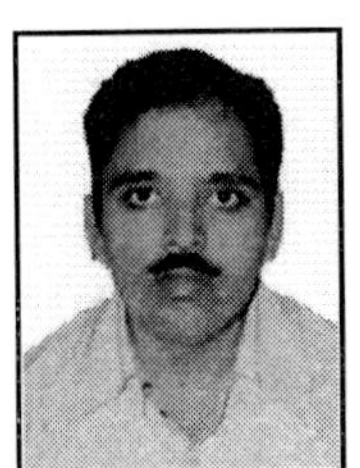

Dr. Sanjeev Kumar is working as Head of Krishi Vigyan Kendra, Nalanda of Bihar agricultural university, Sabour, Bhagalpur. He is throughout First class from Matriculation to Doctorate level. He did his post-graduation (2000) from Allahadad Agricultural Institute, Naini and doctoral degree from National Dairy Research Institute, Karnal, Haryana (2009. He qualified NET in Dairy Engineering and Technology conducted by ICAR. He stared his carrier as Research Associate in Network project at NDRI, Karnal since 2000. Subsequentely, he joined as Training Associate-cum-Jr Scientist at Krishi Vigyan Kendra, Banka under the jurisdiction of Rajendra Agricultural University, Pusa. Further, he selected and contributed as Assistant Professor at Sanjay Gandhi institute of Dairy Technology, Patna under BAU, Sabour and promoted to Assistant Professor (Sr. Scale). He again selected and contributed as Head for KVK, Nalanda under BAU, Sabour since 2012. Dr. Kumar has having experiences in Teaching, Research, Extension and Administration for more than 15 years in different capacity. Dr. Kumar has completed four Research projects as principal investigator (PI) and Co-Pi comprising one externally funded project. He has one patent, more than 50 publications including four Book chapter, Technical Bulletin. Dr. Kumar has been awarded Young scientist award by All India veterinary public health Association in year 2010. Dr Kumar has also been awarded six Best paper/poster awards in International/National Conferences and Seminars.

Dr. Yasin Jeshima, K. Born on 4th February 1979 at Madurai city of Tamil Nadu in a farmer's family, Dr. Yasin Jeshima, K., has obtained her graduation (1999) and post-graduation in Plant breeding and Genetics (2001) from TNAU, Doctorate in Genetics (2008) from KAU and Post-Doctoral Fellowship (2012-2013) from ARO, Volcani, Ministry of Agriculture, Israel; acquired masters in animation. She started her career as ARS-Scientist in 2009; attended a semester course from Weizmann Institute of Science, Rehovot, Israel and three major courses on Intellectual Property from World Intellectual Property Organisation. She is currently working on abiotic stress tolerance of legumes at ICAR-NBPGR and is a PGR faculty at IARI. She has contributed significantly in developing two of the leading salinity tolerant rice varieties. She was a topper and was a recipient of JNU-JRF and ICAR-SRF fellowships. Qualified NET of ICAR-ASRB in Genetics (2003), Plant Breeding (2006), Biotechnology (2008) and Crop Improvement (2009). She has developed a concept known as "SMART" agriculture and strongly in favour of "Merging Plant Breeding with Crop Biotechnology". Participated and organised many workshops and trainings. She is a Bioinformatician and Statistician too, known for her commendable interpretations. She has published more than 15 peer reviewed research papers, 5 books, 6 chapters, 5 practical manuals, 78 gene bank submissions including three publications in science, more than 25 conference presentations and invited lectures. She is a member of top scientific societies in India and abroad. So far she has been conferred with three society awards. She is a recipient of foreign recognition in the form of fellowships and collaborative research projects.

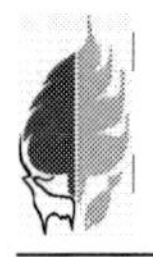

National Academy of Agricultural Sciences

Prof. Panjab Singh
President

Dated: 07.07.2017

Foreword

Lentil is one of the important pulse crops of India. Next only to Canada, India is the second largest producer of lentil in the world. In India this crop is widely grown across states, especially under marginal soil and water resource situations and offers huge potential to grow it on unconventional area epically under rice fallow situation. Lentil being legume food crop, presentation of this book is small step toward providing technologies information for realizing self-sufficiency in food legume production. I am happy to see the book entitled *"Scientific Lentil Production"* is being published by the Society for Upliftment of Rural Economy (SURE) and is authored by a team of ICAR Scientists. This book contains 35 chapters on various topics relating to important aspects ranging from production to post harvest management.

I congratulate all the contributors and editors of this publication. I am sure the scholars, researchers, teachers and all other stakeholders interested in lentil will make use of this useful and informative publication.

With best wishes

(Panjab Singh)

NASC, Dev Praksh Marg, PO Pusa, New Delhi-110012 India

Society for Upliftment of Rural Economy
Varanasi 221 005 (Uttar Pradesh) India

Mrs. Manju Lata Singh
President

File No: SURE/ BL/2017
Dated: 21.09.2017

Foreword

Lentil is the second most important cool season pulse crops after chickpea of India. Lentil plays a vital role in nutritional security to fulfill protein supplementparticularly forvegetarians.

Globally, India is the second largest producer of lentil after Canada. India is one of the promising places with huge potential to grow lentil on unconventional area epically under rice fallow situation. I am happy to share that Society for Upliftment of Rural Economy (SURE) Varanasi is presenting its first ever book entitled "*Scientific Lentil Production*". This is a timely publication on this crop, as year 2016 was celebrated as International Year of Pulses. This book contains 35 chapters on various topics related to lentil right from production to safe storage. This book is edited by a team of ICAR Scientists comprising Dr Anil Kumar Singh, Dr Narayan Bhakta, Dr Manibhushan, Dr UR Sangle, Dr Prem K Sundaram, Dr YJ Khan and Dr Sanjeev Kumar. I must congratulate all the contributors including editors, who took special pain in shaping this important book. Credit also goes to the Society for Upliftment of Rural Economy, Varanasi India and Satish Serial Publishing House, Azadpur, Delhi for timely publication of this valuablebook on lentil.

I am confident enough that this book will be boon for the academicians, researchers, professors and all other stakeholders involved in lentil.

With Due Regards

Manju Lata Singh
(Manju Lata Singh)

List of Contributors

1. Dr. A A Saad, Assistant Professor (Agronomy), Sher-e-Kashmir University of Agricultural Science and Technology, Kashmir, Shalimar, Srinagar (J&K)190 016 India; Email: saad_aa@rediffmail.com; jagannathpathak@yahoo.in
2. Dr. A K Mishra, Head, Division of Agricultural Engineering, ICAR Indian Agricultural Research Institute, New Delhi-110012, India
3. Dr. Abdul Haris A, Principal Scientist (Agronomy) ICAR Research Complex for Eastern Region, P.O. Bihar Veterinary College, Patna 800 014 Bihar, India; Email:abdulharis123@rediffmail.com
4. Dr. Aditya Kumar Singh, Assistant Professor, Agronomy, College of Post Graduate Studies, Central Agricultural University, Barapani, Shillong (Meghlaya) 793 103, India; Email: adityakumar1972@ yahoo.co.in
5. Dr. Ajay Kumar Pandey, Assistant Professor College of Forestry & Hill Agriculture, G. B. Pant University of Agriculture & Technology Hill Campus, Ranichauri, TehriGarhwal-249 199 (Uttarakhand), India; Email: drajay2002@gmail.com
6. Dr. Ajay Veer Singh, Assistant Professor (Microbiology), Department of Biological Sciences, G B Pant University of Agriculture and Technology, Hill Campus, Ranichauri, TehriGarhwal, Uttarakhand-249 199, India; Email: ajaygbpuat@gmail.com
7. Dr. Amarjeet Gupta, Senior Scientist, Directorate of Onion and Garlic Research, Rajgurunagar-410 505, Pune, Maharashtra, India; Email: guptaaj75@yahoo.co.in
8. Dr. Amit Bhatnagar, Department of Agronomy,GB Pant University of Agriculture and Technology PantnagarUttarakhand, India
9. Dr. Amjad Masood , Assistant Professor (Agronomy), Sher-e-Kashmir University of Agricultural Science and Technology, Kashmir, Shalimar, Srinagar (J&K)190 016, India
10. Dr. Anand Kumar, Bihar Agriculture Uuniversity, Sabour, Bihar, India; Email: anandpbgkvkharnaut@gmil.com
11. Dr. Anil Kumar Singh, Principal Scientist (Agronomy), Division of Land and Water Management, ICAR Research Complex for Eastern Region, P.O. Bihar Veterinary College, Patna 800 014 Bihar, India; Email: anil.icarpat@gmail.com
12. Dr. Arbind Kumar Choudhary, Principal Scientist (Plant Breeding), ICAR Research Complex for Eastern Region,P.O. Bihar Veterinary College, Patna 800 014 Bihar, India; Email: akicar1968@gmail.com

13. Dr. Ashok Yadav, International Rice Research Institute , India, New Delhi, India; Email: aky444@gmail.com
14. Dr. Ashutosh Upadhya, Division of Land and Water Management, ICAR Research Complex for Eastern Region, P.O. Bihar Veterinary College, Patna 800 014 Bihar, India; Email: upadhyaya66@yahoo.co.in
15. Dr. B P Bhatt, Director ICAR Research Complex for Eastern Region, P.O. Bihar Veterinary College, Patna-800 014, India; Email: drbpbhatt.icar@yahoo.com
16. Dr. B S Mahapatra, Professor Department Agronomy GB Pant University of Agriculture & Technology PantnagarUttarakhand, India
17. Dr. FA Pir, Assistant Professor (Agronomy), Sher-e-Kashmir University of Agricultural Science and Technology, Kashmir, Shalimar, Srinagar (J&K)190 016, India
18. Dr. H K Singh Associate Professor, Department of Plant Pathology, Narendra Deva University of Agriculture and Technology, Kumarganj-224 229, Faizabad (Uttar Pradesh), India; Email: hksndu@gmail.com
19. Dr. H L Kushwaha, Division of Agricultural Engineering, ICAR Indian Agricultural Research Institute, New Delhi-110 012, India;
20. Dr. J S Mishra, Head, Division of Crop Research, ICAR Research Complex for Eastern Region, P.O. Bihar Veterinary College, Patna-800 014 Bihar, India; Email: jsmishra31@gmail.com
21. Dr. J Veeranna, Division of Agricultural Engineering, ICAR-Indian Agricultural Research Institute, New Delhi-110 012, India; Email: veerucae023@gmail.com
22. Dr. Jaganath Pathak, Subject Matter Specialist, KrishiVigyan Kendra, Burhanpur (MP) 450331, India
23. Dr. Jeshima Khan Yasin, Scientist, ICAR National Bureau of Plant Genetic Resources , Pusa Campus, New Delhi, India- 110 012; Email: yasinlab.icar@gmail.com
24. Dr. Jitendra Kumar Sr. Scientist (Plant Breeding), Indian Institute of Pulses Research, Kanpur (UP) India
25. Dr. K M Singh, Director Extension, Professor & Chairman, Department of Agricultural Economics, Rajendra Agricultural University, Bihar, Pusa-Samastipur-848 125, India; Email: m.krishna.singh@gmail.com
26. Dr. Lal Chand Prasad, Department of Genetics and Plant Breeding, Institute of Agricultural Sciences, Banaras Hindu University, Varanasi (India)-221 005
27. Dr. Lal Singh, Associate Professor (Agronomy), Sher-e-Kashmir University of Agricultural Science and Technology, Kashmir, Shalimar, Srinagar (J&K)190 016, India; Email: drlalsingh@rediffmail.com

28. Dr. Lala I P Roy, Assistant Professor, Agronomy, College of Post Graduate Studies, Central Agricultural University, Barapani, Shillong (Meghlaya) 793 103, India; Email: lalaipray@rediffmail.com
29. Dr. Mandhata Singh, KVK, Buxar, ICAR Research Complex for Eastern Region, P.O. Bihar Veterinary College, Patna-800 014 Bihar, India; Email: mandhataagro@gmail.com
30. Dr. Manibhushan, ICAR Research Complex for Eastern Region, P.O. Bihar Veterinary College, Patna 800 014 Bihar, India; Email: mani_patna2000@yahoo.com
31. Dr. Manoj Kumar, Join Director, Central Potato Research Station Meerut, Uttar Pradesh, India; Email:
32. Dr. MdIdirs, Principal Scientist (Agricultural Entomology), ICAR Research Centre, Darbhanga, Bihar, India; Email: idrism1958@yahoo.co.in
33. Dr. Mukesh Kumar Meena, Soil Scientist, National Research Centre, Bharatpur Rajasthan, India; Email: mukesh.icar@gmail.com
34. Dr. N Janaki Singh, Assistant Professor, Agronomy, College of Post Graduate Studies, Central Agricultural University, Barapani, Shillong (Meghlaya) 793 103, India
35. Dr. N K Gautam, ICAR National Bureau of Plant Genetic Resources, Pusa Campus, New Delhi, India-110 012
36. Dr. Narayan Bhakta Principal Scientist (Plant Breeding), Division of Crop Research, ICAR Research Complex for Eastern Region, P.O. Bihar Veterinary College, Patna 800 014 Bihar, India; Email: bhakta65@rediffmail.com
37. Dr. Naresh Chandra, Senior Scientist, ICAR Research Complex for Eastern Region, P.O. Bihar Veterinary College, Patna 800 014 Bihar, India; Email: nchandra_09rediffmail.com
38. Dr. Nita Singh, ICAR National Bureau of Plant Genetic Resources, Pusa Campus, New Delhi, India-110 012
39. Dr. Nupur Malavika Indian Institute of Pulse Research, Kanpur, Uttar Pradesh-208024, India; Email: malviya.nupur@gmail.com
40. Dr. Pankaj Kumar Yadav, KrishiVigyan Kendra Banka, Bihar, India
41. Dr. Parashanath, BPSAC, Purnea, Bihar, India
42. Dr. Parmeet Singh, Associate Professor (Agronomy), Sher-e- Kashmir University of Agricultural Science and Technology, Kashmir, Shalimar, Srinagar (J&K)190 016, India; Email: singh.parmeet@rediffmail.com
43. Dr. Pramod Maal, Assistant Professor College of Forestry & Hill Agriculture, G. B. Pant University of Agriculture & Technology Hill Campus, Ranichauri, TehriGarhwal-249 199 (Uttarakhand), India

44. Dr. Pravesh Kumar, Assistant Professor (Agronomy), Bihar Agriculture University Sabour, Bhagalpur Bihar, India; Email: praveshkumar.agron@gmail.com

45. Dr. Prem Kumar Sundaram Scientist (Farm Power and Machinery) Division of Land and Water Management, ICAR Research Complex for Eastern Region, P.O. Bihar Veterinary College, Patna 800 014 Bihar, India; Email: prem.k.sundaram@gmail.com

46. Dr. Purshotam Singh, Assistant Professor (Agronomy), Sher-e-Kashmir University of Agricultural Science and Technology, Kashmir, Shalimar, Srinagar (J&K)190 016, India; Email: drpurshotam@gmail.com

47. Dr. R B Singh, Department of Plant Pathology, Narendra Deva University of Agriculture and Technology, Kumarganj-224 229, Faizabad (Uttar Pradesh), India; Email: rbspath.2010@gmail.com

48. Dr. R C Bharatai, Principal Scientist, ICAR Research Complex for Eastern Region, P.O. Bihar Veterinary College, Patna-800 014, India; Email: drrcbharati@yahoo.com

49. Dr. R C Shakywar, Department of Plant Protection, College of Horticulture & Forestry, Central Agricultural University, Pasighat-791 102 (Arunachal Pradesh), India; Email: rcshakywar@gmail.com

50. Dr. R K Singh, Senior Scientist (Seed Technology), Central Potato Research Station, Sahaynagar Patna- Bihar 801506, India; Email: rksingh.cprs@gmail.com)

51. Dr. R K Solanki, Senior Scientist (Plant Breeding), National Research Centre on Seed Spices , Ajmer-305206 Rajasthan, India

52. Dr. R P Maurya, Assistant Professor College of Forestry & Hill Agriculture, G. B. Pant University of Agriculture & Technology Hill Campus, Ranichauri, TehriGarhwal-249 199 (Uttarakhand), India

53. Dr. R P Singh, Professor, Department of Agronomy, IAS, BHU, Varanasi (UP), India

54. Dr. Raghu Raj Singh, ICAR-Directorate of Medicinal and Aromatic Plants Research, Boriavi, Anand-378310, Gujarat, India; Email: raghuraj9271@gmail.com

55. Dr. Rahul Bakade, Scientist, Central Potato Research Station, Patna Bihar, India; Email: rahulbcpri@gmail.com

56. Dr. Ram Chandra, Division of Agricultural Engineering, ICAR-Indian Agricultural Research Institute, New Delhi-110 012, India;

57. Dr. Ravindra Prashad, Department of Genetics and Plant Breeding, Institute of Agricultural Sciences, Banaras Hindu University, Varanasi (India) -221 005; Email: rprasadbhugpb@gmail.com

58. Dr. Roaf A Parray, Division of Agricultural Engineering, ICAR-Indian Agricultural Research Institute, New Delhi-110012, India;
59. Dr. S K Singh, Principal Scientist (Agronomy) Indian Institute of Vegetable Research Varanasi UP, India; Email:skscprs@gmail.com
60. Dr. Sanjeev Kumar Senior Scientist and Program Coordinator, KrishiVigyan Kendra, Harnaut, Nalanda Bihar, India; Email: sanju_kvk@yahoo.co.in
61. Dr. Santosh Kumar Magadum, ICAR National Bureau of Plant Genetic Resources, Pusa Campus, New Delhi, India-110 012; Email: santoshsbm@gmail.com
62. Dr. Santosh Kumar, Scientist (Plant Breeding) Division of Crop Research, ICAR Research Complex for Eastern Region, P.O. Bihar Veterinary College, Patna-800 014 Bihar, India; Email: santosh9239@gmail.com
63. Dr. Sati Shankar Singh, Director Agricultural Technology Application Research Institute, Kolkata, West Bengal, India; Email: sssinghpatna@yahoo.com
64. Dr. Satish D, Lande Division of Agricultural Engineering, ICAR-Indian Agricultural Research Institute, New Delhi-110012, India
65. Dr. Satish Santosh Bornare, Department of Genetics and Plant Breeding, Institute of Agricultural Sciences, Banaras Hindu University, Varanasi (India)-221005
66. Dr. Saurabh Sharama Assistant Professor, PBR Agriculture Degree College, Hardoi (UP) 241 001India; Email: saurabhpdcsrmeerut@ gmail.com
67. Dr. Shachi Shah, Associate Professor, School of Agriculture, Indira Gandhi National Open University, New Delhi, India; Email: sshah@ignou.ac.in
68. Dr. Sharad Kumar Dwivedi Scientist (Plant Physiology) Division of Crop Research, ICAR Research Complex for Eastern Region, P.O. Bihar Veterinary College, Patna-800 014 Bihar; Email: sharad.dwivedi 9736@gmail.com
69. Dr. Shishir Kant Singh, Division of Crop Research, ICAR Research Complex for Eastern Region, P.O. Bihar Veterinary College, Patna-800 014 Bihar, India; Email: shishir.singh68@gmail.com
70. Dr. Shiv Kumar, Principal Scientist,Division of Genetics, Indian Agriculture Research Institute New Delhi 110012, India
71. Dr. Shyam Singh, Division of Plant Pathology, Central Coffee Research Institute, C.R. Station-577117, Chikmaglur (Karnataka), India; Email: shyamsingh.ccri@gmail.com

72. Dr. SK Singh, Head & Principal Scientist (Agronomy), Division of Land and Water Management, ICAR Research Complex for Eastern Region, P.O. Bihar Veterinary College, Patna-800 014 Bihar, India; Email: santprincipal@gmail.com

73. Dr. Sudhanshu Singh, International Rice Research Institute , India, New Delhi, India; Email: sud.singh@irri.org

74. Dr. Sunita Kushwaha, Scientist, KrishiVigyan Kendra Banka, Bihar , India; Email: sunita17kk@rediffmail.com

75. Dr. Sushil Dimree, Associate Professor, Department of Soil Science and Agricultural Chemistry, CS Azad University of Agriculture & Technology, Kanpur-208 002 (UP), India; Email: dimri.astro@ gmail. com

76. Dr. Tapan K Khura, Division of Agricultural Engineering, ICAR Indian Agricultural Research Institute, New Delhi-110012, India; Email: tapankhura@gmail.com

77. Dr. U R Sangle, Senior Scientist (Plant Pathology) Division of Crop Research, ICAR Research Complex for Eastern Region, P.O. Bihar Veterinary College, Patna-800 014 Bihar, India; Email: umeds@rediffmail.com

78. Dr. U S Singh, International Rice Research Institute , India, New Delhi, India

79. Dr. V Venkat Raman, Assistant Professor, School of Agriculture, Indira Gandhi National Open University, New Delhi, India; Email: vvramanan@ignou.ac.in

80. Dr. Vedpraksh , Scientist, Division of Crop Research, ICAR Research Complex for Eastern Region, P.O. Bihar Veterinary College, Patna-800 014 Bihar, India; Email: jyanivedprakash486@gmail.com

81. Dr. Vishram Ram, Associate Professor, Farming System Management, Agronomy, College of Post Graduate Studies, Central Agricultural University, Barapani, Shillong (Meghlaya) -793 103, India; Email: ramvishramcau@gmail.com

82. Dr. Wasim Raza, Assistant Professor (Agronomy), Sher-e-Kashmir University of Agricultural Science and Technology, Kashmir, Shalimar, Srinagar (J&K)190 016, India

83. Dr. Zahida Rashid, Assistant Professor (Agronomy), Sher-e-Kashmir University of Agricultural Science and Technology, Kashmir, Shalimar, Srinagar (J&K)190 016, India; Email: zahida1926@gmail.com

❑❑❑

Contents

Chapter 1

Lentil: Introduction

Anil Kumar Singh, N Bhakta, UR Sangle, Manibhushan, PK Sundaram, S Kumar and Yasin Jeshima Khan

SYNOPSIS

Lentil is probably the first food legume that ever cultivated as evident from the earliest archeological finding from the Paleolithic and Mesolithic layers of Franchthi caves in Greece that dates back to almost 13000 to 9500 years ago is none other than lentil . It is one of the vegetable sources of quality protein for large portion of Indian, which is still vegetarian with respect to their dietary preferences. Fortunately India ranked first in the acreage of lentil with 35.57% of total lentil area of world, and second in the production of lentil with 23.94% of production share of world. However lentil accounted for 7 % share of total Indian pulse production. Uttar Pradesh was ranking first in area and production (41.1 % and 47.2 %) followed by Madhya Pradesh (35.86% and 26.20 %) and Bihar which accounted for 13.9 % of total area coverage and 17.3 % of the country's Lentil production. Inter district analysis of lentil production revealed that district Jalaun of Uttar Pradesh with 4.36% of production has the highest share followed by Patna of Bihar (3.52%). Pulses are the staple source of protein to the majority of Indian population and contribute significantly to the nutritional security of the country. Production of these crops has been stagnant over the years. There is widening gap between demand and supply with about 20 % of the total demand met by imports. Major areas under pulses are in the States of Madhya Pradesh (20.3%), Maharashtra (13.8%) and Rajasthan (16.4). Lentil fixes 30-40 kg/ha atmospheric 'N' in readily available form to the succeeding crop. Associated intercrop (other than legume) also gets benefited by 'N' transfer from lentil roots up to some extent. It also contributes to sustain production system through physical, chemical and biological improvements of soil properties, as a rotation effect.

Keywords: Lentil scenario, agrotechnique recommended varieties, nutrient management, plant protection, bottleneck.

Introduction

India is no longer ship to mouth country thanks goes to policy planner researcher and the most to our farmers who make this dream to happen in our life time. India has traveled a lot on this rough path in respect to agricultural growth and development to match or even exceed to population growth. India has a unique characteristic of having nearly 60% of the population directly or indirectly dependent on agriculture and allied activities. Currently, we are in the mid-way of self-sustaining in pulses production as we are world leader in production, consumption and import as well. India import 2-3 MT of pulses during 2010-11, causing huge hard foreign earning. By the 2050 we will able to sustain our production and we turned to net importer to net exporter for pulses if every things goes as per plan. Another unique feature is its source of livelihood and still not a commercial business (Singh *et al.,* 2013a). *Lens culinaris* (Modik) name has been given by the German botanist Medikus in 1778 (Cubero, 1981). It is one of the most nutritious cool season food legumes and ranks next only to chickpea. It is grown throughout the northern and central India, besides its utilisation as a dal (whole or dehulled grains) in various Indian preparations. Lentil is a valuable human food, mostly consumed as dry seeds (whole decorticated, seed decorticated and split). In Indian sub continent it is mostly consumed as 'Dal' by removal of outer skin and separation of cotyledons, snacks and soup preparation etc. It is easy to cook and easily digestible with high biological value, hence also referred to patient. Dry leaves, stems, empty and broken pods are used as valuable cattle feed. Its bold seeded, attractive shaped grains have high demand for export at premium prices. Lentil, based on the seed size and test weight, is classified into two main groups *i.e.* bold seeded: Includes sub sp. *macro-sperma,* with the test weight of more than 25 g. also known locally as *Masur* or *Malka Masur* and mainly cultivated in Bundelkhand region of UP/ MP and Maharashtra state and (ii) small seeded: Sub sp. *micro-sperma,* test weight, less than 25 g, locally known as *masuri* and primarily grown in Indo Gangtic plains of NEPZ (UP, Bihar, West Bengal and Assam).

Table 1.1 Botany of Lentil

Botanical Name	*Lens culinaris Medikus subsp. culinaris*
Chromosome	2n = 14
Origin	Turkey to South Iran
Synonym/Other name	Masur, Malka (bold seeded), lentille (French) lines (German), Lenteja (Spanish) and Mercimek (Turkish).

Lentil is an herbaceous annual plant mostly erect and bushy type with four to six primary branches, plant height not exceeding 50-60 cm in general. It has a well developed root system including a central tap root with several lateral branches, spreading in all directions. Root nodules, the site of atmospheric 'N' fixation, are mainly concentrated on primary root. The stem is weak and quadrangular and ends of leaflets some time forms tendrils. Inflorescence is a raceme of two to four flowers. Flowers are small and white with blue, violet or pink tinged. Ovary is short with one or two ovules hence, pods are one to two seeded. Anthesis takes place in buds sometimes before opening of flowers in the next morning. Hence, self pollination is a general rule. Pods are one to one and half cm in length with a curved beak. Grains are often light brown in colour with lens shaped.

Lentil is the earliest known and probably first food legume that ever been cultivated as evident from the earliest archeological finding from the Paleolithic and Mesolithic layers of Franchthi caves in Greece that dates back to almost 13000 to 9500 years ago (Zohary and Hopf, 1973; Zohary and Hopf,2000 and Harlan *et al.*, 1976). Lentil is one of the prominent sources of vegetable protein in the South East Asia including India. It is essentially grown as a rainfed crop on the residual soil moisture of preceding crop (rice in-general) (Ali *et al.*, 2012; Singh *et al.*, 2012 and Singh *et al.*, 1995).

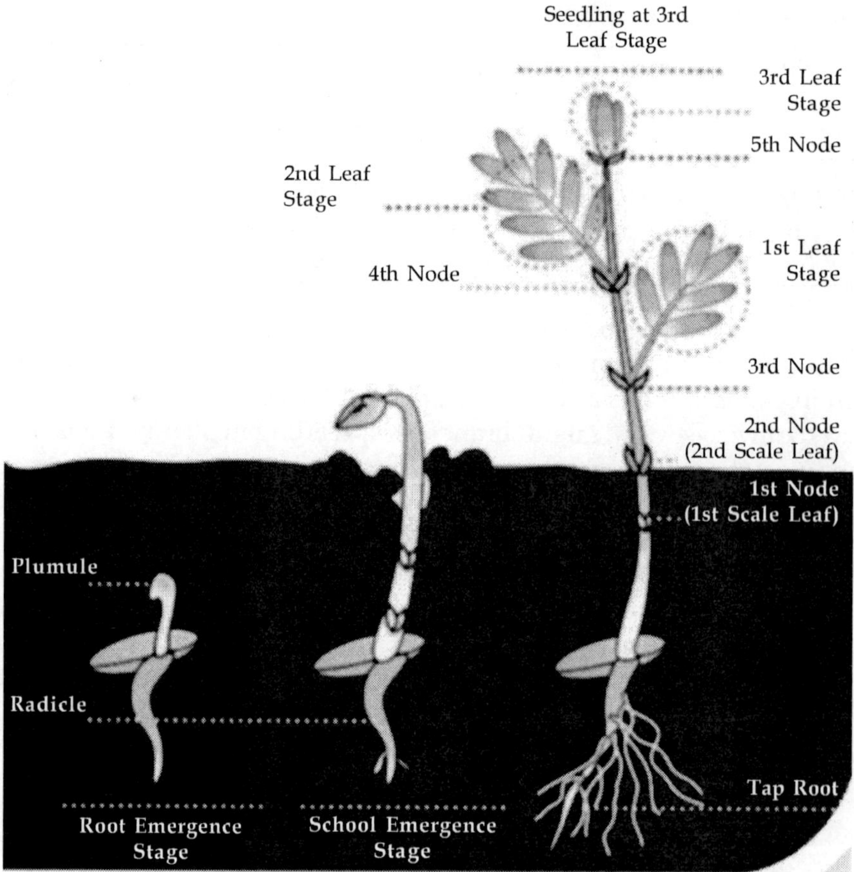

Fig. 1.1 Lentil growth and development (Adopted from Pulse Production Manual, 2000)

Table 1.2 Nutritive value of Lentil

Constituents	Magnitudes
Protein	24-26%
Carbohydrate	57-60%
Fat	1.3%
Fibre	3.2%
Phosphorus	300 mg/100 g
Iron	7mg/100 g
Vitamin C	10-15 mg/100 g
Calcium	69 mg/100g
Calorific value	343
Vitamin A	450 IU

Pulses Scenario in India

Indian agriculture is endowed with the second largest cultivable land area in the world, the largest number of varied agro climatic conditions, 270 days of sunshine, 1120 mm of annual rainfall, rivers crisscrossing the country, 7000 km long coast line, extra ordinary bio-diversity, and abundant labour force. India is the largest producer, consumer, importer and processor of pulses in the world. Ironically, the country's pulse production has been hovering around 14–15 MT, coming from a near-stagnated area of 22–23 M ha, since 1990–91 (Singh *et al.*, 2013). Major areas under pulses are in the States of Madhya Pradesh (20.3%), Maharashtra (13.8%), Rajasthan (16.4), Uttar Pradesh (9.5%), Karnataka (9.3%), Andhra Pradesh (7.9%), Chhattisgarh (3.8%), Bihar (2.6%) and Tamil Nadu (2.9%). Pulse productivity which was 441 kg/ha in 1950 increased up to 689 kg/ha during 2011, registering 0.56% annual growth rate (Table 1.3).

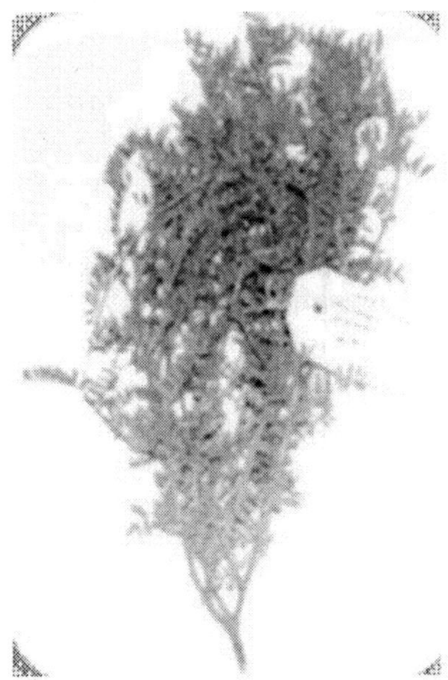

Fig. 1.2 (a) Ideal plant type of cultivar Arun

Fig. 1.2 (b) Plant sample collected from farmer field type of cultivar Arun

Table 1.3 Compound growth rate achieved in pulses production in India during the period 1950-2011

Particular	Pulses
Productivity 1950	441 kg/ha
Productivity 2011	689 kg/ha
Overall compound growth rate (%)	1.011
Compound growth rate (%) Area	0.52
Compound growth rate1 (%) Production	1.27
Compound growth rate (%) Productivity	0.73

Source: Department of Agriculture and Cooperation, Ministry of Agriculture, Govt. of India

Production of pulses in 2008-09 was 14.66 million tons with an average yield of 655 kg/ha. In the year 2013-14 India produced record 19.5 Mt of total pulses. Share of chickpea, pigeonpea, mungbean and urdbean to total production has been worked out about 39, 21, 11 and 10%, respectively. Lentil and field pea accounted for 7% and 5% share of total production. For meeting the demand of the growing population, the country is importing pulses to the tune of 2.5–3.5 MT every year. Strong upward trend in the import of pulses is a cause of concern, since an increase in demand from India has shown to have cascading effect on international prices, thus draining the precious foreign exchange. By 2050, the domestic requirements would be 26.50 MT, necessitating stepping up production by 81.50%, i.e. 11.9 MT additional produce at 1.86% annual growth rates. This uphill task has to be accomplished under more severe production constraints, especially abiotic stresses, abrupt climatic changes, emergence of new species/ strains of insect-pests and diseases, and increasing deficiency of secondary and micronutrients in the soil. This requires a two-pronged proactive strategy, i.e. improving per unit productivity and reducing cost of production. The yield levels of pulses have remained low and stagnant, also area and total production. Number of districts harvesting more than 0.8 or 1 t/ha yield of *kharif* pulses is very small. Situation of *rabi* pulses is better in this regard. The gap between demand and supply has been widening and has necessitated import of pulses of 2.8 million tons in 2007-08.

Table 1.4 All-India Area, Production and Productivity of pulses

Year	Area (Mha)	Production (MT)	Productivity (kg/ha)
1950-51	19.09	8.41	441
1960-61	23.56	12.70	539
1970-71	22.02	11.69	531
1980-81	22.46	10.63	473
1990-91	24.66	14.26	578
2000-01	20.35	11.08	544
2010-11*	26.28	18.09	689

* Fourth Advance Estimates

Department of Agriculture and Cooperation, Ministry of Agriculture, Govt. of India

Perusal of data presented in table 1.5 revealed that Uttar Pradesh contributes significantly to the pulses production and its share to the national pulses security is 21.8% with 3.196 Mt. Other two leading states are Madhya Pradesh and Rajasthan with 19.5% and 13.6% production share of India. Total eight (8) states contributing 90 % of total pulses production (Table1.5).

Table 1.5. State wise contribution in the Indian pulses production

State	Production (Mt)	Production (%)
Uttar Pradesh	3.196	21.8
Madhya Pradesh	2.859	19.5
Rajasthan	1.994	13.6
Maharastra	1.407	9.6
Orissa	1.202	8.2
Bihar	1.041	7.1
Karnataka	0.777	5.3
Haryana	0.748	5.1
Other states	1.437	9.8
Total	**14.66**	**100**

Data presented in table 1.6 reveal the importance of individual legume crop in national pulse production. Chickpea top the list with 26.85 and 38.81 per cent share with respect to area and production in India, followed by pigeonpea and mungbean respectively. Lentil occupies only 4.94 % area and corresponding contribution in national pulse production is 6.96%.

Table 1.6 Contribution of individual pulses in Indian scenario

Name of crop	Area (Mha)	Area (% of Total)	Production (Mt)	Production (% of Total)
Chickpea	5.81	26.85	5.69	38.81
Pigeonpea	3.62	16.73	3.07	20.94
Mungbean	2.98	13.77	1.61	10.98
Urdbean	3.18	14.70	1.46	9.96
Lentil	**1.07**	**4.94**	**1.02**	**6.96**
Other Pulses	6.30	29.11	1.75	11.94
Total	**21.64**	**26.85**	**14.66**	**100.00**

Lentil: Global scenario

India ranked first in the area and second in the production with 35.57% and 23.94% of world area and production respectively (Table 1.7). In case of productivity, India ranked 30th in the world. The highest productivity is recorded in Armenia (2181 kg/ha) followed by China. Canada ranks first with respect to production due to very high level of productivity (1482 kg/ha) as compared to India (697 kg/ha) as well as other lentil producing countries. Global lentil production in 2009 was 4.5 million metric tons, the top producers being Canada, India, Turkey and the United States. Canada is the leading lentil exporting nation. The major importing countries are Spain, Colombia, Egypt, Algeria and Sri Lanka. India, Turkey, Ethiopia, China, Syria, Iran and Bangladesh are among the top ten producers. On average, about 70% of global production is consumed in the developing countries where it is produced. About a quarter of global production is from India, most of which is consumed in India itself. In 2010 the area of lentil worldwide was 4.2 million hectares and the average worldwide yield was 1.1 metric tons per hectare.

Table 1.7. Global ranking of major lentil producing countries with reference to area, production and productivity

Rank	Area (Lakh ha)			Production (Lakh tonnes)			Productivity (kg / ha)	
	Country	Area	% to World	Country	Production	% to World	Country	Yield
I	India	14.73	35.57	Canada	12.78	30.77	Armenia	2181
II	Canada	8.62	20.81	India	9.94	23.94	China	1978
III	Turkey	4.40	10.62	Turkey	5.70	13.72	New Zealand	1866
IV	Iran	2.26	5.45	USA	2.34	5.64	Azerbaijan	1817
V	Nepal	1.89	4.56	Australia	2.10	5.06	Egypt	1770
VI	USA	1.78	4.29	Nepal	1.61	3.87	Australia	1650
VII	Syrian Arab	1.43	3.45	Syrian Arab	1.54	3.70	France	1580
VIII	Bangladesh	1.30	3.14	China	1.38	3.33	Kenya	1520
IX	Australia	1.27	3.07	Bangladesh	1.18	2.84	Canada	1482
X	Ethiopia	0.96	2.33	Iran	1.13	2.73	Argentina	1333
XXX	-	-	-	-	-	-	**India**	**675**
	World	**41.41**		**World**	**41.53**		**World**	**1003**

Source: FAO statistics 2005.

National scenario of lentil

Indian is world largest homeland of vegetarian population and world leader in pulses production and imports large quantities to provide protein supplements to it population. Indian pulse production has been stuck between 14 and 15 MT since mid-nineties, resulting in poor consumption (33g/capita/day) during 2010 (Ali and Gupta, 2012). Lentil occupies 1.47 million ha area in India with a production of 0.95 million tons (Dixit *et al.* 2009). Uttar Pradesh was ranking first in area and production (41.1 % and 47.2 %) followed by Madhya Pradesh (35.86% and 26.20 %) and Bihar which accounted for 13.9 % of total area coverage and 17.3 % of the country's Lentil Production. Further, lentil occupies 0.30 M ha in the Indian IGP, either alone or under rice- lentil sequential cropping system, being second most important cropping system after rice -wheat (Singh *et al.*, 1997 and Singh *et al.*, 2012). Lentil sowing get delayed either due to late harvest of preceding long duration rice crop or due to heavy soil moisture in the field kept for lentil, because it is the only crop which can grow or field is left fallow (Singh and Singh, 2008 and Singh *et al.*, 2012). IGP contributes 66 % of total lentil production from 54.7 % area. Lentil productivity in the Indian IGP is quite satisfactory (839kg/ha) compared to national average (697kg/ha), though it faces several bio-physical constraints (Anonymous, 2011).

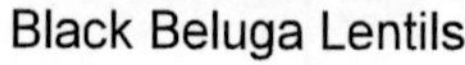

Black Beluga Lentils

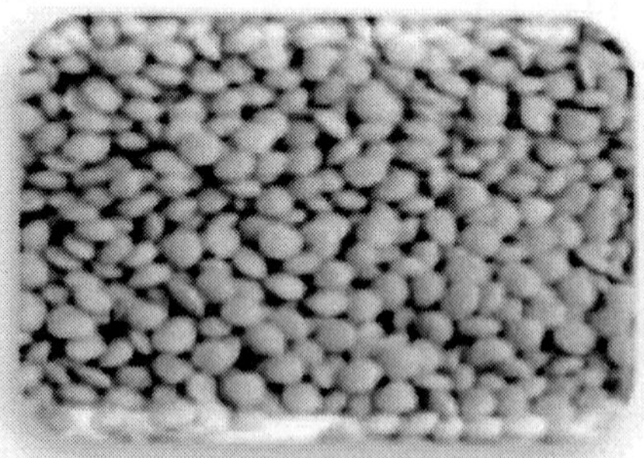

Brown Lentils

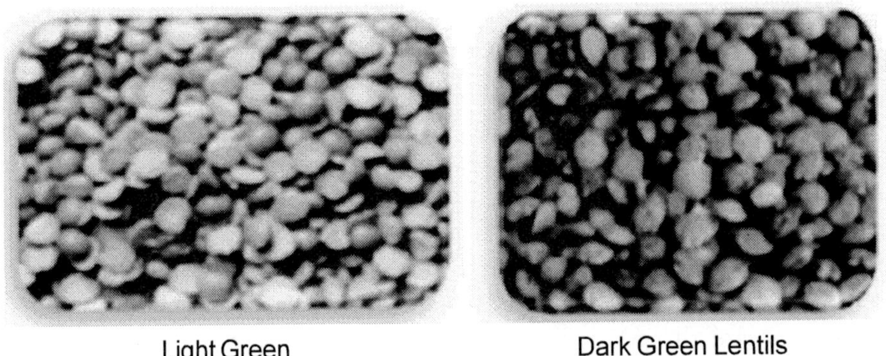

Light Green Dark Green Lentils

Fig. 1.3 Diversity in lentil seed coat colour

Lentil potential realized in top ten districts of Indian (2003-04)

Analysis of the intra-state status of Lentil crop is presented in table 1.9. Inter district analysis revealed that district Jalaun of U.P. with 4.36% of production has the highest share followed by Patna of Bihar (3.52%), Bahraich (3.57%), Barbanki (3.04%) and Hamirpur (3.00%) of U.P. Other districts have less than 3 percent of the country's lentil contribution. District-wise area, production and yield of top ten district of India in respect of production are presented below which contributed 21.58 per cent and 27.86 per cent of area and production of the country. It may be concluded that these districts may be targeted specifically to tap the potential through important inputs and technological back-up with aggressive extension programmes.

Climatic requirements for lentil

Being a winter season crop, require cold climate. Being very hardy in nature, it can tolerate frost and severe winter to a larger extent. The range of cultivation with regard to climate is very wide. It requires cold temperature during vegetative growth and comparatively warm temperature during maturity, with the optimum temperature range of 18-30^{0}C. Unlike Bengal gram, it can thrive well under stress conditions of frost and winter rains, even at flowering and fruiting stage.

Agro techniques for lentil production

The crops leaves a reasonable good amounts of atmospheric 'N' in readily available form (upto 30-40 kg/ha) to the succeeding crop. Associated intercrop (other than legume) also gets benefited by 'N' transfer from lentil roots up to some extent. It also contributes to sustain production system through physical, chemical and biological improvements of soil properties, as a rotation effect. It offers good scope in late vacated paddy fields either as *Utera* or succeeding crop (if soil is workable after paddy harvest) as delayed sowing does not affect as adversely as in chickpea and pea due to its high cold tolerant nature. For this reason, lentil is preferred over gram in the regions having cold winters like plains of North and lower Himalayan Hills. It is also a good substitute of chickpea in areas which may be too dry due to shorter duration. The crop is also used as cover crop to check soil erosion in problem areas. Well drained loamy soils with neutral reaction, are best, though it can also be grown successfully in low lying paddy soils of poorer type, even tolerate moderate alkalinity conditions. Lime treatment (either seed pelleting or soil amendments) is must, prior to its cultivation in acidic soil. Like gram, it also require good aeration for nodule development, achieved by one deep ploughing followed by one cross harrowing. Lentil is generally grown after the harvest of *kharif* crops or also as the sole crop of the year. The most common rotations under sequential cropping as well as intercropping are depicted in table 1.10.

Table 1.8 State wise lentil scenario in India during VIIIth to Xth Plan

States		XIIIth Plan	% to Country	IXth Plan	% to Country	Xth Plan	% to Country
Assam	A	0.14	1.10	0.22	1.52	0.20	1.40
	P	0.08	1.06	0.11	1.20	0.11	1.12
	Y	1373	212	525	79	546	80
Bihar	A	1.71	13.88	1.76	12.39	1.71	11.71
	P	1.39	17.34	1.59	16.85	1.34	13.47
	Y	810	125	903	136	784	115
Chhattisgarh	A	*		0.06	0.40	0.18	1.23
	P			0.02	0.18	0.06	0.56
	Y			119.40	18	308	45
Haryana	A	0.11	0.90	0.09	0.64	0.07	0.46
	P	0.08	0.96	0.07	0.75	0.06	0.59
	Y	697	108	757	114	853	125
Madhya Pradesh	A	4.43	35.86	5.03	35.52	5.14	35.22
	P	2.10	26.20	2.35	25.00	2.59	26.04
	Y	473	73	467	70	503	74

Maharashtra	A	0.11	0.91	0.07	0.52	0.07	0.51
	P	0.05	0.57	0.03	0.34	0.03	0.27
	Y	405	63	430	65	348	51
Rajasthan	A	0.18	1.49	0.33	2.29	0.24	1.68
	P	0.17	2.08	0.37	3.92	0.25	2.49
	Y	878	136	1118	168	1001	147
Uttar Pradesh	A	5.08	41.11	5.83	41.12	5.93	40.64
	P	3.77	47.14	4.31	45.72	4.80	48.38
	Y	741	115	740	112	813	119
Uttarakhand	A	*		0.06	0.43	0.16	1.12
	P			0.04	0.44	0.08	0.84
	Y			270	41	520	76
West Bengal	A	0.49	3.95	0.67	4.69	0.65	4.43
	P	0.33	4.10	0.49	5.18	0.46	4.64
	Y	671	104	724	109	712	104
All India	A	12.35		14.17		14.58	
	P	8.00		9.42		9.93	
	Y	647		664		682	

(A = Area Lakh ha, P = Production Lakh tonnes, Y = Yield kg / ha)

* Newly carved states during 2000

Table 1.9 Top potential districts (2003-04)

Rank	Name of District	State	Area (lakh ha)		Production (lakh tonnes)		Yield (Kg/ha)	
			Area	% to India	Prod	% to India	Yield	YI
I	Jalaun	UP	0.389	2.79	0.452	4.36	1160	156
II	Patna	Bihar	0.289	2.07	0.365	3.52	1260	170
III	Bahraich	UP	0.460	3.29	0.371	3.57	810	109
IV	Barabanki	UP	0.307	2.20	0.315	3.04	1030	139
V	Hamirpur	UP	0.308	2.21	0.311	3.00	1010	136
VI	Lalitpur	UP	0.219	1.57	0.255	2.46	1160	156
VII	Sitapur	UP	0.318	2.28	0.254	2.45	800	108
VIII	Shivbasti	UP	0.234	1.68	0.196	1.89	840	113
IX	Murshidabad	WB	0.246	1.76	0.191	1.84	770	104
X	Kheri	UP	0.235	1.68	0.181	1.74	770	104
	Total above		**3.005**	**21.52**	**2.891**	**27.86**	**962**	**129**
	All India		**13.964**		**10.378**		**743**	

Fig. 1.4 GAP : Well managed lentil showing marked difference of dates of sowing

Fig. 1.5 (a) Poorly managed lentil crop at early growth stage

Fig.e 1.5 (b) Good managed lentil crop at early growth stage

Table 1.10 Sequential and Inter cropping

Sequential cropping		
kharif fallow – Lentil (rainfed areas)	Bajra – Lentil	Groundnut-Lentil
Paddy – Lentil	Cotton – Lentil	Jowar-Lentil
Maize – Lentil		
Inter cropping		
Lentil + Sugarcane (Autumn): two rows of lentil at 30 cm row spacing in between two rows of sugarcane	Lentil + Linseed (2:2)	Lentil + Mustard (2:1)

Soils suitable for lentil production

Indian soil suitable for lentil cultivation

Lentil is raised on light loams and alluvial soils in upper India, and on well-drained, moderately-deep, light-black soils in Madhya Pradesh and Maharashtra. It is also grown on low-lands, poor soils. In Punjab, it is often cultivated on inundated lands. The crop can withstand moderate alkalinity. It cannot tolerates water logging. Soil should be made friable and weed free so that seed could be placed at a uniform depth. In case of light soils, less tillage is required to prepare an ideal seed-bed. In heavy soils, one deep ploughing followed by 2-3 cross harrowing should be given. After harrowing, the field should be levelled by giving a gentle slope for easy irrigation. There should be proper 5 moisture in the soil at the time of sowing for proper germination of seeds. In utera/paira cropping system of rice-lentil in eastern Uttar Pradesh, Bihar, Orissa, Madhya Pradesh, seeds are sown in standing rice crop and therefore no tillage is done.

Loam soil

This soil consists of sand, silt and clay to some extent. It is considered to be the perfect soil. The texture is gritty and retains water very easily, yet the drainage is well. There are various kinds of loamy soil ranging from fertile to very muddy and thick sod. Yet out of all the different kinds of soil loamy soil is the ideal for cultivation. The deep, sandy loam soils high in phosphorus and potassium.

Alluvial soils

This soil belong to soil order entisol, occupy about 15 lakh sq. km of area stretching from the river Satluj in the west to the Brahmaputra valley in the east. They also occur in the valleys of Narmada and Tapi in Madhya Pradesh and Gujarat; Mahanadi in Chhattisgarh and Orissa; Godavari in Andhra Pradesh; Krishna in Karanataka and Andhra Pradesh and Kaveri in Tamilnadu. Along the coast of Kerala these are called coastal alluvium and in the deltas of the Mahanadi, Godavari, Krishna and Kaveri as deltaic alluvium. These soils are mainly derived from the debris brought down from the Himalayas or from the silt left out by the retreating sea. Their colour varies from light grey to ash grey and the texture is sandy to silty loam. These soils are both well drained and poorly drained. At some places these have high textured sub soil while at other places the subsoil is low textured.

These soils have an immature profile in undulating land while the flat lands have a more mature profile. The coastal allu-vial soils are of tidal origin. The two sub-types-(i) Alluvial loam, and (ii) Calcareous-clavey alluvial are identical with the low-lying younger, floodplain alluvium locally known as Khadar and the older alluvium seldom liable to inundation known as Bhangar respectively. "The Khadar occupies the riverine flood plains of the rivers and is enriched by fresh silt deposits every year. Its alluvium is light coloured and poor in calcareous matter consisting of deposits of sand, silt, mud and clay" (Tiwari, 1984, p. 3). A little of Khadar can be assigned to the Upper Pleistocene and most to the recent age (Krishnan, 1960, p. 574). The Bhangar lands are level plains above the normal flood limit. Here the alluvium is of dark colour (pale reddish brown) generally rich in concretions, and nodules impure calcium carbonate or Kankar (Mukerji, 1964, p. 100). In drier areas it also exhibits stretches of saline and alkaline efflorescence's known as reh. The main constituent of Bhangar is clay while at places it gives way to loam and sandy loam (Shafi, 1968, p. 31). It also occupies the Pleistocene terrace dating back to middle and upper Pleistocene periods. The depressions in Bhangar containing clay loam to clay (grey to dark grey in colour) are better suited to rice crops, while the rest is used for other crops. Bhangar soils require recurrent manuring to main-tain their fertility. The Khadar soils, although rich in humus, yield only one crop during rabi season after the retreat of the floods. Alluvial soils are rich in potash, phosphoric acid, lime and organic matter but deficient in nitro-gen and humus contents. These soils are capable of fixing nitrogen very rapidly through leguminous crops. The percentage of phosphoric acid in the oven dry soil varies from 0.005 to 0.02; humus from 0.8 to 0.9; potash from 0.1 to 0.35; and lime less than 7.0.These soils are well suited to irrigation which helps in raising a number of crops like rice, jute, sugarcane, wheat, cotton, maize, oil seeds, fruits and vegetables. These constitute the rice and wheat bowls of India and support large clusters of population.

Black soils

This soil belong to soil order vertisol, locally called regard or black cotton soils, and internationally known as 'tropical black earths' or 'tropical chernozems' have been developed by the weathering of the Deccan lava in majorparts of Maharashtra, western Madhya Prades (Hoshangabad, Narsinghpur, Damoh, Jabalpur, Raisen and Shahdol districts), Gujarat (Surat, Bharuch, Vadodara, Kheda, Sabarkantha and Dang districts), Andhra Pradesh (Adilabad, Warangal, Khammam, Mahbubnagar, Kurnool, Guntur and Karimnagar districts), Karnataka (Bijapur, Dharwar, Gulbarga, Bidar, Belgaum, Raichur, Bellari and Chitradurga districts), Rajasthan (Kota, Bundi, Sawai Madhopur, Bharatpur and Banswara districts), Tamil Nadu (Ramnathpuram, Tirunelvelli, Coimbatore, Madurai and South Arcot districts) and Uttar Pradesh (Jalaun, Hamirpur, Banda and Jhansi districts). According to Krebs the regur soil is essentially a mature soil which has been produced by relief and climate rather than by a particular type of rock. It occurs where the annual rainfall is between 50 cm to 75 cm and the number of rainy days is from 30 to 50. The colour of these soils varies from deep black to light black and chestnut and is dependent on the colour of the mechanical fractions. The black color is attributed to the presence of titaniferous magnetite, compounds of iron and aluminium, accu-mulated humus and colloidal hydrated double iron and aluminium silicate. In general these soils have clay texture, average clay content being 50% and the range being 40-50%. Except in cases where there is stratification, the clay content down the profile is uniform. The structure of these soils is usually clod-dish but occasionally friable. Regard soils are calcar-eous neutral to mild alkaline in reaction, high in carbon exchange capacity and low in organic matter. In general these soils are rich in iron, lime, calcium, potash, aluminium and magnesium carbonates but poor in nitrogen, phosphorus and organic matter. The chemical test near Nagpur yields soluble matter 68.71%, ferric oxide 11.24%, alumina 9.3%, water and organic matter 5.83%, lime 1.81% and magne-sium 1.79%. Black soils are highly retentive of moisture, extremely compact and tenacious when wet, consid-erably contracted developing deep wide cracks on drying and self-ploughing. Black soils are credited with high fertility. These are well suited to legumi-nous crops like cotton, turn and citrus fruits. Other crops include wheat, jowar, millets, linseed, castor, tobacco, sugarcane, safflower, vegetables etc. On the uplands these soils are comparatively less fertile than on the low lands.

Bundelkhand soils

Above an impervious layer of rock that is found at depths of 6 to 15 metres, several kinds and grades of soil are found across Bundelkhand. Broadly, the soils fall into two categories: red soils and black soils. Across Bundelkhand, soils of both categories have poor organic content.Red soil, locally called pathari, are found over granites, from which it is derived, and is seen in shades of brown, yellow or grey, depending on the iron content. The soil is rich in potash, but has poor phosphate and nitrogen constituents. It is hence not suitable for farming; some varieties of trees can however be grown in it.A yellowish, light-coloured variety of red soil, called **parua**, is sandy and has some clay content. It is well aerated and easily accepts water. Parua soil is found across UP Bundelkhand and is suited for cultivation of wheat.Two kinds of black soils are found Bundelkhand. The first, known as **kabar**, resembles black cotton soils of central India. Its colour varies from dark black to grey black and brownish black. It has high clay content; hence it is highly adhesive, retains moisture, and quickly turns dry and into hard blocks. Kabar is found in both Bundelkhand Upland and Bundelkhand Plains (see Bundelkhand Sub-regions). In UP, it is seen in parts of Lalitpur, Jalaun, Banda and Hamirpur. In MP the soil is found across Sagar district and in many other places, such as the Sonar valley in Damoh district, where one commonly found local variety is called mund; there are also varieties with other names. The soil is suitable for cultivation of gram and jowar, but requires careful handling.The second variety of black soil, called **mar**, is what is generally called black cotton soil. It has high clay content and is prone to water logging. The soil has relatively high organic matter content, and hence can be cropped without use of fertilisers. However, like kabar, it requires careful handling. Time for ploughing has to be selected carefully; the soil can be either too dry or it too wet for tilling. Mar is Bundelkhand's most prized soil, and wheat, gram and rice are cropped on it. The soil is found in parts of the districts of Jalaun (Konch tehsil), Hamirpur (Maudaha, Rath tehsils), Jhansi (Mauranipur, Moth) and Banda (Banda tehsil). In small patches, it is found in other districts as well, such as in southern Lalitpur, and parts of Sagar district. Apart from the red and black soils, some varieties of loamy, alluvial soils are found along river banks. Among these, the **rankar** variety is stony and found on sloping surfaces of ravine land. It is a highly degraded kind of coarse soil. However, where irrigation is available, the soil is suitable for cultivation of jowar, til and bajra. A variety of rankar, called moti rankar, is deteriorated mar soil and can be made productive; the soil is hence prized in parts of Tikamgarh and Jhansi districts, where it is found. A variety of rankar in Banda district has stones of various sizes. Dark soils found along river beds, known as kacchar and tari soils, are used to grow water melons and vegetables.

Canadian soils suitable for lentil cultivation

In Saskatchewan, Canada the lentil is best adapted to the Brown and Dark Brown soil zones, but can be grown successfully in the Thin Black soil zone in years without excessive moisture. In the Brown soil zone, lentil performance is best on fallow in medium to fine textured soils, or on stubble using direct seeding techniques. In the Dark Brown, moist Dark Brown and Thin Black soil zones, lentil grown on stubble tends to receive a stress needed to reduce the time to maturity, prevent excessive vegetative growth and reduce the risk of damage from early fall frost. To date, most lentil production in Saskatchewan has been of the large green market class, but red lentil production has recently grown to equal or greater proportion. There is opportunity to expand lentil production into the moist Dark Brown and Thin Black soil zones by increasing red and small green production. These more determinate red and small green varieties are better suited to the cooler moist lentil growing areas of the province. Research is ongoing to determine the best varieties and agronomic practises for lentil production in these areas.

In **India,** it can be concluded that the lentil is best adapted to the all soil types, from sand to clay loam, if there is good internal drainage. Lentil does not tolerate flooded or waterlogged soils, and does best on deep, sandy loam soils high in phosphorus and potassium. Good drainage is required, because even short periods of exposure to waterlogged or flooded field conditions kill plants. A soil pH near 7.0 is best for lentil production. In **Canada,** lentil is best adapted to the Brown and Dark Brown soil zones, but can be grown successfully in the Thin Black soil zone in years without excessive moisture. In the Brown soil zone, lentil performance is best on fallow in medium to fine textured soils, or on stubble using direct seeding techniques. In the Dark Brown, moist Dark Brown and Thin Black soil zones, lentil grown on stubble tends to receive a stress needed to reduce the time to maturity, prevent excessive vegetative growth and reduce the risk of damage from early fall frost. To date, most lentil production in Saskatchewan has been of the large green market class, but red lentil production has recently grown to equal or greater proportion. There is opportunity to expand lentil production into the moist Dark Brown and Thin Black soil zones by increasing red and small green production. These more determinate red and small green varieties are better suited to the cooler moist lentil growing areas of the province.

Seed and sowing

Recommended sowing time for rainfed is First fortnight of October in Central and South India and second fortnight of October in North India; under irrigated condition first fortnight of November in North India and for late sowing first week of December in rice fallows of NEPZ or on fields vacated very late by *kharif* crops under irrigated condition. For small seeded 40-45 kg/ha; bold seeded varieties and, late sown condition 50-60 kg/ha and under utera cropping 60 kg/ha seed is recommended. Treat the seed with thiram or carbendazim @ 2 g/kg of seed before 3 days of sowing followed by dual seed inoculation (culture of *Rhizobium* + PSB, one packet each for 10kg seed. Based on the region, time of sowing and purpose of cultivation, recommendations etc selection of variety from table 11 is recommended. However situation specific varieties for Rice-lentil cropping system are i) *Utera cultivation*: PL-406, PL-639, Arun and, ii) *Late sowings*: PL-406, PL-639, IPL-15, Narendra Masoor-1.

Fig. 1.6 Faba bean: An potential pulse crop of eastern region grown under faba bean + lentil cropping system.

Table 1.11 Recommended lentil varieties/characteristics

Variety	Source	Year of Release/ Notification	Area of adoption Zone/State	Ave. yield (Q/ha)	Duration	Characteristics
Pant L 406	GBPUAT	1980	NWPZ NEPZ	12-16	140	Rust resistant, small seeds
Asha(B77)	BCKV	1980	West Bengal	12-17	122-125	Rust resistant, small seeds
Pant L 639	GBPUAT	1980	NWPZ NEPZ	12-16	140	Rust resistant, small seeds
Vipasha	HPKV	1982	HP	10	170-180	Rust res., bold seeds
LL 56	PAU	1983	Punjab	12-17	150-155	Rust resistant, small seeds
Ranjan	BCKV	1984	West Bengal	15-16	120-125	Small seeds
Mallika(K 75)	CSAUAT	1986	NEPZ & CZ	14	135	Bold seeds
Arun	RAU, Dholi	1986	Bihar	12	130	Bold seed, tol. to rust
L 147	PAU	1988	Punjab	14	140	Small seed Rust resistant
JL 1	JNKVV	1991	MP	8.0	120-125	Early, Tol. to wilt, *Seed bold*
Sapana (LH 84-8)	CCSHAU	1991	NWPZ	15.0	135-140	Tolerant to Rust & Bold Seeded
VL Masoor 4	VPKAS	1991	Uttarakhand	12.5	168	Tol. to wilt & Rust, *Small seeded & black.*
Pant lentil-4 (PL-81-17)	GBPUAT	1993	NWPZ	16.0	140-145	Resistant to Rust & tolerant to wilt.
Lens-4076	IARI	1993	NWPZ CZ	14.0	130-135	Tolerant to wilt & Rust. *Seed bold*
DPL-15 (Priya)	IARI	1995	NWPZ	15-18	130-135	Tolerant to wilt & Rust, ***bold seeded.***
Pusa Vaibhav (L-4147)	IARI	1996	NWPZ	20-24	130-135	Res. to Rust & Tol. to wilt, *small seeded.*
Garima (LH-84-6)	CCSHAU	1996	Haryana	15-20	135-140	Tol. to Rust, wilt & Blight. *bold seeded.*
Narendra Masoor-1	NDAUT	1997	Uttar Pradesh	14.0	125-130	Resistant to Rust & Tol. to wilt.
DPL-62 (Sheri)	IIPR	1997	NWPZ	17.0	130-135	Resistant to Rust & wilt, ***bold seeded.***
Subrata	BCKV	1998	West bengal	12-18	120-125	Tolerant to Rust, ***bold seeded.***
JL-3 Jawahar lentil-3	JNKVV	1999	CZ	15-19	115-120	Tolerant to wilt, *bold seeded.*
VL Masoor 103	VPKAS	2000	Uttarakhand	12-14	1645	Tolerant to Rust, ***small seeded.***
Pant Lentil-5	GBPUAT	2001	Uttarakhand	15-18	135	Resistant to Rust, *bold seeded.*
Noori (IPL-81)	IIPR	2000	CZ	17-18	110-120	Tolerant to Rust, wilt, *bold seeded*
KLS 218	CSAUAT	2005	NEPZ	14-15	125-130	Tolerant to Rust, wilt, small seeded
Malaviya Vishwanath (HUL 57)	BHU	2005	NEPZ	14.0	130	Resistant to rust & wilt, ***small seeded.***
VL-Masoor-507	VPKAS, Almora	2006	J&K, H.P., Uttarakhand, North Eastern Hills	10-12	140-209	Resistant to wilt

Haryana Masar-1 (LH-89-48)	CCSHAU	2006	Haryana	14	138	Moderate resistant to all disease
VL Masoor 125	VPKAS, Almora	2006	Uttarakhand	20		Resistant to wilt
VL Masoor 126 (VL-126)	VPKAS, Almora	2007	Uttarakhand, H.P., J&K and North Eastern Hills	12-13	126-212	Resistant to GM and Moderately resistant to wilt and rust
IPL-406 (Angoori)	IIPR	2007	NWPZ	17	120-155	Resistant to rust and wilt

CZ- (MP, Maharashtra, Chhattisgarh, Gujarat) , SZ- (A.P., Karnataka, Tamil Nadu, Odisha) NEPZ-North East plane Zone (Eastern Uttar Pradesh, Bihar, Jharkhand, West Bengal). NWPZ- North West Plain Zone (Punjab, Haryana, Delhi, West UP & North Rajasthan) Res.= Resistant, Tol.= Tolerant, Mod.= Moderately,

Fig. 1.7 The Need of Hours : Quality seed materials of lentil

Plant nutrient management

On an average, production one ton of grain of high-yielding varieties of rice, remove about 22 kg N, 7 kg P_2O_5, 32 kg K_2O, 5 kg MgO, 4 kg CaO, 1 kg S and 40g Zn from the soil. Similarly for producing one tonne of biomass, lentil crop in general removes about 40 kg N, 3.9 kg P_2O_5, 85 kg K_2O, 8.0 kg S 7.0 kg Ca, 9.0 kg Mg, 35 g Zn from soil (Ahlawat and Ali, 1993). There is a need to ascertain and promote the uses of types of fertilizers required to correct the deficiency of all these nutrients especially zinc and sulphur. Zn deficiency is the most widespread micronutrient disorder in lowland rice and application of Zn along with NPK fertilizer increases the grain yield dramatically in most cases (Deokaran *et al.*, 2014; Singh and Singh, 2008). Deficiency symptoms of Zn in rice crop and sulphur in lentil are frequent and reduction in yield of rice and lentil is often blamed to zinc sulphur deficiency (Singh *et al.*, 2014; Singh and Bhatt, 2015). Being a legume it does not respond to nitrogen except for some types for initial boosting of growth whereas response to potash is inconsistent due to good 'K' supply status of most of the Indian soils. However, phosphorus definitely plays a vital role in root development, nodulation and growth and yield of the crop. General recommendation is 15-20 kg N and 50-60 kg 'P' as basal placement at soil depth of 10-15 cm during sowing/ last ploughing could be met easily through 100 kg DAP/ha .Lentil also responds positively to 'S' (20-40 kg/ha) giving an average nutrient use efficiency of 10-15 kg grain/kg S especially in light textured sandy loam soils of Northern India. SSP is the best source of 'P' followed by Gypsum and 'Pyrite'. Among micro-nutrient, Zn is most critical in intensive Rice-Wheat cropping system areas of Punjab, Haryana, Rajasthan (Eastern) U.P. and Bihar General recommendation is 25 kg zinc sulphate as basal, a foliar spray of 0.5% ZnSO4 + 0.25% lime (5 kg zinc sulphate + 2.5 kg lime in 1000 lt. of water per ha). 'Mo' and 'Fe' are the integral components of enzyme 'nitrogenous' for 'N' fixation. 'Mo' deficiency may create twin deficiency of 'N' and 'Mo'. 'Boron' and 'Mo' is found deficient in acidic soil of Eastern India hence 10 kg borax and 1 kg ammonium molybdate as soil application and foliar spray of 2% each of DAP and 'KCL' at pre flowering and pod development enhance yield by 10-15% along with increasing its ability to resist terminal drought.

Fig. 1.8 Adoption of clean cultivation, key to efficient application of good agronomic practices, leads toward bumper lentil output

INM Strategies

Application of 2-2.5 tonnes 'vermicompost' or 5 t FYM to the *'kharif'* crop in rotation and seed inoculation with efficient strain of Rhizobia takes care of the initial nitrogen requirement and there is no need to apply 'N' as booster (required especially in low fertile and paddy soils). Dual inoculation with 'Rhizobium' and 'PSB' takes care of 'N' as well as reduces 25-30% of phosphorus requirement by making available the initial fixed soil 'P' to the plants. Rhizobium inoculation is must after paddy as it is aerobic bacteria and most of its population dies during flooding and compaction for want of oxygen. In-situ management of rice straw/residues takes care of Zinc and other micronutrient and no need to apply them separately.

Fig. 1.9 Lentil: An important pulse crop of eastern region at pre-flowering stage showing sulphur and zinc treatments effect

Water management

Most critical stage for moisture stress is pod formation followed by flower initiation. In absence of winter rains and where contribution of soil moisture is negligible viz in Central India, two light irrigations may be applied for significant yield improvement.

Fig. 1.10 Managing water efficiently: Key to success for good harvest of lentil at the end

Weed management

Major weeds are *Chenopodium spp.* (bathua), *Fumaria parviflora* (gajri), *Lathyrus aphaca* (chatri matri), *Vicia sativa* (ankari), *Crisium arvense* (kateli), *Melilotus alba* (senji), *Asphodelus enuifolius* (jungli piaji), *Convolvulus avensis, Phalaris minor* and *Avena ludoriciana*. Orobanche, a parasitic weed is also seen as major problem at some places. Similarly *V sativa* adultrate the grain due to its size, shape and colour. One hand weeding/inter-culture at 30 DAS and another at 55-60 DAS, depending upon the intensity of weed infestation, provides efficient soil oxygen environment to rhizobium bacteria along with soil moisture conservation breaking soil capillaries, creating dust mulch (Singh *et al.*, 2015).

Fig. 1.11 Weed free lentil field: Insurance for lest incidence of insect pests and diseases and efficient management, if any

Plant protection measures

Lentil is in general pest free crop under normal condition if proper crop rotation is follows. However pod borer, aphids and Wilt *(Fusarium lentis)* are major insects and disease pests (Singh *et al.*, 2013b). Hence the control measures of all three are listed in Table 12.

Table 1.12 Pest and diseases in lentil and their management

Insect pest/ disease/causal organism	Nature of damage/ symptoms	Control measures
i) Pod borer	The caterpillar defoliates the tender leaves and also bores the green pods and feeds upon the ripening grains.	Cypermethrin (0.02%) or Endosulfan (0.07%) or Monocrotophos (0.04%).
ii) Aphids	Aphids suck the sap and in case of severe damage the growth is suppressed.	Metasystox or Monocrotophos (0.04%).
iii) Wilt *(Fusarium lentis)*	The growth of the plant is checked due to yellowing of leaves, drying of plants. The roots of affected plants remain under-developed and looks light brown in colour.	i) Seed Treatment with Thiram + Benomyl (1:1) x @ 3 g/kg of seed ii) Adopt crop rotation. iii) Use healthy seeds.
iv) Rust *(Uromyces fabje)*	Pink to brown pustules appear on leaves and stems. In severe attack, the affected plants may dry.	i) Grow early maturing/ duration variety. ii) Seed Treatment with Agrosan GN @ 2.5 g /kg seed. iii) Spray the crop with Maneb, Zineb or Ferbam @ 2.5 g/litre of water.

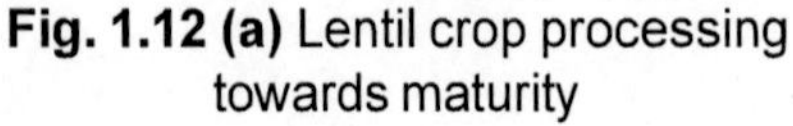

Fig. 1.12 (a) Lentil crop processing towards maturity

Fig. 1.12 (b) Lentil crop at podding stage

Harvesting, threshing, storage and yield

Crop become ready for harvest when leaves begin to fall, stem and pod turn brown or straw in colour and seeds are hard and rattle with 15% moisture inside them. Over ripening may lead to fall of pods as well as shattering and seed cracking if seed moisture fall below 10% due to delay in harvesting. The crop should be allowed to dry for 4-7 days on threshing floor and threshed by manually or bullock/ power drawn thresher. The clean seed should be sun dried for 3-4 days to bring their moisture content at 9-10%. The seed should be safely stored in appropriate bins and fumigated to protect them from bruchids. A good harvest may produced lentil seed of 14-18 q/ha.

Fig. 1.13 (a) Lentil crop at flowering stage

Fig 1.13 (b) Late sown lentil crop at podding stage

Major bottleneck in lentil production

Participatory rural appraisal (PRA), a widely adopted methodology to prioritized list of researchable issue, has been conducted by several workers to identify the important cause of such low productivity (Ali *et al.,* 2012 and Reddy, 2009). PRA techniques revealed several factor responsible for extremely low productivity related to inputs and their efficient management (Singh *et al.,* 2012). An earlier study revealed that area under lentil in mostly predetermined, but as the irrigated area increases, pulses are relocated to rainfed areas and their area is replaced by cereals or some cash crop (Singh *et al.,* 1995). In India, the irrigated area under pulses was only 12 per cent, while under wheat and paddy; it was more than 60 per cent of the total area (Reddy and Reddy, 2010). However, lentil is also grown as a paira crop in the eastern India, which helps in timely planting of the crop even before, the paddy has been harvested (Singh and Singh, 1995). Substantial increases in the production of lentil in the IGP can thus be achieved by fine tuning management aspects and thereby increasing total system productivity and sustainability (Ramakrishna *et al.,* 2000; Reddy, 2009 and Singh *et al.,* 2012). Poor soil and agro-climatic conditions not only compel late sowing of lentil, which leads to reduced length of growing period but also necessitate to sustain cold injuries at early vegetative phase which freeze all biological activities for prolonged period. A sudden rises in temperature after that, not only induces forced maturity but simultaneously invites several biotic stress viz., diseases and insects pests (Ali *et al.,* 2012; Reddy, 2009 and Singh and Singh, 2008). Traditionally lentil sowing is delayed up to last week of November and some time under extreme circumstances it goes up to the first fortnight of December, obviously due to reasons already explained. However, optimum sowing time of lentil is first fortnight of October (Singh *et al.,* 2011 and Ramakrishna *et al.,* 2000). Consequent upon delayed planting, early encounter with severe cold, growth and development of lentil crop gets hampered for a considerable period. Subsequently plants get comparatively less time to complete their lifecycle which, by and large forces maturity (Ramakrishna *et al.,* 2000). In India, normal sown lentil is a medium duration (130-150 days) crop, while under late sown conditions it is forced to complete its life cycle in 105±5 days (Joshi, 1998; Ramakrishna *et al.,* 2000; Reddy, 2009; Singh and Singh, 2008 and Singh *et al.,* 2012). Typically, late sown lentil undergoes three distinct phases and considerable degrees of phenological modifications are bound to happen. Eventually, lentil crop during its early seedling phase grows slowly due to its energy invested in the initial establishment (Singh *et al.,* 2002 and Singh *et al.,* 2012). However, in mid-phase, very insignificant growth and development is observed. This poses serious threat to realization of yield potential due to cold injuries. This phase is very important for creating source of channelizing the energy at later stage. In the last and most important phase lentil faces heat injury, resulting in early onset of reproductive phase, causing imbalance in resources and inputs, biotic stress and forced maturity (Joshi, 1998; Dixit *et al.* 2009; Reid *et al.,* 2011 and Singh *et al.,* 2012). Growth and development of lentil plant is comprehensively controlled by plant growth regulators (PGRs). To provide momentum to the vegetative growth auxins plays a vital role especially under low ambient temperature and cold rhizosphere regime (Singh and Bhatt, 2013). Major bottleneck in lentil Production are summarized here as under:

Fig. 1.14 (a) Lentil crop at harvest stage

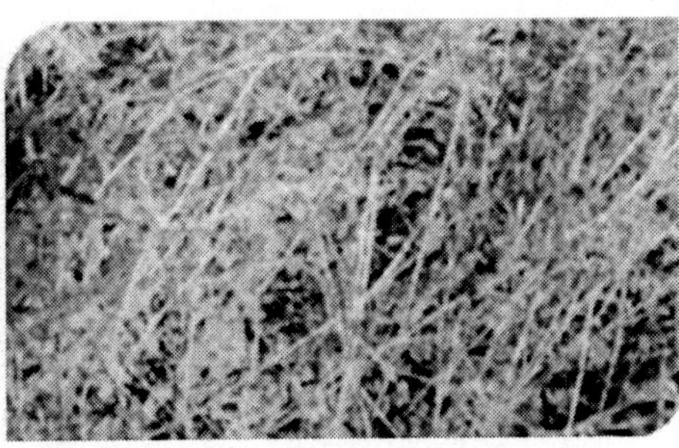

Fig. 1.14 (b) Over mature lentil crop, leads to huge losses

Fig. 1.15 (a) Manual harvesting is in progress

Fig. 1.15 (b) Harvested lentil bundelled and ready for further processing

a) Climatic factors

Lentil is mainly grown under rainfed conditions except in few districts of Uttar Pradesh, Madhya Pradesh, Rajasthan and Bihar. As a consequence area under pulses and their productivity are dependent on amount and distribution of rainfall. Rainfall intensity and distribution leads to vulnerability of lentil to water stress. Occurrence of mid-season cold waves and terminal heat during winter season has also been causing losses to crop productivity of lentil in many regions.

b) Soil related constraints

Lentil crops are generally very sensitive to acidic, saline and alkaline soil conditions. North-western states have extensive areas with high soil pH whereas eastern and north eastern states have chronically acidic soils. The problem has been compounded by rising deficiency of micronutrients such as zinc, iron, boron and molybdenum and that of secondary nutrients like sulphur particularly in traditional pulse growing areas. This emerges to an extent from the fertilizer subsidy policies. Recent incentives to specialty fertilizers ameliorate this stress. On the other hand, shallow and coarse textured soils in north and western states have low water rententivity and require irrigation for supporting a good rabi pulse crop.

c) Input quality and availability related constraints

Nutrient requirement of lentil is much lower than cereals mainly because of biological nitrogen fixation and relatively low productivity levels although lentil crops respond favourably to higher doses of fertilizer nutrients than generally applied or even recommended. But, since lentil are invariably subjected to abiotic stresses leading to sub-optimal nutrient uptake, farmers tend to use low doses of fertilizer nutrients. Further, nutrient use is unbalanced and seldom based on soil-test values. Timely availability of quality chemical fertilizers continues to be a problem in many pulses growing area. Inadequate availability of gypsum or pyrites as a cheap source of sulphur remains a serious impediment in many states/regions. Availability of pesticides (including herbicides) in most of the states has been comfortable but their quality in terms of effectiveness and eco-friendliness has been an issue in spite of a well-designed regulatory mechanism put in place.

d) Pests and diseases

Although lentil crops are prone to many insect pests and seed borne diseases, a major cause of concern as its incidence, if not controlled, devastates the crop. *Fusarium* wilt is wide spread in lentil growing regions. In addition, heavy damage to lentil grain is caused by pests during storage.

Fig. 1.16 Good crop stand of lentil, promised for good harvest at the end of crop cycle

e) Blue bull menace

Lentil is vulnerable to attack by Blue Bulls in the Indo-Gangetic Plains. Because of the widespread menace particularly in Uttar Pradesh, Bihar, Madhya Pradesh, Rajasthan and Chhattisgarh the potential area suitable for taking lentil crops is left uncultivated by the farmers. There is no viable strategy available in the country to effectively manage the menace.

a) Technological constraints

Lentil is grown under varied agro-climatic conditions (soil types, rainfall and thermal regime) in the country. This calls for region specific production technology including crop varieties with traits relevant to prevailing biotic and abiotic stresses. Even biological fertilizers and pesticides used should be based on strains isolated from regions with similar agro-climatic conditions for them to be effective. Our research and development programme in pulses has yet to appreciate and address this issue adequately. Production technology for a lentil crop has to be soil type/region specific. So is true for tillage and seeding device/gadgets.

Fig. 1.17 Blue Bull roaming freely

b) Infrastructural constraints

Rainfall received during maturity of *kharif* pulses, causes losses in yields and grain quality when farmers usually do not have pakka and covered threshing floor. Farmers also lack awareness and means for safe storage of grain/seed of pulses. Many areas are approachable only during fair weather. Warehousing facilities are either inadequate or inaccessible.

c) Credit and marketing related constraints

Farmers engaged in cultivation of lentil are mostly small and marginal. A majority are in areas with poor banking infrastructure.

They have poor resource base and lack risk-bearing capacity. They therefore either lack access to credit or turn defaulters. Delivery of credit to such farmers is also not hastle-free. There is lack of marketing network in remote areas. Procurement of produce by a dedicated agency is virtually non-existent or in-effective.

d) Policy related issues

System of regulating quality of inputs though in place in all the states, needs to be made more effective. Delivery of improved technology, inputs, credits need to be stream lined through appropriate policy interventions. Benefit of crop insurance need to be extended to pulses farmers.

Opportunities in lentil production

A wide spectrum of agro-climatic conditions, favourable thermal regime for almost year round cropping and availability of generally adequate rainfall point to the fact that there is a vast untapped potential for improving productivity of pulses and bringing additional area under pulses. There is overwhelming scientific evidence suggesting a vast gap between farmer's yield of pulses including lentil and front line demonstration plot yield. Further, a large chunk of rice fallow lands can be brought under lentil provided available land and water resources (soil moisture) are scientifically and innovatively managed. Exploitation of promising intercropping systems in rainfed/partial irrigated areas offers a vast opportunity for improving pulse production. Scientific management of "utera" cultivation of lentil in rice based cropping systems. Lentils are environment friendly crops that have the unique ability to fix nitrogen and thereby help improve soil health. Even though, they have low genetic potential in terms of realizing productivity as compared to cereals, they contribute to the environment protection. Opportunity exists in cultivation of pulses not only to increase production of other crops in the cropping system but also entitles pulses growers to claim Payment for Environment Services (PES) through carbon trading or other similar mechanisms.

Strategy for increasing production

The issues assigned to each sub-group were dealt in detail during presentations and subsequent discussion. The strategy and approach suggested to address each issue is narrated below:

Identification of additional area having potential for lentil crops following avenues for area expansion have been suggested.

1. **Utilization of potential area of rice fallow lands**

   The area left un-cropped after kharif rice is estimated to be around 11.65 million ha. The area is primarily rainfed and exists in the states of Bihar, Madhya Pradesh, Chhattisgarh, Orissa, Eastern Uttar Pradesh, West Bengal and Jharkhand. About 25% of this area has potential for supporting a rabi pulses after rice depending on soil type and depth. Thus, the 3 to 4 million ha additional area can be brought under rabi pulses. Assuming an average productivity of 600 kg/ha, the area can produce 1.8 to 2.4 million tons of pulses. Farmers need to be encouraged through various incentives and region specific extension strategy for cultivation of pulses in the identified districts. Necessary technological back up in terms of suitable short-duration varieties, nutrient application rates and other agronomic practices should come from local research stations. SAUs/KVKs may be mandated to conduct field demonstrations on pulses in rice fallow lands and train field staff and farmers participating in demonstrations.

2. **Replacement of low productivity crops with lentil and other pulses**

   About 5 lakh ha area of upland rice, 4.5 lakh ha area of millets and 3 lakh ha area under barley, mustard and wheat can be brought under rabi pulses. Rabi pulses such as lentil and chickpea should replace mustard, barley and wheat. If possible, harvested rain-water should be used for rabi crop establishment.

3. **Promotion of intercropping and utera cultivation of pulses**

   There are a good number of promising intercropping systems for pulses developed by Zonal Agricultural Research Stations. Farmers in rainfed states (Gujarat, Madhya Pradesh, Chhattisgarh, Maharashtra, Karnataka, and Andhra Pradesh) are familiar with some of them as they have been practicing them in traditional ways.

   The approach to be followed by the rainfed states should include:

   i) Identification of districts and promising intercropping systems for each agro-climatic zone and setting of area coverage targets.

ii) Conduction of field demonstrations on intercropping with farmer's active participation and comparing returns with sole cropping system.

iii) Ensuring availability of seed of pulse varieties recommended for intercropping.

iv) Demonstration of suitable seeding devices (animal drawn and tractor drawn) for simultaneously planting of main and intercrop components.

v) Seed-minikits of pulses may be given to farmers opting for intercropping only.

vi) KVKs at districts level should be involved in training of farmers and field demonstration of production technology.

4. **Promotion of utera cultivation of lentil and field pea in rice fields**

Utera cultivation is a practice commonly and traditionally followed in tribal regions of Madhya Pradesh, Chhattisgarh, Bihar, Jharkhand and can contribute to pulse production ensuring additional income to tribal rice growers.

5. **Rainfed areas developed as watersheds**

Multiple ministries/departments and agencies have been involved in the growth and development of watersheds with an array of watershed schemes. Three schemes of Drought Prone Area programme (DPAP), Desert Development Programme (DDP) and Integrated Wasteland Development Programme (IWDP) of Ministry of Rural Development have now been merged into Integrated Watershed Management Programme (IWMP). National Watershed Development Programme for Rainfed Areas (NWDPRA), River Valley Projects (RVPs), Catchment Area Programme in Flood Prone Rivers, Shifting cultivation scheme of Ministry of Agriculture, Hill Area Development Programme (HADP) and Western Ghats Development Programme (WGDP) of Planning Commission; various afforestation programmes and other national and externally aided projects (EAPs) are other programmes being implemented for developing land and water resources in rainfed areas.

Increasing crop productivity

Opportunities for increase in crop productivity exist in the form of new varieties of seeds developed for recording higher yields and through better crop management practices that make the pulses cultivates of more efficient. Strategies and approaches relating to the two aspects: promotion of quality seed varieties and promotion of efficient package of practices are discussed as under.

Promotion of quality seeds

Main issue relating to promotion of quality seeds is the availability of seed of promising varieties to the farmers in adequate quantities and in time. To increase supply of quality seeds, following measures are recommended.

a) Seed replacement/multiplication strategy

Use of good quality/certified seed in pulses has generally been low. Seed replacement Rate (SRR) estimated for the year 2006-07 was only 10.41%. Efforts made through various Government Sponsored Programmes such as Integrated Scheme of Oilseeds, Pulses, Oilpalm and Maize (ISOPOM), National Food Security Mission, (NFSM), Seed Village Programme etc have been successful in raising SRR of pulses to 22.5% by the year 2010-11.

Fig. 1.18 Lack of suitable genotypes and their quality seed is primary constraint for economic lentil production

Primary constraint for economic lentil production

Identification of best agronomic practices

Agronomic practices that have major impact on productivity of pulses include tillage, crop geometry, plant population, planting method and time, nutrient and water (rainwater and irrigation) management, seed treatment (with fungicides) and crop-specific bacterial cultures, weed management and plant protection. Crop-specific recommendations based on applied and adaptive research findings generated in different agro-climatic regions are developed by Zonal Agricultural Research Stations. The same are usually presented and finalized during Research-Extension Interface held between SAUs/ICAR Institutes and State Department of Agriculture and allied departments/agencies twice in a year. The recommendations that emerge are passed on to farmers for adoption. Thus availability of sound region, and crop-specific agronomic practices is adequate however, low Crop-wise regions for seed production with enhanced SMR and quality are indicated below.

Crop Region/State

In Uttar Pradesh, Madhya Pradesh, (Bundel- khand region) rate of adoption of improved agronomic practices has been a major constraint. Promising agronomic practices applicable across pulse growing regions are suggested as below:

a) Agronomic practices

Wide spread deficiency of zinc and sulphur in major pulse growing states and boron deficiency in acid soils of eastern and north eastern states has necessitated use of sulphur containing fertilizers and zinc sulphate as a source of zinc. Sulphur application @ 20-40 kg/ha (through gypsum, SSP) at sowing and zinc sulphate @25-50 kg/ha once in two years effectively address the problem and tend to maximize crop productivity. Correction of soil pH has a major role in nutrient and water use efficiency and consequently on crop yield. Use of gypsum in western states and liming in eastern and parts of southern states at the rates prescribed by SAUs/ICAR research centres located in the region is strongly recommended. Nitrogen requirement of pulses is much higher than that of cereals. However, most of the requirement is met through biological N-fixation. It is, therefore important that farmers are encouraged to adopt agronomic practices that facilitate N-fixation. Minimizing magnitude and duration of moisture stress period for lentil cropping season. Weed infestation has been observed to cause heavy yield losses in lentil pluses. Use of chemical herbicides particularly during *kharif* season needs to be promoted through incentives and appropriate extension strategy as frequent rains and too wet soil conditions do not allow mechanical/ manual weeding. A number of cost-effective herbicides are available in the market. Local research stations can provide accurate recommendation in this regard. Technologically advanced private sector companies with a wide distribution network must be involved in pesticide and herbicide propagation for yield expansion.

b) Pest surveillance mechanism and pest management practices

It is important that region specific advisories are issued for guiding pulse growers on pest control. This calls for an effective pest surveillance mechanism to be put in place at district level. National Centre for Integrated Pest Management (NCIPM) has come out with a model for pest surveillance in cotton, pigeon pea, chickpea and other crops and tested the same in Maharashtra state. The model is being demonstrated in some selected NFSM-pulses states. The model has met with notable success. As a consequence, other states such as Orissa, Gujarat are keen to adopt. The key features of the mechanism followed are as below Demonstration of IPM module in farmers' participatory mode.

i) Capacity building of farmers, Subject Matter Specialists and Extension functionaries.

ii) Conduction of roving as well as fixed plot surveys on weekly basis so as to provide real time data base for use by National Pest Reporting and Alert System established at NCIPM.

iii) Periodic release of pest advisories using electronic media.

iv) Ensuring that the advices are complied with through provisioning of the needed biological or chemical pesticides.

It is therefore suggested to adopt the model with defined priorities.

c) Innovative method of better dissemination and adoption of agronomic practices

Agronomic practices to be disseminated have to be not only region/agro-climatic zone-specific but should also match the resource-base of the farming community. Similarly, extension strategy to be followed should take into account the prevailing socio-economic status of farmers. It is however widely observed that training of farmers and field demonstrations of improved production practices are effective extension methods. Timely availability of credit and critical inputs facilitates adoption of a practice which has been popularized. Innovative ways of institution building that aggregates the produce of scattered lentil farmers and links them up with the businesses for better quality of inputs and for efficient marketing of the produce need to be found. Building farmer-producer organization discussed separately is an example of this approach. More such approaches using Non-Government Organizations and private sector should be tried. Producer Companies and other strategic PPP models need to be followed with priority.

Fig. 1.19 Line sowing, adoption of better cultural practices leads to good crop harvest with least attack of various pests

d) Mechanization in pulses

Lentils are grown in different agro climatic regimes and soil conditions can vary. In many soils mechanisation is essential to raise productivity. Adoption of many scientifically sound and economically viable agronomic practices requires use of certain farm machines/ implements. Very good examples are deep ploughing, ridge planting, line sowing, inter-culture operations etc. Besides, mechanization contributes to timeliness of operations, reduces cost of production and improves resource (water, energy and inputs) use efficiency. Another important aspect on farm mechanization is that it needs to be promoted on 'service' approach. Considering small holding of the farmers, custom hiring of the machines is the only viable option for increasing the reach. In this context, example of 'Haldhar' program of Madhya Pradesh Government is a good practice that subsidizes the farmers to the extent of Rs. 2000/- per hectare for deep ploughing of their lands.

Government can facilitate farmers' access to the recommended machines by empanelment of the vendors for the services against a fixed fee/charges. This way instead of subsidizing capital of individual farmers, the subsidy amount on service charges would generate demand for capital creation that could be served from institutional finance and ensure more optimal utilization of assets. The following suggestions are worth consideration for promotion of farm mechanization.

i) Make an assessment of operations that if mechanized will contribute to crop productivity, resource use efficiency and reduction in cost of cultivation in each agro-climatic region.

ii) Make an assessment of availability of farm power.

iii) Identify farm implements/machines based on the above.

iv) Organize demonstrations of identified farm implements/ machines to pulse growers.

v) Involve prominent manufacturers of farm implements/machines for manufacturing, establishing supply outlets of identified machines/implements and providing after sales service.

vi) Provide financial assistance for purchase of implements/ machines.

e) Post harvest handling of grains for reducing losses

i) Mechanical threshing needs to be promoted by providing incentives for purchase of threshers.

ii) Procurement of pulses grains by Govt. authorized organizations will considerably reduce the need for storage at farmers level.

iii) Small Dal-mills should be popularized and promoted through various incentives.

iv) Private sector should be encouraged to establish' Dal Mill's in rural areas/districts with large acreage under pulses on the pattern of sugar mills. Private companies need to be involved in processing, packing and marketing of pulses. The public sector procurement agencies are severely handicapped for funds and expertise in this area.

f) Expansion of irrigation using resource conservation technologies

Pulses crops are invariably grown under moisture stress which leads to sub-optimal productivity levels. Scientific scheduling of irrigation, an estimate of quantity of water to be applied and deployment of water saving irrigation methods can lead to enhanced yield, higher water and nutrient use efficiency and larger area coverage under irrigation. Use of sprinkler irrigation has enormous potential for saving irrigation water and expanding area under irrigation. The method has gained popularity in many districts with limited water resources. Drip irrigation attention of the policy makers, administrators, social workers, as it has assumed social, economic and ecological dimensions. The problem was discussed at length in a Brain Storming Session chaired by Agriculture Commissioner and some useful recommendations emerged. It is suggested that a committee consisting of senior officers from Department of Agriculture, Department of Forestry and wild Life, and Department of Animal Husbandry may be constituted with a mandate to come out with practicable recommendations including policy interventions in a time bound manner as the issue is too serious to be ignored or left unattended for long, and fertigation hold promise for widely spaced crops like pigeonpea. These devices can expand irrigation area by 30-50%.

g) Control of damage by blue bull

The damage caused to pulses crops by blue bull has been on the rise in the extent and magnitude. The problem has become so acute that area of pulses in general and summer mungbean in particular has witnessed drastic reduction in the states of Punjab, Haryana, Rajasthan, Uttar Pradesh and Gujarat. Although, the problem has been in existence for decades, no socio-economically viable control measures have been evolved and implemented. Pulse growers continue to suffer heavy economic losses.

Pulses control order

Regarding stock limits, the concerned Food and Supply Department of State Government who is implementing pulses control order need to be approached. Different states have different license requirements and stock limits imposed on pulses. Only Madhya Pradesh does not have a stock limit imposed on domestic pulses. Gujarat has done away with control order license but has a stock limit. Due to these reasons, none of the big trading companies deal in domestic pulses. Applying through the right department in every state for EC license as well as obtaining the EC license is a complex and time consuming exercise. States like Delhi are not issuing Pulses Dealer license for over one year. All channel partners have to also apply for licenses. Generally, the official time limits vary from 1-2 months, while the actual time taken is much longer. New methods for marketing should be devised to supplement some of the shortfalls in specific pulses crops. For example, Yellow Dal is being aggressively promoted by Ministry of Consumer Affairs through publicity campaign that not only succeeded in introducing the split field pea for consumption as dal but also promoted its nutritive and culinary benefits in preference to the gram dal or toor dal. New avenues need to be explored to manufacture dal from other vegetative nutritive base that meets the culinary and nutritive requirement of vast majority of Indians.

National Pulses Development Board

Sustainable availability of the pulses in general and kharif pulses like Red Gram, Black Gram and Green Gram in particular in the context of limited global marketable surplus and increasing demand makes it important to ensure that these crops get special attention. It is necessary to view various functions under the pulses supply chain from the farmers' field to the consumers' plate under a single integrating umbrella to bring in more efficiency and develop a holistic approach to pulses development. For production, marketing, processing, retailing and consumption, information relating to the specialized functions performed by different departments, agencies and private sector could be collected, collated and analysed by a central institution for knowledge exchange to bring about efficiency gains in the supply chain and for timely policy and programmatic responses so as to improve the availability of pulses in the market. It is proposed that on the lines of Coconut Development Board, National Pulses Development Board may be constituted as a Central knowledge bank on pulses. This one stop administrative structure on pulses would encompass knowledge exchange on production, markets, trade, processing, manufacturing and consumption. It could be an autonomous body of expertise covering all aspects of pulses. Mandate and the functions of the Board could be worked out separately. Board can be given an initial corpus for performing the mandated functions. A revenue model would need to be evolved for sustaining its efforts.

Summary

Forgone discussion it is clearly visualized that, this is the high time to frame policies for oilseeds and pulses in general and with special focus to this region. In the technology front, good agronomic practices (GAP) their different components like upgraded cultivars having potential to perform better under change climate condition, ensuring optimum plant stand through adjustment in spacing and thinning, site specific nutrient management (SSNM), efficient weed management, protective irrigation need-based plant protection, etc. have profound influence on the productivity of different oilseed and pulses. The improved technology packages were also found to be economically attractive. There is need to adopt the all component of advocated technology as a unit not to choose few of them at will , which were leading to several complication soil health hazards and poor response of technology as compared to demonstrated through FLD etc. Moreover adoption of numerous modules of technology was low, highlighting the necessity for better diffusion. For efficient and smooth realization of set goal steps as mentioned below may be taken on priority basis (1) Encouraging accelerated adoption of present technology for bridging the yield gap (2) pulses and oilseed grower are poorer among the poor farming community hence institutional support is inevitable, which not only boost seed replacement rate but quality production as well (3) there must be provision for life-saving irrigation in pulse and oilseed growing pockets on priority basis(4) guaranteeing availability of critical inputs viz., seed fertilizer, pesticides (5) gradual mechanization for pulse and oilseeds production (6) public–private partnership for sustaining chain of producer to consumer and to minimizing post-harvest losses due to better-quality machines for harvesting, threshing and more importantly (7) policy support for value chain for pulse is need of the hour.

References

Ahlawat IPS and Ali M. 1993. Fertilizer management in food crops. In: Nutrient Management for Pulses - A Review (Tandon HLS Eds.). pp. 114-138. FDCO, New Delhi.

Ali M and Gupta S. 2012. Carrying capacity of Indian agriculture: pulse crops *Cur. Sci.* 102(6): 874-881.

Ali RI, Awan TH, Ahmad MM, Saleem U and Akhtar M. 2012. Diversification of Rice-Based Cropping Systems to Improve Soil fertility, Sustainable Productivity and Economics. *J. Animal & Plant Sciences* 22(1): 108-112.

Anonymous. 2011. Agricultural Statistics At a Glance 2011, Department of Economics and Statistics, Department of Agriculture and Cooperation , Ministry of Agriculture, Govt. of India.

Cubero, JI. 1981. Origin, taxonomy and domestication In: Lentils (eds C.Webb & G.C. Hawtin). Commonwealth Agricultural Bureau, Slough, England.

Das TK. 2008. Weed Science: Basics and Application. Jain Brothers, New Delhi, 901pp.

Deokaran, Singh SB and Ram Kewal. 2014. Effect of zinc and boron application on Yield of lentil and nutrient balance in the soil under indo-gangetic plain zones. *Journal of AgriSearch* 1 (4): 206-209.

Dixit GP, Katiyar PK, Singh BB and Shivkumar. 2009. Lentil varieties in India. AICRP on MULLaRP. Indian Institute of Pulses Research, Kanpur.

Guilfoyle T and Hagen G. 2001. Auxin response factors. *J. Plant Growth Regul.* 10: 281-291.

Harlan, JR, de Wet JMJ and Stemler ABL .1976. Plant domestication and indigenous African agriculture. In: Harlan JR, de Wet JMJ and Stemler ABL (eds) Origin of African plant domestication. Mouton, The Hogue, Netherlands.

Joshi PK.1998. Performance of Grain Legumes in the Indo-Gangetic Plain in Residual Effects of Legumes in Rice and Wheat Cropping Systems of the Indo-Gangetic Plain. In: Legumes in Rice and Wheat Cropping Systems of the Indo-Gangetic Plain. (Kumarrao JVDK, Johansen C and Rego TJ Eds.). pp. 3-13. ICRISAT, Patancheru, Andhra Pradesh.

Ramakrishna A, Gowda CLL and Johansen C.2000. Management factors affecting legumes production in the Indo-Gangetic Plain. In: Legumes in rice and wheat cropping systems of the Indo-Gangetic Plain - constraints and opportunities. (Johansen C, Duxbury JM, Virmani SM, Gowda CLL. Eds.). pp. 156-165. ICRISAT, Patancheru, Andhra Pradesh.

Reddy AA and Reddy GP. 2010. Supply Side Constrains in Production of Pulses in India: A Case Study of Lentil. *Agricultural Economics Research Revie.* Vol. 23 January-June 2010 pp 129-136.

Reddy AA. 2009. Pulses Production Technology: Status and Way Forward. *Economic & Political Weekly*.44(52): 73-80.

Reid DE, Ferguson BJ, Hayashi S, Lin YH and Gresshoff PM. 2011. Molecular mechanisms controlling legume autoregulation of nodulation. *Annals of Botany* 108: 789-795.

Singh AK and Bhatt BP 2015. Late sown lentil performance in response to foliar application of zinc. *Bangladesh Journal of Botany* 44(1): 125-128.

Singh AK and Bhatt BP. 2013. Effects of foliar application of zinc on growth and seed yield of late-sown lentil. *Indian J.Agril. Sci.* 83 (6): 622-626.

Singh AK and Singh NP. 2008. Yield and uptake of primary nutrients by large seeded varieties of lentil under varying seed rates in normal and late sown conditions. *Journal of Food Legumes* 20(2): 187-189.

Singh AK, Bhatt BP, Singh KM, Kumar Abhay, Manibhushan, Kumar Ujjawal, Chandra Naresh and Bharati RC 2013b. Dynamics of powdery mildew (*Erysiphe trifolii*) disease of lentil influenced by sulphur and zinc nutrition. *Plant Pathology Journal* 12(2): 71-77.

Singh AK, Manibhushan, Bhatt BP, Singh KM and Upadhyaya A.2013a. An Analysis of Oilseeds and Pulses Scenario in Eastern India during 2050-51. *Journal of Agril. Sci.* 5(1): 241-249.

Singh AK, Meena, MK Bharati RC. 2011. Sulphur and Zinc Nutrient Management in Rice Lentil Cropping System. International Conference on "Life Science Research for Rural and Agricultural Development. 27-29 December, 2011, CPRS Patna (Bihar). pp. 66-67.

Singh KM and Singh RKP. 1995. An Economic Analysis of Lentil Cultivation in N-E Alluvial Plains of Bihar. *Economic Affairs* 40(3): 157-163.

Singh P, Singh L, Singh P, Lone BA, Qayoom S, Ahmad L, Ganai MA, Kanth RA and Fayaz A. 2015. Response of lentil (*Lens culinaris* Medik) and weeds to different weed management practices under temperate conditions. *Journal of Agrisearch* 2(1): 72-74.

Singh SK, Varma SC and Singh RP. 2002. Integrated nutrient management in rice and its residual effect on lentil. *Indian J. Agric. Res.* 36(4): 286-289.

Zohary D and Hopf M. 2000. Domestication of plants in the old world: The origin and spread of cultivated plants in West Africa, Europe and the Nile valley. Oxford University Press New York USA.

Zohary D and Hopf M.1973. Domestication of pulses in the old world. *Science* 182: 887-894.

Citation

Singh AK, Bhakta N, Sangale UR, Manibhushan, Sundaram PK, Kumar S and Yasin JK. 2018. Lentil: Introduction. In: Scientific Lentil Production. Singh AK, Bhakta N, Sangale UR, Manibhushan, Sundaram PK, Kumar S and Yasin JK. (eds.) Society for Upliftment of Rural Economy Varanasi, India. Pp. 1-44.

❑❑❑

Chapter 2

An Introduction to Taxonomy, Floral Biology and Genetics of Lentil

RK Solanki, Jitendra Kumar and RK Singh

SYNOPSIS

Lentil (*Lens culinaris* M.), is one of the ancient food legume crop. Lentil has originated from the South Western Asia region. The primitiveness of lentil is visible in Indian ancient literature. Lentil is mentioned as *masura* (Pillow in sanskrit) in Indian language. Taxonomically, the specific status of *Lens* species or sub-species is still confusing. However, based on studies, the genus *Lens* consists of six main species. The crop is classified under the subfamily *Papilionoideae* of legume, having four clades. Cross-ability among the species has been established for defining subspecies, i.e. *L. odemensis* with *L. culinaris* ssp. *culinaris* and *L. culinaris* ssp. *orientalis* has helped to differentiate *L. nigricans* ssp. *nigricans* and ssp. *ervoides* which lack free cross-ability with *culinaris*. The two subspecies of Group-I have similar morphology and also share very high homology for nucleotide sequences, they are cross compatible with *odemensis* but results partially fertile seeds. Two subspecies of cultivated lentil including *microsperma* and *macrosperma* has been described on the basis of seed size. Most of traits in lentil are found to be controlled by single genes (growth habit, foliage colour, leaf shape, tendril, plant height, earliness, flower per peducle, seed coat colour, seed hardness, rust resistance, ascochyta blight resistance etc.). Cotyledon colour is governed my two genes whereas, seed size and winter hardiness is said to be governed by polygenes.

Keywords: Floral biology, Genetics, *Lens culinaris* Medik, Morphology, Taxonomy.

Introduction

The scheme of crop genetic improvement relies on the basic information of the crop. The taxonomy, origin, floral biology and genetics underlying the genome of any crop suggest about the framework which has to be implied for boosting the expression of plant agro-morphological traits that contribute to high economical yield. The *prima facia* objective is to understand the variability available and strategy for its utilization. Basic genetic information of any crop guide's about the breeding methods that can be adopted. In the present chapter, authors have placed their sincere efforts to develop an understanding of the crop genetical behaviour and have simultaneously explained about the crop taxonomical classification for suggesting the near wild parents available for pre-breeding efforts. A recombination plan depends upon the floral biology and genetics of the trait. Henceforth the information furnished will definitely help scientists working on lentil genetical improvement. Lentil (*Lens culinaris* M.), is one of the ancient food legume crop (Deokaran *et al.*, 2014). After chickpea it is the second most important cool season grain legume (Singh *et al.*, 2011). At Global level, India ranks first in area and second in production, the other countries growing lentil are Ethiopia, North Africa and to lesser extent in southern Europe, USA, Australia and Canada. Being a grain legume, it has the ability to fix atmospheric nitrogen and do carbon sequestration to improve soil health. The seeds are very rich source of dietary protein (24-28%) having abundant lysine, making it a good supplement with cereals for a balanced diet. Seeds also contain mineral and vitamins useful for human health, moreover straw is also suitable for animal feeding. The annual production of lentil has gone three times in the last five decades, as in 1971 it was 1.05 MT and in 2009 it has reached 3.6 MT (FAO, 2011). This bounce has happened with the gradual and concerted efforts of scientists by developing high yielding varieties and improved production technology, no doubt the area under lentil cultivation has also increased with time. In India it occupies acreage of about 1.49 million ha producing about 1.03 MT. The global productivity of the crop is about 1016 kg/ha compared to Indian which is quite low of about 741 kg/ha. The major factors which affects productivity potential are drought, cold, salinity, water logging etc. which need to be addressed with the recent molecular breeding tools.

Taxonomy

The scientific classification of lentil is reported to be:

Kingdom	:	Plantae
Unranked	:	Angiosperms
Order	:	Fabales
Family	:	Fabaceae
Subfamily	:	Faboideae
Tribe	:	Vicieae
Genus	:	Lens
Spices	:	Culinaris (Medikus)

The crop is further classified under the subfamily *Papilionoideae* of legume, having four clades (Gepts *et al.,* 2005), *viz.,* haseoloids/ millettioid, galegoids/hologalegina, dalbergioid (mostly tropical) and genistoid (Moretzsohn *et al.,* 2007). Important tropical and temperate food legume comes of the world comes under Haseoloids/millettioid and galegoids/hologalegina clades sharing a common ancestor. The further bifurcation of the clade is to Loteae and inverted repeat loss subclades (i.e., presence of a lost copy of the inverted repeat in the chloroplast genome of most angiosperms). Alfalfa (*M. truncatula*), chickpea (*Cicer arietinum*), faba bean (*Vicia faba*), lentil (*Lens culinaris*) etc., many cool-season legumes species belong to the latter subclade. *Medicago truncatula* shares a high level of genome synteny with Lentil (Phan *et al.,* 2007) and pea at nucleotide level (Weeden *et al.,* 1992).

Origin, Centre of Diversity and Gene Pool

Lentil has originated from the South Western Asia region (Turkey-Syria-Iraq) as per archaeological evidences, a pattern of distribution of wild and cultivated species overlapping in the Turkey-Cyprus region also supports the South-Western Asia or Mediterranean area as the centre of origin of the cultivated species *Lens culinaris* (Cubero, 1981). In course of time, in these regions few populations of *orientalis sp.* might have been naturally selected resulting in evolution of cultivated species (Cubero *et al.,* 2009). One second theory which has developed concurrently, suggesting the South-West Asian border region between Afghanistan, India, and Turkistan possibly a center of origin is also there due to presence of highest proportion of endemic varieties (Barulina, 1930). But by deciphering more information this border region was explained as secondary center of diversity. Molecular and biochemical evidences confirm the domestication of cultivated species from the subspecies *orientalis,* on nuclear basis the cytosolic glutamine synthetase gene intronic region variation and two paralogous genes coding for Bowman Birk protease inhibitors also shows that lentil may have derived from a specific stock of the ssp. *Orientalis* (Sonnante *et al.,*2003).

The primitiveness of lentil is visible in Indian ancient literature. Lentil is mentioned as *masura* (Pillow in sanskrit) in Indian language. The citation of lentil can be seen in *Rigveda* (8000 B.C), *Yajurveda* (7000 B.C), *Brahadaranyaka* (5500 B.C), *Charaka* (700 BC), *Susruta* (400 BC) and *Kautilya* (321–296 BC). In the modern world, German botanist and physian Medikus was the first to describe Lentil scientifically as *Lens culinaria* in 1787 (Cubero, 1981). *Lens* genus primary have chromosome complement of 2n=14 with self-pollination behaviour. All the species belonging to *Lens* genus shows wide variations in chromosomal rearrangements affecting both intraspecific and interspecific recombination (Ladizinsky, 1979b), based on which gene pools are grouped.

Taxonomically, the specific status of *Lens* species or sub-species is still confusing. However, based on studies, the genus *Lens* consists of six main species (*L. culinaris, L. odemensis, L. ervoides L. nigrcans, L. tomentosus* Ladiz. and *L. lamotti*) and two sub species *L. culinaris* ssp. *culinaris* and *L. culinaris* ssp. *orientalis* (wild progenitor of cultivated species) (Ladizinsky 1993, Van Oss *et al.,* 1997). *L. orientalis* is clearly observed as the wild progenitor of cultivated species and hence it is considered as sub species of cultivated species *L. culinaris* (Harlan and De Wet, 1971).

Crossability among the species has been established for defining subspecies, i.e., *L. odemensis* with *L. culinaris* ssp. *culinaris* and *L. culinaris* ssp. *orientalis* has helped to differentiate *L. nigricans* ssp. *nigricans* and ssp. *ervoides* which lack free cross-ability with *culinaris* (Ladizinsky *et al.,* 1984). However, on basis of flower characters like pollen and pistil morphology, close relationship between *L. odemensis and L. nigricans* has been established and further the two species showed relationship with *L. culinaris* subsp. *orientalis.* Whereas, *L. ervoides* isc loosely related to all these taxa (Fratini *et al.,* 2006). The specific status of *L. tomentosus* (Ladizinsky, 1997) is still debatable as it has been found to be separated from *L. culinaris* ssp. *orientalis,* which is evident on the basis of tomentose as opposed to glabrous pods and by the presence of a small, asymmetrical and minute satellite chromosome. Based on morphological, biochemical and molecular markers, *L. odemensis* and *L. tomentosus* are considered as subspecies of *L. culinaris* (Ferguson *et al.,* 2000). Based on the cross-ability three groups are described (Table 2.1). Besides, a number of other studies based on SDS-PAGE, ITS and cytoplasic DNA markers have also helped to draw phylogenetic relationship among various *Lens* species (Zimniak-Prybylaska, 2001, Mayer and Bagga 2002, Sonnante *et al.,* 2003, Galasso, 2003, Duran *et. al.,* 2004, Fratini *et al.,* 2004, Fratini and Ruiz 2006 and Fratini *et al.,* 2006). All these efforts have been discussed and summarized earlier by Cubero *et al.,* 2009.

Table 2.1 The Cross-ability groups of lentil species

Group	Species	Reference
I	*L. culinaris.* ssp. *culinaris* and ssp. *Orientalis, L. Odemensis*	Hancock, 2004
II	*L. nigricans, L. ervoides* and *L. lamottei* *L. nigricans, L. ervoides* and *L. lamottei, L. Odemensis*	
III	*L. tomentosus*	

The two subspecies of Group-I have similar morphology and also share very high homology for nucleotide sequences, they are cross compatible with *odemensis* but results partially fertile seeds (Ladizinsky *et al.,* 1984). Group I and Group II (*L. nigricans)* crossing results in irregular meiotic events producing non-viable seeds in 1st filal (F_1) generation (Ladizinsky *et al.,* 1984 & 1985). Group I (*L. culinaris*) and Group II (*L. ervoides*) hybrids can be raised by embryo rescue (Ladizinsky *et al.,* 1984). Reports suggest that up to nine percent success rate is there for developing interspecific hybrid by embryo rescue between *L. culinaris* ssp. *culinaris* and *L. nigricans, L. ervoides* and *L. odemensis.* Based on the above information's *L. odemensis* is placed in secondary gene pool and *L. nigricans,* and *L. ervoides* are placed in the tertiary gene pool by Ladizinsky, 1993. However, Cubero *et al.* (2009), based on similar observation, suggested to put *L. odemensis* along with *L. nigricans, and L. ervoides* in secondary gene. However, further study on hybridization is still needed to establish place of *L. tomentosus* and *L. lamottei* in secondary or tertiary gene pool.

Two subspecies of cultivated lentil including *microsperma* and *macrosperma* has been described on the basis of seed size. Based on geographical distribution, 6 different regional groups or *greges* (i.e. *pilosae, subspontanea, aethiopicae, europeae, asiaticae* and *intermediate*) were also considered within former (*micropserma*) subspecies (Barulina, 1930). Distributions of Barulina's *greges* and wild lentils have better explained evolution of cultivated species and its varietal *facies* in lentil. Three *greges* having only a distinct character are restricted to very concrete regions: *pilosae* to the Indian subcontinent (a strong pubescence), *aethiopicae* to Ethopia and Yemen (pods with a characteristic elongated apex) and *subspontanea* to the Afghan regions closer to the Indian subcontinent (very dehiscent pods, purple coloured before maturity). However, above unique characters of each *greges* are shown together with a cluster of primitive characteristics of closely related to *orienatalis*. The distribution of *subspontanea* also overlaps with wild species *orienatalis* and both *orienatalis* and *culnaris* forms are found together in South Turkey & North Syria. Thus *orienatalis* has played an important leading role in the evolution of eastern small seeded lentils. While wild species *L. ervoides* spreading to southwards and overlapping with short calyx *aethiopicae* suggests its contribution in evolution of this small seeded *greges*. However, other *microsperma* and *macrosperma* lentils overlap to greater and lesser extent with known wild lentils are clearly intermixed. However, easy cross compatibility of *Lens odemensis*, with *culnaris* could have generated the genetic raw material for western lentils having bigger seeds with more number of large leaflets and calyx teeth longer than corolla. Spreading of *nigricans* and *ervoides* to westwards also has their role in evolution of western lentil, because there is probability of survival of some crosses in natural environments despite their cross incompatibility with cultigens due to hybrid embryo abortion. Thus, *orienatalis* and *odemensis* forms are most likely candidates as companion weeds of the cultigens, and *microsperma* and macrosperma have evolved simply through disruptive selection (Cubero *et al.*, 2009).

Fig. 2.1 (a) Green coloured lentil **Fig. 2.1 (b)** Brown gray coloured lentil

Floral biology and Plant morphology

Lentil flowers are complete type having typical papilionaceous structure (Æ; female k ; K (5); c1+2=(2);A(9)+1;G1; leg.). Flower size is small 4-9 mm long. Petal colour may be pink, purple, light purplish blue pale blue or white. Flower wings and keel are generally white; often wings may have a violet-blue tinge. The calyx tube is divided near its base into five narrow pointed lobes that are curved over and beyond the corolla. The stamens are diadelphous (9 + 1) with upper vexillary stamens. The ovary contains one or two ovules (sometimes three ovules) and it terminates in a short curved style. The style is pilose on the inner side and the stigma is slightly swollen and glandular. Lentil flowers are cleistogamous offering self pollination, but to some extent out-crossing (<0.08%) is reported, may be due insects (Muehlbauer *et al.*, 2002). However a recent report suggests that the degree of out crossing in lentil ranges from 0.06 to 5.12% depending upon the cultivar, year and location. Moreover, outcrossing rate of individuals within cultivars can also occur up to 22.2% (Horneburg, 2006).

Plant of lentil is an annual bushy herb with slender, soft haired and highly branched stem. Leaves are alternate, compound and pinnate with one to eight pairs of subsessile or sessile, ovate, elliptical or lanceolate leaflets, ending with or without a tendril. Height varies from 15 to 75 cm (Duke, 1981). Ample amount of variation has been found in growth habit, plant may be either single stem erect to semi-erect in shape or compact type with heavy branched to low bushy type. However, this expression depends mainly on genotypes, but environmental conditions also interact to contribute for expression of these characters (Saxena and Hawtin, 1981). Blooming starts after a defined period of vegetative growth depending upon the genotype (Fig. 2.2). Flowering occurs in acropetal form that is lower nods mature first, while younger nodes may carry immature pods or flowers. Period of flowering may vary from 10 to 40 days depending upon genotype and environment. The number of peduncles per plant ranges from 10 to 150 depending on the genotype (Malhotra *et al.*, 1974). Generally, one to four flowers have been observed on each peduncle, but sometimes as much as seven flowers have also been found. Pods are oblong, laterally compressed bulging over the seeds, 6-20 mm long and 3.5 to 11 mm wide. The number of pods per penducle varies from one to four and in rare cases, six has also been found. However, frequency of one pod per peduncle is also reported (Muehlbauer, 1974). Lentil seeds are typically lens shaped and with a diameter of 2-9 mm (Barulina, 1930). The large seeded types generally have a diameter of 6 to 9 mm. For medium and small seeded types, a diameter range of 6-9 mm and 5-6 mm, respectively has been observed. Seeds may be globose (diameter: thickness ratio = 2.5:4) or flattened (diameter: thickness ratio= 1.5:2.5). The 100-seed weight ranges from 1.05 to 8.55 g depending upon the genotype, management practices, environment, etc. (Barulina, 1930; Tullu *et al.*, 2001). The seed coat colour also varies (Fig. 2.1); it may be green, green-brown, light red with black speckles (Duke, 1981). Cotyledons may be red/orange, yellow or green. Depending on the genotypes, the seed shows 3-4 week dormancy period after the harvest.

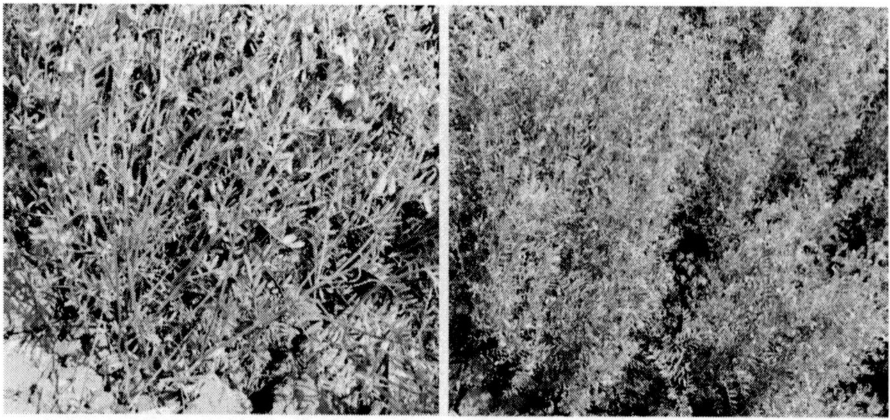

Fig. 2.2 (a) Lentil crop at podding stage

Fig. 2.2 (b) Lentil crop at flowering stage

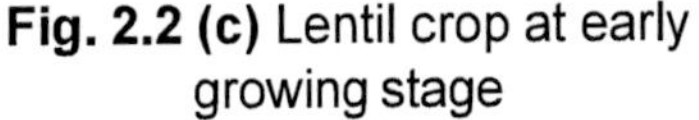

Fig. 2.2 (c) Lentil crop at early growing stage

Fig. 2. 2 (d) Lentil crop at grand growth stage

Trait genetics

Genetics is any trait suggests the way for its further improvement. The inheritance pattern of any trait is worked out by knowing the expression in the segregating generations. A true recombinant of two diverse parents will offer segregating population for the diverse trait, based on which genetics of any trait is deciphered. In lentil both at national and international level tremendous efforts have been made to generate knowledge about the genetical behaviour of many traits. Studies shows that pod dehiscence and seed dormancy are the two main traits which have contributed more towards domestication process and both of them are found to be governed by single recessive genes. Most of traits in lentil are found to be controlled by single genes (growth habit, foliage colour, leaf shape, tendril, plant height, earliness, flower per peducle, seed coat colour, seed hardness, rust resistance, ascochyta blight resistance etc.). Cotyledon colour is governed my two genes whereas, seed size and winter hardiness is said to be governed by polygenes (Table 2.2).

Conclusion

Lentil (*Lens culinaris* M.), is classified under the subfamily *Papilionoideae* of legume, having four clades. Cross-ability among the species has been established for defining subspecies, i.e., *L. odemensis* with *L. culinaris* ssp. *culinaris* and *L. culinaris* ssp. *orientalis* has helped to differentiate *L. nigricans* ssp. *nigricans* and ssp. *ervoides* which lack free cross-ability with *culinaris*. The two subspecies of Group-I have similar morphology and also share very high homology for nucleotide sequences, they are cross compatible with *odemensis* but results partially fertile seeds. Two subspecies of cultivated lentil including *microsperma* and *macrosperma* has been described on the basis of seed size. Most of traits in lentil are found to be controlled by single genes (growth habit, foliage colour, leaf shape, tendril, plant height, earliness, flower per peducle, seed coat colour, seed hardness, rust resistance, ascochyta blight resistance etc.). Cotyledon colour is governed my two genes whereas, seed size and winter hardiness is said to be governed by polygenes.

Table 2.2 Genetic studies for morphological traits in lentil

Gene Action	Traits	References
Monogenic dominant gene	Light green foliage	Kumar, 2002
	Oval leaflet shape	Kumar *et al.*, 2006 unpublished data
	Leaflet number	Kumar *et al.*, 2005a
	Plant pubescence	Vandenberg and Slinkard, 1989; Emami 1996, Kumar 2002.
	Plant height	Tahir et al, 1994
	Flower colour	Lal and Srivastava, 1975, Ladizinsky, 1979a, Kumar 2002, Wilson and Hudson, 1978; Gill and Malhotra 1980
	Two-flower per peduncle	Gill and Malhotra, 1980, Emami 1996 and Kumar 2002
	Pod dehiscence	Ladizinsky 1979a, 1985, Vaillancourt and Slinkard 1993,
	Seed coat colour black > brown > grey > green	Vandenberg and Slinkard, 1990
	Tannin in seed coat	Vaillancourt et al 1986
	Seed coat spotting	Ladizinsky 1979, Vandenberg and Slinkard 1990, and Emami 1996.
	Hard seededness	Ladizinsky 1985, Vaillancourt and Slinkard, 1992
	Fusarium wilt	Eujayl *et al.* (1988)
	Rust resistance	Kumar *et al.* (1997)
	Ascochyta blight resistance	Ford *et al.* (1999)
Monogenic recessive gene	Growth habit	Emami and Sharma 1999
	Red foliage	Ladizinsky, 1979, Emami and Sharma, 1996a, Vandenberg and

		Slinkard, 1989, Havey and Muehlbauer, 1989, Emami 1996
	Acute leaflet shape	Kumar et al., 2006 unpublished data
	Tendril formation	Vandenberg and Slinkard 1989, Vaillancourt and Slinkard, 1992, Emami, 1996.
	Earliness	Sarker et al. 1998, 1999
	Three flower per peduncle	Gill and Malhotra, 1980, Emami 1996 and Kumar 2002
	Three flower per peduncle	
Incomplete dominance	Stipule size	Kumar, 2002
	Pod size	Kumar et al 2005b
Two dominant	Orange cotyledon	Emami and Sharma 1996 a, b, c, Vaillancourt and Slinkard 1993
genes	Yellow cotyledon colour	
Two recessive genes	Green cotyledon colour	
Duplicate dominant gene	Rust resistance	Lal *et al.* (1996)
Polygenic	Seed size	Abbo *et al.*, 1991
with partial dominance	Winter hardiness	Kahraman *et al.*, 2004
Polygenic		

References

Abbo S, Ladizinsky G and Weeden NF. 1991. Genetic-analysis and linkage study of seed weight in lentil. *Euphytica* 58: 259-266.

Barulina H. 1930. Lentil of the USSR and other countries. *Bulletin of Applied Botany, Genetics and Plant Breeding* 40: 265-304.

Cubero JI, Perez de la Vega M and Fratini R. 2009. Origin, phylogeny, domestication and spread. In: Erskine, W., Muehlbauer, F.J., Sarker, A. and Sharma, B. (eds) *The Lentil: Botany, Production and Uses.* CABI, UK, pp. 13-33.

Cubero JI.1981. Origin, domestication and evolution. In: Webb, C. and Hawtin, G.C. (eds) Lentils. Commonwealth Agricultural Bureau, Slough, UK, pp. 15-38.

Duke JA. 1981. Handbook of legumes of world economic importance. *Plenum,* New York, pp. 317-338.

Duran Y, Fratini R, Garcia P and Perez de la Vega M. 2004. An interspecific genetic map of *Lens. Theor Appl Genet* 108: 1265-1273.

Emami MK and Sharma B. 1996a. Inheritance of brown leaf pigmentation in lentil (*Lens culinaris*). *Indian Journal of Genetics and Plant Breeding* 56(3): 362-365.

Emami MK and Sharma B. 1996b. Digenic control of cotyledons colour in lentil (*Lens culinaris*). *Indian Journal of Genetics and Plant Breeding* 56(3): 357-361.

Emami MK and Sharma B. 1996c. Confirmation of digenic inheritance of cotyledon colour in lentil (Lens culinaris). *Indian Journal of Genetics and Plant Breeding* 56(4): 563-568.

Emami MK and Sharma B. 1999. Linkage between three morphological markers in lentil. *Plant Breeding* 118(6): 579-581.

Emami MK. 1996. Genetic mapping in lentil (*Lens culinaris* Medik.). PhD thesis, Division of genetics, Indian Agricultural Research Institute, New Delhi, India.

Eujayl I, Erskine W, Bayya B, Baum M and Pehu E. 1998. Fusarium vascular wilt in lentil: inheritance and identification of DNA markers for resistance. *Plant Breeding* 117: 497-499.

FAO. 2011. Production Statistics. Food and Agriculture Organization, Rome, Italy.

Ferguson ME, Maxted N, Van Slageren M and Robertson LD. 2000. A reassessment of the taxonomy of *Lens* Miller (Leguminosae, Papilionoideae, Vicieae). *Botanical Journal of the Linnean Society* 133: 41-59.

Ford R, Pang ECK and Taylor PWJ. 1999. Genetics of resistance to ascochyta blight (*Ascochyta lentis*) of lentil and the identification of closely linked RAPD markers. *Theretical and Applied Genetics* 98: 93-98.

Fratini R and Ruiz ML and Perez de la Vega M. 2004. Intra specific and inter sub specific crossing in lentil (*Lens culinaris* Medik.) *Canadian Journal of Plant Science* 84: 981-986

Deokaran, Singh SB and Ram Kewal. 2014. Effect of Zinc and Boron application on Yield of Lentil and Nutrient Balance in the Soil under Indo-Gangetic Plain Zones. *Journal of AgriSearch* 1(4): 206-209.

Fratini R and Ruiz ML. 2006. Interspecific hybridization in the genus *Lens* applying *in vitro* embryo rescue. *Euphytica* 150: 271-280.

Fratini R, Garcia P and Ruiz ML. 2006. Pollen and pistil morphology, *in vitro* pollen grain germination and crossing success of *Lens* cultivars and species. *Plant Breeding* 125: 501-505.

Galasso I. 2003. Distribution of highly repeated DNA sequences in species of the genus *Lens* Miller. *Genome* 46: 1118-1124.

Gepts P, Bavis WD, Brummer CE, Shoemaker RC, Stalker TH, Weeden NF and Young ND. 2005. Legumes as a Model Plant Family. Genomics for Food and Feed Report of the Cross-Legume Advances through Genomics Conference. *Plant Physiology* 137: 1228–1235

Gill AS and Malhotra RS.1980. Inheritance of flower colour and flower number per inflorescence in lentils. *LENS Lentil Experimental News Service* 7: 15-19.

Hancock JF. 2004. Plant Evolution and the origin of crop species, 2nd edn. CABI, Wallingford, UK.

Harlan JR and De Wet JMJ. 1971. Towards a rational classification of cultivated plants. *Taxon* 20: 509-517.

Havey MJ and Muehlbauer FJ. 1989. Linkages between restriction fragment length, isozyme and morphological markers ion lentil. *Theoretical and Applied Genetics* 77(3): 395-401.

Horneburg B. 2006.Outcrossing in lentil (*Lens culinaris*) depends on cultivar, location and year, and varies within cultivars *Plant Breeding* 125: 638-640.

Kahraman A, Kusmenoglu I, Aydin N, Aydogan A, Erskine W and Muelbauer FJ. 2004. Genetics of winter hardiness in 10 lentil recombinant inbred line population. *Crop Science* 44: 5-12.

Kumar V, Singh BM and Singh S. 1997. Genetics of lentil resistance to rust. *Lens Newsletter* 24: 23-25.

Kumar Y, Mishra SK, Tyagi MC and Sharma B. 2005a. Linkage between genes for leaf colour, plant pubescence, number of leaflets and plant height in lentil (*Lens culinaris* Medik.). *Euphytica* 145: 41-48.

Kumar Y, Mishra SK, Tyagi MC and Sharma B. 2005b. Inheritance of pod size in lentil (*Lens culinaris* Medik.). *Journal of Lentil Research* 2: 31-33.

Kumar Y. 2002 Inheritance and linkage of genes for morphological traits in lentil (*Lens culinaris* Medik.). PhD thesis, Charan Singh University, Meerut, India. p. 127.

Ladizinsky G, Braun D, Goshen D and Muehlbauer FJ. 1984. The biological species of the genus *Lens. Botanical Gazette* 145: 253-261.

Ladizinsky G. 1979a. The genetics of several morphological traits in the lentil. *Journal of Heredity* 70: 135-137.

Ladizinsky G. 1979b. The origin of lentil and its wild gene pool. *Euphytica* 28: 179-187.

Ladizinsky G. 1985. The genetics of hard seed coat in the genus Lens. *Euphytica* 34: 539-543.

Ladizinsky G. 1993. Wild lentils. *Critical Reviews in Plant Science* 12: 169-184.

Ladizinsky G. 1997. A new species of *Lens* from south-east Turkey. *Botanical Journal of the Linnean Society* 123: 257-260.

Lal C, Sharma S and Chahota RK. 1996. Inheritance of rust resistance in lentil. *Indian Journal of Genetics and Plant Breeding* 56(3): 350-351.

Lal S and Srivastava RS. 1975. Inheritance of flower colour in lentils. *Indian Journal of Genetics and Plant Breeding* 35(1): 29-30.

Malhotra RS, Singh KB, and Singh JK, 1974. Genetic variability and genotype-environment interaction in lentil. *Journal of Research Punjab Agricultural University* 10: 17-21.

Mayer MS and Bagga SK, 2002. The phylogeny of Lens (*Leguminosae*): new insight from ITS sequence analysis. *Plant Systematics and Evolution* 232: 145-154.

Moretzsohn M, Bertioli SL, Guimarães P, Madsen L, Fredslund J, Hougaard B, Schauser I, Sandal N, Stougaard J, Tabata S and Bertioli D. 2007. Can legume synteny be useful in guiding the introgression of wild genes into cultivated peanut? *Lotus Newsletter* 37: 95-96.

Muehlbauer FJ and McPhee. 2005. Lentil (*Lens culinaris* Medik.). In: Singh, R.J. and Jauhar, P.P. (eds) *Genetic Resources, Chromosome Engineering, and Crop Improvement: Grain Legumes*. Vol.1. CRC Press, Boca Raton, Florida, USA, pp. 219–230.

Muehlbauer FJ, Summerfield RJ, Kaiser WJ, Clement SL, Boerbomm CM, Welsh-Maddux M M and Short RW. 2002. Principles and Practices of Lentil Production.United States Department of Agriculture (USDA) Agriculture Research Service (ARS). Available at http://www.ars.usda.gov/is/np/lentils/lentils.htm (accessed on 28 february 2008).

Muehlbauer FJ. 1974. Seed yield components in lentil. *Crop Science* 14: 403-406.

Phan HTT, Ellwood SR, Hane JK, Ford R, Materne M and Oliver RP. 2007. Extensive macrosynteny between *Medicago truncatula* and *Lens culinaris* ssp. *culinaris*. *Theotical and Applied Genetics* 114: 549-558.

Sarker A, Erskine W, Sharma B and Tyagi MC. 1999. Inheritance and linkage relationship of days to flower and morphological loci in lentil (*Lens culinaris* Meikus subsp. culinaris). *Journal of Heredity* 90(2): 270-275.

Sarker A, Erskine W, Sharma B and Tyagi MC.1998. Genetics of flowering in lentil: a new major gene and polygenes. In: *Abstracts of the Third European Conference on Grain Legumes,* Valladolid, Spain, p. 465.

Saxena MC and Hawtin GC. 1981. Morphology and growth patterns. In: Webb, C. and Hawtin, G.C. (eds) *Lentils*. Commonwealth Agricultural Bureau, Slough, UK, pp. 39-52.

Singh, AK, Meena, MK Bharati RC. 2011. Sulphur and Zinc Nutrient Management in Rice Lentil Cropping System. International Conference on "Life Science Research for Rural and Agricultural Development. 27-29 December, 2011, CPRS Patna (Bihar).pp. 66-67.

Tahir M and Muehlbauer FJ. 1994. Gene mapping in lentil with recombinant inbred lines. *Journal of Heredity* 85(4): 306-310.

Tullu A, Kusmenoglu I, Mc Phee KE and Muehlbauer FJ. 2001. Characterization of core collection of lentil germplasm for phenology, morphology, seed and straw yields. *Genetic Resources and Crop Evolution.* 48: 143-152.

Vaillancourt CE, Slinkard AE and Reichert RD. 1986. The inheritance of condensed tannin concentration in lentil. *Canadian Journal of Plant Science* 66: 241-246.

Vaillancourt RE and Slinkard AE. 1993. Inheritance of new genetic markers in lentil (*Lens* Miller). *Euphytica* 64: 227-236.

Van Oss H, Aron Y and Ladizinsky G. 1997. Chloroplast DNA variation and evolution in the genus *Lens* Mill. *Theoretical and Applied Genetics* 94: 452-457.

Vandenberg A and Slinkard AE. 1989. Inheritance of four new quantitative genes in lentil. *Journal of Heredity* 80: 320-322.

Vandenberg A and Slinkard AE. 1990. Genetics of seed coat colour and pattern in lentil. *Journal of Heredity* 81: 484-488.

Wilson VE and Hudson LW. 1978. Inheritance of lentil flower colour. *Journal of Heredity* 69: 129-130.

Zimniak-Przybylska Z, Przybylska J and Krajewski P. 2001. Genetic resources, *Lens*, Lentil, SDS-PAGE, seed globulins. *Journal of Applied Genetics* 42: 435-447.

Citation

Solanki RK, Kumar J and Singh RK. 2018. An Introduction to taxonomy, floral biology and genetics of lentil. In: Scientific Lentil Production. Singh AK, Bhakta N, Sangale UR, Manibhushan, Sundaram PK, Kumar S and Yasin JK. (eds.) Society for Upliftment of Rural Economy Varanasi, India. Pp. 45-60.

❑❑❑

Chapter 3

Lentil Breeding for Improving Productivity using Classical and Molecular Approaches

Ravindra Prasad, Lal Chand Prasad and Satish Santosh Bornare

SYNOPSIS

Lentil is cool season annual legume crop. The genus *Lens* Miller comprises of five annual species of which only *Lens culinaris* Medik is cultivated. As per historical record of released varieties, majority of them are selected from germplasm is evident that pure line selected among local adopted Germplasm lines has been very successful. Lentil is strictly self pollinated crop due to cleistogamy and dehiscence of anthers is before the flowers opens. Varietal hybridization followed by application of pedigree, bulk breeding methods or their modifications are now increasingly being used. Major breeding objectives are to develop high yielding varieties with bold seed, containing high protein content with less cooking time and resistance to various biotic and abiotic stresses.

Keywords: Crop improvement, classical breeding, molecular approaches, lentil.

Introduction

Lentil (*Lens culinaris*, 2n = 2x =24) is low input pulse crop grown mainly in our country, Pakistan, Egypt, Greece, Italy, in Mediterranean basis (Ram and Singh, 2003). In India, Uttar Pradesh, Madhya Pradesh, Bihar and West Bengal are the states were 75 % acreage is concerned (Jeswani and Baldev, 1990). It was one of the earliest cultivated crops in the world with archeological evidence from the early Stone Age. India leads the world in lentil production, and most of the lentils are consumed by India and its neighbors. Currently, Canada and Turkey are the world's largest lentil exporters. Historically, lentil was widely used in India, Southwest Asia, and the Mediterranean areas in the form of split lentil (dhal) and it is still an important source of dietary protein in these areas. Lentil contains approximately 22% protein, and is high in fiber, vitamin A, calcium, starch, iron, phosphorus, copper and manganese (Singh *et al.*, 2015b). While the lentil seed is used mainly as food, the straw can also be used as a high quality animal feed or as a source of organic material for soil improvement. It can be used to diversify and lengthen the crop rotation, which reduces disease pressure in other crops and has weed control advantages. Lentil also improves soil tilth and reduces the requirement for nitrogen fertilizer. Lentil prices vary, but net returns are often higher than those obtained from wheat. On the prairies, lentil yields range between 450 and 2500 lb/ac (500 and 2800 kg/ha), with an average yield of about 1345 kg/ha. Average concentration of major nutritional components in lentil i.e., protein 22 per cent (in seed), 6.4 per centage (in straw) for livestock rations.

History

The Lentil crops may have been one of the first agricultural crops grown more than 8,500 years ago. Production of this cool season annual crop spread from the Near East to the Mediterranean area, Asia, Europe and finally the Western Hemisphere. The crop has received little research attention to improve its yield and quality. It grows well in limited rainfall areas of the world (Ladizinsky, 1979).

Species (cultivated and wild species): The genous *Lens* Miller comprises of five annual species of which only *Lens culinaris* Medik (= *L. esculenta* Moench) is cultivated, while the four wild species namely *L. montbretii*, *L. nigricans*, *L. ervoides* and *L. orientalis* are delicate, small flowered annuals distributed over South West Asia and Mediterranean (Zohary, 1972).

Stages of development in Lentil

i) Vegetative growth stages

The lentil is much branched, softly pubescent, light green, annual herbaceous plant with a slender stem and branches (Saxena and Hawtin, 1981). The branching habit of lentil contrasts with the better development main stem axis of other food legumes such as pea, common bean etc. where descriptions of vegetative growth stages have focused on the development of the main stem. In lentil, it is necessary to use the primary branches. Collectively the primary branches carry 52% of the pods compared with an average of only 17 % born on the main stem. The basal primary branches also carry an average of 17% of the total pod yield, but is of most importance in the measurement of reproductive growth stages. It carries the nodes that are most frequently occupied by the pods. Both vegetative and reproductive growth measurements on lentil focus on the basal primary branch (Fig. 3.1).

Fig. 3.1 Line sown lentil crop at early growth stage

Fig. 3.2 Line sown lentil crop at grand growth stage

ii) Reproductive growth stages

Flowering in the lentil is in indeterminate, occurring from axillary bud on the main stem and branches. It proceeds acropetally from lower to higher nods. A single plant may have open flowers at high nodes and full pods in lower nodes at same time (Fig. 3.1 to 3.4). The developmental stages of reproductive growth are described in the Table 3.1.

Crossing

Lentil is strictly self pollinated crop due to cleistogamy and dehiscence of anthers is before the flowers opens. Wilson and Law in 1972 reported less than 8.0% occasional cross pollination through thrips or other small insects. Wilson (1972) reported successful crossing in green house. Crossing was best on young vigorous plants when the relative humidity was above 50%, the temperature between 15-25 °C in 12-15 hours of light and 9-12 hours of darkness. When the RH dropped below 35%, the percentage of successful manual crosses decreased sharply. Lentil flowers continue to self-pollinate at the lower RHs, but pollen for crossing is difficult to gather and transfer. The buds (one half-three fourth the length of calyx lobes) are held between the thumb and the forefinger with the suture of the keel facing the operator. Special care is taken not to bend or twist the peduncle. Sharp-pointed forceps are used to carefully remove one or two calyx lobbes nearest the suture side of the keel. The wings and keels are removed individually. The standard is either or folded away from the stigma. The stamens are removed either by grasping individual filaments with forceps and breaking them free from the stamina column or by pulling them outward. Manual pollinations are made immediately after emasculation. Pollen is selected from the vigorous flowers as soon after another dehiscence as feasible. The keel is opened with forceps so that the stamina column and the pistil could be removed as a unit. This brush like unit is used to transfer pollen to the exposed stigma of emasculated flowers. The pollen loded pistil and the anthers are brushed against the trichoid side of the stigma. After pollen transfer the standard of the maternal flower is returned to its original positions around the pistil.

Table 3.1 Developmental stages in lentil crops

Stage designation	Abbreviated stage title	Descriptions
R_1	First bloom	One open flower at any node.
R_2	Full Bloom	Open flowers on node 10-13 of the basal primary branches.
R_3	Early pod	Pods on nodes 10-13 of the basal primary branch visible.
R_4	Flat pod	Pods on nodes 10-13 of the basal primary branch have reached full length and are largely flat. Seeds till less than half of the pod area but can be felt as a bump between the fingers.
R_5	Full seed	Seeds in any single pod on nodes 10-13 of the basal primary branch are swollen and completely fill the pod cavity.
R_6	Full pod cavities	All the normal pods on nodes 10-13 of the basal primary branches are swollen and completely fill the pod cavity.
R_7	Physiological maturity	The leaves start yellowing and 50 % of the pods have turned yellow.
R8	Full maturity	The 90 % of pods on the plant are golden brown.

Source: Erskine *et al.*, 1990

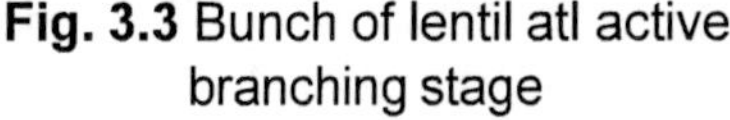

Fig. 3.3 Bunch of lentil atl active branching stage

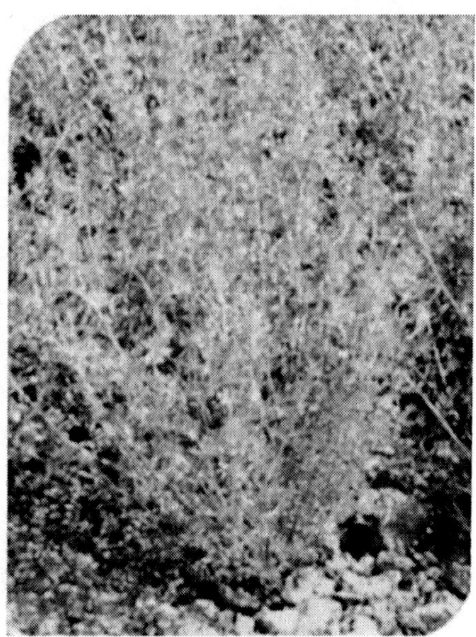

Fig. 3.4 Lentil plant at early podding stage

Breeding objectives

In lentil crops major breeding objectives are

1. To develop high yielding varieties.
2. To developed variety with bold seed, containing high protein content with less cooking time.
3. Early maturing
4. Resistance to various diseases such as rust, and ascochyta blight.
5. Resistance to insects such as pod borer, cutworm, aphids etc.
6. Resistance to shattering
7. Tolerance to drought etc.

Breeding procedures

Classical/traditional/conventional approaches

As per historical record of released varieties, majority of them are selected from Germplasm is evident that pure line selected among local adopted Germplasm lines has been very successful. Varietal hybridization followed by application of pedigree, bulk breeding methods or their modifications are now are increasingly being used (Singh *et al.*, 2015a). The trends are in favour of Bulk and single seed descent (SSD) method where segregating populations are handled in bulk till F_5/F_6 generation, preferably with introduction off-season nurseries and single plant selections are made in F_5/F_6 to be grown as individual plant progeny rows. The bulk harvest of selected rows (each row to be separately bulk) are evaluated in yield trials. While advancing the generations, it should be insured that the natural spread of prevalent disease is adequate. Mutation breeding can also be applied. A detail review on mutation breeding in pulse crop has been written by Singh (1991b). According to this germination and survival at maturity in lentil where unaffected at 4 and 8 kR (Krad dose of gamma rays). However these characters were markedly affected at 12-32 kR. According to Singh (1991a) a series of induced morphological mutants listed in lentil including crinkled leaf, boat shape leaf, lendril leaf multiflorate/sterile crumple petals, late flowering, semi fertile with long peduncle, genetic male sterile, cytoplasmic male sterile, seed colour etc. Mutation breeding would be advantageous if carried out with one–two specific objectives. The doses less than LD 50 are GR 50 could be desirable to obtain economically useful mutants. The mutant generations may be handled as per the requirement of breeding objectives.

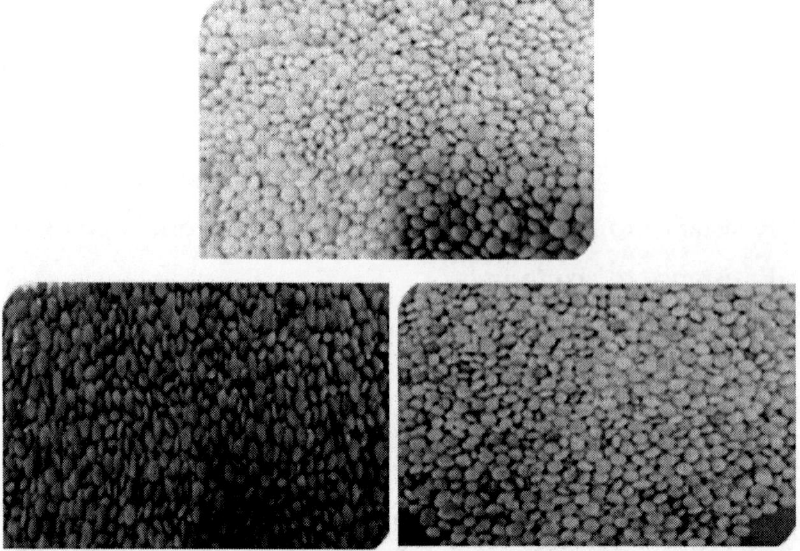

Fig. 3.5 Variability in seed size and seed coat clour in lentil

Molecular/non-conventional approaches

Recently combined approaches viz., conventional and markers (morphological, biological and molecular) are widely being used to improve the lentil crops for trait of interest (Frederick *et al.*, 2006). Gottlieb, 1981 emphasised that biochemical markers are powerful tools for the assessment of genetic variability within and between plant populations. In continuation DNA marker namely nuclear restriction fragment length polymorphism (RFLP) have recently received much attention as plant breeding tool (Soller and Beckmann, 1983) and as an estimator of phylogenetic relationship (Sharma *et al.*, 1996 and Song *et al.*, 1988). Further, Beckmann and Soller (1983) viewed that Isozymes (biochemical marker) and RFLP have several advantages over morphological markers e.g., multiple allelic forms, co dominance, and absence of pleiotropic effects on economically important traits. Havey and Muehlbauer (1989) evaluated thirty accession of lentil including wild species for restriction fragment length polymorphism (RFLP) using ten relative low-copy-number probes selected from partial genomic and cDNA libraries of lentil and concluded that the wild lentil *L. culinaris* and ssp. *orientalis* and *L. culinaris* ssp. *odemensis* showed greatest variability for restriction fragment length and were closely positioned to domestic lentil. Sharm *et al.* (1996) applied AFLP and RAPD techniques on 54 lentil accession lines including wild species to know the actual differences among them and phylogentic information as well. Based on finding they have reported that the level of variation detected within the cultivated lentil with AFLP analysis indicates that it may be a more efficient technology than the RAPD analysis for construction of genetic linkage maps between carefully chosen lentil accessions. Tar'an *et al.* (2003) used the molecular markers and studied recombinant inbred lines developed for resistance to Ascochyta blight disease caused by the fungus *Ascochyta lentis* Vassilievsky in lentil and draw their conclusions regarding genetic of said disease which is badly reduce both seed yield and seed quality of the crop.

Lentil varieties

There are two types of cultivars/ varieties commonly called as masur or malk masur (bold seeded) and masuri (small seeded) are grown in India (Fig. 3.5). The bold seeded cultivars are grown primarily in Bundelkhand region of Uttar Pradesh, Madhya Pradesh and in Maharastra. The small seeded cultivars are popular in Bihar,

Eastern Uttar Pradesh, West Bengal and Assam. Varietal improvement programme of lentil in India was initiated in 1924 when single plant selections were made in locally collected mixed seed lots. Some of these earlier developed pure line selections were NP1, NP47 (IARI), T8, T38 (Uttar Pradesh), L12 (Punjab), BR25 (Bihar) and B77 (West Bengal). Subsequently the lentil breeding programme got boost under All India Coordinated Pulse Improvement Project in 1970s and several important varieties were developed at Pantnagar, IARI, and PAU Ludhiana etc. Some selective state wise recommended lentil varieties are given below (DPR, 1988 and Singh *et al.*, 2008).

Uttar Pradesh: Type 36, Type 8, Pant 406, Pant L639, VL Massor, Malika (K75), Lens 4076, PL 77-2, L-4-12.

Bihar: BR 25, NO 26, Pant L 406, PL 639, Malika (K75), PL 77-2, Pant L 209.

Madhya Pradesh: T36, B 77, JLS-1, PL 639, Malika (K75)

Haryana: Pant L 639, Lens 4076, Pant L 406, L9-12

Panjab: L9-12, Pant L 406, PL 639, Lens 4076, LL 56

Delhi: Pusa 4, Pusa 6, PL 639, Lens 4076

Gujrat: PL 639, Malika (K 75), Lens 4076

Himachal Pradesh: L9-12, Vipasha.

Jammu & Kashmir: T36, L9-12

West Bengal: C31, S 177 (B 177), S 235, Ranjan (B256), Asha (B77).

North Eastern Hill Region: BL 25, T 36, Pant L 406, PL 639, B 77 (Asha).

Maharastra: PL 639, Malika (K75), Lens 4076, L 9-12.

Assam: L 9-12, Asha (B77).

Orrisa: B 77 (Asha).

Rajasthan: L9-12

Resistance to disease and insects

Rust, wilt and ascochyta blight are the serious disease of lentil in India, Chile, Ethiopia and Pakistan (Erskine, 1985). The sources of resistance against these diseases as compiled and given in table**3.2.**

Table 3.2 The sources of resistance against these diseases

Disease	Sources of resistance
Rust	UPP 173, UPP 175, EC10, LW 54, L 32-1, Bombay 77, TT 2, T36, L 15-10, L9-12, Pant L209, Pant L 220, Pant L 370, Pant L 406, Pant L 533, Pusa 5, Pusa 2, Pant L 8, UP 6, PL 208
Fusarium wilt	L 48, JL 33, LP 145, LP 199, Pusa 6, Pusa 9, Pant L 209, Pant L 220, P 43, UPL 1
Ascochyta blight	LG 169, LG 170, LG171, LG172, LG173, LG174, LG175, LG176, LL26, LL27, LL56, LL57,
Resistance to insects (Pod borrer, Cutwarms and Aphids)	P202, P239, P248, P262, P 772, P803, P927, P942

Above described varieties could be used in the process under breeding programme for crop improvement.

Conclusion

Lentil is cool season annual legume crop. The genous *Lens* Miller comprises of five annual species of which only *Lens culinaris* Medik is cultivated. As per historical record of released varieties, majority of them are selected from germplasm is evident that pure line selected among local adopted Germplasm lines has been very successful. Lentil is strictly self pollinated crop due to cleistogamy and dehiscence of anthers is before the flowers opens. Varietal hybridization followed by application of pedigree, bulk breeding methods or their modifications are now are increasingly being used. Major breeding objectives are to develop high yielding varieties with bold seed, containing high protein content with less cooking time and resistance to various biotic and abiotic stresses.

References

Beckmann J and Soller M.1983. Restriction fragment length polymorphisms in genetic improvement: methodologies, mapping and costs. *Theor. Appl. Genet.* 67: 35-43.

Erskine W, Muehlbauer FJ and Short RW. 1990. Stages of development in lentil. *Expl. Agric.* 26: 279-302.

Erskine W. 1985. Prospective in lentil breeding. IN Proc. Int. Workshop on faba beans, kabuli chickpeas and lentils in the 1980s, pp. 91-100. Saxena, Verma, Aleppo, Syria: ICARDA.

Frederick J. Muehlbauer, Seungho Cho, Ashutosh S, Kevin Clarice E, McPhee, Coyne J, Rajesh PN and Rebecca F. 2006. Application of biotechnology in breeding lentil for resistance to biotic and abiotic stress. *Euphytica* 147: 149–165.

Gottlieb L. 1981. Electrophoretic evidence and plant populations. *Prog. Phytochem.* 7: 1-46.

Havey M and Muehlbauer F .1989. Linkages between restriction fragment length, isozyme, and morphological markers in lentil. *Theor. Appl. Genet.* 77: 395-401.

Jeswani LM and Baldev B .1990. Advances in Pulse production technology. ICAR, New Delhi.

Ladizinsky G. 1979. The origin of Lentil and its wild gene pool. *Euphytica* 28: 179-187.

Ram HH and Singh GB. 2003. Crop Breeding and Genetics. pp. 209-221 Kalyani Publishers, New Delhi.

Saxena MC and Hawtin GC.1981. Morphology and growth patterns, pp. 39-52. In C. Webb and G.C. Hawtin (eds.) Lentils. Farnham, U.K. Commonwealth Agricultural Bureaux.

Sharma SK, Knox MR and Ellis THN. 1996. AFLP analysis of diversity and phylogeny of *Lens* and its comparison with RAPD analysis. *Theor. Appl. Genet.* 93: 751-758.

Singh AK, Pedapati A. and Manibhushan 2015a. Assessment of exchange of crop in view of change climate and International treaties. *Environ. Biol.* 36: 309-317.

Singh AK, Manibhushan, Chandra Naresh and Bharati RC. 2008. Suitable crop varieties for limited irrigated conditions in different agro climatic zones of India. *Int. J. Trop. Agri.* 26(3-4): 491-496.

Singh AK, Singh SS, Prakash V, Kumar S and Dwivedi SK. 2015b. Pulses Production in India: Present Status, Bottleneck and Way Forward. *Journal of Agrisearch* 2(2): 75-83.

Singh. 1990b. Mutation breeding in pulse crops. Proc. Golden Jubilee Celebration-Symp. Grain Legumes.

Singh.1990a. Genetics and breeding of pulse crops. pp. 217-238. Kalyani Publishers, New Delhi, Ludhiana. Feb. 9-11, IARI, New Delhi; Indian Soc. Genet. Plant. Breed. pp. 227-245.

Soller M and Beckmann J. 1983. Genetic polymorphism in varietal identification and genetic improvement. *Theor. Appl. Genet.* 67: 25-33.

Tar'an B, Buchwaldt L, Tullu A, Banniza S, Warkentin TD and Vandenberg A. 2003. Using molecular markers to pyramid genes for resistance to ascochyta blight and anthracnose in lentil (*Lens culinaris* Medik). *Euphytica* 134: 223–230.

Wilson VE .1972. Morphology and technique for crossing *Lens esculenta* Moench. *Crop. Sci.* 12: 131-2312.

Wilson VE and Law AC .1972.Natural crossing in *Lens esculenta* Moench. *Crop. Sci.* 12: 142-143.

Zohary D. 1972. The wild progenitor and the place of the cultivated lentil: *Lens culinaris. Eco. Bot.* 26: 326-332.

Citation

Prasad R, Prasad LC and Bornare SS. 2018. Lentil breeding for improving productivity using classical and molecular approaches. In: Scientific Lentil Production. Singh AK, Bhakta N, Sangale UR, Manibhushan, Sundaram PK, Kumar S and Yasin JK. (eds.) Society for Upliftment of Rural Economy Varanasi, India. Pp. 61-71.

□□□

Chapter 4

Plant Genetic Resource and Evaluation of Lentil (*Lens culinaris* Medik. ssp. *Culinaris)*

Yasin Jeshima Khan, Santoshkumar Magadum, Anil Kumar Singh, NK Gautam and Neeta Singh

SYNOPSIS

Assessment of genetic diversity in lentil is desirable for prospective future breeding activities, in terms of broadening and maintaining the diversity of the genetic base, improving opportunities for selection of improved genetics and cultivar identification. Yield stability is an interesting feature of today's lentil breeding programs, due to the high annual variation in mean yield, particularly in the arid and semi-arid areas. One of the major objectives of lentil breeding strategy in the world today is to develop cultivars with important agricultural characters, as well as high yielding varieties with drought tolerance. Adequate variability for all these traits is known to exist within the genetic resources of the crop. This suggests developing crop varieties with higher yield that include environment-specific responses to stress. Genomics-assisted breeding (GAB), that integrates the use of genomics resource tools.

Keywords: Biotic Stresses, Genetic Resources, Genomics-assisted breeding Evaluation.

Introduction

Species and cross compatibility

The genus Lens has seven closely related taxa (Ferguson and Erskine 2001). Lens culinaris Medik. ssp. culinaris is the only cultivated species and other species remain as wild relatives, L. orientalis, L. tomentosus, L.odemensis, L. lamottei, L. ervoides and L. nigricans in the genus. Lens orientalis is the presumed progenitor of cultivated L. culinaris and these two species are cross compatible to produce completely fertile progenies. From the stand point of cross compatibility to directly involve in breeding, the other Lens species are also genetically compatible through hybridization and embryo rescue. L. ervoides and L. nigricans are considered to be in the secondary gene pool (Ladizinsky *et al.*, 1984; Muehlbauer and McPhee 2005). The cultivated species of lentil (Lens culinaris Medkus) has two races viz., microsperma and macrosperma. The former has smaller rounded seeds of 2 to 6 mm diameter, yellow or orange cotyledon and testa of various colors from pale yellow to black, plants are dwarf, more pigmented, having smaller leaves and leaflets compared to macrosperma types. Macrosperma has large more flattened seeds with yellow cotyledons and pale green testa which may be speckled, 6 to 9 mm in diameter larger leaflets and pods and more robust vegetative growth. Introgression of genes between microsperma and macrosperma groups of lentil from closely related species like L. culinaris sub spp. orientalis and use of biotechnological tools, wherever necessary, have been recommended by various workers to broaden the genetic base of this crop. Direct selection for grain yield through important component traits such as pods/plant, seeds/pod, number of clusters/plant, harvest index, biological yield, early maturing and breeding for better yield quality have been suggested. Realizing the significance of plant germplasm, special efforts have been made by the national bureau to collect the genetic resources of lentil from different states of India. About 700 accessions were characterised and evaluated for various agro-morphological traits, using the lentil minimal descriptor and two catalogues have been published (Kumar *et al.*, 1983; Negi *et al.*, 1996). Lentil (Lens culinaris Medikus) seeds are a rich source of protein in many developing countries and they are becoming increasingly popular in developed countries as they are perceived as a healthy component of the diet (Tullu *et al.*, 2013). Lentil is grown in 1.4 million ha in India, out of which 86-90% is grown in Uttar Pradesh, Madhya Pradesh and Bihar. Genetic base of the presently cultivated varieties and losses due to the biotic and abiotic factors of lentil plant breeding (Tickoo *et al.*, 2015). One of the major objectives of lentil breeding strategy in the world today is to develop cultivars with important agricultural characters, as well as high yielding varieties with drought tolerance. The success of lentil as a crop depends greatly on the ability to produce economically-viable and yields potential of lentil germplasm conservation (Babayeva *et al.*, 2014). Grain legumes are second into human and animal diets after cereals and occupy an important place in the world stable food and economically important of malnutrition of lentil. The primary source of high quality dietary protein, grain legumes are important protein deficiency and malnutrition prevailing among poor in developing countries, as well as the contributing significantly to global food and nutritional value of lentil (Sharma *et al.,* 2013). Wild species which are crossable to cultivated pea, lentil, and chickpea have been collected and maintained in major germplasm collections throughout the world. The primary, secondary, and tertiary gene pools of the cool season food legumes represent potential genetic diversity that may eventually be exploited in cultivated types to overcome biotic and abiotic stresses (Muehlbauer *et al.,* 1994). The status of PGR conservation, documentation and utilization of germplasm conservation. Promote the PGR utilization including core development, pre-breeding, gene prospecting, and allele mining, public private partnerships. Major issues and future thrust areas in the field of plant genetic resources (Varaprasad and Sivaraj 2010). Plant genetic resources in conservation and utilization of crop production and productivity (Varaprasad and Sivaraj 2010). Average yields are small due to biotic and abiotic constraints and to the fact that lentil is generally produced on marginal lands that are relatively dry and lack fertilizer inputs and irrigation (Vega *et al.,* 2011). In contrast with tropical legumes that has been largely studied leading to the identification of their associated rust resistance genes and closely linked markers that are readily used in marker-assisted back crossing (Biales et al., 2013). The present productivity of these pulse crops is not only low as compared to potential yield realized in the experimental plots mainly because of prevalence of biotic and abiotic stresses legumes. In order to combat these biotic stresses mainly diseases and of course to attain higher yield levels, a large number of improved varieties have been developed through different breeding methods mainly pure line selection and hybridization utilizing in specific diversity (Shivkumar *et al.,* 2004). With the use of molecular techniques it would now be possible to hasten the transfer of desirable genes among varieties and to introgress novel genes from related wild species. Polygenic characters which were previously very difficult to analyze using traditional plant breeding methods, would now be easily tagged using molecular markers (Mohan *et al.,* 1997). This collective group of genotypes demonstrated valuable germplasm traits under stress and may therefore serve as source of useful genes in breeding lentils for drought tolerance (Babayeva *et al.,* 2014).

The present study, the genetic diversity of 505 lentil cultivars and landraces obtained from different geographical regions (Lombardi *et al.*, 2014). The genetic effects including genetic main and genotype × environment (GE) interaction effects for grain yield of eighteen lentil genotypes were studied with fourteen non parametric stability statistics (Sabaghnia *et al.*, 2014). This population was used to construct an extensive genetic linkage map in lentil using deferent types of DNA markers. The large number of unlinked markers (37) rejects the need for a large number of markers to evenly cover the lentil genome compared with most food legumes, lentil has a relatively large genome (Arumuganathan *et al.*, 1991). Lentil is a self-pollinating diploid (2n = 14 chromosomes) annual cool season grain legume produced as a high protein food source throughout the world. It has a large genome size (4,063 Mb) (Arumuganathan and Earle 1991). Several lentil genome maps are available and recent progress towards a consensus map has been made by employing robust locus markers that are derived from the model legume *Medicago truncatula* and other legume genomes (Ford *et al.*, 2007). Molecular tools have rather recently been employed by several research teams for assistance with breeding by understanding the genetic basis of many traits and for selection against major production constraints such as susceptibility to important foliar fungal disease. Consequently, much of this chapter will focus on the development of molecular markers and the identification of gene sequences associated with resistance to fungal foliar pathogens, as well as the development of advanced technologies such as genetic transformation and transcript profiling (Ford *et al.*, 2007). A lentil consensus map will comprise a set of robust and transferable markers from which genetic distance can be measured and compared across different genetic backgrounds. This will enable the tracking of gene recombination events for the building of superior genotypes (Ford *et al.*, 2007).

Major Biotic Stresses

Working with various species of food legume crops for the last few decades, food legume scientists around the world have identified several biotic agents, which substantially limit crop productivity and cause yield instability. Some of these agents pose a serious threat to the existence of the crops as well. Due to yield instability and high-risk cultivation caused by various biotic stresses, farmers in many countries have abandoned the cultivation of some food legumes, thus posing a severe threat to the sustainability of whole crop production systems. Although several authors have reported on the effects of numerous major and minor biotic factors, the key biotic stresses, which substantially affect food legumes, are presented in Table 4.1.

Table 4.1 Important biotic factors in Lentil

Fungal diseases	Insects	Viruses
Parasite weeds		
Rust	Aphids	Faba bean necrotic
Orobanche		
Fusarium wilt	Sitona weevil	Yellow virus
Cuscuta		
Root rots	Bruchids	Pea enation mosaic
Ascochyta blight	virus	
Stemphylium blight	Bean leaf roll virus	

Available Genetic Resources

Availability of resistance genes(s) among germplasm is the key building block of a crop improvement program. Sources of resistance in the primary gene pool and ability to identify resistant germplasm are keys to the breeding for resistance to biotic stresses. Success depends on the availability and level of resistance among cross compatible germplasm and the development of efficient screening techniques. The international centers involved in food legume research have collected and conserved substantial amount of germplasm in their gene banks. For example, 17, 258 accessions of *desi* chickpea and 13,548 accessions of pigeonpea have been conserved at ICRISAT, India (Sardana *et al.,* 2004); 12,421 *kabuli* chickpea, 10,520 lentil and 10,649 faba bean accessions have been conserved at ICARDA, Syria (ICARDA, 2005); 54,906 accessions of common bean at CIAT, Columbia; 16,654 accessions of cowpea at IITA, Nigeria; and more than 5,000 accessions at AVRDC, Taiwan. Besides, some national institutions, such as NBPGR, India; IPSR, UK; USDA, USA; VIR, Russia, and PGRI, Pakistan have a rich collection of pulses germplasm. Over 60,000 accessions of 26 food legumes have been assembled from 56 countries at NBPGR, India (Sardana *et al.,* 2004). A large number of germplasm accessions have been evaluated for different biotic stresses and sources of resistance have been identified and utilized in the development of improved varieties for the majority of the stresses.

Major Achievements in Biotic Stress Resistance

Various national and international breeding programs have put considerable effort son host plant resistance, they have made progress in identifying resistance sources and their use in genetic enhancement. In some instances, resistant varieties have been released and are being cultivated by farmers. Among the biotic stresses, diseases caused by fungal pathogens are the most devastating yield-limiting factors of lentil. Of them, Fusarium wilt, rust, and Ascochyta blight are globally important diseases (Bayaa *et al.*, 1994), as they cover a wide range of environment. Screening techniques for these diseases have been developed and sources of resistance identified. Sources of resistance to *Fusarium* wilt have been identified through rigorous screening in a wilt-sick plot at ICARDA, Tel Hadya. Screening of all new germplasm and breeding lines is done systematically. Recently, 34 new sources of resistance were identified from 1500 accessions of diverse origin and included in the breeding program at ICARDA (Sarker *et al.*, 2004). Sources of resistance have been used in cross breeding programs for the development of new genotypes. Several national programs have identified numerous lines and have released them as varieties (Table 4.2). Lines resistant to rust have been reported in India, Bangladesh and Ethiopia (Khere *et al.*, 1979; Agarwal *et al.*, 1993; Bakr 1993; Bejiga *et al.*, 1998). Rust screening is being done at two hot spots in India. Some lines have been identified, through rigorous and repeated screening in these sites, as highly resistant: ILL 4605, -7199, -7204, -7215, -7217, -7219, -7670, -7683, -7717, -7949,-8008, -8076, -8077, -358, -5244, -5604, -5755, -6258, -6465, -7193, -7517, and -7537 (Mishra *et al.*, 2005). By using these resistance sources, lines have been developed through backcross breeding and bulk pedigree method, where single plant selection is done at hot spots. Breeding and selection have, thus, led to the development of rust resistant varieties in Bangladesh, Ethiopia, and Morocco (Table 4.2). High level resistance to Ascochyta blight have been identified for foliar as well as seed infection in Australia (Nasir and Bretag 1997). Sources of resistance have also been reported in Canada (Slinkard and Vandenberg 1995; Tay 1989), India (Singh *et al.*, 1982; Sugha *et al.*, 1991), New Zealand (Cromey *et al.*, 1987; Russel 1994), Pakistan (Iqbal *et al.*, 1990). Some of the highly resistant lines with desirable agronomic traits and wide adaptation found in ICARDA are: ILL 358, 5244, 5604, -5588, -5684, 5755, -6258,-6465, -7193, -7517 and -7537. Source of resistance have been reported in wild relatives (Bayaa *et al.*, 1994; Nasir and Bretag

1997). Sources of resistance to minor diseases and viruses have also been reported (Agarwal *et al.*, 1993; Bakr 1993; Makkouk *et al.*, 2001; Tullu *et al.*, 2005). Adequate variability for all these traits is known to exist within the genetic resources of the crop (Wild and cultivated), allowing manipulation through plant breeding. By contrast, several other important traits, such as resistance to pea leaf weevil (*Sitona* sp.) and aphids, and the parasitic weed broomrape (*Orobannche* sp.) are not currently addressable by breeding due to a lack of genetic variation among the germplasm evaluated.

The next frontier is to ensure resistance to multiple stresses prevailing under local conditions, understanding the genetics of resistance, and applying appropriate breeding methodologies to assemble the resistant gene in a gene-pyramiding program. Test of allelism provides clues to identifying diverse genes conferring resistance to a particular disease. Backcross breeding has been applied for major genes, and multiple convergent crosses are applied in case of polygenetically inherited traits in combining desirable resistant genes in segregants. A number of disease resistant varieties have been developed by national programs from ICARDA-supplied germplasm/ segregating population (Table 4.2).

Legumes the third largest family of flowering plants, are vitally important in agriculture and the environment. Because of their capacity for symbiotic nitrogen fixation, legumes provide a substantial fraction of all nutritional protein and reduce the need for agricultural chemicals. Legumes are an old and diverse plant family comprising nearly 18,000 species in essentially all terrestrial habitats (Cannon *et al.*, 2006). Molecular markers are the most powerful genomic tools to increase the efficiency and precision of breeding practices for crop improvement. Progress in the development of genomic resources in the leading legume crops of the semi-arid tropics (SAT), namely, chickpea (*Cicer arietinum*), pigeon pea (*Cajanus cajan*) and groundnut (*Arachis hypogaea*) (Varshney *et al.*, 2012). These trait-associated robust markers along with other genomic resources including genetic maps and genomic resources will certainly accelerate crop improvement programmes in the SAT legumes (Varshney *et al.*, 2012). Development of mapping populations using identified genetic resources for trait mapping and marker-assisted breeding is the subsequent important task. The suitable genotypes for developing the mapping population should be genetically diverse and with contrast phenotype for the trait of interest (Saxena *et al.*, 2010). The global production of grain and forage legumes is about 300 million metric tons (Vance *et al.*, 2000). Which are grown on about 190 million hectares. Legumes are a rich source of proteins, vitamins, minerals and dietary fibre (Duranti and Gius 1997). The importance of common bean as a crop legume, very little expressed sequence tag (EST) information is publicly available. Development of molecular tools and to provide an initial platform for comparative functional genomics (Graham *et al.*, 2006).

Table 4.2 Disease resistant cultivars developed from ICARDA-supplied germplasm and released by the national programs

Disease Country	**Cultivar**
Rust	
Bangladesh	Barimasur-2, Barimasur-4
Ethiopia	Ada'a, Alemaya, Gudo, Assano, Alemtina, Teshale
Morocco	Bakria, Hamria, Bichette, Zaria, Chouia, Abda
Pakistan	Masoor-93, Masoor-2002
Fusarium wilt	
Syria	Idlib-2, Idlib-3, Idlib-4
Lebanon	Talya-2, Rachayya, Hala
Turkey	Firat-87, Syran-96
Ethiopia	Ada'a, Alemaya, Assano, Alemtina, Teshale
Iraq	IPA-98
Ascochyta blight	
Pakistan	Masoor-93
Australia	Nugget, Cassab, Boomer, Nipper
Stemphylium blight	
Bangladesh	Barimasur-4
Botrytis blight	
Pakistan	Masoor-93
and stem rot	
Australia	Boomer, Nipper

Leguminous plants are able to grow in the soil/substrate without any combined nitrogen due to the fixation of atmospheric nitrogen by symbiotic nodule bacteria. In rhizobia, legumes make a large contribution to the global nitrogen balance in natural and agricultural eco systems. Nitrogen fixation occurs within special plant organs, root nodules in some associations stem nodules are also formed. Development of these organs represents a well-organized process based on the tightly coordinated expression of specialized symbiotic plant and bacterial genes. The legume nodules provide an ecological niche for bacteria, as well as structure for metabolic/signal exchange between the partners and for the control of symbionts by the hosts (Zhukov *et al.,* 2013). Recent developments have rapidly increased the rate of gain for biomass yield, largely by increasing the focus and intensity of selection and improving the choice of germplasm and selection methods (Casler, 2012). All the legume crops are known for their symbiotic behaviour various genomics and genetics resources have been developed in model legume (Kulwal, 2009). The use of genomics technologies in chickpea breeding programs has been limited, since available genomic resources were not adequate and limited polymorphism was observed in the cultivated chickpea for the available molecular markers. Remarkable progress has been made in the development of genetic and genomic resources in recent years and integration of genomic technologies in chickpea breeding. Marker-assisted breeding is currently being used for improving drought tolerance and combining resistance to diseases (Gaur *et al.,* 2012). Genomics crop improvement various the applications of genomics identifying the breeding process, specific genomics approaches (Ishitani *et al.,* 2014).

Legumes more than 650 genera and 16,000 species, legumes are secondary only to grasses in economic importance in world agriculture. The central role of this plant family in agriculture is not surprising, given the abundant levels of protein found in legumes, along with their enormous contribution to soil fertility. Legumes are unique in their ability to fix atmospheric nitrogen the formation of root nodules in which they house symbiotic bacteria, known as rhizobia. Legumes have developed through evaluation of, the ability to accumulate remarkable levels of proteins, a property that is both biologically important and agricultural significant. Grain legumes can contain up to two to three times more protein s than cereals in forages legumes, such as lucerne and clovers, the protein concentrations in leaves higher than in other plant families. The use of nitrogen-fixing legumes to produce plant proteins results in a substantial decrease in the consumption of fossil fuels, and thereby lowers the agricultural contribution to global warming. Legumes also decrease the need for nitrogen fertilizer for the following crop of the rotation. Legumes are thus extremely important in the development of sustainable agriculture systems. They found that there had been successful testing of technologies but that the "rate of adoption has not significantly contributed to rural livelihoods". Land allocated to legumes was too small to make a significant impact and, where intercropping was carried out, a strong focus was maintained on maize yields, which were often very low (Abberton, 2010). Legumes have been an integral part of agricultural for thousands of years and remain a key component of sustainable agricultural systems. This report describes the significant progress in symbiosis research resulting from legume genomics and functional of genomics (Udvardi, 2002). Legumes are the primary sources of dietary proteins in many developing countries were protein hunger and malnutrition is widespread (Upadhyaya *et al.,* 2011).

Genomic resources provide the starting point for understanding the unique traits present in the given crop and also tools for implementation of molecular breeding for the development of high varieties (Varshney *et al.,* 2013). Genetic markers are tools that can facilitate molecular breeding, even in species lacking genomic resources (Kim *et al.,* 2015). Plant genomics for the purpose of this research paper the investigation and advancement of new crop varieties by studying the whole genome of a plant species and sequencing nucleotides to find constituent genes (Boettiger *et al.,* 2012). Plant breeding has the latest innovations in biology and genetics to enhance the crop improvement of plant breeding. (Stephen *et al.,* 2004). The use of plant genetic resources (PGR) in crop improvement, followed by adoption, cultivation and consumption or marketing of the improved cultivars by farmers, the most sustainable methods to conserve valuable genetic resources and simultaneously to increase the agricultural production and food security (Haussmann *et al.,* 2004). The model is very important for develop studies the pathogenesis of these parasitic plants in legumes (Pérez de Luque *et al.,* 2007). Advances in plant genomics research are opening a new plant breeding where the linkage of genes to specific traits to more efficient and predictable breeding in future (Kulwal *et al.,* 2011). Large extent an under exploited crop where relatively large genetic gains can likely be made with only modest investments in both applied plant breeding and molecular genetics, new opportunities for improving cowpea exist by the emerging genomic tools and knowledge gained through research on other major legume crops (Timko and Singh, 2008). This suggests developing crop varieties with higher yield that include environment-specific responses to stress. Genomics-assisted breeding (GAB), that integrates the use of genomics resource tools (Rajeev *et al.*, 2013). Characterization of genetic diversity and population structure of lentil landraces was also performed made for germplasm conservation and breeding programs (Lombardi *et al.*, 2014). Genomic information may also assist the discovery of genomic regions that contribute to the underlying genetic variance of legumes of plant breeding programs (Escriche and Racio, 2011). First, the adoption of two genetic model legumes, Medic ago truncatula and Lotus japonicas, and second, the application of modern methods in functional genomics (*e.g.* transcriptomic, proteomic and metabolomic analyses) (Stacey *et al.*, 2015). Plant breeding and genetics are powerful tools for increasing plant productivity through development of improved varieties. The rapid progress of plant genomics in recent years has opened new possibilities in targeted breeding of specific traits, and provides a powerful approach to sustainable crop production. Plant genomics, in combination with genetics and breeding, has a particularly crucial role to play in ensuring food security to the rapidly growing world population (Ranjan *et al.*, 2012). In addition, Low levels of genetic variability and limited genomic resources have been serious impediments to pigeon pea crop improvement through modern breeding approaches (Pazhamala *et al.*, 2015). Moreover, because comparative genomic studies have revealed the limits of conservation between rice and the other cereal genomes, genomic resources and programs have been developed for maize, sorghum, wheat, and barley to serve as the foundation for future genome sequencing and the acceleration of genomic based improvement of these critically important crops (Feuilleta and Eversole, 2007).

Germplasm Conservation

Extensive collections of lentil germplasm now exist in various genebanks around the world. This germplasm including wild Lens species has been used in plant introduction strategies and in efforts to widen the potential sources of increasing genetic diversity in the breeding programmes. The World Information and Early Warning System on plant genetic resources for food and agriculture of the FAO lists a total of 58,404 lentil accessions worldwide (FAO, 2010). In India, the responsibility of maintaining the *ex situ* collections on long-term basis is entrusted to the National Genebank at NBPGR. The National Genebank operates in the network mode to have an effective linkage with other organizations, engaged in lentil improvement programmes and to make conserved PGR easily accessible to plant breeders. The conservation activities were initiated in 1980s with storage and frequent regeneration of germplasm and strengthened with the provisions of cold storage facilities at NBPGR headquarters under the Indo-UK (Overseas Development Agency) project. The Indian Institute of Pulse Research (IIPR), Kanpur, designated as National Active Germplasm Site (NAGS), and 7 NBPGR Regional Stations were also equipped with medium-term seed storage modules (+4°C) and, thereby, a network was established. A total of 2,758 accessions are conserved as base collections in long-term storage facility at National Genebank at NBPGR, New Delhi, which are maintained at -18°C. It includes indigenous collection of 2,081 accessions from the different states and 677 exotic collections including accessions of 4 wild species. In addition, a duplicate set of accessions (comprising 7,712 from ICARDA, Syria), is also maintained at the National Genebank. Further, some 14,329 accessions of lentil which may also include some duplicates are held as active collections at various research organizations major one being the IIPR, Kanpur, a NAGS for pulses which holds over 5000 accessions of lentil. The genetic resources of lentil have a wide spectrum of variability of both qualitative and quantitative nature. Availability of information on conserved germplasm resources is key to its utilization. Sources of desirable agronomic traits like early maturity, dwarf types, good plant height, plant vigour, high biomass, profuse branching, high peduncle number, bold seed, high pod number, long pod, broad pod, high yield, more seeds/pod (>5) and biotic stress like, rust and pod borer resistance are available in National Genebank. The germplasm conserved as a base collection is assigned a national identity number, dried to seed moisture of around 5+2 per cent at 15°C and 15% relative humidity. Un-treated and pest free seeds (2000-4000) meeting international standards, with (>85%) seed viability are transferred to long-term storage. Base collections are monitored for seed viability, quantity, etc., at an interval of 10 years from storage. The base collections are not meant for distribution and are generally used for regeneration of active collections to support crop improvement activities. For conservation of active collections, seed moisture is brought down to 8-10%. The active collections are conserved in medium-term storage maintained at 4°C and 35-40% relative humidity. In active collections also, the seeds are periodically monitored for viability and quantity at an interval of 5 years. As per the Global Crop Diversity Trust (2008); FAO (2010) approximately, 58,407 lentil accessions exists in various collection worldwide, of which 11,125 are conserved at the International Centre for Agriculture Research in Dry Areas (ICARDA), Syria, 5,254 at Australian Temperate Field Crops Collection (ATFCC) Australia, 3000 at Seed and Plant Improvement Institute (SPII), Iran, 3,182 at National Centre for Genetic Resources Preservation (NCGRP, formerly, NSSL, USA, 2556 at N.I. Vavilov All-Russian Scientific Research Institute of Plant Industry (VIR), Russia.

Registered germplasm

Many promising genotypes which have specific desirable attributes but do not show the required yield superiority to get released or do not have direct commercial value can be registered with NBPGR as the nodal institute. The 3 registered lentil genetic stocks are : (1) IC 296883 (LH-92-27) with INGR no.02 006 for its multiple resistance to *Meloidogyne incognita, M. javanica,* Botrytis grey mould and lentil pod borer, (2) IC 296884 (LH-90-54) with INGR no. 02 005 for its bright dark green leaves, (3) INGR 08 026 for resistance to rust. Information on genetic stocks conserved in the National genebank can be viewed here.

(http//www.nbpgr.ernet.in/IRCG%20search/index.htm)

References

Abberton M. 2010. Enhancing the role of legumes: potential and obstacles. Grassland carbon sequestration: management, policy and economics 11:177-187.

Agarwal SC, Singh K and Lal SS. 1993. Plant protection of lentil in India. Proceeding of the Seminar on Lentil in South Asia. pp. 147-167. 11-15 March, 1991, New Delhi, India.

Arumuganathan K and Earle ED. 1991. Nuclear DNA content of some important plant species. *Molecular Biology Reports* 9: 208–221.

Arumuganathan K and Earle ED. 1991. Nuclear DNA content of some important plant species. *Plant Molecular Biology Reporter* 9: 208-218.

Babayeva S, Akparov Z, Damania A, Izzatullayeva V, Aslanova G and Abbasov M. 2014. Genetic diversity for drought tolerance in lentils from central asia and the caucasus: caclentil. *Albanian Journal of Agricultural Sciences* 13(2): 1-8.

Bakr MA. 1993. Plant protection of lentil in Bangladesh. Proceedings of the Seminar on Lentil in South Asia. pp. 177-186. 11-15 March, 1991, New Delhi, India.

Bayaa B, Erskine W and Hamdi A. 1994. Response of wild lentil to *Ascochyta fabae* f. sp. *Lentis* from Syrai. *Genetics Resources and Crop Evolution* 41: 61-65.

Bejiga G, Tadesse N and Erskine W. 1998. We fixed rust, now wilt. *ICARDA Caravan* 9: 12.

Biales D, Maria M, Molina R and Sillero JC. 2013. Identification of pre and post-haustorial resistance to rust (Uromyces Viciae-Fabae) in lentil (*Lens culinaris*) germplasm. *Plant Breeding* doi:10.1111/Pbr.12096 2013.

Boettiger S, Anthony V, Booker K and Starbuck C. 2012. Public Private Partnerships in Plant Genomics for Global Food Security. GATD Foundation.

Cannon SB, Sterck L, Rombauts S, Sato S, Cheung F, Gouzy J, Wang X, Mudge J, Vasdewani J, Schiex T, Spannagl M, Monaghan E, Nicholson C, Humphray SJ, Schoof H, Mayer KF, Rogers J, Quétier F, Oldroyd GE, Debellé F, Cook DR, Retzel EF, Roe BA, Town CD, Tabata S, Van de Peer Y and Young ND. 2006. Legume genome evolution viewed through the *Medicago truncatula* and *Lotus japonicas* genomes. *Proceedings of the National Academy of Sciences U.S.A.* 103(40): 14959-14964.

Casler MD. 2012. Switchgrass Breeding, Genetics, and Genomics. Switchgrass, Green Energy and Technology. Monti A (Ed). Springer-Verlag London.

Duranti M and Gius C. 1997. Legume seeds: protein content and nutritional value. *Field Crops Research* 53: 31-45.

Erez De La Vega P, Fratini MRM and Muehlbauer FJ. 2011. Genetics, genomics and breeding of cool season grain legumes. pp. 89-150. Vaga MP, Torres AM, Cubero JI and Kole C (Eds). *Science Publishers, CRC Press, Enfield.*

Escriche NI and Recio OG. 2011. Promises, pitfalls and challenges of genomic selection in breeding programs. *Spanish Journal of Agricultural Research* 9(2): 404-413.

Feuilleta C and Eversole K. 2007. Physical mapping of the wheat genome: A coordinated effort to lay the foundation for genome sequencing and develop tools for breeders. *Israel Journal of Plant Sciences* 55: 307–313.

Ford R, Mustafa B, Inder P, Shaikh R, Materne M and Taylor P. 2007. Lensomics: advances in genomics and molecular techniques for lentil breeding and management. Lentil: An ancient crop for modern times, pp 275-289. Yadav SS, McNeil DL and Stevenson PC (Eds). Springer Publication, Netherlands.

Gaur PM, Jukanti AK and Varshney RK. 2012. Impact of Genomic Technologies on Chickpea Breeding Strategies. *Agronomy* 2: 199-221.

Graham MA, Ramírez M, Valdés-López O, Lara M, Tesfaye M, Vance CP and Hernandez G. 2006. Identification of candidate phosphorus stress induced genes in *Phaseolus vulgaris* through clustering analysis across several plant species. *Functional Plant Biology* 33: 789-797.

Haussmann BIG, Parzies HK, Prester T, Susic Z and Miedaner T. 2004. Plant genetic resources in crop improvement. *Plant Genetic Resources* 2(1): 3–21.

Iqbal SM, Bakshsh A and Malik RA. 1990. Identification of resistant sources for Ascochyta blight in lentil. *Lens* 17: 26-27.

Khare MN, Agrawal SC and Jain AC. 1979. Diseases of lentil and their control. Jawaharlal Nehru Krishi Vishwa Vidyalaya. Jabalpur, Madhya Pradesh India, 29 pp.

Kim JH, Lee C, Hyung D, Jo YJ, Park JS, Cook DR and Choi HK. 2015. Platform for designing cross-species intron-spanning genic markers linked with genome information of legumes. *Plant Methods* 11:30. DOI 10.1186/s13007-015-0074-6.

Kulwal PL, Thudi M and Varshney RK. 2011. Genomics Interventions in Crop Breeding for Sustainable Agriculture. Encyclopedia of Sustainability Science and Technology. Springer. New York.

Kulwal PL. 2009. Recent event on legume genomics and genetics. *International Journal of Integrative Biology* 6: 38-40.

Lombardi M, Materne M, Noel OI, Rodda M, Hans DD, Anthony TS, John WF and Kaur S. 2014. Assessment of genetic variation within a global collection of lentil (*Lens culinaris* Medik.) cultivars and landraces using SNP markers. *BMC Genetics* 15: 150.

Makkouk K, Kumari S, Sarker A and Erskine W. 2001. Registration of six lentil germplasm lines with combined resistance to viruses. *Crop Science* 41: 931-932.

Mohan M, Nair S, Bhagwat A, Krishna TG, Yano M, Bhatia CR and Sasaki T. 1997. Genome mapping, molecular markers and marker-assisted selection in crop plants. *Molecular Breeding* 3: 87-103.

Moose SP and Mumm RH. 2008. Molecular plant breeding as the foundation for 21st century crop improvement. *Plant Physiology* 147: 969–977.

Muehlbauer FJ, Kaiser WJ and Simon CJ. 1994. Potential for wild species in cool season food legume breeding. *Euphytica* 73: 109-114.

Nasir M and Bretag TW. 1997. Resistance to *Ascochyta fabae* f. sp. Lentis in *lentil* accessions. *In:* Program and Papers, International Food Legume Research Conference, Adelaide, Australia.

Pazhamala L, Saxena RK, Singh VK, Sameerkumar CV, Kumar V, Sinha P, Patel K, Obala J, Kaoneka SR, Tongoona P, Shimelis HA, Gangarao NVPR, Odeny D, Rathore A, Dharmaraj PS, Yamini KN and Varshney RK. 2015. Genomics assisted breeding for boosting crop improvement in pigeonpea (*Cajanus cajan*). *Frontiers in Plant Science* 6: 1-12.

Ranjan A, Ichihashi Y and Sinha NR. 2012. The tomato genome: implications for plant breeding, genomics and evolution. *Genome Biology* 13: 167.

Russell AC. 1994. "Rajah" lentil (*Lens culinaris* Medik.) *New Zealand Journal of Crop and Horticultural Science* 22: 469-470.

Sabaghnia N, Karimizadeh R and Mohammad M. 2014. Graphic analysis of yield stability in new improved lentil (*Lens culinaris* Medik.) genotypes using nonparametric statistics. *Acta Agriculturae Slovenica* 103: 113-127.

Sardana S, Upadhyaya HD and Dhilon BS. 2004. Genetic Resources of Pulse Crops-present status and future plans. *In:* Pulses in new perspective: Proceedings of the National Symposium on Crop Diversification and Natural Resource Management, 20-22 December 2003, Indian Institute of Pulse Research, Kanpur, India.

Sarker A, Bayaa B, El-Hassan H and Erskine W. 2004. New sources of resistance to fusarium wilt in lentil. (*Lens culinarsi* Medikus subsp. *Culinaris*). *Journal of Lentil Research* 1: 30-33.

Saxena RK, Saxena KB, Kumar RV, Hoisington DA and Varshney RK. 2010. SSR-based diversity in elite pigeonpea genotypes for developing mapping populations to map resistance to fusarium wilt and sterility mosaic disease. *Plant Breeding* 129: 135-141.

Sharma S, Upadhyaya HD, Varshney RK and Gowda CLL. 2013. Pre breeding for diversification of primary gene pool and genetic enhancement of grain legumes. Frontiers *in Plant Science* 4: 1-14.

Shivkumar, Gupta S, Chandra S and Singh BB. 2004. How wide is the genetic base of pulse crops. Proceedings of pulses in new perspectives. pp. 211-221. *Indian Institute of Pulses Research, Kanpur.*

Shugha SK, Kappor SK and Singh BM. 1995. Management of chickpea wilt with fungicides. *Indian Phytopathology* 48: 27-31.

Singh G, Singh K, Gill AS and Brar JS. 1982. Screening of lentil varieties/ lines for blight resistance. *Indian Phytopathology* 35: 678-679.

Slinkard AE and Vandenberg A. 1995. Lentil. Harvest of Gold, The History of Field Crop Breeding in Canada, Slinkard AE and Knott DR (Eds). pp 191-196. Saskatoon, Canada, University Extension Press, University of Saskatchewan.

Stacey G, Libault M, Brechenmacher L, Wan J and May GD. 2006. Genetics and functional genomics of legume nodulation. *Current Opinion in Plant Biology* 9: 110–121.

Tay J. 1989. Inheritance of resistance to ascochyta blight in lentil. M.Sc. thesis. Department of Crop Science and Plant Ecology, University of Saskatchewan, Canada.

Tickoo JL, Sharma B, Mishra SK and Dikshit HK. 2005. Lentil (*Lens culinaris*) in India: Present status and future perspectives. *Indian Journal of Agricultural Sciences* 75(9): 539-62.

Timko MP and Singh BB. 2008. Cowpea a multifunctional legume. Genomics of Tropical Crop Plants. pp 227-258. Moore PH and Ming R (Eds). Springer, New York.

Tullu A, Buchwaldt L, Warkentin T, Taran B and Vandenberg A. 2013. Genetics of resistance to anthracnose and identification of AFLP and RAPD markers linked to the resistance gene in PI 320937 germplasm of lentil (*Lens culinaris* Medikus). *Theoretical and Applied Genetics* 106: 428.

Tullu A, Taran B, Breikreutz C, Buchwaldt L, Baninza S, Warkentin T and Vandenbger A. 2005. A QTL for resistance to ascochyta blight maps to the same linkage group containing the gene for anthracnose resistance in lentil. Proceedings of the International Edible legume conference, Durban, South Africa, 17-21 April 2005. pp 22.

Udvardi MK. 2002. Legume genomes and discoveries in symbiosis research. *Genome Biology* 3(9): reports 4028.1–4028.3.

Upadhyaya HD, Dwivedi SL, Ambrose M, Ellis N, Berger J, Smykal P, Debouck D, Duc G, Dumet D, Flavell A, Sharma SK, Mallikarjuna N, Gowda CLL. 2011. Legume genetic resources: management, diversity assessment, and utilization in crop improvement. *Euphytica* 180(1): 27-47.

Vance CP, Graham PH and Allan DL. 2000. Biological nitrogen fixation. Phosphorus: a critical future need in Nitrogen Fixation: From Molecules to Crop Productivity. pp 506-514. Pedrosa FO, Hungria M, Yates MG and Newton WE (Eds). Kluwer Academic Publishers, Dordrecht, The Netherlands.

Varaprasad KS and Sivaraj N.2010. Plant genetic resources conservation and use in light of recent policy developments. *Electronic Journal of Plant Breeding* 1(4): 1276-1293.

Varshney RK, Kudapa H, Roorkiwal M, Thudi M, Pandey MK, Saxena RK, Chamarthi SK, Mohan SM, Mallikarjuna N, Upadhyaya H, Gaur PM, Krishnamurthy L, Saxena KB, Nigam SN and Pande S. 2012. Advances in genetics and molecular breeding of three legume crops of semi-arid tropics using next-generation sequencing and high-throughput genotyping technologies. *Journal of Biosciences* 37: 811–820.

Varshney RK, Roorkiwal M, and Nguyen HT.2013. Legume genomics: from genomic resources to molecular breeding. *The Plant Genome* 6: 1-7.

Zhukov VA, Shtark OY, Borisov AY and Tikhonovich IA. 2013. Breeding to Improve Symbiotic Effectiveness of Legumes, Plant Breeding from Laboratories to Fields. pp 167-207. Andersen SB (Ed). In Tech Publishers, Croatia-European Union.

Citation

Yasin JK, Magadum Santoshkumar, Singh Anil Kumar, Gautam NK and Singh Neeta. 2018. Plant Genetic Resource and Evaluation of Lentil (*Lens culinaris* Medik. ssp. *Culinaris).* In: Scientific Lentil Production. Singh AK, Bhakta N, Sangale UR, Manibhushan, Sundaram PK, Kumar S and Yasin JK. (eds.) Society for Upliftment of Rural Economy Varanasi, India. Pp. 73-90.

□□□

Chapter 5

Genetic Improvement of Lentil for Fusarium wilt Resistance

AK Choudhary and Shiv Kumar

SYNOPSIS

Lentil is one of the major components of low-input agriculture in the sub-tropics including parts of Indian sub-continent. It plays an important role in alleviating protein malnutrition for millions of vegetarian peoples of these regions. The production and productivity of lentil is severely constrained by Fusarium wilt (FW). FW has been reported to cause 100% yield losses, if it affects the crop in the seedling stage. However, recent advances in the identification of races of the pathogen, screening techniques to identify resistant genotypes, genetic dissection and mapping and tagging of resistance genes through molecular markers have resulted in the release of several FW resistant varieties the world over. This has not only narrowed the gaps between potential and realized yields, but also minimized yearly fluctuations in production and productivity of lentil.

Keywords: Fusarium wilt, screening techniques, gene tagging, wilt resistance, molecular markers, gene pyramiding.

Introduction

Pulses form an important component of typically low input agriculture in sub-tropical regions. Lentil (*Lens culinaris* Medikus subsp. *culinaris*) is one of the major pulse crops in these areas. Resource poor small and marginal farmers hardly provide better-than-average-management practices to this crop. The crop is also exposed to harsh, erratic and unpredictable climatic conditions of these regions in addition to biotic stresses imposed by several pathogens and insect pests. This has not only slowed down its productivity growth, but also caused yield instability leading to shifts into other crops mainly cereals. Among the biotic stresses, Fusarium wilt (FW) disease is considered as the most severe yield limiting factor in lentil. The FW affects the lentil from seedling to the stage of maturity, causing heavy losses both in quantity and quality. Chemical and cultural control measures are not very effective; however, resistance breeding has been very efficient, and many wilt rust resistant varieties in lentil have been released and adopted by farmers in sub-tropical and semi-arid (SAT) regions (Choudhary *et al.*, 2013).

Importance of wilt disease in lentil

Among several biotic stresses affecting stability of production, FW is the most important. This vascular wilt, which is caused by *Fusarium oxysporum* f.sp. *lentis*, is a widespread disease of lentil with its report of occurrence from as many as 26 countries in South Asia, Sub-Saharan Africa and West Asia and North Africa (WANA) regions. It was first reported from Hungary, and later on from many countries including India, USA, Czechoslovakia, USSR, France, Turkey, Syria, Myanmar and Pakistan, Nepal, Ethiopia and Egypt. The disease is known to cause economic yield losses in parts of WANA region, Sub-Saharan Africa and South Asia (Erskine *et al.*, 1994). In India, FW is the major factor limiting lentil production in the states of Uttar Pradesh, Madhya Pradesh, Himachal Pradesh, Bihar, West Bengal, Assam, Rajasthan, Haryana and Punjab (Agrawal *et al.*, 1993; Chaudhary *et al.*, 2009; 2010). Yield losses due to FW in lentil depend on the crop stage at the time of infection (Khare *et al.*, 1979), environment and crop variety (Table 5.1). Wilt incidence at seedling stage can lead to a complete crop failure whereas at adult stage (flowering and podding) infection, the plants are able to produce some grain yield that could be shriveled. Wilt incidence as high as 50-78% has been reported in some fields of Madhya Pradesh (Khare *et al.*, 1979; Agrawal *et al.*, 1993). In west Asian countries like Syria, the yield losses range from 5-72% (Bayaa *et al.*, 1986). There is a strong correlation between wilt incidence and grain yield, estimating 8.8% yield loss for every 10% wilt incidence (Erskine and Bayaa, 1996).

Table 5.1 Economic losses due to Fusarium wilt in lentil

Stage of infection	Economic loss (yield/value)	Country	Reference
Seedling to pod filling stage	50-78%	India	Khare *et al.* (1979), Agrawal *et al.* (1993)
Seedling stage	100%	India	Khare (1981)
Stage not specified	29.98%	India	Kumar and Bourai (2012)
Stage not specified	5-72%	Syria	Bayaa *et al.* (1986)
Stage not specified	8.8% (for 10% wilted plants)	Syria	Erskine and Bayaa (1996)

Pathogenic variability

Though *F. oxysporum* f. sp. *lentis* (FOL) is host specific, presence of genetic variability has been reported on the basis of reactions in host genotypes and pathogen morphology and cultural characters (Kannaiyan and Nene, 1978; Belabid *et al.*, 2004; Taheri *et al.*, 2010). Khare *et al.* (1975) reported eight strains of FOL, whereas Kannaiyan and Nene (1978) established seven strains. However, no variation in virulence/aggressiveness was detected among these Indian strains that could play a major role in breaking the resistance of existing genotypes of lentil. Belabid *et al.* (2004) studied virulence and vegetative compatibility of 32 Algerian isolates of FOL, and grouped them as a single race (Table 5.2). However, these isolates differed in their aggressiveness on susceptible lines. Study on 333 isolates from different states of India revealed 43 cultural and morphological groups (Chaudhary, 2008). On the basis of disease reactions against 7 lentil differentials, these isolates were grouped into three clusters. Similarly, variability analysis of 24 isolates collected from north eastern Indo-Gangetic plains using 40 RAPD and 12 SSR primer pairs revealed two sub-populations with little genetic variations (Datta *et al.*, 2009).

Table 5.2 Races/variants of *F. oxysporum* f. sp. *lentis* (FOL)

Race/variant	Reported from	References
Eight strains	India	Khare et al. (1975)
Seven strains	India	Kannaiyan and Nene (1978)
Single race	Algeria	Belabid et al. (2004)
Three classes (based on 43 cultural and morphological groups)	India	Chaudhary (2008)
Two groups	India	Datta et al. (2011)

Screening techniques to identify wilt resistant genotypes

The initial step to utilize host plant resistance (HPR) relates to the development of reliable and reproducible disease screening techniques to evaluate large numbers of germplasm accessions and breeding materials. Effective and efficient screening for resistance to soil borne pathogens such as *Fusarium* spp. calls for simulation of natural soil and environmental conditions and uniform inoculum load across all the plants of test genotypes to discriminate between resistant and susceptible genotypes. In general, screening under field and controlled conditions (green house and laboratory conditions) has been suggested to identify resistant genotypes for FW resistance in lentil (Kraft *et al.*, 1994; Alessandro *et al.*, 2006).

Field screening

The most common and widely used method for screening of FW resistant genotypes has been the wilt sick plot (WSP) method. The main advantage of WSP method is that it allows screening of a large number of genetic materials under field conditions (Infantino *et al.*, 2006). Typical disease symptoms are the main criteria for evaluating breeding lines and establishment of WSPs, while the re-isolation of the causal organism is a confirmatory test. In WSP method, inoculum load needed to get the typical wilt symptoms can vary with race/variant, environmental conditions, crop and its maturity groups and ecotypes (e.g., *Macrosperma* and *Microsperma* types of lentil) type. The procedures of the "field screening" for WR are almost similar in all the pulses including lentil. The details of field screening have been described in lentil (Kumar *et al.*, 2010). It involves planting of test genotypes along with a susceptible cultivar, which serves as an indicator line or the susceptible check after every 2-4 test entries to monitor uniformity of the inoculum in the plot (Fig. 5.1). The widely used susceptible checks are 'ILL 4605'. In addition, resistant genotype(s) (e.g., ILL 5588) should also be planted after every 10 rows to monitor if there are other pathogens that can confound the wilt reaction. For screening a large number of gemplasm lines against FW, WSPs have been developed at ICARDA and NARS systems of countries where lentil is a major crop. In India, WSPs have been created for lentil at IIPR, Kanpur, and at selected major centres of the AICRP on MULLaP. Field screening of germplasm and cultivars for wilt resistance has been carried out widely for so many years, which has resulted in the identification of a number of wilt resistant genotypes and cultivars.

Screening under controlled conditions

Field screening, although widely used, has been criticized because many edaphic and climate factors are not under control, and involvement of other soil borne fungal pathogens and nematodes has also been noticed. Screening under controlled conditions in glasshouse is suggested to confirm the results of WSP method. This is particularly important for inheritance and molecular mapping and tagging studies using a well characterized race of the pathogen. Moreover, pathogenic diversity studies can be done under the controlled conditions that support genotypic information.

Greenhouse and laboratory screening

The greenhouse screening technique (multiplication of inoculum, raising of seedlings of pigeonpea in autoclaved soil, root dipping in inoculum and transplanting in pots filled with autoclaved soil and assessing disease incidence) as developed for pigeonpea can also be applied for screening FW resistant accessions of lentil (Haware and Nene, 1994). The sick pot screening technique (Nene and Kannaiyan, 1982) can also be refined and utilized for scoring wilt reaction in lentil (Choudhary *et al.*, 2013).

In lentil, inoculum density of 10^6 conidia ml^{-1} is generally used to inoculate seedlings. Different inoculation methods such as seeding surface of disinfected lentil seeds in the soil infested with the pathogen grown on autoclaved millet or other grains (10% w/w) and inoculating by pouring spores grown on PDA near the roots of 15-day old seedlings in pots are used (Riccioni *et al.*, 2003). Roots of 10-

day old seedlings grown on sterilized sand can be dipped in a spore suspension with concentration of 10^5 conidia ml^{-1}. The wilt reaction in terms of severity of incidence can be evaluated after 7-10 days of inoculation. Fusaric acid (FA), one of the toxins produced by the *Fusarium*, is used as the selective agent to screen chickpea genotypes in laboratory. Concentration of FA to inhibit 50% pollen tube growth differs for resistant, late wilting and susceptible types, which has also been validated by molecular markers. This selective agent can be revalidated for lentil as well. However, despite many limitations, field screening is still a widely used technique to discriminate between resistant and susceptible genotypes of lentil for FW owing to operational simplicity and economy of labour. Nonetheless, it should be used for preliminary screening only. Resistant genotypes must be confirmed for resistant reaction under controlled screening condition. Rapid discrimination between resistant and susceptible genotypes may be performed *in vitro* by using selective agents such as FA, which has already been used in tissue culture studies to select the wilt resistant variants in banana (Matsumoto *et al.*, 1995) and pigeonpea (Pandey *et al.*, 1995). However species/race specificity of FA produced by the pathogen needs to be investigated further. With the availability of diagnostic PCR based molecular markers, resistance to FW can be established without subjecting germplasm and segregating generations for phenotyping in wilt-sick plot. For this purpose, molecular markers that are closely linked with WR genes are required.

Fig. 5.1 Screening of lentil germplasm in wilt sick plot at Tel Hadya, Syria. The picture shows highly susceptible check 'ILL 4605' along with resistant lines.

Genetics of wilt resistance

Development of wilt-resistant varieties is the major objective in the pulse breeding program to ensure stability in production and productivity (Choudhary, 2010). The accomplishment of the objective more often becomes difficult due to evolution of new races and co-existence of more than one pathotype at any one location. The transfer of FW resistant genes from the donors to an otherwise high-yielding genotype requires knowledge about the inheritance and genetics of wilt resistance. Only limited inheritance studies have been carried out to know the genetics and inheritance pattern of WR in lentil. Five independent genes have been reported to confer resistance to FW in lentil (Kamboj *et al.*, 1990). Based on allelism test, two duplicate genes and two complementary genes have been identified, imparting WR in the variety 'PL 234' and in 'JL 446' and 'PL 286', respectively. However, only a single dominant gene has been reported to control WR in the crosses made at ICARDA (Abbas, 1995). Eujayl *et al.* (1998) also recorded monogenic inheritance for WR in 'ILL 5588' and designated the gene as *Fw* (Table 5.3).

Tagging of resistance gene(s) through molecular markers

Though simply inherited, the transfer of WR to locally adapted cultivars has been difficult due to linkage drag and difficulty in accurate phenotyping under field screening because of uneven concentration of inoculum and presence of different races/pathotypes of *Fusarium* spp. (Choudhary *et al.*, 2013). Therefore, tagging of WR gene(s) through molecular markers is highly desirable. Only limited progress has been achieved towards tagging of resistance gene(s) in lentil (Table 5.4).

Table 5.3 Inheritance of FW resistance in lentil

Fusarium race/ variant	Number and nature of WR gene	Gene symbol	Remarks	Reference
Strain not specified	Five genes	-	Independent genes	Kamboj *et al.* (1990)
Strain not specified	Duplicate genes	-	Resistance in PL 234	Kamboj *et al.* (1990)
Strain not specified	Two complem-entary genes	-	Resistance in JL 446 and PL 286	Kamboj *et al.* (1990)
Strain not specified	Monogenic dominant gene	-	ICARDA experiments	Abbas (1995)
Strain not specified	Monogenic dominant gene	*Fw*	Resistance in ILL 5588	Eujayl *et al.* (1998)

Table 5.4 WR genes tagged in lentil

Fusarium race	Gene tagged	Marker identified	Distance (cM)	Linkage group	Refer-ence
Strain not specified	*Fw*	RAPD marker OPK-15_{900} OP-BH_{800} and OP-$D15_{500}$OP-C0465o	10.8	Coupling phase Repulsion phase	Eujayl *et al.* (1998)
Strain not specified	WR gene	SSR59-2B	8.0		Hamw-ieh *et al.* (2005)
Strain not specified	WR gene	AFLP p17m30710	3.5		Hamw-ieh *et al.* (2005)

Eujayl *et al.* (1998) identified RAPD marker OPK-15_{900} linked with *Fw* gene at a distance of 10.8 cM and established its linkage with the RAPD markers OP-$B17_{800}$ and OP-$D15_{500}$ in coupling and OP-$C04_{650}$ in repulsion phase. These arbitrary markers can be made more useful by converting them into locus-specific sequence characterized amplified region (SCAR) markers for marker-assisted screening and selection. Subsequent study identified one SSR and AFLP markers that were linked with *Fw* gene at 8.0 and 3.5 cM, respectively

(Hamwieh *et al.*, 2005). However, WR genes present in the Indian germplasm are yet to be mapped. Efforts are underway to develop mapping populations involving 'Precoz' and 'Sehore 74-3' as the susceptible and 'PL2' and 'IPL406' as the resistant parents. For developing mapping populations without any segregation distortion, molecular markers have been very useful in establishing hybridity of F_1 plants (Solanki *et al.*, 2010). New RILs have been developed at ICARDA involving parents from different geographical regions for mapping race-specific resistance genes.

Conventional and molecular breeding for wilt resistance in lentil

Lentil is a highly self-pollinated crop. As FW resistance appears to be simply inherited, conventional breeding methods used in autogamous crops such as backcross and recombination breeding should be equally effective for breeding wilt resistant varieties. Simple field screening in WSPs and selection has resulted in the identification and release of a number of FW resistant donors and varieties, respectively in lentil (Table 5.2). Recombination breeding, a selection-crossing-selection cycle which consists of controlled crossing between agronomically superior genotype(s) and wilt resistant donor(s) followed by pedigree selection or its various modifications in the segregating generations has been the most utilized breeding approach for incorporating WR in these three pulse crops. The bulk pedigree method has been the preferred method at ICARDA in which targeted crosses are advanced under disease-free conditions as bulks up to F_4 generation, and the selected single plant progenies (F_5) are grown in the wilt-sick plot. Plant progenies with resistant reaction are further evaluated in WSP and in normal field as preliminary screening nursery (F_6), preliminary yield trial (F_7), and advanced yield trial (F_8). Finally, the elite lines with WR, high yield and other desirable traits in different genetic backgrounds are included in Lentil International Fusarium Wilt Nursery (LIFWN) and other yield nurseries for multi-location testing in the targeted countries. In addition to genetically fixed elite lines and germplasm, segregating populations are also made available to the national programs for selection in the local wilt-sick plot and agro-climatic conditions. Systematic utilization of resistant sources such as 'ILL 5883', 'ILL 5588', 'ILL 4400' and 'ILL 590' at ICARDA has resulted in the development of a wide spectrum of FW resistant varieties for cultivation in different countries (Table 5.5). Some of the prominent wilt resistant varieties are 'Idleb 2', 'Idleb 3', 'Idleb 4' and

'Ebla 1' in Syria; 'Talya 2', 'Rachayya' and 'Hala' in Lebanon; 'Firat 87' and 'Syran 96' in Turkey; 'Ada', 'Alemaya', 'Assano', 'Alemtina' and 'Teshale' in Ethiopia; 'Kimiya' in Iran and 'IPA 98' in Iraq. In India, national program has released several wilt resistant varieties, and prominent among them are 'L 4147', 'Pant L 406', 'Pant L 4', 'Pant L 639', 'Priya', 'Seri', 'JL 3', 'Noori', and 'VL 507' (Pandya *et al.*, 1980; Singh *et al.*, 1994; Rahman *et al.*, 2009).

With increasing information on host-pathogen interaction, genetic variation in the pathogen and temporal variation in pathogenicity, more efficient screening and breeding methods would be required for improving WR in lentil (Fig. 5.2). For example, early and late wilt reactions are noticeable in genotypes. These host reactions may be under the control of different genetic systems. Combining them together through marker assisted selection (MAS) may be essential for stable resistance. Similarly, resistant sources identified in wild species require allelism test to establish their genetic relationship with resistance gene already identified in the cultivated germplasm and, if found alien, these should be introgressed in the cultivated germplasm for durable resistance. However, transfer of desirable alleles is not so simple because of difficulty in efficient tracking for desired and non-desired alleles in breeding lines. This problem can be overcome by advanced-backcross QTL based breeding (AB-breeding) as it is the most suitable for introducing novel alleles from wild relatives to the cultivated species cultivars or varieties in a controlled manner (Tanksley and Nelson, 1996). Furthermore, establishment of pathogenic races in FOL will require search for race-specific resistance genes and their pyramiding in superior genotypes. However, it is difficult through recombination breeding approach by selecting desirable plants on the basis of phenotype. Marker-assisted gene pyramiding can be used to combine in a single genotype the desirable WR genes as well-established tight association between markers and target traits has already been reported in chickpea and lentil (Kumar *et al.*, 2011). Recently, gametophytic selection for WR has been reported to be effective in chickpea for developing wilt resistant genotypes in a short period (Ravikumar *et al.*, 2013). They have demonstrated the effectiveness of gametophytic selection in two populations segregating for wilt resistance using molecular markers linked to H1 and H2 locus for WR in chickpea. The same may also be tried for lentil.

Table 5.5 Important varieties/donors of lentil for FW resistance

Resistant variety/donor	Country	Reference
Pant L 406	India	Pandya *et al.* (1980)
Pant L 4	India	Singh *et al.* (1994)
Pant L 639, Priya, Seri, JL 3, Noori, VL 507 L 4147	India	Rahman *et al.* (2009)
IPL 306	India	IIPR (2012)
IPA 98	Iraq	Rahman *et al.* (2009)
Adaa, Alemaya	Ethiopia	Sarker and Erskine (2002)
Firat 87, Syran 96	Turkey	Rahman *et al.* (2009)
Talya 2, Rachayya, Hala	Lebanon	Rahman et al. (2009)
ILL 5883, ILL 5588,ILL 4400, ILL 590	Syria	Erskine *et al.* (1994)
Idleb 2, Idleb 3, Idleb 4, Ebla 1	Syria	El-Ashkar *et al.* (2003; 2004a; 2004b)
ILL 6256	Nepal	Joshi and Maharjan (2003)

Therefore, identification and incorporation of new WR genes in breeding programs, and development of genotypes with multiple combinations of WR genes will remain a continuous activity for sustainable production of lentil. Molecular markers offer a viable option to accelerate breeding progress through indirect selection for WR in segregating generations without actual phenotyping in the wilt-sick plot. Marker-assisted introgression of WR gene(s) is possible only when locus specific co-dominant markers tightly linked with the WR gene(s) are identified. Presently, the most tightly linked marker with WR gene '*Fw*' in lentil is AFLP marker p17m30710 (3.5 cM) followed by SSR marker SSR59-2M (8 cM) and RAPD marker OPK-15_{900} (10.8 cM). However, their distance from the gene of interest '*Fw*' does not provide confidence for use in marker-assisted screening and selection. Therefore, there is a need to develop more locus-specific co-dominant markers such as SSR, ESTs, CAPS, and SNPs in the map of lentil at the closer proximity (<1 cM) with WR gene(s). It will make MAS an essential component in resistance breeding to develop FW resistant varieties in lentil.

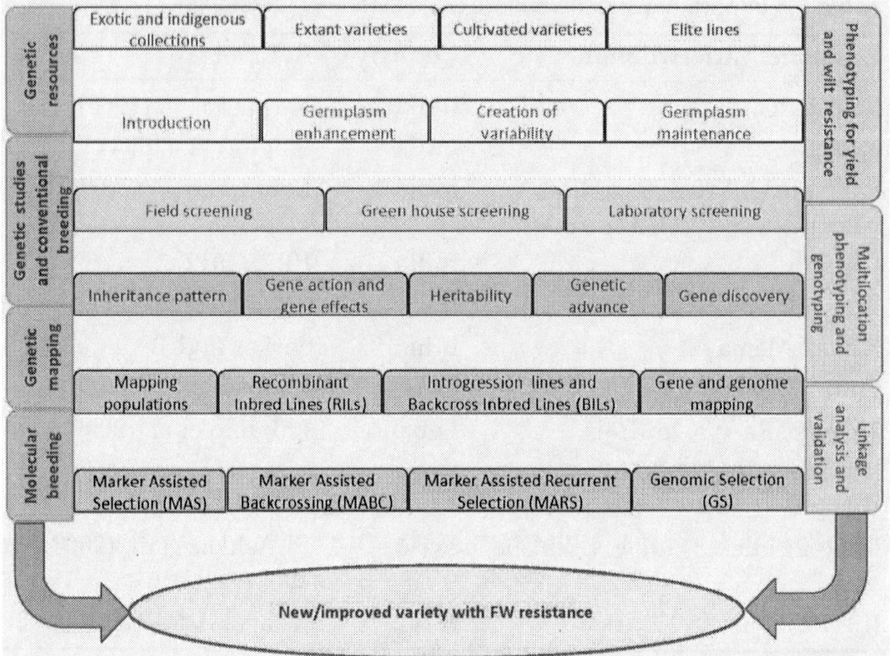

Fig. 5.2 Integration of conventional and molecular breeding for FW resistance in lentil

SUMMARY

Considerable progress has been made during the last three decades in characterizing pathogenic variability of FOL, identifying resistant sources for FW, establishing genetics of wilt resistance and incorporating WR gene(s) into the improved cultivars of lentil. The rapid adoption of these resistant varieties may prevent yield losses due to FW, reduce the gaps between potential and realized yield and bring about stability in production. However, many milestones have still to be achieved. For example, there is need to establish distinct race specificity. Further, there are many discrepancies regarding set of differentials used to classify the races. Since WR in lentil is governed by major resistance genes, there is a need to develop an improved differential set for lentil wilt to reduce ambiguity in variant/race determination. The major factor that has lead to discrepancy in molecular marker studies relates to the application of different phenotyping methods and disease scoring scales (Tullu, 1996). Therefore, it seems appropriate to standardize uniformly applicable phenotyping method and disease scoring scale along with permanent mapping. Similarly, it seems imperative to combine in a single genotype multiple WR genes through MAS to make it resistant to multiple FOL races. The use of FA as a selective agent for scoring wilt reaction needs further investigation as it (fusaric acid) may not be the sole factor resulting in the development of wilt disease in pulse crops including lentil. In lentil, lack of precise knowledge on the existence of pathogen race is the major hindrance to develop durable resistant cultivars for different regions. In spite of good progress in breeding wilt resistant varieties, its impact could not be demonstrated in farmers' fields due to their susceptibility to other soil borne pathogens causing root rot diseases (e.g., collar rot and dry and wet root rots in lentil). Due to lack of efficient screening techniques, stable resistance for these related pathogens could not be identified, and thus remain the major breeding goals in Asia and Africa. Therefore, there is an urgent need to identify resistance genes for these soil-borne pathogens and incorporate them in wilt susceptible cultivars for visible impact in farmers' fields. Studies are also needed to ascertain that incorporation of WR should not accompany susceptible reaction for other diseases such as rust or *Ascochyta* blight. Besides, further studies are required especially in lentil to establish pathogenic races using an international differential set which is not yet available. Efforts are underway in lentil to develop a common differential set for pathogenicity test which can distinguish different FOL isolates into pathogenic races. This is the pre-requisite for generating information on geographical distribution of races and efficient deployment of race specific resistance genes in lentil cultivars for durable resistance. However, many gaps still exist in our knowledge on the influence of environmental parameters on disease progression which is very crucial for controlling the disease by cultural practices. Preliminary studies indicate that morphological and anatomical characters as well as biochemical constituents of roots of lentil do play an important role in disease reactions, and thus influencing the wilt incidence. However, there is no information on the underlying mechanism of wilt resistance. Marker-assisted breeding for FW resistance in lentil is very limited, partly because WR can be easily identified in field and laboratory. However, these markers can be strategically used to avoid combined effect of other soil-borne pathogens and genotype x environment interactions, and in identification of race-specific resistance genes and their pyramiding in a single cultivar.

References

Agrawal SC, Singh K, Lal SS (1993). Plant protection of lentil in India. In: W. Erskine, MC. Saxena, eds., Lentil in South Asia, ICARDA, Aleppo, Syria, pp. 147-165.

Alessandro I, Mohamed K, Luca R, Clarice JC, Kevin EM, Niklaus JG (2006). Screening techniques and sources of resistance to root diseases in cool season food legumes. *Euphytica* 147: 201–221.

Abbas A (1995). Variation in some cultural and physiological characters and host/pathogen interaction of Fusarium oxysporum f. sp. lentis and inheritance of resistance to lentil wilt in Syria. PhD Thesis, Faculty of Agriculture, University of Aleppo, Aleppo, Syria.

Bahl PN, Lal S, Sharma BM (1993). An overview of the production and problems of lentil in South Asia. In: W. Erskine, MC. Saxena, eds., Lentil in South Asia, ICARDA, Aleppo, Syria, pp. 1-10.

Bayaa B, Erskine W, Khoury L (1986). Survey of wilt damage on lentil in Northern Syria. *Arab J. Plant Prot.* 4: 118–119.

Bayya B, Kumari SG, Akkaya A, Erskine W, Makkouk KM, Turk Z, Ozberk I (1998). Survey of major biotic stresses of lentil in Southeast Anatolia, Turkey. *Phytopathol. Medit.* 37: 88-95.

Belabid L, Baum M, Fortas Z, Bouznad Z, Eujayl I (2004). Pathogenic and genetic characterization of Algerian isolates of Fusarium oxysporum f. sp. lentis by RAPD and AFLP analysis. *Africa. J. Biotech.* 3: 25-31.

Chaudhary RG (2008). Pathogenic variability in Fusarium oxysporum f. sp. lentis. *Pulses Newslett.* 19: 3.

Chaudhary RG, Dhar V, Singh RK (2009). Association of fungi with wilt complex of lentil at different crop growth stages and moisture regimes. *Arch. Phytopathol. Plant Prot.* 42: 340-343.

Chaudhary RG, Saxena DR, Dhar V, Singh RK, Namdev JK (2010). Prevalence of wilt root rot and their associated pathogens at reproductive phase in lentil. *Arch. Phytopathol. Plant Prot.*, 43: 996-1000.

Choudhary AK (2010). A wilt resistant line 'IPA 204' of long-duration pigeonpea (Cajanus cajan). *Indian J. Agric. Sci.* 80: 907-909.

Choudhary AK, Kumar Shiv, Patil BS, et al. (2013). Narrowing yield gaps through genetic improvement for Fusarium wilt resistance in three pulse crops of the semi-arid tropics. SABRAO J. Breed. Genet. 45: 341-370.

Choudhary AK, Raje RS, Datta S, Sultana R, Ontagodi T (2013). Conventional and molecular approaches towards genetic improvement of pigeonpea for insects resistance. *American J. Plant Sci.* 4: 372-385.

Claudius GR, Mehrotra RS (1973). Wilt and root rot disease at lentil (Lens esculenta) at Sagar. *Indian Phytopathol.* 26: 268-273.

Datta S, Chaudhary RG, Rai R, Singh RK, Dhar V (2009). Molecular diversity and pathotype analysis of Fusarium oxysporum f. sp. lentis population from Northeastern Gangetic Plain region of India. *SABRAO J. Breed. Genet.* 41: 1-11.

Datta S, Choudhary RG, Shamim Md, Singh RK, Dhar V (2011). Molecular diversity in Indian isolates of Fusarium oxysporum f.sp. lentis inciting wilt disease in lentil (Lens culinaris Medik). *African J. Biotech.* 10: 7314-7323.

El-Ashkar F, Sarker A, Erskine W, Bayaa B, El-Hassan H, Kadah N, Karim BA (2004a). Registration of 'Idlib-3' Lentil. *Crop Sci.* 44: 2261.

El-Ashkar F, Sarker A, Erskine W, Bayaa B, El-Hassan H, Kadah N, Karim BA (2004b). Registration of 'Idlib-4' Lentil. *Crop Sci.* 44: 2261–2262.

El-Ashkar F, Sarker A, Haddad N, Bayaa B, El-Hassan H, Erskine W (2003). Registration of 'Idlib-2' Lentil. *Crop Sci.* 43: 728–729.

El-Morsy GA, Abou-Zeid NM, Hassanein AM (1997). Identification of Fusarium wilt caused by Fusarium oxysporum and pathogen variability in faba bean, lentil and chickpea crops in Egypt. *Egyptian J. Agric. Res.* 75: 551-564.

Erskine W, Bayaa B (1996). Yield loss, incidence and inoculum density associated with vascular wilt of lentil. *Phytopathol. Medit.* 35: 24-32.

Erskine W, Tufail M, Russell A, Tyagi MC, Rahman MM, Saxena MC (1994). Current and future strategies in breeding lentil for resistance to biotic and abiotic stresses. *Euphytica* 73: 127-135.

Eujayl I, Erskine W, Bayaa B, Baum M, Pehu E (1998). Fusarium vascular wilt in lentil: Inheritance and identification of DNA markers for resistance. *Plant Breed.* 117: 497–499.

FAO (2012). Food and Agricultural Organization of the United Nation, FAO Statistical Database, http://faostat.fao.org.

Fleischmann A (1937). Observations on lentil wilt. *Pfianzenbau* 14: 49-56.

Hamwieh A, Udapa SM, Choumane W, Sarker A, Dreyer F, Jung C, Baum M (2005). A genetic linkage map of lentil based on microsatellite and AFLP markers and localization of Fusarium vascular wilt resistance. *Theor. Appl. Genet.* 110: 669–677.

Infantino A, Kharrat M, Riccioni L, Coyne CJ, McPhee KE, Grunwald NJ (2006). Screening techniques and sources of resistance to root diseases of food legumes. *Euphytica* 147: 201-221.

Jimenez-Diaz RM, Crino P, Halila MH, Mosconi C, Trapero-Casas A (1993). Screening for resistance to Fusarium wilt and Ascochyta blight in chickpea. In: KB. Singh, MC. Saxena, eds., Breeding for Stress Tolerance in Cool Season Food Legumes, ICARDA, Wiley-Sons Pvt. Co. UK, pp. 77-95.

Jimenez-Diaz RM, Trapero-Casas A, Cabrera de la Colina J (1989). Races of Fusarim oxypsourm f.sp. ciceris infecting chickpeas in Southern Spain. In: EC. Tiamos, C. Beckmann, eds., Vascular wilt Diseases of plants (Vol. H28), NATOASI Series, Springer-Verlag, Berlin, Germany, pp. 515-520.

Joshi S, Maharjan R (2003). Evaluation of Lentil genotypes to Fusarium wilt. In: Proceedings of 25th National Winter Workshop (Grain legumes), pp. 33-35.

Kamboj RK, Pandey MP, Chaube HS (1990). Inheritance of resistance to Fusarium wilt in Indian lentil germplasm (Lens culinaris Medic.). *Euphytica* 50: 113-117.

Kannaiyan J, Nene YL (1978). Host range of Fusarium oxysporum f.sp. lentis, Sclerotinia rolfsii and Rhizoctonia solani isolated from lentil. *LENS* 5: 12.

Kannaiyan J, Nene YL (1978). Strains of Fusarium oxysporum f. sp. lentis and their pathogenicity on some lentil lines. *LENS* 5: 8-10.

Karki PB (1993). Plant protection of lentil in Nepal. In: Lentil in South Asia (eds., W. Erskine and M.C. Saxena), ICARDA, Aleppo, Syria, pp. 187-191.

Khare MN (1981). Diseases of lentil. In: Lentils (eds. Webb, C., Hawtin, G.,). Farnham Royal, England, UK, Commonwealth Agricultural Bureaux, pp. 163-172.

Khare MN, Agrawal SC, Dhingra OD, Kushwaha LS (1975). Variability in the growth of eight strains of Fusarium oxysporum f. sp. lentis on different solid media. India. *Phytopathol.* 28: 126-128.

Khare MN, Agrawal SC, Jain AC (1979). Lentil diseases and their control. *Tech. Bull.*, JNKVV, Jabalpur, India, pp. 29.

Kraft JM, Haware MP, Jimenez-Diaz RM, Bayaa B, Harrabi M (1994). Screening techniques and sources of resistance to root rots and wilts in cool-season food legumes. *Euphytica* 73: 27-39.

Kumar J, Choudhary AK, Solanki RK, Pratap A (2011). Towards marker-assisted selection in pulses: a review. *Plant Breed.* 130: 297-313.

Kumar S, Kumar J, Singh S, Ahmed S, Chaudhary RG, Sarker A (2010). Vascular wilt disease of lentil: a review. *J. Lentil Res.* 4: 1-14.

Kumar S, Sarker A, Erskine W, Geletu B, Ahmed S, Bayya B, El-Hassan H (2010). Valuable donors for Fusarium wilt resistance in lentil. In: Legumes for Global Health, Proceedings of the V International Food Legumes Research Conference and VII European Conference on Grain Legumes, 26-30 Apr 2010, Antalya, Turkey, pp. 37.

Matsumoto K, Barbosa ML, Souza LAC, Teixeira JB (1995). Race-1 fusarium wilt tolerance on banana plants selected by fusaric acid. *Euphytica* 84: 67-71.

Moreau (1978). Diseases and pests of lentil. Institute National de Vulgarisation pour les Fruits, Legumes at Champignons, pp.95-100.

Pandey RN, Pawar SE, Bhatia CR (1995). Effect of culture filtrate of Fusairum udum and fusaric acid on wilt susceptible and resistant pigeonpea cultivars. *Indian Phytopathol.* 48: 444-448.

Pandya BP, Pandey MP, Singh JP (1980). Development of Pant L 406 lentil resistant to rust and wilt. *LENS* 7: 34-37.

Ravikumar RL, Chaitra GN, Choukimath AM, Soregaon CD (2013). Gametophytic selection for wilt resistance and its impact on the segregation of wilt resistance alleles in chickpea (Cicer arietinum L.). *Euphytica* 189: 173-181.

Riccioni L, Giambattista GD, Valvassori M, Malta R, Porta-Puglia A (2003). Fungi associated with root rot of lentil landrace from Sicily. *J. Plant Pathol.* 85: 300.

Sarker A, Erskine W (2002). Lentil production in the traditional lentil world. In: Proceedings of Lentil Focus (ed. Brouwer, J.B.,) Horsham, Victoria, Australia. Pulse Australia, Sydney, pp. 35-40.

Singh IS, Singh JP, Singh AK, Chauhan MR (1994). Pant lentil 4: A high-yielding, rust, wilt and blight resistant variety for the north-western plains of India. *LENS* 21: 8-9.

Solanki RK, Singh S, Kumar J (2010). Molecular markers assisted testing of hybridity of F1 plants in lentil. *J. Food Leg.* 23: 21-24.

Srivastava SK, Singh SN, Khare MN (1984). Assessment of yield losses in some promising gram cultivars due to Fusarium wilt. *Indian J. Pl. Prot.* 12: 125-126.

Taheri N, Fallahati R, Jafarpour B, Bagheri AR, Jahanbaghsh V (2010). Pathogenic and genetic characterization of Fusarium oxysporum f. sp. lentis by RAPD and IGS analysis in Khorasan province. *World Appl. Sci. J.* 9: 239-244.

Tanksley SD, Nelson JC (1996). Advanced backcross QTL analysis: a method for the simultaneous discovery and transfer of valuable QTLs from unadapted germplasm into elite breeding lines. *Theor. Appl. Genet.* 92: 191–203.

Tullu A (1996). Genetics of Fusarium wilt resistance in chickpea. Ph D Thesis, Washington State University, Pullman, U.S.A.

Ujevic I, Kovacikova E, Urosevic B (1965). Botrytis parasitica and Fusarium oxysporum as dangerous parasites of lentil. *Polnohospodarstve* 11: 584-594.

Vasudeva RS, Srinivasan XV (1952). Studies on the wilt disease of lentil (*Lens esculenta* Moench.). *Indian Phytopathol.* 5: 23-32.

Wilson VE, Brandsberg J (1965). Fungi isolated from diseased lentil seedlings in 1963—64. *Plant Dis. Rep.* 49: 660-662.

Citation

Choudhary AK and Kumar S. 2018. Genetic improvement for fusarium wilt resistance in lentil. In: Scientific Lentil Production. Singh AK, Bhakta N, Sangale UR, Manibhushan, Sundaram PK, Kumar S and Yasin JK. (eds.) Society for Upliftment of Rural Economy Varanasi, India. Pp. 91-108.

□□□

Chapter 6

Enhancing the Lentil Production Through Improved Varietal Technology

Pravesh Kumar, AK Singh, RP Singh, BS Mahapatra and Amit Bhatnagar

SYNOPSIS

Lentil plant was one of the first crops domesticated in the world. Humans started eating lentils before they invented ceramics. More than 20 varieties of lentil have been released for cultivation in different agro-ecological zones of India. Most of them are small seeded. Recently some bold seeded varieties like WBL 58, Lens 4076, K75, DPL 62 and Sehore74-3 have been developed. Some of the genotypes have resistance/tolerance to rust like Pant L 639, Pant L 4076, DPL 32, Sapna. Several improved varieties were also released by state variety release committee. Apart from this a brief review were also presented about the prominent lentil cultivars grown in different lentil growing countries presently.

Keywords: Characteristic feature, Exotic varieties, Indian varieties, Green varieties, Red varieties, Lentil

Introduction

According to a Hindu proverb, "Rice is good, but lentils are my life." The plant was one of the first crops domesticated in the world. Humans started eating lentils before they invented ceramics. They are also considered as nutritional dynamite; contains 112 grams of protein per pound as compared to 59 grams per pound for T-bone steak. In 1971 Frances Moore Lappe' in "*Diet For A Small Planet*" had proposed a very complicated formula for combining the various proteins found within the plants to get "protein complementarity" as all eight "essential" amino acids in one meal (Actually, there are 23 essential amino acids, but the human body produces 15 of them on its own). The theory states that in order to get "complete" protein in a vegetarian diet, you must eat beans at the same time as cereals or grains, nuts or seeds, rice, or dairy products in certain important proportions. Now, the usability of bean protein has been in question since 1971. Still pulses are cultivated in the country on more than 12 per cent of total cultivated area and they constitute more than 4 per cent of the output of crop sector in value terms. The seed replacement rate is still (30%) lower than wheat and rice. Lentil contains 24-26% good quality protein. Based on the region, time of sowing and purpose of cultivation, recommendations etc selection of variety from table 7 is recommended. However situation specific varieties for Rice-lentil cropping system are i) *Utera cultivation* - PL-406, PL-639, Arun and, ii) *Late sowings* – PL-406, PL-639, IPL-15, Narendra Masoor-1. Although there are several varieties of lentils grown throughout the world, the following varieties are the most common (Singh *et al.*, 2008 and Singh *et al.*, 2015).

Fig. 6.1 Lentil variety Arun, at grand growth stage

A. Indian Varieties

More than 20 varieties of lentil have been released for cultivation in different agro-ecological zones of India (Fig. 6.1). Most of them are small seeded. Recently some bold seeded varieties like WBL 58, Lens 4076, K75, DPL 62 and Sehore74-3 have been developed. Some of the genotypes have resistance/tolerance to rust like Pant L 639, Pant L 4076, DPL 32, Sapna etc. Characteristics features of some of the varieties are given below.

PL 01: Released from GBPUA&T Pantnagar and resistant for rust, wilt, stem rot and ascochyta blight.

PL 02: This variety is also released from GBPUA&T Pantnagar and resistant for rust, wilt, stem rot and ascochyta blight.

Pant L 4 (PL 81-17): Developed from cross UPL 175 x (PL 184 x P 288), released in 1993 for North- West plain zone, semispreading, foliage green, seeds grey mottled and small (1.9 g/1O0 seeas1: resistant to rust, and wilt/root rot diseases, matures in 135-140 days and average yield 15-16 q/ha. This variety is suitable for North Western Zone Punjab, Haryana, Delhi and Western U.P.

Pant L 5: Released from GBPUA&T Pantnagar in 2001. Plant height is 45-50 cm and has the synchronized and medium maturity period. Test weight is 28 g. Maturity period is 135 days and average yield is 15-18 q/ha. This is resistant to rust and suitable for timely sown conditions of Uttar Pradesh.

PL-6 (PL 02): Released in 2009 from GBPUA&T Pantnagar and matures in 125-145 days. This is cross of Pant L 4 x DPL 55.Small seeded variety has the yield potential of 16-18 q/ha. Resistant for rust and suitale for Uttarakhand State.

Pant L 7 (PL 024): This variety is also released in 2009 from GBPUA&T Pantnagar. A cross of L 4076 x DPL 15 is large seeded variety. Matures in 125-145 days and has the yield potential of 16-18 q/ha. Suitale for growing in Uttarakhand State.

Pant L 8 (Pant L 063): A cross of DPL 59 x IPL 105 released from GBPUA&T Pantnagar in 2010. Recommended for release and cultivation in the states of Punjab, Haryana, plains of Uttarakhand, Western Uttar Pradesh, Delhi and Rajasthan as a pure crop, suitable for timely sown rainfed conditions in *rabi* season. Average plant height is 40 cm and plants are semi erect with dark green foliage; mottled grey seed coat with pink cotyledon colour. Flowers are bluish white in colour. Moderately resistant to rust wilt and pod borer. This variety has medium maturity (135 days) and average yield is 15 q/ha.

Pant L 209: This variety matures in 145-150 days. Plants are semi-spreading in nature and seeds are medium in size and light pink in colour. Yield potential is about 25 q/ha. This variety is resistant to rust and wilt and suitable for the hilly regions of Himanchal Pradesh. It can even be grown after late harvesting of rice.

Pant L 234: This variety matures in 130-135 days. Seeds are attractive and large in size. Yield potential is about 15-20 q/ha. This variety is resistant to rust and wilt and also some extent to drought.

Pant L 406: A selection from germplasm line P 495, released in 1979, widely adapted, green foliage, seeds grey mottled and small

(1.8 g/ 100 seeds). Plants are large and semi-spreading in nature. This is resistant to rust and tolerant to wilt/root rot diseases, suitable for North-West and North-East plains and can be sown normal as well as in late conditions. Maturity period is 140-150 days. It can yield up to 25-30 q/ha.

Pant L 639: Developed from cross L 9-12 x T 8, notified in 1982, widely adapted, matures in 135-140 days, semi-spreading in nature and have light green foliage. This variety takes medium time for maturity and resistant to rust, wilt and blast, suitable for North- West and North-East plains. Seeds are medium in size and brown in colour and can resist drought up to some extent. The yield potential is 18-20 q/ha.

Type 1 to Type 66: Bushy spreading plants and small seeded. Maturity period of these varieties are 120-150 days. Test weight varies from 16-28 g and yield potential from 6-25 q/ha. They are found suitable for whole country.

Type 8: Crop maturity period is 120-125 days. Plants are dwarf and spreading in nature. Seeds are pinkish brown in colour. Yield potential is about 16-20 q/ha. This is susceptible to rust.

Type 36: This variety is released from Kanpur by selection method and can be grown in all the states. Maturity period is 130-135 days. Plants are dwarf and spreading in nature. Seeds are pinkish brown in colour. The yield potential is 18-20 q/ha. Susceptible to wilt but can tolerate the drought.

Pusa 1: Maturity period of crop is 100-110 days. Suitable for growing in Bihar, Uttar Pradesh, Madhya Pradesh, Bihar and Orissa. The uktimate yield potential is 15-20 q/ha.

Pusa 4: This is matured in 130-135 days. Plant height is 40 cm and test weight is 18-20 g. Seeds are small in size. It has the yield potential of 20-25 q/ha and recommended for U. P. Bihar and West Bengal. Can tolerate the drought up to some extent.

Pusa Masur-5 (L-4594): Released for NCR Delhi in 2006. Suitable for irrigated and rainfed situations and found resistant to rust. It has medium growth habbit, small seeded and orange cotyledon. Yield potential is 17 q/ha.

Pusa 6: This is also matured in 130-135 days. The height of plants are 35 cm and seeds are smaller than Pusa 4. Yield potential of this variety is 20-25 q/ha. It is recommended for U. P., Punjab, Haryana, Delhi and Himanchal Pradesh.

Pusa 830: Maturity period of this variety is 120-125 days. Seeds are large in size and the average yield potential is 25-30 q/ha. It can be sown up to the end of November but for getting higher yield it should be sown up to the end of November.

Pusa 4076 (Shivalik): This is the only improved variety of Malka lentil. Plant height is 50-60 cm and erect in nature. Seeds are large in size. Comparatively resistant to wilt and rust than Desi lentil. In Northern India the sowing of seed can be done up to last October while in Central India up to last December. It is recommended in all the regions where Malka lentil is grown. The maturity period of this variety is 130 days and average yield is 25-28 q/ha.

Table 6.1 Characteristic feature of some improved varieties of lentil

Variety	Year of release/ identification	Average yield (tones/ha)	Duration (days)	100 seed weight (g)	Area of adoption	Special features
Pant L 406	1980	1.5-2.0	140	1.9	NEPZ, NHZ, NWPZ, NEPZ, NHZ, UP	Rust resistant
Pant L 639	1982	1.5-2.0	135-140	1.8	NEPZ, CZ CZ	Rust resistant
Malika	1986	1.2-1.5	135-140	2.8	NWPZ, Uttarakhand	Bold seeded
Lens 4076	1986	1.2-1.6	135-140		hills	
LL 147	1987	1.2-1.6	145-150		CZ	Rust resistant
VL Masoor 4	1989	1.2-1.5				
					NEPZ	
Sehore 74	1988	1.0-1.5	120-125	2.8		Bold seeded,
					NWPZ	Drought tolerant
WBL 58	1990	1.5-2.0	120-125	2.0		Bold seeded
					CZ	
Sapna	1991	1.5-2.0	125-130		NWPZ	Bold seeded Rust resistant
JLS 1	1991	1.2-1.5	120	2.5	NWPZ	Short duration
Pant L 4	1993	1.5-2.0	130-140	1.8	NEPZ	Rust and wilt resistant
DPL 15	1995	1.5-1.8	125-140	2.7		Rust resistant
DPL 62	1996	1.0-1.2	130 140	3.3		Bold seeded

Table 6.2 State wise list of improved varieties of lentil

States	Varieties
Uttar Pradesh	Pant L 209, Pant L 234 Pant L 406, Pant L 639, HUL 57, DPL 62 (Sheri), IPL 81 (Noori), Narendra Masoor 1, IPL 406, L 9-12, Pusa 4, Pusa 6, Type 8, Type 36, V. L. Masoor 1
Bihar	HUL 57, WBL 77, Pusa 4, Pusa 6, Arun (PL 77-12) Pant L 406, Pant L 639, Type 36, B. R. 25, B. R. 26, L 9-12
Madhya Pradesh	Pant L 639, IPL 81 (Noori), JL 3, IPL 406, J. L. S. 1, L 9-12
Haryana	Pusa 4, Pusa 6, DPL 62 (Sheri), IPL 406, Pant L 639, L 9-12
Punjab	DPL 62 (Sheri), Pant L 4, LH 84-8, LL 147 Pant L 639, L 9-12, Pusa 6
Delhi	DPL 62 (Sheri), LH 84-8, Type 36, Pant L 639, Pusa 4, Pusa 6, L 9-12
Gujarat	IPL 81, JL 3
Himachal Pradesh	HUL 57, VL 507
J&K	Shalimar Masoor 1, HUL 57, VL 507
West Bengal	Ranjan (B 256), Asha (B 77), HUL 57, WBL 77, KLS 218
NEH Region	HUL 57, DPL 62
Maharashtra	IPL 81 (Noori), JL 3
Assam	Asha (B 77), HUL 57, WBL 77, KLS 218
Orissa	WBL 77, B 77 (Asha), HUL 57
Rajasthan	Pant L 639, Type 36, IPL 406, DPL 62, L 9-12,

Pusa Vaibhav (L 4147): Released from IARI, New, Delhi in 1996, suitable for cultivation in North Western Plain Zone with rainfed and irrigated conditions. Variety has medium tall growth habit with small pubescent leaves. It is small seeded variety. It is resistant to wilt and rust and matures in 120-125 days. Seeds are small with orange cotyledons. It has yield potential of 20-25 q/ha.

B 17: Matures in 120-125 days and having the yield potential of 18-20 q/ha.

B.R. 25: The cropping period of this variety is 125-130 days. This is selected from local varieties of Bihar. Plants grow straight and bushy in nature. Average yield potential is 20-25 q/ha. Suitable for Southern Bihar and Chhota. Nagpur region of India.

B.R. 26: The growing period and yield potential of this variety is similar to B.R. 25. This is also selected from local varieties of Bihar. Plants are long and straight growing. Suitable for Northern Bihar area.

B 77 (Asha): High-yielding, small seeded and medium duration wilt resistant variety was notified in year 1980. Recommended for rabi season and late sown conditions in rainfed areas.

B 177 (S 177): Maturity period of this variety is 125-130 days but it matures in only 90 days in coastal areas of Orissa. Seeds are medium sized and brownish-gray in colour. Test weight is 18 g yield potential is 15-20 q/ha. Protein content is 29%.

B 235: Plants are compact and branched and having small seeds. It matures in 116-120 days with average yield of 16-18 q/ha and 28% protein content.

B 256 (Ranjan): Mutant of B 77 (Asha), notified in 1982 for West Bengal, spreading plant type, flowers white, seeds mottled and small (1.8 g/100 seeds), average yield 12 q/ha.

Shekhar Masoor-2 (KLB-303): Plant height is 40-45 cm, green foliage, bold, flattened pods, mottled gray, bold seeded; days of flowering: 55-62 days and days of maturity: 128 days.

Sekhar Massor-3 (KL-320): Plant height is 40-45 cm, green foliage, bold flattened pods, mottled gray, bold seeded; days of flowering: 55-60 days; and days to maturity is 128 days.

S 256: Plants are of spreading nature which contains small seeds. It matures in 125-130 days and the yield potential is 15-20 q/ha. Test weight is 18 g. Recommended for growing in Bengal state.

C 31: Tall, spreading and seeds are deep ash in colour. Matures in120-125 days, Test weight is 19 g and yield potential is 16 q/ha. This variety is adapted in West Bengal and Bihar regions.

No. 26: Bushy, violet flowers and seeds are ash in colour. Maturity period of this variety is 125-130 days and test weight is 25 g. Yield potential is 20 q/ha. Adapted in Bihar region.

No. 27: This variety has similar characteristics as No. 26 but comparatively longer maturity period of 135-140 days.

Arun (PL 77-12): Mutant of BR 25, notified in 1986 for Bihar, matures in 115-120 days, medium bold seeded, tolerant to rust and wilt diseases, average yield 12-13 q/ha.

JA 4: High-yielding and medium duration wilt resistant variety.

JKM 7: High-yielding and medium duration wilt resistant variety.

JKM 189: High-yielding and medium duration wilt resistant variety.

NA 3: High-yielding and medium duration wilt resistant variety.

J.L.S. 1: This variety is developed by selection method from local varieties of M.P. and long and straight growing in nature. Yield potential is 10-15 q/ha and suitable for cultivation in central regions.

JL 1: Bold seeded variety matures in 120 days and the yield potential is 15 q/ha. It is wilt resistant.

JL 2: large seeded wilt resistant variety.

J.L. 3 (SLC 06): A selection from local germplasm of M.P., released in 1999 for CZ, erect plant type, dull green foliage, tolerant to wilt disease, matures in 115-120 days, seeds grey mottled and bold (3.0 g/100 seeds) and has the yield potential of 14 q/ha.

VL 1: A selection from hill land race, notified in 1983 for hills of Uttar Pradesh (Now Uttarakhand), matures in 165 days, light green foliage, seed coat black, seeds small (1.9 g/100 seeds), tolerant to rust and wilt diseases and average yield 11 q/ha.

VL 3: Maturity period is 160-165 days and found suitable for growing in hilly areas of Uttar Pradesh and other states. Tolerant to aschochyta blight disease.

VL 4: Released from VPKAS, Almora in 1991. Plants are dwarf, bushy and seeds are small black. Maturity period is 165-170 days. Test weight is 20 g and average yield is 18.5 q/ha. It is tolerant to rust and wilt and can be grown in J&K and Uttarakhand states.

VL 103: Maturity period of this variety is 170-175 days. Recommended for cultivation in hills of Uttarakhand. Yield potential is 20-22 q/ha.

VL-125: Plant height is30-35 cm; small seeded and black in colour. Maturity period is 160-165 days and the yield potential is 18-20 q/ha. Tolerant to wilt and root rot and adapted in hills of Uttarakhand (Up to mid-altitude).

VL-126: Released from VPKAS, Almora in 2006 and suitable for NHZ. Small seeded, resistant to rust, matures in 160-170 days and yield potential is 13 q/ha.

VL 129: Plant height is 24-29 cm and brown seeded. Maturity period is 148-155 days. Average yield is 10-12 q/ha. Resistant to root rot and wilt and protein content is 24.6%. Adapted in hills of Uttarakhand.

VL 133: Released from VPKAS, Almora in 2011 and adapted in Uttarakhand state. Yield potential is 11 q/ha and resistant to wilt, root rot and rust.

VL 507: A large-seeded variety evolved through selection from ILL 7978 has been identified for cultivation in Jammu and Kashmir, Uttarakhand, Himachal Pradesh and hilly regions of north-eastern zone states. It has an average yield of 12 q/ha in the North Hills Zone with 100-seed weight of 3.32 g. It has showed resistance against wilt.

VL Masoor 514: The growing period of this variety is 140 days. Average yield is 10 q/ha. It is resistant to lentil rust and recommended for Uttarakhand.

NDL-1: This variety is developed at ICARDA and has the yield potential of 18 q/ha.

NDL 94: A new bold-seeded lentil variety NDL 94, a pedigree selection from DLG 105 × PL 406, has been identified for eastern Uttar Pradesh, Bihar, West Bengal and Orissa. It has small seeds with an average yield of 15 q/ha. The variety possesses resistance to rust.

Azad Masoor 1 (KLS 218): Duration – Released from CSAUA&T Kanpur in 2005 is resistant to rust, and suitable for NEPZ. Matures in 120-125 days and average yield is 13 q/ha.

WBL 58 (Subrata) : Developed from cross JLS 2 x T 36, released in 1998 for West Bengal, semi-spreading, green foliage, tolerant to rust, matures in 120-125 days, average yield 12-13 q/ha.

WBL-77 (Moitree): Matures in 125 days and the average yield is 8-9 q/ha. Small seeded, resistant to rust and suitable for growing in Eastern Uttar Pradesh, Bihar, Jharkhand, West Bengal and Assam regions.

DPL 15 (Priya): Developed from cross PL 406 x L 4076, released in 1995 for NWPZ, semi-spreading, green foliage, matures in 135-140 days, seeds bold (2.7 g/100 seeds), resistant to rust, tolerant to wilt/ root rot diseases, average yield 14-15 q/ha.

DPL 58: This variety is released from IIPR, Kanpur and resistant for rust and wilt.

DPL 62 (Sheri) : Developed from cross JLS 1 x LG 171, released in 1997 for NWPZ, Haryana, Delhi and Western U.P., semi-erect, foliage green, less hairy, resistant to rust and wilt diseases, matures in 130-135 days, seed grey mottled and bold (3.4 g/100 seeds) and average yield is 17 q/ha.

IPL-81 (Noori): Developed from cross K 75 x PL 639, released in 2000 for Central Zone like M.P. and Maharashtra, semi-spreading, green foliage, tolerant to rust and wilt, matures in 135-140 days, seeds grey mottled and bold (2.7 g/100 seeds) and average yield is 20-22 q/ha.

IPL 313: Another high yielding variety IPL 313 (average yield 12 q/ha) and large seeded (2.6 g/100 seed wt.) developed from a three way cross [(ILL 7659 x DPL-58) x KL 178] is suitable for normal sown conditions in UP.

IPL 315: The lentil variety has been developed from PL 4 x DPL 62 and has the average yield of 1487 kg/ha. It has yielded 13.2% more than the best check variety DPL 62. The variety is resistant to rust and moderately tolerant to wilt and has been identified for the states of Madhya Pradesh, Chhatisgarh and Bundelkhand region of U.P.

IPL 316: Developed from the cross of Sehore 74-3 x DPL 58. Average yield is 13 q/ha. The variety has been identified for release in M.P. Chhatisgarh, Maharashtra and parts of Rajasthan.

IPL 406 (Angoori): It is a cross of DPL 35 x EC 157634/382 from IIPR, Kanpur. Seeds are extra large tolerant to rust and suitable for growing in Punjab, Haryana, Rajasthan, Delhi and western Uttar Pradesh.

Bombay 18: Matures in 130-140 days and recommended for Maharashtra, Punjab and Haryana. The yield potential is 10-12 q/ha.

Garima (LH 82-6): Suitable for sowing in Haryana, matured in 137 days and has the yield potential of 14.5 q/ha.

Narendra Masoor-1 (NDL-91-1): Released in 1996 and yield potential is 18-22 q/ha. Maturity period is 130-135 days. Resistant to rust and tolerant to wilt and root-rot diseases.

Narendra Masoor-2 : This variety released in 2002 and can produce up to 22-25 q/ha yield. Maturity period is 125-130 days.

Resistant to rust, wilt and root rot diseases. Suitable for both irrigated and rainfed conditions.

NDL 94: A new lentil variety NDL 94, a pedigree selection from DLG 105 × PL 406, has been identified for Eastern Uttar Pradesh, Bihar, West Bengal and Orissa. It has small seeds with an average yield of 1.52 tonnes/ha. The variety possesses resistance to rust.

Haryana Masoor 1 (LH 89-48): Released from CCSHAU, Hissar, Haryana in 2006 and recommended for growing in the state. Matures in 138 days and the average yield is 14 q/ha. It is moderately resistant to all the diseases.

Lentil KLS 218: This small-seeded variety has a yield advantage of 27% over the best check in the North-Eastern Plains Zone and is fairly tolerant to rust; it is identified for the North-Eastern Plains Zone.

L 9-12: This is developed in Punjab and recommended for cultivation in most of the states of India. Maturity period of this variety is 150 days. Plants are semi-spreading in nature and seeds are medium in size with spotted brownish colour (Kattthai). It has the yield potential of 18-22 q/ha and can resist the drought up to considerable extent.

L 830: The maturity period of this variety is 120-125 days which is found suitable for cultivation in North Western Plain Zone. Yield potential is 8-12 q/ha.

L 4076: A cross of Pant L 639xPant L 234, released in 1984 from IARI, New Delhi. Bold seeded variety matures in 145-150 g. Test weight is 32 g and yield potential is 24 q/ha. It is rust and wilt resistant and found suitable for NWPZ and CZ.

Lens 4076 : Developed from a cross of Pant L 234 x Pant L 639, released in 1993 for North-West plain and Central zones, semi-spreading plant type, deep green foliage, seeds bold (28 g test weight), grey mottled, resistant to rust, average yield 22 q/ha. Maturity period is 130-140 days.

L 4147: Small seeded variety released for cultivation in 1996 for NWPZ (Punjab, Haryana, Delhi, West UP.) Yield potential is 18-24 q/ha and maturity period is 130-135 days. Resistant to rust & tolerant to wilt.

(L 4596): The plant of lentil variety L4596 is dwarf with erect growth habbit. It has green cotyledons and small seeds (2.33g/100 seeds). This variety matures in 125 days. yield potential is 21 q/ha.

L 4603: Released from IARI, New Delhi and found resistant for rust and root rot.

L 4632: Extra bold seeded variety suitable for cultivation in North Western Plain Zone.

LL 56: Released from PAU, Ludhiana in 1983. This variety is suitable for Delhi region and matures in 145-155 days. Small seeded variety having many branches and pods. Test weight is 20 g and yield potential is 18 q/ha. It is rust resistant and found suitable for Punjab and Haryana states.

LL 147: Maturity period of this variety is 135-140 days. Yield potential is 12-16 q/ha. It is resistant to rust.

LL 664: Released from PAU, Ludhiana and resistant for pod borer.

LL 699: Early in flowering. Average yield is 14.2 q/ha. Maturity period is 145 days. Moderately resistant to rusts and blight and tolerant to pod borer.

LL 931: The cropping period of this variety is 146 days. It is resistant to lentil rust but tolerant to pod-borer. Suitable for timely sown irrigated conditions of Punjab.

EC 1: Released from Dhaulakuon and resistant for rust.

Shalimar Masoor 1: Plant is erect type having lush green foliage white flower and amber pink seed colour.

Vipasha: Released from HPKVV, Palampur in 1982. Maturity period is 160-170 days. Test weight is 28 g and yield potential is 18 q/ ha. This medium bold seeded variety is resistant to rust and found suitable for H.P. J&K and Uttarakhand states.

Malika (K 75) : A selection from Bundelkhand local, released in 1986 for NEPZ and CZ, semi -spreading, deep green foliage, seeds bold (3.0 g/100 seeds), average yield 12-14 q/ha in NEPZ and 10-12 q/ha in CZ.

Sapna (LH 84-8) : Developed from cross L 9-12 x JL 2, released in 1991 for NWPZ, semi-spreading, green foliage, matures in 135-140 days, seeds bold (2.7 g/100 seeds), tolerant to rust disease, average yield 14-15 q/ha.

Malaviya Vishwanath (HUL 57): Released in 2005 and suitable for Eastern and Central U.P., Bihar, Jharkhand, West Bengal and Assam. Matures in 112-120 days have the yielding ability of 14.0 q/ ha found resistant to rust.

Table 6.3 Disease resistant/tolerant varieties

Disease	Varieties
Rust	Pant L 406, Pant L 639, Pant L 4, L 4076, Sapna, Priya, Sheri, IPL 81, Vaibhav, LL 56, LL 147
Wilt	Pant L 406, Pant L 639, Pant L 4, L 4076, Priya, Sheri, Jawahar Masoor 3, Vaibhav, IPL 81

Table 6.4 Sate wise recommended lentil varieties

State	Small seeded	Bold seeded
Jammu & Kashmir	Pant L 406, Pant L 639	
Himachal Pradesh	Pant L 406, Pant L 639	Vipasha
NEH	Pant L 406, Pant L 639	
Punjab	LL 56, LL 147, Pant L 4, Vaibhav	L 4076, Sapna, Priya, Sheri
Haryana	Pant L 4, Vaibhav	L 4076, Sapna, Garima, Priya, Sheri
Delhi	Pant L 4, Vaibhav	L 4076, Sapna, Priya, Sheli
Uttaranchal	VL Masoor 1, VL *Masoor* 4	
Uttar Pradesh (West)	Pant L 4, Vaibhav	L 4076, Sapna, Priya, Sheri
Uttat Pradesh (East)	Pant L 406, Pant L 639	Malika
Uttar Pradesh (Bundelkhand)	-	Malika, L 4076, JL 3, Noori
Bihar & Jharkhand	Pant L 406, Pant L 639, Arun	Malika
West Bengal	Pant L 406, Pant L 639, Ranjan	Malika, Subrata
Madhya Pradesh & Chhattisgarh	-	JL 1, Malika, L 4076, Jawahar Masoor 3, Noori
Maharashtra	-	Malika, L 4076, Noori, Jawahar Masoor 3
Rajasthan	-	Malika, L 4076, Jawahar, Masoor 3, Noori

Fig. 6.2 Lentil crop at peak flowering and podding stage

B. Exotic Varieties

Canada

1. **Green varieties:** The majority of lentil varieties grown in Canada have green seed coats and yellow cotyledons (seed leaves). Green lentil is divided into three market classes in Canada e.g. large green, medium green and small green.

Large green varieties: Most of the large green varieties require early seeding because they are relatively late maturing and indeterminate. They produce tall plants which can be prone to lodging, and are susceptible to botrytis (grey mould) infestations in high rainfall areas. but maturity can be much earlier in dry years, much later in cool wet years.

Source: Govt. of Saskatchwan, 2013

Laird: is a Chilean variety of lentil. It is considered by many lentil importers and consumers to be the prima donna of lentil.

It is the most widely-grown variety, and is considered an extra-large seeded type. It requires 100 days to reach maturity and has some resistance to ascochyta blight. It is the tallest variety available. The release of this large green variety by the Crop Development Centre in Saskatoon in 1980 was the single most important factor contributing to the rapid expansion of the lentil industry in Canada.

Table 6.5 Red lentil varieties

Variety	Seed coat Colour	Seed Size (g/1,000 seeds)	Seeding Rate (lbs/ac)	Height (cm)	Days to flower	Maturity Rating	Ascochyta Blight Resistance	Anthracnose Race 1 Resistance
Extra Small Red								
CDC Impala	Grey	31	35-40	30	51	Early	Good	Good
CDC Imperial CL	Grey brown	30	35-40	30	49	Early	Good	Good
CDC Redbow	Grey	32	35-40	30	49	Early	Good	Good
CDC Robin	Brown	30	35-40	30	49	Early	Good	Good
CDC Rosebud	**Tan	31	35-40	30	50	Early	Good	Good
CDC Rosetown	Grey	31	35-40	31	52	Early	Good	Good
Small Red								
CDC Blaze	Grey	34	40-45	30	47	Early	Good	Poor
CDC Imax CL	Grey	45	50-55	35	51	Early medium	Good	Fair
CDC Impact CL	Grey	34	40-45	30	47	Early	Good	Poor
CDC Maxim CL	Grey	40	45-50	34	51	Early medium	Good	Good
CDC Redberry	Grey	42	45-50	34	50	Early	Good	Good
CDC Redcoat	Grey	41	45-50			Early	Good	Good
CDC Red Rider	Grey	45	50-55	34	52	Early medium	Good	Fair
CDC Rouleau	Grey	37	40-45	33	52	Medium	Good	Good
Large Red								
CDC KR-1	Grey	56	65-70	37	52	Medium	Good	Good

*Co-op and Regional Trials in Saskatchewan since 1995. Comparisons to CDC Milestone. **Potential new market class.
CL Clearfield tolerant variety, Maturity ratings: Normal maturity range in days based on May 1 seeding is E=100, VL=110

C Glamis: is an ascochyta blight resistant large-seeded Chilean type lentil to replace Laird. CDC Glamis has out-yielded Laird in trials. It has a slightly smaller seed than Laird which may present a marketing problem in some countries.

CDC Sedley: is a large green lentil with moderate ascochyta blight resistance and medium maturity.

CDC Improve: is a new large green lentil with tolerance to clearfield herbicides. It is very similar to CDC Sedley in its physical attributes.

CDC Plato, CDC Grandora, CDC Greenland, and CDC Sovereign are ascochyta blight resistant large green lentil varieties. **CDC Impower** is also a large green variety. Varieties such as **CDC Gold, CDC Royale, CDC Matador** and **Indianhead** are grown in small volumes and are sold into niche markets.

a) Medium green varieties

CDC Richlea: is a medium-sized Chilean-type similar in size and color to the Chilean lentil grown in the USA. It is a high yielding variety that is susceptible to both ascochyta blight and anthracnose.

CDC Meteor: is a high yielding, ascochyta resistant, medium green variety with similar seed and plant characteristics as CDC Richlea.

CDC Impress CL: is a Clearfield variety with herbicide tolerance. It is very similar to CDC Meteor; however, it has marginally better resistance to Race 1 anthracnose, but yields are lower.

CDC Vantage: is an ascochyta resistant medium-sized Chilean type lentil to replace CDC Richlea. It has out-yielded CDC Richlea in trials. CDC Vantage has similar seed characteristics and is slightly shorter than CDC Richlea.

CDC Impress and CDC Imigreen are some other medium green varieties grown in Canada.

b) Small green varieties

Eston: is a Persian variety of lentil. Eston lentil is small-seeded, matures about 10 days earlier than Laird, and out-yields Laird by about 7%. Eston lentil can be less popular because it is shorter than Laird and is more susceptible to ascochyta blight.

CDC Milestone: is an ascochyta blight resistant small-seeded Persian type lentil to replace Eston. CDC Milestone has out yielded Eston in trials. Characteristics of the plant and seed are very comparable to Eston. so CDC Milestone should be well accepted in the market-place.

CDC Redwing: is a small-seeded red lentil with a pale green seed coat suitable for the red, split market. It is resistant to ascochyta blight, but susceptible to anthracnose.

CDC Milestone: is a high-yielding, ascochyta blight resistant, small green variety.

CDC Viceroy: is high yielding and resistant to ascochyta blight and Race 1 anthracnose.

CDC Invincible and CDC Peridot are other small green varieties. All other green lentil varieties grown in Canada are susceptible to anthracnose.

2. Red varieties

Red lentil varieties have brown, grey, tan or pale green seed coats with red cotyledons. Seed coat colour has some influence on red lentil quality, particularly the ability to withstand weathering in years with prolonged periods of wet weather at maturity. Under those conditions, varieties with grey seed coats show less weathering damage. Because most red lentil is dehulled before consumption, the suitability of new red lentil varieties for secondary processing such as dehulling and splitting is of utmost importance. Dehulling and splitting yields in some processing plants are higher for more thick seeds which may be more desired for specific markets. Red lentil accounts for more than half the world trade of the crop and is sold in either whole seed or split form. Processing facilities capable of de-hulling and splitting red lentil suitable for the world market have been built in Saskatchewan. Plant breeding goals for red lentil include the development of well-adapted, high yielding varieties that meet market needs. Canada has recently become an important producer and exporter of red lentil.

Crimson: is a red lentil developed in the USA at Washington State University. Crimson is highly susceptible to lodging before harvest, particularly in seasons when the crop is short.

CDC Blaze: has a grey seed coat, is resistant to ascochyta blight, and has similar seed traits compared to Crimson, with improved yield and lodging tolerance.

CDC Rouleau: have grey seed coats and are resistant to both ascochyta blight and Race 1 anthracnose.

CDC Red Rider: have a larger seed size compared to other red lentil varieties.

CDC Rosetown: have a brown seed coat, smaller, more rounded seed and resistant to aschochyta blight and anthracnose race 1.

CDC Robin: also have a brown seed coat, smaller, more rounded seed and resistant to both diseases.

CDC Redberry: has shown good ability to resist lodging under a wide range of conditions. CDC Redberry also have comparatively a larger seed size than other red lentil varieties.

CDC Impact: may mature earlier compared to other varieties because the plants may not be injured when treated with herbicide.

CDC Imperial: is grown in Canada and differ in resistance to lodging when compared to CDC Impact.

CDC Impala: (extra-small red) CDC Impala is similar to CDC Rosetown, but is tolerant to the Clearfield herbicides.

CDC Imperial: (extra-small red) CDC Imperial has the same yield, disease and quality attributes as CDC Robin, but is tolerant to the Clearfield herbicides.

CDC Maxim: (small red) CDC Maxim is similar to CDC Redberry, but is tolerant to the Clearfield herbicides.

CDC Impact:(small red) CDC Impact has the same yield, disease and quality attributes as CDC Blaze, but is tolerant to the Clearfield herbicides.

CDC Imax: is another small red variety grown in this region.

Other varieties

Small quantities of French green or dark speckled market class lentil varieties, such as **CDC LeMay**, are produced in Saskatchewan. Varieties such as **CDC Gold**, **Red Chief**, and **Indianhead** (called Beluga in the United States) are produced in small volumes in western Canada and are sold into niche (specialty) markets.

Australia

Large red lentils

PBA Jumbo: PBA Jumbo is the highest yielding large seeded red lentil and a direct replacement for Aldinga. It is suited to most current lentil growing areas where it has consistently yielded around 15% higher than Aldinga although attaining larger seed size is more likely in medium to high rainfall zones. PBA Jumbo is moderately susceptible to BGM and this disease will need to be managed in disease prone areas. It has resistance to foliar and seed ascochyta blight, which are both improved over Aldinga. Plant type and lodging susceptibility is similar to Aldinga and like this variety PBA Jumbo is well suited to no-till, inter-row sowing into standing residue. PBA Jumbo has improved tolerance to soil boron and salinity over Aldinga and Nugget. It has a seed size and shape similar to Aldinga (20% larger than Nugget) but with a grey seed coat and like Aldinga is well suited to the postharvest removal of small broadleaf weeds seeds. Milling quality is improved over Nugget and it is well suited to premium large red split markets such as those in Sri Lanka.

Medium red lentils

PBA Ace: PBA Ace (evaluated as CIPAL803) is a disease resistant, vigorous growing, mid flowering and maturing line with broad adaptation. It will provide an alternative to Nugget in all regions and PBA Flash and PBA Blitz in favourable lentil growing areas where crop topping and drought tolerance are not major priorities. It has a good disease resistance profile with resistance to ascochyta blight and moderate resistance to botrytis grey mould. PBA Ace has a grey seed coat colour with a medium seed size and the highest yielding variety in long term in all regions of Sothern Australia and Victoria but due to its later maturity than PBA Bolt, PBA Blitz and PBA Flash is likely to be better suited to areas where mid maturing varieties are favoured. Some level of shattering has been observed under some conditions in PBA Ace at maturity but it is unlikely to cause significant yield loss, yield data includes sites where some shattering may have occurred in this variety.

PBA Blitz: Early maturing red lentil variety will reduce the impact of climatic stresses such as drought and heat and help to avoid financial losses in poor years. PBA Blitz is suited to all current lentil areas with particular adaptation to shorter-season areas where its combination of mid to early flowering, early maturity, high yield, good disease resistance and medium to large seed size will improve lentil reliability and economics of production. PBA Blitz is the best option where crop topping and/or delayed sowing are practiced. It has a good level of early vigour and an erect plant type. PBA Blitz is a medium sized (larger than PBA Flash and Nugget) with a grey coloured seed coat. PBA Blitz has a low level of "pale coat Blitz" seeds which still have red cotyledons and are a natural part of the genetic make-up of the variety and do not affect the splitting or cooking characteristics of the variety.

PBA Bolt: This variety will reduce the impact of climatic stresses such as drought and heat. PBA Bolt is resistant to seed ascochyta blight and has similar salt and boron tolerance to PBA Flash. It is also resistant to lodging and has the best combination of pod drop and shattering resistance among all varieties. PBA Bolt is a short season red lentil variety intended as a replacement for PBA Flash. Mid flowering but early to mid maturing lentil with excellent lodging resistance at maturity and high yield in drought years and dry areas. It will provide an alternative to PBA Flash in all areas where ascochyta blight, harvestability and drought tolerance are major issues but particularly in the mallee areas of Victoria where it has consistently performed well. PBA Bolt is moderately susceptible to botrytis grey mould. It has a grey seed coat colour with a medium seed size.

Table 6.6 Lentil variety agronomic guide (may vary with sowing time and location)

	Variety	Grain type	Seed coat	Seed size	Flowering	Maturity	Lodging	Shatt-ering	Sali-nity	Bo-ron
Small seed	PBA Bounty	Red	Grey	Small	Mid/Late	Mid	S	R	MI	I
	PBAHerald XT	Red	Grey	Small	Mid/Late	Mid/Late	MR	MR	I	I
	Nipper	Red	Grey	Small	Mid/Late	Mid	MR	MR	MT	I
	Northfield	Red	Tan	Small	Mid	Mid	MS	MR	I	I
Medium seed	PBA Ace	Red	Grey	Small	Mid	Mid	MRMS	R	I	I
	PBA Blitz	Red	Grey	Med/Large	Early	Early	MS	MR	I	I
	PBA Bolt	Red	Grey	Med	Early/Mid	Early/Mid	R	R	MI	MI
	PBA Flash	Red	Green	Med	Early/Mid	Early/Mid	MR	MR	MI	MI
	Nugget	Red	Grey	Med	Mid/Late	Mid/Late	MS	MR	I	I
Large seed	PBA Jumbo	Red	Grey	Med/Large	Mid	Mid	S	MR	I	MI
	Aldinga	Red	Green	Med/large	Mid	Mid	S	R	MI	I
Green lentil	Boomer	Yellow	Green	Large	Mid	Mid/Late	S	MR	I	I
	Tiara	Yellow	Green	Very large	Very late	Very late	MS	*	*	

R = Resistant, MR=Moderately Resistant, MS=Moderately Susceptible, S=Susceptible, VS=Very Susceptible, I=Intolerant, MT= Moderately Tolerant, MI= Moderately Intolerant. # Ratings relative to Digger. Can flower and mature relatively earlier if sown early.

PBA Flash: This is also red lentil and early maturing variety will reduce the impact of climatic stresses such as drought and heat. PBA Flash is a high yielding red lentil with a green seed coat and medium seed size. It is well suited to shorter season and lower yielding lentil growing areas where its higher yield and earlier maturity improves reliability of yield. Earlier maturity also makes PBA Flash better suited to crop topping than Nugget although caution is still required with this practice due to seasonal variation in weed and crop maturity. It is moderately susceptible to foliar ascochyta blight and may require foliar fungicide sprays prior to flowering and at podding to control this disease under high intensity situations or in disease prone environments. PBA Flash is susceptible to botrytis grey mould but has improved tolerance to boron and salinity compared to Nugget. PBA Flash has improved seed splitting characteristics over Nugget and is well suited to medium sized red lentil grain markets.

Aldinga

Released in 1994, Aldinga has medium-large seed with good milling characteristics. Aldinga is moderately resistant to foliar infection by ascochyta blight but moderately susceptible to seed infection and botrytis grey mould and lodges more than other varieties. It has been a preferred variety for some farmers using no till systems.

Nugget

Released in 1999, Nugget is a mid season variety with a medium size seed and a grey seed coat. It has been the benchmark for lentil production and marketing but is gradually being superseded. It is moderately resistant to foliar infection by ascochyta blight but moderately susceptible to seed infection. Nugget is moderately resistant to botrytis grey mould.

Table 6.7 Optimum sowing time in some regions of Australia

Sowing time	Varieties
2nd-4th week of May	Aldinga, Boomer, Nugget, PBA Ace, PBA Bolt, PBA Bounty, PBA Blitz, PBA Flash, PBA Jumbo
3rd week of May-3rd week of June	Aldinga, Boomer, Nipper, Northfield, Nugget, PBA Ace, PBA Bolt, PBA Bounty, PBA Blitz, PBA Flash, PBA Herald, PBA Jumbo, Tiara

Small red lentils

PBA Herald XT: First herbicide tolerant lentil variety PBA Herald XT released in 2011, will aid growers in increasing lentil production in southern Australia through helping to overcome current production limitations. The widespread availability of PBA Herald XT now provides growers with the first lentil variety with improved tolerance to imazethapyr and flumetsulum along with reduced sensitivity to some Group B herbicide residues. PBA Herald XT also has a high level of disease resistance and will provide an alternative to all current varieties where improved weed control is required however its plant type and yield is only similar to that of Nipper. It is a mid to late flowering and maturing lentil and similar to Nipper in many agronomic characteristics, including short height, seed shape, grey seed coat colour and grain yield. It has very good disease resistance including resistance to ascochyta blight, and moderate resistance to botrytis grey mould. Limited herbicide tolerance testing suggests PBA Herald XT, like Nipper, is more sensitive than Nugget, PBA Blitz and PBA Flash to high rates of group C herbicides such as metribuzin and simazine and caution is urged with the application of these products particularly on variable soil types. PBA Herald XT has a very low level of seeds with a black seed coat typically at frequencies of 0.1% or less. These seeds still have red cotyledons and are a natural part of the genetic make-up of the variety and do not affect the splitting or cooking characteristics of PBA Herald XT.

PBA Bounty: PBA Bounty is a high yielding small seeded red lentil that is broadly adapted and suited to most lentil growing environments particularly the southern mallee region of Victoria. PBA Bounty is moderately resistant to ascochyta blight, moderately susceptible to botrytis grey mould and has improved tolerance to salinity compared to Nugget. PBA Bounty has a prostrate plant growth habit early in the season compared with all other varieties. PBA Bounty produces a small round seed that is slightly larger than Nipper.

Nipper: Nipper has good levels of resistance to both ascochyta blight and botrytis grey mould and is a low cost, low disease risk option in disease prone areas but yields less than recently released varieties. Nipper has a small seed size similar to Northfield but grey in colour. Nipper, like Northfield, flowers later than Nugget but often matures earlier than Nugget. Nipper is favoured for the practice of early sowing in Southern Australia due to the combination of a lower biomass production and a high level of disease resistance. Nipper is more sensitive to metribuzin than other varieties and caution is required to avoid application when conditions are conducive to damage. An old mid season variety with a small rounded seed and a tan seed coat that has often attracted a premium price. Northfield is resistant to foliar and seed infection by ascochyta blight but it is susceptible to botrytis grey mould. It is better adapted to the Wimmera than the Mallee region. It was released in 1996.

Table 6.8 Disease rating for current lentil varieties

Variety	Ascochyta blight foliar	Ascochyta blight seed/pod	Botrytis grey mould (BGM)
Small seed			
PBA Bounty	MR	MR	MS
PBA Herald XT	R	R	MR
Nipper	MR	R	R
Northfield	MR	R	S
Medium seed			
PBA Ace	R	R	MR
PBA Blitz	MR	MR	R
PBA Bolt	MR	R	MS
PBA Flash	MS	MR	MS
Nugget	MRMS	MS	MRMS*
Large seed			
PBA Jumbo	MR	R	MS
Aldinga	MR	MS	MS
Green lentil			
Boomer	MR	MS	MR*

R = Resistant, RMR = Resistant to moderately resistant, MR = Moderately resistant, MRMS = Moderately resistant to moderately susceptible, MS = Moderately susceptible, MSS = Moderately susceptible to susceptible, S = Susceptible, SVS = Susceptible to very susceptible, VS=Very susceptible

* Nugget and Boomer have a tendency for greater vegetative growth and height, which can predispose them to botrytis.

Large green lentils

Boomer: Boomer is a large seeded high yielding green lentil released as Australia's first adapted green variety for all lentil growing areas. It has superior seed size to Matilda (approximately 30% larger) making it more competitive than this variety with overseas varieties in world export markets. It can produce prolific amounts of growth making it prone to lodging under favourable growing conditions. It has improved resistance to ascochyta blight and botrytis grey mould compared to Matilda. Boomer will be best adapted to medium rainfall areas of Southern Australia with mild finishing conditions to ensure that the large seeds can fully form. Ascochyta blight must be managed during podding to avoid disease blemish on the seed coat. Boomer is moderately susceptible to shattering at maturity (more susceptible than current varieties). Delayed harvest under some conditions can result in shattering and/or loss of green seed colour resulting in downgrading in this variety. The current Australian lentil varieties Digger, Cobber, Matilda, Aldinga and Northfield, are much better adapted to conditions in Western Australia than old varieties such as Laird and Callisto, mainly because of their earlier flowering time. Nevertheless, these varieties were selected and released in the eastern states in the early 1990's and are still considered late flowering for the low rainfall regions of Western Australia. Two new varieties selected in Western Australia from germplasm introduced from ICARDA Syria were released in 1998 (Cassab and Cumra, 1998) with even earlier flowering and high seed yield in dry environments. Lentil variety evaluation trials continued in 1998 at various sites. Over all sites, Cassab produced 18% more seed yield than Digger.

Table 6.9 Seed yields (kg/ha) from lentil variety evaluation trials

Variety	Average yield (kg/ha)	% Digger
Digger	624	100
Aldinga	557	89
Cassab	733	118
Cobber	674	108
Cumra	473	76
Matilda	603	97
Northfield	431	69
Raja	477	76
ILL61	288	46
ILL7127	511	82
ILL7180	516	83
ILL7220	633	102
ILL7536	432	69

At present, lentil growers in Western Australia are advised to select Cassab or Digger, which are widely adapted. The early flowering and maturing variety Cumra has shown a specific adaptation to short season low yielding situations and hence, should be grown in the low rainfall regions. Cumra is generally taller than Digger or Cassab, which makes harvesting easier, but is more susceptible to ascochyta blight. Fortunately, this disease is unlikely to be a major problem in low rainfall areas.

Bangladesh

Uthfala: Early maturing, small-seeded, semi-dwarf type, with determinate growth pattern, and good podding intensity. It was selected from a landrace, Pabna, and released in 1991 with national accession number BLL 79694. It has an average yield of 13-15 q/ha, and is less susceptible to rust and Stemphylium blight than the local cultivars. Stem pigmentation is absent at the seedling stage, but is light green at the late-vegetative stage. Leaves are dark green with slight pubescence. Leaf size is medium with a dark green, short petiole and rachis that form no tendrils. Its flower is white. Seed color is ash and testa is dotted. The seeds are smooth seed surface with red cotyledon. The average 100-seed weight is 1.6 g. Its recovery percentage (splitted/dehulled quantity from whole seed) is 74%. The variety requires 16 minutes to cook, which is liked by the consumers. Its seed contains 27.8% protein. The variety is suitable for sole cultivation in all lentil-growing areas in Bangladesh.

Bari masur2: Released in Bangladesh in 1993 for its high yield and wide adaptability with a high level of resistance to rust. It was developed through single plant selection from a cross ILL 4353 (India) ´and ILL 353 (Mexico) made at ICARDA. The variety has an average yield of 18 q/ha. Barimasur-2 is a semi-erect and medium stature cultivar with plant height of 40 cm. Leaves are pubescent and grey-green, with narrow terminal leaflets and ending in a short tendril. Stems are light green, flowers white, seed coat light gray without pattern, and cotyledon bright orange. Test weight is 15 g. Matures in 110 days, which is similar to Uthfala and traditional landraces. The major achievement is that it is the first rust resistant lentil variety that provides stability in yield. Seed of Barimasur-2 retains 89% kernel content after dehulling, but actual recovery after splitting by traditional manual method is 76%, compared with 73% for local cultivars. It requires 15 minutes of cooking time, with 52% solid dispersion. Its seed protein content is 28.3%. The variety is performing well in Natore, Rajshahi and Pabna districts in both as sole and inter-cropping with sugarcane.

Barimasur-3: was released in Bangladesh in 1996 for its stable and higher yield, medium seed size, and resistance to rust. It was developed through a national hybridization program from a cross between BLL 7966 (Indian) and Pabna Local (landrace). Average seed yield is 20 q/ha, medium in height (42-44 cm), and semi-erect with more basal primary branches. The leaves are dark green with broad leaflets (without tendrils), flowers are white, and pods and leaves turn to straw color, while the stem remains green at maturity. Test weight is 25 g, which is much bigger than the local varieties. The farmers like the variety for to its bigger seed size. Seeds of Barimasur-3 has 76% recovery with traditional manual dehulling method. It takes 21 minutes to cook and shows 52% solid dispersion. Dehulled seed contains 26.9% protein. Adoption of Barimasur-3 has been seen in Rajbari, Faridpur and Kustia areas, mostly as sole and mixed cropping with mustard and wheat.

Barimasur-4: was released in Bangladesh in 1996 for its stable and high yield and its resistance to rust and stemphylium blight. It was developed from the cross ILL 5888 and ILL 5782 made at ICARDA in Syria specifically for Bangladesh. The female parent (Uthfala) was an improved lentil variety developed through pure line selection from a landrace in Bangladesh. Barimasur-4 produced an average mean seed yield of 23 q/ha. Due to its wide adaptability, the cultivar is recommended for all lentil-growing areas in Bangladesh. Barimasur-4 is an erect cultivar of medium height (40-42 cm) with long fruiting branches. The stems have anthocyanin pigmentation, the leaves are light green with narrow leaflets and rudimentary tendril. The flower is bluish-purple, and the pods, leaves and stems turn a light straw color at maturity. Most leaflets are shed at 100% pod maturity. The seed coat is reddish-grey, dotted, and cotyledon is light orange. Test weight is about 17 g and matures in 116 days. Barimasur-4 has high levels of resistance to lentil rust and *Stemphylium* blight diseases. It is the only variety grown in Bangladesh that has combined resistance against these two major diseases. Seeds of the cultivar have 89% kernel content, but produce 78% head dahl. It takes about 17 minutes to cook and shows solid dispersion of 54%. It contains 28.5% protein. Due to erect growth habit, more plants can be accommodated per unit area. It is a potential cultivar for intercropping with sugarcane, and mixed cropping with mustard. The cultivar is popular all over lentil-growing areas of Bangladesh except Rajbari and Faridpur.

CWANA region: Some varieties of this region with their characteristics are listed in table 6.10.

Conclusion

Central variety release committee of India released more than 20 varieties of lentil for cultivation in different agro-ecological zones. Recently some bold seeded varieties like WBL 58, Lens 4076, K75, DPL 62 and Sehore74-3 have been developed. Important lentil cultivars grown in different lentil growing countries presently were also listed along with major/specific characters if any. This will give a glimpse of modern lentil cultivar and there potential as of now, making a ready reckon for selection of appropriate variety under particular situations.

Table 6.10 Improved lentil varieties released for highland cultivation in CWANA region

Country	Varieties	Characteristics	Planting season
Afghanistan	ILL 5582 ILL 7180	High yield, wide adaptation High yield, erect, moderate winter-hardy	Early spring
Azerbaijan	Arzoo *	Bold seed, high biomass, erect, winter-hardy	Winter
Iran	Giacharan	High seed yield, moderate winter-hardy, erect, resistant to wilt	Early spring
Uzbekistan	Oltindon *	High seed yield, erect, early, winter-hardy	Winter
Georgia	Pablo	High biomass, erect, wide adaptation, winter-hardy	Winter
Pakistan	Shiraz-96	Winter-hardy, bold seed, high yield	Winter
Kazakastan	ILL 6434 *	High biomass, bold seed, erect	Spring
Morocco	Hamria Bichette	High seed yield, moderate winter-hardy, resistant	Winter

		to rust, Ascochyta blight and wilt diseases	
Turkey	Kafkas	Winter-hardy, high seed yield, erect, wide adaptation	Winter
	Ozbek	Winter-hardy, high seed yield, erect	Winter
	Cifci	Winter-hardy, high seed yield, large seed	Winter
	Meyvici-2001	Large seed, wilt resistant, erect	Spring
	Ali Dayi	Erect, high yield, wilt resistant	Spring

* In pre-release stage

References

Macmurray L. 2013. Lentil variety sowing guide 2013. Sowing guide 2013. 42-45 p.

Majumdar DK. 2011. Pulse Crop Production : Principles and Technologies. 567 p. PHI Learning Pvt. Ltd.

Punia SS, Baldev R, Dheer M, Jain NK, Koli NR and Khedar, OP. 2011. Variability and Association Studies in Land Races of Lentil Collected From South-Eastern Rajasthan. *Academia Arena* 3(4): 46-51.

Punia SS, Baldev R, Koli NR, Verma P and Ranwha BR. 2014. Hyper-variable spontaneous genetic variation for earliness, seed characters and other yield-contributing traits in lentil (*Lens culinaris* Med.). *Current Science* 106(1): 75-83.

Singh AK, Manibhushan, Chandra Naresh and Bharati RC. 2008. Suitable crop varieties for limited irrigated conditions in different agro climatic zones of India. *Int. J. Trop. Agri*. 26(3-4): 491-496.

Singh AK, Singh SS, Prakash V, Kumar S and Dwivedi SK. 2015. Pulses Production in India: Present Status, Bottleneck and Way Forward. *Journal of Agrisearch* 2(2): 75-83.

Singh C, Singh P, Singh RB. 2007. Modern techniques of raising field crops. 583 p. Oxford and IBH Publishing, Pvt. Ltd.

Tickoo JL, Shari B, Mishra SK and Dikshit HK. 2005. Lentil (*Lens cuslinaris*) in India: Present status and future perspectives. *Indian Journal of Agricultural Sciences* 75(9): 539-567.

Citation

Kumar P, Singh AK, Singh RP, Mahapatra BS and Bhatnagar A. 2018. Enhancing the Lentil production through improved varietal technology. In: Scientific Lentil Production. Singh AK, Bhakta N, Sangale UR, Manibhushan, Sundaram PK, Kumar S and Yasin JK. (eds.) Society for Upliftment of Rural Economy Varanasi, India. Pp. 109-138.

□□□

Chapter 7

Modern Seed Production Technology for Improving Lentil Production

RK Singh, SK Singh, Rahul R Bakade and Manoj Kumar

SYNOPSIS

The quality seed of any crop has been recognized as an important input in modern agriculture and is considered essential for increasing crop production. The lentil seed production demands trained manpower and also timely and sufficient inputs for maintenance of yield and quality of the produce. One has to follow the general as well as specific crop standards for production of quality seed. With the proper care and management one can harvest 20-25 quintal of lentil seed. It offers a good opportunity to get profit from seed production. Comparatively it is better than many pulse crops as the crop is a bit hardy and there is less attack of insect pests which deter the successful production of the crop.

Keywords: *Lens culinaris*, production, plant protection and quality seed.

Introduction

Lentil (*Lens culinaris* Medic.) is a winter season crop which thrives well in cool conditions. This is a self pollinated crop having chromosome number 2n=14. It belongs to family leguminoseae and grown throughout India during winter season. Owing to its hardiness, particularly for drought, lentil is the most popular pulse crop in the rainfed areas of eastern and central parts of the country. It is grown both as upland crop and also under *paira* or *utera* conditions in rice fallows under residual moisture and without much additional inputs. The range of its adoption to climate is very wide and can be grown at 3500 meter altitude in north-western India. In India, two types of lentils are commonly recognized in trade, comparatively bold seeded is known as masur or malka masur and small seeded is known as musri. Although, Indian lentils are more polymorphous and as a group they are generally shorter, more pigmented and have smaller pods, leaves and leaflets. A large variability exists with regards to seed size, which is almost continuous and it is really difficult to make any distinct demarcation. Howard and Khan (1942) had reported that the large seeded types of lentil generally belong to the black cotton soil areas (Madhya Pradesh, Maharashtra etc.) whereas in the Indo-Gangetic areas (Bihar, Eastern Uttar Pradesh, Assam etc.), the corresponding varieties were very small seeded. It is reveating to note that most of the large seeded varieties are late in maturity and small seeded varieties are early. Generally large seeded types are poor yielders. The lentil seeds are flat and round in shape. Langer and Hill (1991) described two seed types: (i) the large seeded cultivars with large flat pods bearing flattened seeds 6-9 mm in diameter with yellow to orange cotyledons and large white or occasionally blue flowers; and (ii) the small seeded cultivars having seeds 3-6 mm in diameter with small convex pods, convex cotyledons and small violet-blue or pink flowers. The lentil follows hypogeal mode of germination wherein the cotyledons are remained inside the soil. The good quality seed production of any crop is an exacting task that requires trained manpower coupled with some standards fixed for a specific crop. There is need to follow general as well as specific crop requirements specified for the lentil crop in the process of seed production.

Seed production

Climatic requirement

It requires cold climate. It is very hardy and can tolerate frost and severe winter to a great extent. Unlike grams, it remains unaffected by rain at any stages of its growth, including flowering and fruiting. It requires cold temperature during its vegetative growth and warm temperature at the time of maturity. The optimum temperature for growth is 18-30 ºC.

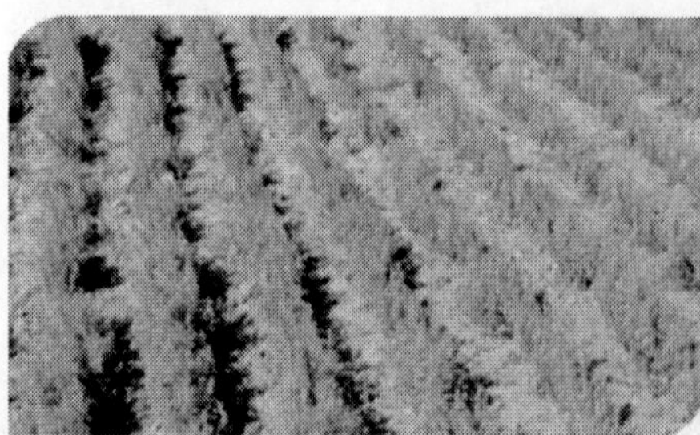

Fig. 7.1 Line sown lentil crop at early grand growth phase

Fig. 7.2 Line sown lentil crop at grand growth phase

Land requirement

One should select such a field for lentil seed production on which the same crop was not grown in the previous season unless the crop in the previous season was of the same variety and certified, in order to avoid volunteer plants. The crop can be raised on a variety of soils, such as the light loam and alluvial soils of the Indo-Gangetic plains and on the black cotton soils of Maharashtra and Madhya Pradesh. In Bengal, it thrives on clay soils while in Assam the crop grows on light loamy soils. The field should be well drained and free from weeds, disease inoculums and other undesirable materials. It should be well leveled having sufficient organic matter content. Lentil crop should not immediately follow *Lathyrus* or *Vicia spp.* as it is difficult to differentiate between the young plants of each of three species during early roguing.

Land preparation

On heavy soils, one deep ploughing followed by two to three cross harrowings should be given. In the case of light soils less tillage is needed to prepare an ideal seedbed. After harrowing, the field should be leveled. Efforts should be made to make the soil friable and weed-free so that seeding could be done at a uniform depth (Singh *et al.*, 2012a).There should be sufficient moisture in the soil at the time of sowing for better and uniform germination.

Time of sowing

The most suitable time for sowing of lentil crop is the middle of October in the northern India.

Source of seed

Always obtain required class of seed from reliable sources like research institutes of ICAR, Agricultural Universities, State Farm Corporation of India, National Seed Corporation or established private companies which are approved by concerned certification agency.

Selection of varieties

Improved high yielding varieties adapted to that area and resistant to various diseases should be selected for seed production. If the variety is not well adapted to that particular area, there is danger of accumulation of developmental variation within the variety (Singh *et al.*, 2008).

Method of sowing

The crop should be sown in lines as this facilitate intercultural operations, field inspections etc. smoothly. The seed should be sown at a depth of 2-3 cm. Deep sowing may result into uneven and less emergence leading to poor crop stand and finally yield (Singh *et al.*, 2013a).

Spacing

Row to row distance: 25-30 cm

Plant to plant distance: 05-10 cm

Seed rate

Small seeded varieties: 25-30 kg/ha

Large seeded varieties: 35-40 kg/ha

Fig. 7.3 Dark green coloured lentil seed

Fig. 7.4 Brown coloured lentil seed

Seed treatment

i) Fungicide treatment: Bavistin + Thiram (1:1) at the rate of 2.5 g/ kg seed.

ii) Rhizobium culture: *Rhizobium* strains like L 1-77, L 21-83 can be used for inoculation for the lentil seeds. First of all sticker solution is prepared by dissolving 50 g sugar/jaggery in 500 ml water which is boiled for 15 minutes before use. When it becomes cool, add the *rhizobium* culture and mix it properly. This slurry is further mixed to the quantity of seed enough for one acre sowing. Allow the seed to dry in the shed before sowing. The fungicide treatment must be followed by *rhizobium* inoculation.

Fertilizer application

It is always safe and economic to have soil test before application of fertilizer so that correct amount of fertilizer may be calculated and applied to get maximum return as the soil fertility varies from field to field. On medium to low fertile soils, application of 20-40 kg nitrogen and 60-75 kg phosphorus per hectare is recommended. To overcome zinc deficiency a spray of 0.5 per cent zinc sulphate plus 0.25 per cent lime solution is recommended (Singh *et al.*, 2012b).

Irrigation

One to two irrigations are required for growing the crop. The first irrigation should be given at 45 days of planting and second at pod filling stage, if needed. Avoid more irrigations as it may affect the crop performance adversely.

Interculture operations

One to two weddings / hoeings in the early stages of crop growth is necessary to keep an effective control over weeds (Singh *et al.*, 2013b). Lasso or Toke E-25 sprayed @ 4 liters in 1000 liters of water per hectare on the soil immediately after sowing but not later than 40 hours after sowing, control weed population effectively. Basalin 0.75 kg a.i. per ha can also be used as pre-sowing spray to control the weeds.

Plant protection

There are no serious pests that infest this crop all over the growing areas. However, infestation of aphids, hairy caterpillars and pod borers may occur.

Aphid (*Aphis craccivora* Koch.)

Both adult and nymphs suck the sap of the plants. Although the insects are minute, considerable damage is caused on account of their large numbers. In the cases of severe infestation, the plant withers and succumbs. The insect can be controlled by spraying of 0.04 per cent monocrotophos at the appropriate time. Foliar spray of 0.03.% Dimethoate (30 EC) or 0.03% Phosphomidon is also effective for control of aphids. Apart from this, natural enemies like lady bird beetel and syrphid can also be used to control the population of aphids. Saxena *et al.* (1970) observed a large number of grubs and adults of lady bird beetel, *Coccinella septempunctata* Lin. and groups of syrphid, *Ischiodan scutellaris* Lin. preying upon a heavy population of aphid (*Aphis craccivora*).

Cut worm

This insect is polyphagous in nature and observed in many crops which are cultivated in winter season. During the day, the insect remains hidden under the debris, soil clods and cracks in the fields and become active during night. The caterpillar cuts plants or its branches and in severe infestation the loss is enormous. The damage may not be so serious due to actual feeding by the caterpillar but much great loss is caused on account of its habit of cutting either the entire plant or its branches.The insect can be controlled by applying 5% Aldrin or Heptachlor dust at the rate of 20-25 kg/ha mixed in soil to control the insect.

Pod borer (*Etiella zinckenella* Treitschke)

This is a common pest of cold weather crops and does considerable damage to many pulses including lentil. The caterpillar damages to the pods. The activity of the pest starts when the flowers appear and pods formed. The crop which matures late either because of late sowing or due to long duration is generally more severely damaged by the insect. To control the insect, spray the crop with 0.04% monocrotophos emulsion whenever insect appear to control them. Besides, spraying of 0.1% lindane emulsion and 0.25% carbaryl is also effective in controlling the insect.

Diseases management

Wilt

The disease was reported by Dey (1948) causing considerable damage to all cultivars. Vasudeva and Srinivasan (1952) reported losses upto 67% in some lentil lines at Delhi whereas Khare *et al.* (1979)

reported more than 50% losses due to wilt under field conditions. The disease appears in November increases from February to April but causes maximum damage in March. The leaves of affected plants turns yellow, curl upwards and plants later dry up. The causal organism is *Fusarium oxysporum* Schlect. *F. sp. Lentis* is similar to wilt pathogen of pea but is pathogenic only to *Lens culinaris.* The fungus is seed and soil borne. Ujevic *et al.* (1965) isolated the pathogen from seeds obtained from wilt plant. To control the disease it is always advised to keep the field clean and follow a three year crop rotation. This will help in reducing the disease incidence. Kannaiyan and Nene (1974) as well as Agarwal and Khare (1975) reported that seed treatment with Benlate (0.3%) reduced wilt and correspondingly increased the yield.

Collar rot

It is an important disease in parts of Madhya Pradesh, Uttar Pradesh and Rajasthan as reported by Jain *et al.* (1960), Pavgi and Upadhyay (1967) and Mathur and Deshpande (1968), respectively. The disease is caused by *Sclerotium rolfsii* Sacc. The pathogen attacks the seedlings at the collar region. The tissues at this region get infected, discoloured and stem becomes weak. Under favourable conditions, the pathogen kills the plant outright while under less favourable conditions the leaves turn pale, tender twigs droop and the plant dries up within one or two weeks after infection. White mycelium can be seen on the infected collar region. The young plants are more susceptible to this disease. To control the disease it is suggested to plant only resistant varieties and mortality of seedlings can be reduced by delayed planting until mid November. Khare *et al.* (1974) reported that seed treatment with Thiram plus PCNB (1:1) at the rate of 2.5 g/kg seed reduced the incidence of collar rot disease.

Root rot

This is caused by *Rhizoctonia solani* Kuhn and *Rhizoctonia bataticola* (Taub) Butler (Shukla *et al.,* 1972 and Kannaiyan and Nene 1973). Colour of the leaves of infected plants gradually changes from dull green to reddish brown. Roots turn brown to dark-brown in colour and ultimately dry up. There is complete rotting of finner roots. The sclerotia of the fungus are small dot like structures and develop on the infected roots.The disease can be controlled by seed treatment with PCNB (Brassicol) at the rate of 2 g/kg of seed before sowing.

Rust

It is caused by *Uromyces fabae* (Pers.) de Bany. Brown circular patches can be seen on leaves in low lying areas of the field under high humidity conditions towards the end of January. Such patches serve as foci of infection for the spread of the disease. In severe cases of infection, plants dry up completely even before the formation of the seed; if formed the seed size is appreciably reduced. To control the disease, it is suggested to properly process the seed during processing to remove diseased plant-bits. The affected plant trash should burn in order to reduce the disease pathogen inoculums. Khare *et al.* (1979) reported that protective spray with Dithane M-45 (2500ppm) at 10-12 days interval, starting immediately on the appearance of the disease were quite effective for the control of the disease.

Ascochyta blight

The disease is internally seed borne and perpetuates through infected seed and diseased plant debris. It is caused by *Ascochyta pisi* Lib. In India this was first isolated by Sattar (1933) form Palampur (H.P.) and by abroad by Mukica (1955) in Chile. Circular tan to dark-brown spots appears on leaves, stems and pods. These spots have small, dark-brown to black pycnidia on them in concentric rings. The diseased spots are more conspicuous on pods than on leaves and stems. To control the disease, always use good quality seeds for sowing and must treat the seed with systemic fungicide before sowing.

Powdery mildew

It is common in Madhya Pradesh and the disease may occur at any growth stage of the plant but is more severe at flowering stage. White powdery growth occurs on the leaves, stems and pods. The disease begins on the lower side of leaves, covers both the surfaces of leaves, stems and pods under conditions of heavy infection. The disease can be controlled by dusting of sulphur dust @ 20-25 kg/ha. Spraying of Karathane 0.1% at 15 to 20 days interval during flowering is also effective in controlling the powdery mildew disease.

Roguing

Roguing is a regular and continuous process in the seed crop which is done to remove the off type plants, plants of other varieties, diseased plants, weed plants particularly *Lathyrus spp* and *Vicia spp.*

These undesirable plants should be removed from the field whenever notices as delay removal may affect the quality of the seed. There should be minimum three roguings in the seed field. The roguing should be plan in such a way that it should precede every field inspection. Careful and intensive roguing is required at flowering to remove the contaminants, if any.

Field Inspection

The field inspections are done by the representatives of concerned seed certification agency of the area. A minimum of two inspections are made, the first before flowering and second at flowering and fruit stage. The field crop must meet the prescribed standards fixed for the various contaminants in order to get the crop certified otherwise it will be liable for rejection.

Field Standards

Isolation

The crop is self pollinated and the dehiscence of anthers take place in the bud itself sometimes before the opening of the flower the next morning. Therefore, the lentil seed field should be isolated from the contaminants shown in column 01 of the table below by the distances specified in column 02 and 03:

Harvesting and threshing

Harvesting should be done when plants become yellow and seed resistant to pressure when pressed between the fingers. At this particular stage the lower pods are usually brown in colour. The plants may be pulled manually by hand or cut and collected in small bundles to dry. Care should be taken to not dump the just harvested crop in a big heap for longer as it will spoil the crop due to generation of excessive heat inside the heap.

Threshing is done by flailing with sticks and seed is cleaned by winnowing. The threshers are also available which can be used for large scale harvesting. After winnowing the seed should be dried to nine per cent moisture before storage. Precaution should be taken to avoid mechanical mixtures during harvesting and threshing operations and also contact of seeds with moist soils.

Seed yield

The average seed yield varies from 20 to 25 q/ha from a good crop of lentil with proper management.

Lot size

The recommended lot size for lentil is 100 quintals per lot. Quantities greater than these needs to be divided into different lots accordingly

Aperture Size for grading of lentil seed

Following screen aperture size is recommended for lentil seed processing:

Top screen 7.00r

Bottom screen 3.20s, 4.00r and 4.75r

(r = round holes, s= slotted holes)

Table 7.1 Isolation distance and specific requirements for lentil seed production

Contaminants	Minimum distance (meter)	
	Foundation seed crop	Certified seed crop
Fields of other varieties	10	05
Fields of the same variety not confirming to the varietal purity requirements for certification	10	05
Specific requirements		
Off types (Maximum permitted %)*	Foundation seed crop	Certified seed crop
	0.10	0.20

* Maximum permitted at the final inspection.

Table 7.2 Standards for lentil seeds in India

Particulars	Foundation seed	Certified seed
Pure seed (Minimum)	98.0%	98.0%
Inert matter (Maximum)	2.0%	2.0%
Other crop seed (Maximum)	05 per Kilogram seed (By number)	10 per Kilogram seed (By number)
Weed seeds (Maximum)	10 per Kilogram seed (By number)	20 per Kilogram seed (By number)
Other distinguishable varieties (Maximum)	10 per Kilogram seed (By number)	20 per Kilogram seed (By number)
Germination including hard seeds (Minimum)	75%	75%
Moisture (Maximum)	9.0%0%	

Conclusion

Lentil is one of the most important *Rabi* pulses. It is a self pollinated crop hence, seed production is comparatively easy. This crop is comparatively hardy and needs less water. The attack of insect pest is also very low or occasional and place specific. On an average 20-25 quintal seed per hectare can be obtained under proper care and management conditions along with adopting the above mentioned steps and standards.

Reference

Agrawal SC and Khare MN. 1975. Effect of fungicides on microconidia formation of *Fusarium oxysporum F. sp. lentis. Sci. and Cult.* 41: 280-281.

Dey PK. 1948. *Plant Pathology. Adm. Rep.* Agric. Deptt. U.P. 46: 43-46.

Howard GLC and Khan AR. 1924. Studies in Indian oil seeds. No.2 Mem. Deptt. of Agric. India. Bot. Ser., xii 4: P183.

Jain AC, Nikam BG, Kulkarni SN and Sharma OP.1960. Fungi of Gwalior and Indore region. *The Vikram*. 4: 181-187.

Kannaiyan J and Nene YL. 1973. A new root rot disease of lentil. *Curr. Sci.* 42: 257.

Kannaiyan J and Nene YL. 1974. Effect of fungicidal seed treatment on wilt disease of lentil. *India J. Pl. Prot.* 2: 80-85.

Khare MN, Agrawal SC and Jain AC. 1979. Diseases of lentil and their control. *Tech Bull.* Deptt. *Plant Path.* JNKV, Jabalpur, India. 01-29.

Mathur BN and Deshpande BN. 1968. Collar rot of lentil. *Indian Phytopath.* 21: 455-456.

Mujica RF. 1955. Studies on Sclerotiniosis. *Agricultura Tech. Santiago.* 15: 64-74.

Pavgi MS and Upadhyay R. 1967. Root rot disease of lentil in Uttar Pradesh. *Sci. and Cult.* 33: 75-76.

Sattar A. 1933. A comparative study of the fungi associated with blight diseases of certain cultivated leguminous plants. *Trans. Brit. Mycol. Soc.* 18: 276-301.

Saxena HP, Sircar P and Phokela A. 1970. Predation of *Coccinella septempunctata* Lin and *Ischiodon scutellaris Fabricius* on *Aphis craccivora* Koeh. *Indian J. Ent.* 32(1): 105-106.

Shukla TN, Ahmad ZA and Garg SK. 1972. Root rot of lentil. *Indian Phytopath.* 25: 584-585.

Singh AK, Bhat BP, Sunram PK, Kumar S, Bharati RC, Chandra N and Rai M.2012b.Study of Site Specific Nutrients Management of Cowpea Seed Production and Their Effect on Soil Nutrient Status. *Journal of Agril.* Sci. 4(10): 191-198.

Singh AK, Bhatt BP, Sundaram PK, Gupta AK and Singh Deepak. 2013a. Planting geometry to optimize growth and productivity faba bean (*Vicia faba* L.) and soil fertility. *J. Environ. Biol.* 34(1): 117-122.

Singh AK, Kumar P and N Chandra. 2013b. Studies on yield production of mung bean (*Vigna radiate*) sown at different dates. J. *Environ. Biol.* 34: 1007-1011.

Singh AK, Manibhushan, Chandra Naresh and Bharati RC. 2008. Suitable crop varieties for limited irrigated conditions in different agro climatic zones of India. *Int. J. Trop. Agri.* 26(3-4): 491-496.

Singh AK, Singh D, Singh AK, Gade RM and Sangle UR. 2012a. Good Agronomic Practices (GAP) - An efficient and eco-friendly tool for sustainable management of plant diseases under changing climate scenario. *J. Plant Disease Sci.* 7(1): 1-8.

Ujevic I, Kovacekova E and Urosevic B. 1965. *Botrytis parasitica* and *Fusarium oxysporum* as dangerous prasites of lentil. *Polnohospodarstve.* 11: 584-595.

Vasudeva RS and Srinivasan KV. 1952. The wilt disease of lentil (*Lens esculenta* Moench.) *Indian Phytopath.* 5: 23-32.

Citation

Singh RK. Singh SK. Bakade RR and Kumar M. 2018. Modern Seed Production Technology for Improving Lentil Production. In: Scientific Lentil Production. Singh AK, Bhakta N, Sangale UR, Manibhushan, Sundaram PK, Kumar S and Yasin JK. (eds.) Society for Upliftment of Rural Economy Varanasi, India. Pp. 139-151.

❑❑❑

Chapter 8

Commercial Classification of Lentil

Pravesh Kumar, Anil Kumar Singh and Sushil Dimree

SYNOPSIS

Lentil (*Lens culinaris* Medikus) is a short-statured, annual, self-pollinated, high value crop species. Lentil (*Lens culinaris*) is a free-standing legume divided into two sub-species; the cultivated variety ssp. *L. culinaris* and its wild relative ssp. *L. oriental. It* was first grown in south west Asia about 7,000 B.C. It is best adapted to the cooler temperate zones of the world, or the winter season in Mediterranean climates. Lentil colors range from yellow to red-orange to green, brown and black. Lentils also vary in size and sold in many forms, with or without the skins, whole or split. Red, white and yellow lentils are decorticated, i.e. they have their skins removed. The cultivated species of lentil *(Lens culinaris* Medikus) has two races, viz *microsperina* and *macrosperma.*

Keywords: Black lentils Chilean green lentils, Dark green lentils, Market class, Red chief lentils, Seed size.

Introduction

Lentil (*Lens culinaris* Medikus ssp. *culinaris*) is an important food legume of the Indian subcontinent, the Middle East, Northern Africa and North America. India produces more than 1 million tonnes (mt) of lentil from an area around 1.6 m ha. The Indian productivity is about 25% lower than the global average (1053 kg/ha). Most of the lentil varieties in India have been developed mainly through pureline selection and intraspecific hybridization. Lentil (*Lens culinaris L.*) was first grown in south west Asia about 7,000 B.C. It is best adapted to the cooler temperate zones of the world, or the winter season in Mediterranean climates (Punia *et al.*, 2011) Lentil has restricted root system that is only somewhat resistant to high temperatures and drought. They cannot withstand flooding, water-logging, or soils with high salinity. Lentils work well in rotation with cereals such as spring and durum wheat. It is considered to be the most efficient crop in fixing atmospheric nitrogen through *rhizobium*. They have the ability to fix nitrogen from the air which can then be used by other crops in following years. Lentils are vulnerable to both *ascochyta* blight and anthracnose. To reduce the risk of these blights, lentils should be only seeded in the same field once every four years. Lentil can be used as a green manure crop. Lentils are used almost exclusively for human consumption (Tickoo *et al.*, 2005). Lentils are often used as a meat extender or as a meat substitute because of the high protein content and quality. They can be used in soups, stews, salads, casseroles, and vegetarian dishes. Lentils can also be used for livestock feed, however, only small amounts of low grade lentils are used for feed. In southern Asia, split red lentils are used in curries. Lentil flour is added to cereal flour to make breads, cakes and baby foods. In addition, lentils are often used as a meat extender or substitute because of the high protein content and quality. They are celebrated for their low cost, excellent nutritional content and long shelf life. These features have made lentils a staple food in many cultures across the globe.

Fig.8.2 Dark brown coloured lentil seed

Fig.8.1 Dark green coloured lentil seed

Plant type

Lentil (*Lens culinaris* Medikus) is a short-statured, annual, self-pollinated, high value crop species. Lentil (*Lens culinaris*) is a free-standing legume divided into two sub-species; the cultivated variety ssp. *L. culinaris* and its wild relative ssp. *L. orientalis*. Lentils are a small, round, lens shaped legume that come in many colors and sizes (Tickoo *et al.*, 2005).

Botanical Name	*Lens culinaris Medikus subsp. culinaris*
Chromosome	2n = 14
Origin	Turkey to South Iran

Growth habits

Lentil plants are slender, semi-erect annuals with compound leaves (4 to 7 pairs of leaflets) with a tendril at the tips. Plants normally range from 12 to 20 cm tall, the taller plants resulting from cool growing season temperatures, good moisture and good fertility. Plants can have single stems or many branches depending upon the population in the field. Flowering begins on the lowest branches, gradually moving up the plant and continuing until harvest. Each flower produces a short pod containing one or two lens-shaped seeds. Flowers can be white, lilac or pale blue in color and are self-pollinated. At maturity plants tend to lodge because of their weak stems. Lentil produced in North America has larger seeds than that from India and the Near East. The seeds (2 to 7 mm in diameter) come in colors of tan, brown, or black and some varieties produce purple or black mottled seeds. Lentil seed number varies from 15,600 to 100,000 seeds/lb.

Types

Different kinds of lentils exist, including a variety of large and small lentils. Lentil colors range from yellow to red-orange to green, brown and black. Lentils also vary in size and sold in many forms, with or without the skins, whole or split. Red, white and yellow lentils are decorticated, *i.e.* they have their skins removed. One variety of yellow "lentils", Chana, is in fact not a lentil, but is actually made from the kernels of chickpeas. Split Pigeon peas (either green or yellow) that are actually considered pulses, which include peas and beans are also sometimes erroneously referred to and sold as lentils (Macmurray, 2013). The urad bean is also referred to as a "black lentil".

- Brown/spanish pardina
- French green/puy lentils (dark speckled blue-green)
- Green
- Black/beluga
- Yellow/tan lentils (red inside)

- Red Chief (decorticated yellow lentils)
- Eston Green (small green)
- Richlea (medium green)
- Laird (large green)
- Petite Golden (decorticated lentils)
- Masoor (brown-skinned lentils which are orange inside)
- Petite crimson/red (decorticated masoor lentils)
- *Macachiados* (big Mexican yellow lentils)

Classification

A. According to seed size
B. According to seed weight
C. According to colour
D. According to locality
E. According to main market class
F. According to hull
G. According to cooking time

A. **According to seed size:** The cultivated species of lentil *(Lens culinaris* Medikus) has two races, viz *microsperina* and *macrosperma.*

Microsperma: These types small rounded seeds, 2-6 mm in diameter, yellow or orange cotyledons, testa colour ranging from pale yellow to black and 1000-seed weight up to 25 g. They are mostly grown in India, Nepal, Pakistan Afghanistan, Bangladesh and Egypt. About 65% of lentil area in India is under the varieties of this group, grown in Punjab, Haryana, Uttar Pradesh and Bihar.

Macrosperma: These are large size seeds, 6-9 mm in diameters which are more flattened, yellow cotyledons, pale green testa and 1000-seed weight of 25-50 g. They are generally found in Mediterranean, Spain, Italy, Greece, North and Latin America. In India, about 35% of area is under this group, with about 40% of the total production. These are grown mostly in Madhya Pradesh and adjoining districts of Uttar Pradesh.

B. **According to seed weight:**

For both types of lentils, the seeds are lens-shaped which the source of the name lentil is. Seed coat colours range from clear to green, brown, grey, blotched purple or black. The cotyledons can be yellow, red or green (Punia *et al.,* 2014).

Chilean/large seeded type: The large seeded Chilean type has a seed size that averages 50 grams or more per 1000 seeds.

Persian/small seeded type: The small seeded Persian type has a seed size that averages 40 grams or less per 1000 seeds.

C. According to colour:

Brown lentils

Fig. 8.3 Brown lentils

This is the most basic variety of lentil. These small, brown lentils hold their shape well during cooking but may become mushy if over cooked. This variety cooks quickly (in about 20 minutes) and requires no soaking like most other beans. With a subtle and earthy flavor, this variety is good for soups, stews, salads, pilafs, and more. Common varieties are Spanish Brown, German Brown, or Indian Brown.Italy grows its own brown lentils which remain firm when cooked, and are used for side dishes, soups, and pasta dishes. Italians often simply cook the lentils in broth with pancetta and fennel, and serve them drizzled in good olive oil. In Italy, it is considered good luck to eat lentils on New Year's Day, as the tiny discs are reminiscent of coins and promise a good year fiscally (Fig.8.3).

Light green lentils

Fig. 8.4 Light green lentils

These light green lentils are quite large (about a quarter inch in diameter) and tend to be slightly flatter than other varieties. These lentils are very flavorful, remain fairly firm, and retain their shape with cooking. This makes them ideal for salads and pilafs as they will not easily break down with mixing or stirring (Fig.8.4)

Dark green lentils

These lentils are also called as French green lentils, or Puy lentils, are smaller and darker than regular green lentils (light green lentils) and appear slightly speckled on the surface. This variety of lentil remains firm and requires an extended cooking time of approximately 40 minutes (Fig.8.5).

Fig. 8.5 Dark green lentils

Yellow lentils

Yellow lentils are sweet and nutty and have a reddish interior flesh. These lentils break down quickly with cooking and therefore are better suited for dishes that require thickening. Yellow lentils are popular in Indian "dal" recipes where the lentils are cooked down to a smooth puree (Fig.8.6).

Fig. 8.6 Yellow lentils

Red lentils

Red lentils are yellow lentils that have been hulled and split. They are a light red to orange color, small in size, and create a very smooth puree when cooked. These are the sweetest and nuttiest of the lentils. They're somewhere in the middle in terms of cooking time and are usually done in about 30 minutes. They tend to get mushy when cooked through, so they're perfect for Indian dals and other curries, or for thickening soups. A few varieties are Red Chief and Crimson, and you'll often find them in Indian or Middle Eastern markets labeled as masoor (red lentils) or channa (yellow lentils) (Fig.8.7).

Fig. 8.7 Red lentils

Black lentils

Fig. 8.8 Black lentils

Black lentils are small and quite round in shape. They are often called beluga lentils because they resemble beluga caviar when cooked. The blackest and tiniest lentils you find are usually Beluga lentils, which have a rich and deeply earthy flavor. Although the hull is dark black in color, the flesh is light and creamy. Black lentils are available whole, hulled, or hulled and split (Fig.8.8).

D. **According to locality:**

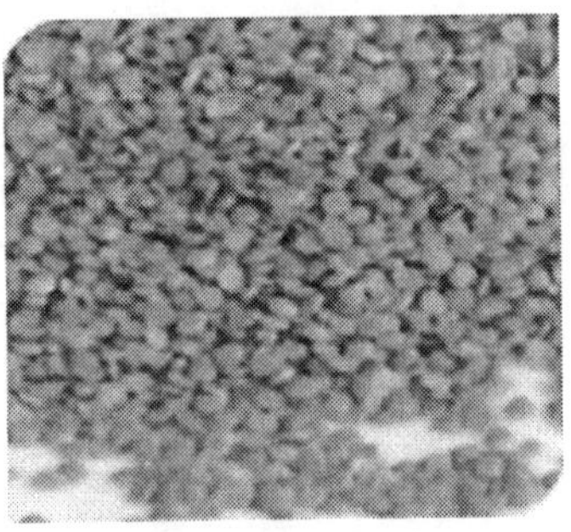

Fig. 8.9 Red chief lentils

Red chief lentils: (Washington, Idaho)Perhaps the prettiest of the lentils, Red chief lentils are tiny and orange hued. You may not find Red chief lentils in typical grocery stores, but they are often available in health food stores or specialty food markets. Ask for local Red chief lentils; imported Persian red lentils are not as tender as Red chiefs and take twice as long to cook. Red chiefs are sold whole and are quite a bit larger than Persian red lentils, which are usually sold split. Red lentils do not hold their shape very well during cooking, so it is important to match them to recipes in which a mash or puree is desired (Fig.8.9).

Chilean green lentils

Fig. 8.10 Chilean green lentils

(Washington, Idaho)Despite the misleading name, this is the khaki-tan hued lentil that many of us associate with the Northwest, as until recently it was the most commonly available variety. Beautifully flavored, Chilean Green lentils hold their shape and color well during cooking (Fig.8.10).

Black beluga lentils: (Montana)

These black beauties are the cavier of lentils. Black beluga lentils are even smaller than Red chief lentils, slightly more rounded, and just as striking in their own way. Softening to a rich dark brown, they glisten when cooked, hold their shape, and have a warm, nutty flavor that lends itself to a wide range of spices (Fig.8.11).

Fig. 8.11 Black beluga lentils

French green or Du puy lentils: (Idaho, Montana)

French Green (also called Du Puy) lentils are varigated shades of lovely dark, muted green. If you didn't know that the photograph above pictured entils, you might think it was a collection of gorgeous beach pebbles. French Green lentils have a rich flavor and remain firm when cooked. The best lentils by far are the French Puy lentils, once grown only in France, but now also in North America. These pretty grey-green lentils are sometimes called Blue lentils. They are sold skin-on, remaining chewy and keeping their colour when cooked, making them ideal for cold salads or as an ingredient in other grain dishes (Fig.8.12).

Fig. 8.12 French green or Du puy lentils

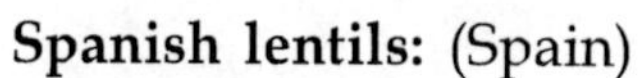

Spanish lentils: (Spain)

Spanish Lentils are medium-sized and a soft, grey-brown hue. Smaller and rounder than Green Chilean lentils, they are said to hold their shape better during cooking, although yet to confirm this (Fig.8.13).

Fig. 8.13 Spanish lentils

Fig. 8.14 Umbrian lentils: (Italy)

Umbrian lentils: (Italy)

Umbrian lentils are beautifully varigated in shades of golden pale orange, tan, and pale green. I have yet to cook these, so will report back later on cooked flavor and texture(Fig.8.14).

E. **According to main market class:** The two main market classes are-

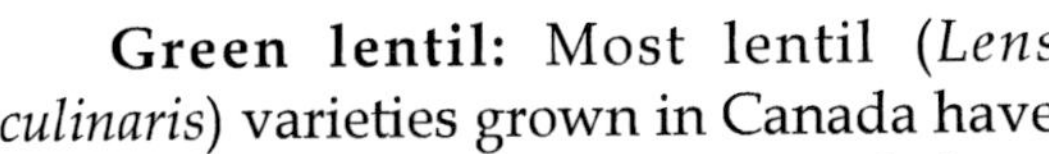

Green lentil: Most lentil (*Lens culinaris*) varieties grown in Canada have green seed coats and yellow cotyledons (seed leaves). Green lentils are divided into three market classes in Canada: large green, medium green, and small green. Large green lentils now include Laird, Glamis, Sovereign, and Grandora varieties; medium green lentils include Richlea and Vandage varieties; and small green lentils include Eston and Milestone varieties. Green lentil is usually marketed as whole seed.

Red lentil: Canada has recently become an important producer and exporter of red lentils. Red lentil varieties typically have brown to pale green seed coats with red cotyledons. The red lentil production consists of the varieties Crimson and Redwig as well as Blaze, Redcap, and Robin. There are also other varieties of lentils that are grown in small volumes in Canada and are sold to niche markets. Red lentil is marketed as whole seed or in dehulled and split form. The majority of world lentil production and trade is in red lentil.

F. **According to Hull:** Classified on the basis of presence and absence of hull/seed coat on the seed.

Hulled: e.g. green and brown lentils.

Dehulled: such as red and some black lentils

G. According to cooking time:

Short period: Shorter for small varieties with the husk removed, such as the common red lentil.

Moderate period: Lentils with husk remain whole takes moderate period for cooking.

Long period: The best, most delicate lentils are the peppery french green lentils. These hold their shape well, but take longer to cook than other lentils.

Point to be consider while going for lentil production

Lentil is a cool season crop with a relatively shallow root system (0.6 metres) and is moderately resistant to high temperature and drought. Lentil is classified into two groups by seed size. The large seeded type has a seed size that averages 50 grams or more per 1000 seeds. The small-seeded type has a seed size which averages 40 grams or less per 1000 seeds. Early lentil production in Saskatchewan focused on the large green market class.Green lentil market classes typically have yellow cotyledons with green seed coats and are described as large, medium, and small. Red lentil varieties typically have grey seed coats with red cotyledons and are described as extra small, small, and medium. Varieties with the Clearfield® trait are a recent development. This trait allows use of imidazolinone herbicides that cause injury to conventional lentils. For red lentil, buyers prefer seed moisture content to be 13 per cent to improve the quality and efficiency of dehulling and splitting processes. Lentil seed at 14 per cent moisture or less and 15°C or less can be stored safely for up to 40 weeks. If heated grain drying is needed, air temperatures should not exceed 45°C to preserve germination, and the sample should not be dried more than four to five percentage points per pass through the dryer. Lentil seed is susceptible to increased chipping and peeling if handled in temperatures colder than -20°C.

Conclusion

Lentil is one of the most nutritious cool season food legumes and ranks next only to chickpea. Lentil is a valuable human food, mostly consumed as dry seeds (whole decorticated, seed decorticated and split). Lentil, based on the seed size and test weight, is classified into two main groups i.e. **bold seeded**: Includes sub sp. *macro-sperma,* with the test weight of more than 25 g. also known locally as *Masur* or *Malka Masur* and mainly cultivated in Bundelkhand region of UP/ MP and Maharashtra state and (ii) **small seeded**: Sub sp. *micro-sperma,* test weight, less than 25 g, locally known as *masuri* and primarily grown in Indo Gangtic plains of NEPZ (UP, Bihar, West Bengal and Assam). Beside this lentil is also classified based on According to seed size, seed weight, colour, locality, main market class, hull and cooking time.

References

Macmurray L. 2013. Lentil variety sowing guide 2013. Sowing guide 2013. 42-45 p.

Punia SS, Baldev R, Dheer M, Jain NK, Koli NR and Khedar, OP. 2011. Variability and Association Studies in Land Races of Lentil Collected From South-Eastern Rajasthan. *Academia Arena* 3(4): 46-51.

Punia SS, Baldev R, Koli NR, Verma P and Ranwha BR. 2014. Hyper-variable spontaneous genetic variation for earliness, seed characters and other yield-contributing traits in lentil (*Lens culinaris* Med.). *Current Science* 106(1): 75-83.

Singh AK, Manibhushan, Chandra Naresh and Bharati RC. 2008. Suitable crop varieties for limited irrigated conditions in different agro climatic zones of India. *Int. J. Trop. Agri.* 26(3-4): 491-496.

Tickoo JL, Shari B, Mishra SK and Dikshit HK. 2005. Lentil (*Lens ccslinaris*) in India: Present status aaad future perspectjves. *Indian Journal of Agricultural Sciences* 75(9): 539-567.

Citation

Kumar P, Singh AK and Dimree S. 2018. Commercial Classification of Lentil. In: Scientific Lentil Production. Singh AK, Bhakta N, Sangale UR, Manibhushan, Sundaram PK, Kumar S and Yasin JK. (eds.) Society for Upliftment of Rural Economy Varanasi, India. Pp. 153-163.

❑❑❑

Chapter 9

Agronomic Management of Lentil for Getting Higher Production

Aditya Kumar Singh, Anil Kumar Singh, Lala IP Ray, Vishram Ram and N Janki Singh

SYNOPSIS

Lentil is one among most nutritious pulse crops of India grown since time immemorial and ranking next to chickpea among Rabi pulses. It is generally grown as a rainfed crop on the residual soil moisture of preceding crop especially rice. Lentil leaves reasonable good amounts of biologically fixed nitrogen. For *utera* sowing, only small seeded varieties should give preference as their chances to hold in standing rice is less and come in contact with soil early due to higher surface resulted in higher germination percentage and establishment of a good lentil crop. Phosphorus (P) deficiency in soil is usually the key factor for poor yield of pulses in all type of soils as they have higher phosphorus requirement. Deficiency of micronutrients especially zinc (Zn) and S is one of the limiting factor for obtaining derived response to applied P in pulses. It is moderately drought tolerant crop and requires 200 mm of water depending on soil and climate. Lentil is suffered badly by severe competition from weeds during early crop growth as crop growth is very slow during this period. The period from 30 - 60 days after sowing is most crucial for completion with weeds. A well-managed crop yields approximately about 1.8-2.0 tones grain and 3.0-4.0 tons of straw per hectare. The clean seed should be sun dried for 3-4 days to bring their moisture content at 9-10%.

Keywords: Lentil, Nipping, Plant nutrient, Water Management, Weed Management, Yield.

Introduction

Lentil (*Lens culinaris,* Medick) is one of the man's oldest food crops. It is one among most nutritious pulse crops of India grown since time immemorial and ranking next to chickpea among rabi pulses. It is one of the prominent sources of vegetable protein in the South East Asia including India. It is generally grown as a rainfed crop on the residual soil moisture of preceding crop especially rice in-general (Singh and Singh, 1995 and Ali *et al.,* 2012). In India small seeded lentil is commonly known as *Masur however,* in many places of central region bold seeded lentil is also known as Malka. It is known by different names in various countries as lentille (French) linse (German), Lenteja (Spanish) and Mercimek (Turkish) (Ali and Mishra, 2000). It contributes significantly to sustainability of food, feed and farming system. Its grains contain high amount of digestible protein (upto 25%), macro and micronutrients (particularly iron and zinc) and vitamins, thus providing nutritional security to consumers. Its cultivation improves soil health by adding nitrogen, carbon and organic matter, thus provides sustainability to the cereal based cropping systems. Lentil is a valuable human food, mostly consumed as dry seeds (whole decorticated, seed decorticated and split). In Indian sub-continent it is mostly consumed as 'Dal' by removal of outer skin and separation of cotyledons, snacks and soup preparation etc. Its grains also used in preparation of snacks and soup and *Dal-moth* etc. It is easy to cook and easily digestible with high biological value, hence also referred to patient. Lentil straw including dry leaves, stems, empty and broken pods are used as valuable cattle feed. Lentil straw is a valuable animal feed as it has high digestibility, protein, calcium and phosphorous compared to wheat straw and is highly palatable as well. Its bold seeded, attractive shaped grains have high demand for export at premium prices.

Deep rooting characteristics, leaf shading ability, capacity to fix atmospheric nitrogen (N) and solublizing phosphorus (P) in association with VAM make lentil one among the most efficient nutrient recycling agents in nature. Biological efficiency of soil is likely to improve when pulses are included in the cropping system. Lentil leaves reasonable good amounts of biologically fixed nitrogen (upto 30-40 kg/ ha) in readily available form to the succeeding crop (Ahlawat *et al.,* 1981). Associated non leguminous intercrop also gets benefited by nitrogen transfer from lentil roots up to some extent. It also contributes to sustain soil productivity through physical, chemical and biological improvements in soil properties, as a rotation effect. It has a potential to cover the risk of dry land agriculture and also used as cover crop to check the risk of soil erosion in problem areas. This crop offers good scope in late vacated paddy fields either as *Utera* or succeeding crop (if soil is not workable after paddy harvest) as delayed sowing does not affect as adversely as in chickpea and pea due to its high cold tolerant nature (Jeswani and Baldeb, 1990 and Ali and Mishra, 2000). By this reason, this crop is preferred over other *rabi* pulses in the regions having cold winters like plains of north and lower Himalayan Hills. It is also a good substitute of chickpea in areas where availability of residual moisture in soil is limited due to its very short duration. The crop is also used as cover crop to check soil erosion in problem areas.

Origin and distribution

Lentil as a cultivated crop is of very ancient origin. Its native home is regarded to be the countries in eastern Mediterranean littoral, such as Asia Minor, Greece, Egypt, from where it extended eastwards to India (Aykroid and Doughty, 1964). The valleys in the Hindukush Mountains in Asia are also claimed as its native home (Decandole, 1904). Its cultivation dates back to pre-historic times in these regions. It was later introduced in Europe, Africa and America. Probably it is the first food legume that ever been cultivated as evident from the earliest archeological finding from the Paleolithic and Mesolithic layers of Franchthi caves in Greece and dates back to almost 13000 to 9500 years ago (Cubero, 1981 and Zohary and Hopf, 2000). Thus, its cultivation dates back to beginning of agriculture itself. It was known to ancients in Egypt and Greece. It had spread to Europe, India and China and now it is introduced and cultivated in most sub- tropical and warm temperate regions.

Status of lentil production

Pulses are the second most important group of crops after cereals. Globally, lentil occupies 3.3 million ha area with annual production of 2.82 million tones and shares only 5% of the total area under pulses. India ranked first in the area and second in the production with 36 and 21 per cents of global area and production, respectively. Highest lentil productivity in the world is recorded from Armenia (2181 kg/ha) followed by China while India ranked 30th however, Canada ranks first in production due to very high level of productivity

(1482 kg/ha) as compared to India (Tiwari, 2008). Indian is the homeland of world largest vegetarian population and world leader in pulses production even though it imports large quantities of pulses to provide protein supplements to its burgeoning population. In India, lentil is the second most important winter pulse crop after chickpea. The area, production and productivity of lentil during 2011-12 were 1.56 million hectares, 1.06 million tones and 678 kg/ha, respectively while the growth rate during 2007-08 to 2012-13 for the same was observed as 0.1, 3.8 and 4.0 percents, respectively (Anonymous, 2013). Being a cool season crop, its production base is mainly confined to northern and central India. Uttar Pradesh is ranking first both in area and production (36.5 and 48.1 percents, respectively) followed by Madhya Pradesh (39.7 and 21.7 percents, respectively) and Bihar (10.9 and 16.4 percents, respectively) and jointly these states share about 87 and 86 percents of country's lentil area and production respectively however; highest lentil productivity (1333 kg/ ha) is reported from Rajasthan (Anonymous, 2013). Further, lentil occupies 3 lakh ha area in Indo Gangetic Plains under rice- lentil sequential cropping system, being second most important cropping system after rice –wheat (Ramakrishna *et al.*, 2000 and Singh *et al.*, 2013). The Indo Gangetic Plains of north India shares about 55 percent of national lentil area and contribute 66 percent to national production with satisfactory level of productivity (839 kg/ ha) compared to national average (697 kg/ha), even though they faces several bio-physical constraints (Singh and Singh, 1995 and Singh *et al.*, 2013).

Botanical description

Lentil (*Lens culinaris* Medick.) is an annual herbaceous plant mostly erect and bushy type with four to six primary branches. Plant height does not exceed 50-60 cm, in general. It has a well-developed root system including a central tap root with several lateral branches, spreading in all directions. Root nodules, the site of biological fixation of atmospheric 'N', are mainly concentrated on primary root however, some of them also spread to lateral roots. The stem is weak and quadrangular while ends of leaflets some time form tendrils. Inflorescence is a raceme of two to four flowers. Flowers are small and white with blue, violet or pink tinged. Ovary is short with one or two ovules hence, pods are one to two seeded. Anthesis takes place in buds sometimes before opening of flowers in the next morning. Hence, self-pollination is a general rule. Pods are flattened and broad about one to one and half cm in length with a minute curved beak. The seeds are lens shaped, light red, brown, grey and of tan speckled with back. Germination is of hypogeal type and the plant has 14 chromosomes (2n=14) (Singh *et al.*, 2008). *Lens* is a latin world which refers to the shape of the lentil seeds

Production technology

The average national yield of lentil (675 kg/ ha) is much below than the global average (845 kg/ ha because of poor crop management and often late planting. The depleting soil moisture conditions and absence of irrigation water, lead to the withering up of the crop towards maturity resulting in poor yield. Owing to its hardiness, particularly for drought, lentil is most popular pulse crop in the rainfed areas of eastern and central parts of the country. It is grown both as upland crop and also under paira or *utera* condition in rice fallows under residual moisture and without much additional inputs (Jeswani and Baldev, 1990). It has assessed that availability of quality seeds along with critical input in time can uplift about 30% production even with present varieties. However, lentil is also grown as a paira crop in the eastern India, which helps in timely planting of the crop even before, the paddy has been harvested (Singh *et al.*, 2008 and Singh *et al.*, 2013). Substantial increases in the production of lentil in the IGP can thus be achieved by fine tuning management aspects and thereby increasing total system productivity and sustainability (Ramakrishna *et al.*, 2000 and Singh *et al.*, 2013).

Soil requirement

In India lentil is grown on a variety of soils ranging from sand dunes in Rajasthan to deep vertisols in Maharashtra. However, well drained loamy soils with neutral in reaction are best for lentil. Lentil is also being grown successfully in low lying paddy soils of poorer type, even tolerate moderate alkalinity conditions. Poor soil condition not only compel late sowing of lentil, which leads to reduced length of growing period but also necessitate to sustain cold injuries at early vegetative phase which freeze all biological activities for prolonged period. A sudden rises in temperature after that, not only induces forced maturity but simultaneously invites several biotic stress viz., diseases and insects pests (Ali *et al.*, 2012). Lime treatment (either seed pelleting or soil amendments) is must, prior to its cultivation in acidic soil. Correction of soil pH has a major role in nutrient and water use efficiency and consequently on crop yield. Use of gypsum in western states and liming in eastern and parts of southern states at the rates prescribed by SAUs/ICAR research centers located in the region is strongly recommended (Nigam *et al.*, 2004 and Singh *et al.*, 2013).

Climatic requirements for lentil

Being a winter season crop, lentil requires cold climate. The plant is hardy and can tolerate frost and severe winter to a great extent. In areas where winter are sufficiently cold, particularly in northern plains and on the lower Himalayas, lentil is preferred over chickpea due to its relatively better tolerance to frost and winter rains, even at flowering and fruiting stage. When temperature reaching at freezing point, during flowering, the crop is damaged severely due to cold injury (Jeswani and Baldeb, 1990). The range of lentil cultivation with regards to climate is very wide and extends from sea level upto an altitude of 3500 m in North West hills. It requires cold temperature during vegetative growth (18-30°C) and comparatively warm temperature during maturity, with the optimum temperature range of 25-30°C. The growth of lentil crop is adversely affected at temperatures above 27°C. Occurrence of mid-season cold waves and terminal heat during winter causes losses to crop productivity of in many lentil growing regions. Hence, it is grown as a winter season crop in semi-arid tropics.

Selection of appropriate varieties

Availability of high yielding varieties insulated against major biotic stresses had increased the scope of growing lentil in cereal based cropping systems both in rainfed and irrigated areas. In India, normal sown lentil is a medium duration (130-150 days) crop, while under late sown conditions it is forced to complete its life cycle in 105±5 days (Jeswani and Baldeb, 1990; Ali *et al.*, 1993; Joshi, 1998 and Ramakrishna *et al.*, 2000). Typically, late sown lentil undergoes three distinct phases and considerable degrees of phenological modifications are bound to happen. Eventually, lentil crop during its early seedling phase grows slowly due to its energy invested in the initial establishment however, in mid-phase, very insignificant growth and development is observed (Singh *et al.*, 2002). This poses serious threat to realization of yield potential due to cold injuries. This phase is very important for creating source of channelizing the energy at later stage.

In the last and most important phase lentil faces heat injury, resulting in early onset of reproductive phase, causing imbalance in resources and inputs, biotic stress and forced maturity (Jeswani and Baldeb, 1990; Joshi, 1998 and Dixit *et al.* 2009). Further, availability of short duration and fertilizer responsive varieties offers great promise for introducing lentil in new as well as existing cropping system.

Based on the seed size and test weight the varieties released for commercial cultivation in India and abroad are broadly classified in two main groups :

i) **Bold seeded**: It includes sub sp. *macro-sperma,* with test weight of more than 25 g (Fig.9.1). Locally it is also known as *Masur* or *Malka Masur.* In India this type of lentil is mainly cultivated in Bundelkhand regions of UP and MP and in Maharashtra state. They have large size seeds with 6-9 mm in diameter which is more flattened. Cotyledons are yellow with pale green testa. Globally this group is concentrated in Mediterranean, Spain, Greece, Canada and North and Latin America.

ii) **Small seeded**: It includes sub sp. *micro-sperma,* with test weight of less than 25 g (Fig.9.2). Locally this type of lentil is also known as *masuri* and primarily grown in Indo Gangtic Plains of NEPZ (UP, Bihar, West Bengal and Assam). This group of lentil dominated for Utera cropping in rice fields and most suitable under delayed planting. Seeds are round with 2-6 mm in diameter. Cotyledons are yellow or orange in colour however, testa colour ranging from yellow to black. These are predominantly grown in India, Nepal, Afganistan, Bangladesh, Ethiopea and Egypt.

Based on the region, time of sowing and purpose of cultivation, the bold and small seeded varieties recommended for different regions/ states of the India. Some situation specific varieties are :

i. *Utera cultivation* - PL-406, PL-639, Arun

ii. *Late sowings* – PL-406, PL-639, IPL-15, Narendra Masoor1

In general, small seeded lentil varieties are preferred for late sowing conditions however, some of the bold seeded varieties also perform better in normal as well as in late sown conditions i.e., Narendra Masoor 1 while DPL15 perform well only in late sown condition of *tarai* regions of Uttrakhand (Singh and Singh, 2002, Tiwari *et al.,* 2008; Dixit *et al.,* 2009.

Fig. 9.1 Bold seeded lentil

Fig. 9.2 Small seeded lentil

Cropping systems

Lentil is grown under different cropping system such as mono cropping, double cropping, relay (*utera*) cropping and mixed or intercropping. The choice of particular system depends on annual rainfall and its distribution, edaphic conditions, irrigation potential and domestic needs and trade requirements. Lentil is more or less shade loving crop so it performs well as intercrop with other crops. Growing lentil with many other rabi season crops is also practiced as a insurance to avoid complete crop failure due to epidemic of any insect pest or diseases and meeting most of the family needs from the limited available land. Lentil is more or less shade loving crop so it performs well as intercrop with other crops. Sowing of lentil seeds by mixing with mustard was quite common during recent past however, now a day's lentil is grown in a definite row arrangement with other crops *viz.*, two rows of lentil after every four rows of wheat or barley or four rows of lentil in between two rows of widely spaced mustard. In north - eastern plains, lentil is also grown as *utera* crop after rice. The seeds broadcast in standing crop of rice just before its harvest (Jeswani and Baldeb, 1990 and Singh *et al.*, 2008).

Crop establishment

Seed is a vital input for realizing high yield. Healthy seeds provide good plant stand and establishment. Seeds should be procured from authorized agencies and change after every 3-4 years to avoid their degradation through varietal mixture and out crossing. Certified seeds meet all the requirement of quality seeds (Nigam *et al.*, 2004).

Field preparation

Tillage requirement for a successful lentil crop depends on soil type, moisture content, preceding crop and weed infestation before planting. Since lentil is generally grown on residual soil moisture under rainfed conditions, effort should be made to conserve late shown showers. The soil of lentil fields should be made friable and weed free so that seeding could be done at a uniform depth. On heavy soils, one deep ploughing followed by one- two cross harrowing should be given. However, light sandy soils require less tillage for better seeding and soil moisture conservation (Ali and Mishra, 2000 and Singh *et al.*, 2008). After harrowing the lentil field should be leveled by given a gentle slope for ensuring uniform irrigation. However, for *utera/paira* cropping of rice-lentil in eastern Uttar Pradesh, Bihar, Orissa, Madhya Pradesh, cropping no tillage is done and seeds are broad casted in standing rice crop about 8-10 days prior to its harvest. There should be proper moisture in the soil at the time of sowing for proper germination of seeds. Lentil requires good aeration for nodule development that could be achieved by one deep ploughing followed by one cross harrowing (Jeswani and Baldeb, 1990 and Ali *et al.*, 1993). Soil should be made friable and weed free so that seed could be placed at a uniform depth. There should be proper moisture in the soil at the time of sowing for proper germination of seeds.

Sowing time

Timely planting is the key factor in realizing full potential of improved varieties of lentil. Recommended sowing time for rainfed regions of Central and South India is first fortnight of October and second fortnight of October in North India (Singh *et al.*, 2008 and Ramakrishna *et al.*, 2000). However, prevailing temperature is the major limitation for October sowing of lentil as if temperature is more than 25^{0}C, chances of wilt and root rot will be more that may results in poor germination and under plant population. Under irrigated condition of *Indo Gangetic Plains* of northern India first fortnight of November is ideal sowing time for lentil. Delay in planting causes reduction in yield and the magnitude of reduction is large after15th November in rainfed areas and after 15 December in irrigated areas especially in rice fallows of NEPZ (Ali *et al.*, 1993; Ali and Mishra, 2000, Ali, 2009 and Singh *et al.*, 2013). However, late sowing by the time of first week of December in the rice fallows of *Indo Gangetic* North Eastern plains or in fields vacated very late by kharif crops is quite common under irrigated condition. Lentil sowing get delayed either due to late harvest of preceding long duration rice or poor working condition of soils due to its high moisture content either by flooding (Singh and Singh, 2008 and Singh *et al.*, 2013). Consequent upon delayed planting, lentil encounters with severe cold in early vegetative phase and its growth and development gets hampered for a considerable period (Singh *et al.*, 2012). Further, the crop gets comparatively less time to complete its lifecycle due to forces maturity (Ramakrishna *et al.*, 2000). The reduction in yield could be minimized up to certain extent by relatively closer spacing, use of higher seed rate and selecting the varieties those have fast initial vigour and shorter in duration, tolerant to low temperature during early vegetative stage and high temperature stress during reproductive phase with high harvest index (Jeswani and Baldeb, 1990; Ali and Mishra, 2000 and Singh *et al.*, 2008). In Madhya Pradesh, sowing of the crop near 15 October was the more advantageous while delay in sowing after 30th October affected the crop adversely. In West Bengal, second fort night of October to first fortnight of November is reported as ideal sowing time for lentil while in Punjab and Haryana, highest crop yield was obtained by sowing the crop before 30th October (Das, 2005). In central zone, where moisture is a limitation, early planting in mid-October is recommended to ensure proper germination. In north-west plains zone, end of October and in north-east zone, second fortnight of November is the optimum time of sowing under adequate moisture (Jeswani and Baldeb, 1990 and Ali *et al.*, 1993).

Seed treatment

Treat the seed with thiram or carbendazim @ 2 g/kg of seed before 3 days of sowing followed by inoculate the seeds with dual culture of *Rhizobium* and phosphorus solublizing bacteria (PSB). One packet weighing about 250g of both the culture is sufficient for 8-10 kg seeds, and thus for one ha sowing a farmer will need about 4-5 packets of both the culture. These cultures are also available with State department of Agriculture at subsidized rates of Rs. 10/packet. PSB bio fertilizer is also known as *Phosphotika*.

The seeds of lentil should be inoculated with dual culture of Rhizobium and PSB by the following simple method:

a) Prepare jaggery solution by dissolving about 50-60 g jaggery in 400-500 ml water by boiling it.

b) Constantly stir the solution with a stick during boiling.

c) Once jaggery is completely dissolved, let it cool down at normal room temperature.

d) Open one packet each of Rhizobium and PSB in the jaggery solution and mixed it properly to make homogenous cultural slurry.

e) This solution is sufficient for 10 kg seeds only.

f) Pour this cultural slurry over the heap of seeds to be treated and mix the seeds homogenously.

g) Place a thin coating of the culture over every seeds

h) Spread the seeds over clean floor/plastic sheet/on clean gunny bag for drying under shade.

The seeds should not be exposed to direct sunlight before sowing because it will decrease the population and activity of Rhizobium and PSB (Jeswani and Baldeb, 1990 and Nigam *et al.*, 2004).

Seed rate and plant spacing

An ideal seed rate for normal sown small seeded crop is 40-45 kg/ha whereas for late sowing or when bold seeded varieties are used, the seed rater should be increased 50-60 kg per hectare (Singh *et al.*, 2008). For *utera* sowing, only small seeded varieties should give preference as their chances to hold in standing rice is less and come in contact with soil early due to higher surface resulted in higher germination percentage and establishment of a good lentil crop. Higher seed rate upto 50-60 kg/ha for utera as well as zero till sowing of lentil should be used as in both the cases germination is poor either by retaining many seeds on rice stover or not placing all the seeds at suitable soil depth with sufficient moisture. When lentil is grown as an intercrop with wheat/barley, mustard or any other rabi crop, seed rate should be adjusted according to area available for lentil sowing. Generally 15-20 kg seeds for small varieties and 20-25 kg seeds for large seeded variety are required for sowing lentil in one ha of land in intercropped system. Under normal planting, lentil should be sown at 30 cm row spacing. However, row spacing should be reduced to 22.5 cm to 20 cm in case of late planting after second fortnight of December. Jeswani and Baldev (1990) and Ali *et al.* (1993) reported that lentil could be sown as late as December but seed rate need to increase upto 40 kg ha^{-1} and row to row spacing should be 20-25 cm (Ali and Mishra, 2000).

Method of sowing

The seeds of lentil are sown either as broadcast or in rows at a narrow spacing of 22.5 cm to 30 cm with the help of seed drill (Fig.9.3 and 9.4). Sowing the seeds in furrow made by country plough is a very common method in rainfed regions. However, in irrigated conditions, sowing in rows made by animal or power drawn seed cum ferti drill is now very common. Seed cum ferti drill helps to place the seeds at uniform spacing besides simultaneously dropping the fertilizers at right place and right depth for their maximum efficiency. In lowland paddy areas where working with farm machinery is not possible due to high moisture content or very delay in sowing is expected because of long duration rice cultivars broadcasting the lentil seeds 8-10 days before rice harvest as *utera* cropping is the best option. Lentil can also be sown in standing rice stubbles with the help of zero seed drill developed by PAU, Ludhiana, GBPUAT, Pantnagar and many other such research institutes. Zero seed cum ferti drills place the seed and fertilizers at right place and right depth and cover the seeds in single operation without any preparatory tillage and avoid much delayed sowing of lentil due to various reasons.

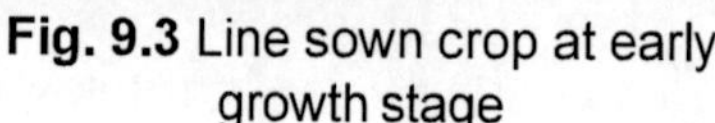

Fig. 9.3 Line sown crop at early growth stage

Fig. 9.4 Line sown crop at later growth stage

Integrated management of plant nutrient supply

Among various constraints identified for low productivity, inadequate and imbalance mineral nutrition play key role in exploiting the genetic potential of pulses since they are grown on marginal and sub marginal lands which are impoverished in soil nutrients with negligible fertilizer use as most of the pulse growing farmers are resource poor. Though nutrient requirement of lentil is much lower than cereals primarily due to biological nitrogen fixation and relatively low productivity levels although lentil crops respond favourably to higher doses of fertilizer nutrients than generally applied or even recommended (Fig.8.5) However, since lentil is invariably subjected to abiotic stresses leading to sub-optimal nutrient uptake and farmers tend to use low doses of fertilizer nutrients. Further, nutrient use is unbalanced and seldom based on soil-test values. Pulses in general, removes 30-50 kg N, 2-7 kg P, 12-30 kg K, 3-10 kg Ca, 1-5 kg Mg, 1-3 kg S, 200-500 g Mn, 5g B, 1g cu and 0.5g Mo (Ravichandran, 2009).

Generally lentil is grown without any fertilizer on residual soil fertility. The nitrogen (N) requirement is largely met through BNF by Rhizobium however, about 40% of pulses growing regions have low to medium population of native rhizobium. Further, in early stages of plant growth when rhizobium is not well established, a starter dose of 10-20 kg N ha^{-1} is necessary for better crop establishment (Jeswani and Baldeb, 1990). The methods, time and rate of N to be applied without any adverse effects on symbiotic N fixation, plant growth, economy, yield and soil fertility is important (Jeswani and Baldeb, 1990).

Phosphorus (P) deficiency in soil is usually the key factor for poor yield of pulses in all type of soils as they have higher phosphorus requirement. Hence, phosphorus application to pulses becomes the basis to fertilize them. For lentil the recommended P level varies from 40-60 kg/ha depending on the soil type and yield potential of the variety (Ali and Mishra, 2000, Ganeshmurthy *et al.*, 2003, Ali, 2009 and Ravichandran, 2009). It also responds to potassium (K) where the soils are deficient or medium in K status. The application of K regulates the utilization of other nutrient in plant system. When soil application of fertilizer is not possible especially in rice fallow, foliar fertilization of 2% DAP and 1% KCl is recommended at 30 and 45 days after sowing , respectively (Ravichandran, 2009). Lentil also responds to sulphur (S) application upto 40 kg/ha giving an average nutrient use efficiency of 10-15 kg grains /kg S especially in light textured sandy loam soils of Northern India. For supply of both P and S, single super phosphate (SSP) is the best source while for S alone, gypsum and pyrite are the more economical, easily available and efficient source (Ahlawat and Ali, 1993).

Fig. 9.5 Supplementing sulphur and zinc: Key to optimize lentil production

Deficiency of micronutrients especially zinc (Zn) and S is one of the limiting factor for obtaining derived response to applied P in pulses (Ali and Mishra, 2000 and Singh and Bhat, 2013). Pulses crops are often exposed to multi-micronutrient deficiencies causing yield reductions to the tune of 10-40 percent depending on the severity and stage of occurrence. In light textured, low organic matter and calcareous soils deficiency of boron (B) and iron (Fe) also occurs. Unless correcting measures through the application of these nutrients are taken, response to applied P cannot be fully realized. By investing small amount in correcting the deficiency of these nutrients farmers can get a quantum jump in their profits. Very encouraging response to application of S and Zn is found with cost: benefit ratio of 10-21%. As B and molybdenum (Mo) are required by rhizobia, their application increases the yield of several pulse crops besides increasing the biological n fixation. The mean response to applied Zn and B varied from 160-460 and 60-420 kg/ha respectively (Ali and Mishra, 2000). In rice - lentil crop sequence, deficiency of Zn in rice and S in lentil are frequent and reduction in yield of rice and lentil is often blamed to zinc sulphur deficiency (Singh *et al.*, 2002 and Singh *et al.*, 2008). Among micro-nutrient, Zn is most critical when lentil is grown after rice in intensive Rice based cropping areas of Punjab, Haryana, Rajasthan (Eastern) U.P. and Bihar where application of 25 kg zinc sulphate as basal or a foliar spray of 0.5% ZnSO4 + 0.25% lime (5 kg zinc sulphate + 2.5 kg lime in 1000 lt. of water for one ha area) is a general recommendation. Molybdenum (Mo) and Iron (Fe) are the integral components of enzyme 'nitrogenase' responsible for 'N' fixation. Hence, Mo deficiency may create twin deficiency of 'N' and 'Mo'. Boron is important for proper grain filling. Both Mo and B are found deficient in acidic soil of Eastern India hence about 10 kg borax and 1 kg ammonium molybdate should be applied in the deficient region as soil application. A foliar spray of 2% each of di-ammonium phosphate (DAP) and potassium chloride (KCl) at pre flowering and pod development stages, respectively enhance lentil grain yield by 10-15% along with increasing its ability to resist terminal drought (Ali and Mishra, 2000 and Singh *et al.*, 2008).

Method of fertilizer application is a non- monetary input which influences growth and consequently the crop yields. Pelleting of seeds with superphosphate, sodium molybdate and soaking of seeds with 8% $MnSO_4$ solution results in higher yield of blackgram and green gram in deltaic region (Ravichandran, 2009). The time gap between the vegetative and reproductive stage in pulses is small and has low portioning efficiency. To meet the increasing demand of nutrients at this stage, foliar application of macro and micro nutrients had been proved beneficial. Foliar applied nutrients usually penetrate the leaf cuticle or stomata and enter the cells facilitating easy and rapid utilization of nutrients. Numerous evidences have indicated the beneficial effect of foliar application of nutrients by influencing the crop both in quantity and quality. The subject of feeding plants though foliage has gained much importance in recent years as an efficient and economical way of supplementing the macro and micro nutrients, because it avoids losses by fixation in soils. It is advantageous over soil application in specific occasions in certain crops *viz.*, deficiencies of Fe, Zn and B are unaviodable in calcareous soils due to precipitation of Zn and Fe as insoluble carbonates and Ca-B antagonism. In soils with high indigenous P supply, Zn and Fe deficiency are quite common. Some of the other nutrient antagonisms commonly encountered are Cu-Mo, Cu-Fe, K-Mg etc. On such occasion, soil application of zinc sulphate or borax may not be useful. Thus, specificity of crops, soil conditions and specific occasions in certain crops in rice fallows warrants foliar application of nutrients (Ravichandran, 2009).

A low external input nutrient supply strategy for getting optimal yield of lentil could be summarized as under-

- Application of 2-2.5 t 'vermicompost' or 5 t FYM to previous 'kharif' crop in rotation takes care about initial nitrogen requirement and no need to apply 'N' as booster dose (required especially in low fertile and puddle paddy soils).
- Dual inoculation with 'Rhizobium' and 'PSB' takes care of 'N' as well as reduces 25-30% of phosphorus requirement by making available the initial fixed soil 'P' to the plants.
- Rhizobium inoculation is must after paddy as these bacteria are aerobic in nature and most of their population dye during flooding and compaction for want of oxygen.
- In-situ management of rice straw/residues takes care of Zinc and other micronutrient and no need to apply them separately (Tiwari, 2008).

Nipping

In case lentil is sown with the seeds of old indeterminate type of varieties with luxuriant plant growth, the shoot tips are nipped to arrest apical dominance, encourage branching, uniform flowering and thereby increasing the crop yield (Das, 2005).

Water Management

Lentil can be raised successfully with the moisture conserved in the soil during the monsoon season. It is moderately drought tolerant crop and requires 200 mm of water depending on soil and climate. Lentil requires 1-2 irrigations depending upon the rains during the growing season (Fig.9.6). First irrigation is applied 6 weeks after sowing and the second at flowering or pod formation stage. Scientific scheduling of irrigation, an estimate of quantity of water to be applied and deployment of water saving irrigation methods can lead to enhanced yield, higher water and nutrient use efficiency and larger area coverage under irrigation. Use of sprinkler irrigation has enormous potential for saving irrigation water and expanding area under irrigation (Das, 2005). However, lentil is mainly grown under rainfed conditions on residual soil moisture except in few districts of Uttar Pradesh, Madhya Pradesh, Rajasthan and Bihar. As a consequence area and productivity of lentil is dependent on amount of residual soil moisture and consequently faces severe moisture stresses during flowering to grain filling. The moisture stress during this critical period resulted in severe reduction in lentil productivity.

Fig. 9.6 Efficient management of water leads to better lentil output

Jeswani and Baldev (1990) reported that modern varieties of lentil respond to irrigation and pre sowing irrigation is must if residual soil moisture is limiting. They also reported that in case of failure of rainfall during rabi season, one or two irrigations are needed, one before flowering and the other at pod filling stage. Ravichandran (2009) also reported the possibilities of quantum jump in productivity can be achieved by applying water during this critical period. Sprinkler irrigation can be used for most efficient use of scare irrigation water especially in central zone of India (Das, 2005). It can tolerate drought to some extent however, wherever irrigation facility is available, one to two irrigations particularly in absence of winter rains should be provided for getting higher yield. First irrigation should be given at 45 days after sowing and second if needed, at pod filling stage. However, more irrigation may affect the crop performance adversely (Ali and Mishra, 2000, Singh *et al.*, 2008 and Ali, 2009). In absence of winter rains and where contribution of soil moisture is negligible *viz.*, in Central India, two light irrigations may be applied for significant yield improvement. However, water stagnation due to poor drainage in heavy soils can results in heavy loss in production on account of low plant stand and increased incidence of diseases (Ravichandran, 2009). Three irrigations applied at 4-6 leaf stage, branching and pod filling produced higher grain yield of lentil on light textured soil reported at New Delhi (Vankateshwarlu and Ahlawat, 1993) while at Hisar, two irrigations, one at 45 DAS and another at early pod filling stage gave maximum grain yield and net profit. At Kanpur, irrigation at early pod filling stage was found more critical (Das, 2005).

Weed Management

Weeds compete with the crops for essential nutrients, space, soil moisture and light, and thus considerably reduce the yields by 40-60%. Lentil is suffered badly by severe competition from weeds during early crop growth as crop growth is very slow during this period. The period from 30-60 days after sowing is most crucial for completion with weeds. Therefore, a weed free period of early 45-60 days is important as thereafter, weeds could not do much harm since the inter row spaces will be covered by crop plants (Fig.9.7). This could be achieved either through mechanical or chemical means. Major weeds are *Chenopodium spp.* (bathua), *Fumaria parviflora* (gajri), *Lathyrus aphaca* (chatri matri),*Vicia sativa* (ankari), *Crisium arvense* (kateli), *Melilotus alba* (senji), *Asphodelus enuifolius* (jungli piaji), *Convolvulus avensis, Phalaris minor* and *Avena ludoriciana* (Singh *et al.*, 2008). Orobanche, a parasitic weed is also seen as major problem at some places. Similarly *V. sativa* adultrate the grain due to its size, shape and colour. Narrow rows of these crops make mechanical weeding out of question, although hand weeding is possible in case of thin density and sporadic weed problem. Lentil is mostly cultivated as dryland crop where weed growth is limited and requires only one manual weeding to obtain acceptable crop yield. One hand weeding/ inter-culture at 30 DAS and another at 55-60 days after sowing, depending upon the intensity of weed infestation, provides efficient soil oxygen environment to Rhizobium bacteria along with soil moisture conservation breaking soil capillaries, creating dust mulch. This coupled with timely weed control by pre-emergence herbicide (pendimethalin@1.5 kg a.i./ha etc.) and one hand weeding after 25 DAS can enhance pulse productivity in the range of 15-32% (Ravichandran, 2009).

Use of herbicides under such conditions is advantageous as the not only econmical but also provides timely protection. However, some important considerations have to be borne in mind; the herbicides besides being selective to lentil should not be harmful to the soil microflora that build up soil fertility and to the nodule- inducing organisms which fix atmospheric N and the grains which would be consume as 'Dal' should not contain any residue that may prove toxic to man and animal besides no toxic effect for subsequent crop in rotation. Keeping all these in view, the following three categories of herbicides are recommendations for weed management in lentil after preparing a solution of 500-600 l of water for one ha area.

Fig. 9.7 Weed management: Way to maintain healthy and optimum plant population

a) Pre-plant incorporation: fluchloralin (0.72 l *a.i.* Jeswani and Baldev, 1990).

b) Pre-emergent: metolachlor, metribuzin and pendimethalin (0.5kg a.i./ha), linuroon (0.5-1.0 kg a.i./ha), trifluralin (0.5-1.0 kg *a.i.*/ ha) and isoproturon (0.5kg a.i./ha) (Jeswani and Baldev, 1990)

c) Post emergence quizalofop-ethyl (40-50g/ha) and clodinafop-propargyl(60g/ha) (Gupta, 2011).

Plant protection measure

The major diseases of lentil are rust in northern plains and wilt in central zones. Use of resistant varieties is helpful in controlling both the disease. However, insects –pests are not a major problem in lentil. However, pod borers at the time of pod formation defoliates the plant, bores the green pods and feed upon the ripening grains and reducing the yield and quality of lentil grains. The caterpillars of cutworm are another pest of minor importance which cut the plant during night and drag the removed plant parts in soil. In some years aphid population also rises due to increasing temperature during January or February, and causes heavy damage. Integrated pest management strategy involving use of resistant varieties, disease free seeds, manipulation of cultural practices, and application of biological and chemical control methods will help in overcoming the losses. To prevent seed borne diseases pre sowing seed treatment with fungicides with bavistin, carbendazim or thiram @ 2g kg^{-1} of seeds is to be resorted.

Further, resource conservation technology in planting, irrigation and manuring offer great promise for improving pulse production and reduce the cost of production. These technologies need to be popularized extensively in pulse growing regions. A brief about the major insect pest and diseases and their control measures are given in table9.1.

Table 9.1 Management of major insect pests and diseases of lentil

Insect pest/ disease/causal organism	Nature of damage/ symptoms	Control measures
Insect - pest		
1. Pod borers	The caterpillar defoliates the tender leaves and also bores the green pods and feeds upon the ripening grains	Cypermethrin (0.02%) or Endosulfan (0.07%) or Monocrotophos (0.04%).
2. Aphids	Aphids suck the sap and in case of severe damage the growth is suppressed	Metasystox or Monocrotophos (0.04%).
Diseases		
1. Wilt *(Fusarium lentis)*	The growth of the plant is checked due to yellowing of leaves, drying of plants. The roots of affected plants remain under-developed and looks light brown in colour.	i) Seed treatment with Thiram + Benomyl (1:1) @ 3 g/kg of seed. ii) Adopt crop rotation. iii) Use healthy seeds
1. Rust (*Uromyces fabje*)	Pink to brown pustules appear on leaves and stems. In severe attack, the affected plants may dry.	Grow early maturing/ duration variety. ii) Seed Treatment with Agrosan GN @ 2g/kg seed. iii) Spray the crop with Maneb, Zineb or Ferbam @ 2.5

Harvesting

Lentil in general matures in 120-130 days and gets ready for harvesting from February – March depending on the time of sowing Harvesting at appropriate time is necessary to avoid post-harvest losses.

The maturity of lentil varies considerably from region to region even for the same date of planting on account of atmospheric temperature, humidity, photoperiod, and moisture in soil profile etc. Crop become ready for harvest when leaves begin to fall, stem and pods turn brown or straw in colour and seeds are hard and rattle with 15% moisture inside them. Over ripening may lead to fall of pods as well as shattering and seed cracking if seed moisture fall below 10% due to delay in harvesting (Singh, *et al.*, 2008). In India harvesting is generally done with sickles (Fig.9.8). However, in Canada, Turkey and other advance countries mechanical harvesting-cum threshing is done by power operated combine machine.

Fig. 9.8 Timely harvest, can save post harvest losses

Fig. 9.9 Bundled lentil crop ready for threshing

Threshing

The harvested plants are carted to threshing floor (Fig.8.9) and are allowed to dry for 4-7 days on threshing floor and threshed by manually or bullock/power drawn thresher. The seeds should be cleaned by winnowing and dried in hot, sunny air to reduce their moisture content for their safe storage.

Yield

A well-managed crop yields approximately about 1.8-2.0 tones grain and 3.0-4.0 tones of straw per hectare. The average lentil yield varies from 15-20 quintals per hectare under better management conditions in soils having better fertility status, good water holding capacities and endowed with protective irrigation. However, it ranges only 5-10 quintals under dry rainfed regions of central India. A well-managed crop yields about 1.8-2.0 tonnes grain and 3.0-4.0 tonnes/ha of straw.

Storage

The clean seed should be sun dried for 3-4 days to bring their moisture content at 9-10%. The seed should be safely stored in appropriate bins and fumigated to protect them from bruchids.

Conclusion

Lentil being an environment friendly crop has the unique ability to fix nitrogen and thereby help in improving soil health. Even though, they have low genetic potential for realizing productivity as compared to cereals, their contribution in protecting the environment make them an inevitable component of sustainable cropping system. A large chunk of rice fallow can be brought under lentil provided available land and water resources (soil moisture) are scientifically and innovatively managed by adapting improved agronomic practices. The agronomic practices that have major impact on productivity of lentil includes field preparation, planting time, seed treatment with fungicides and crop-specific bacterial cultures for ensuring optimum plant population, integrated plant nutrient supply, water management and avoiding undue comptetion between crop and weeds. Further, updation of *Utera* cultivation practices traditionally followed by tribal farmers in the states of Madhya Pradesh, Chhattisgarh, Bihar and Jharkhand can help to enhance national pulse production besides giving some additional income to tribal rice growers. Explring lentil cultivation in promising intercropping of rainfed/partial irrigated areas further expands the opportunity for enhancing lentil production. Finally, farmers need to encourage through various incentives and region specific extension strategy in the identified potential districts through demonstration of improved technology, providing subsidized inputs, credit support and extending crop insurance through appropriate policy interventions.

References

Ahlawat IPS, Singh A and Saraf CS.1981. Effect of winter legumes on nitrogen economy and productivity of succeeding cereals. *Experimental Agriculture* 17: 57-62.

Ahlawat IPS and Ali M. 1993. Fertilizer management in food crops. In: Nutrient Management for Pulses - A Review (Tandon HLS, ed.). FDCO, New Delhi. pp. 114-138.

Ali M. 2009. Pulses. In: Handbbok of Agriculture (ed.). Indian Council of Agricultural Research, New Delhi. pp. 1087-1119.

Ali M and Mishra JP. 2000. Technology for winter pulses. Indian Institute of Pulses Research, Kanpur (India).

Ali M, Saraf CS, Singh PP, Rewari, BB and Ahlabat, IPS. 1993. Agronomy of lentil in India. In: Lentil in South East Asia (Erskine W. and Saxena, M.C., eds). ICARDA, Allepo, Syria.

Ali RI, Awan TH, Ahmad MM, Saleem U and Akhtar M. 2012. Diversification of rice-based cropping systems to improve soil fertility, sustainable productivity and economics. *J. Animal & Plant Sciences* 22(1): 108-112.

Anonymous. 2013. All India area, production and yield of lentil in India. In: Agricultural Statistics at a Glance-2013. Ministry of Agriculture, Directorate of Economics and Statistics, New Delhi.

Aykroid WR and Doughty J. 1964. Legumes in human nutrition. F.A.O. Nutritional Studies, No.19, Rome.

Cubero JI. 1981. Origin, taxonomy and domestication In: Lentils (eds C. Webb & G.C. Hawtin). Commonwealth Agricultural Bureau, Slough, England.

Das PC. 2005. Lentil. *In*: Pulse Crops of India. Kalyani Publishers, Ludhiana. Pp: 126-150.

De Candolle. 1904. Origin of cultivated plants. Kagun Paul, Trench & Co. London.

Dixit GP, Katiyar PK, Singh BB and Shivkumar. 2009. Lentil varieties in India. AICRP on MULLaRP. Indian Institute of Pulses Research, Kanpur.

Ganeshamurthy AN, Srinivas Rao CH, Singh KK and Ali M. 2003. Management of phosphorus for high pulse productivity in different agro-climatic regions of India. *Fertilizer News* 48(5): 23-42.

Gupta OP. 2011.Weed management practices in field crops and vegetables. In: Modern Weed Management (4th edn.). Agribios India, New Delhi. Pp. 369-432.

Jeswani LM and Baldev B. 1990. Advances in pulse production technology. Publications and Information Division, Indian Council of Agricultural Research, New Delhi. Pp. 190.

Joshi PK.1998. Performance of grain legumes in the Indo-Gangetic plain. In: Legumes in rice and wheat cropping systems of the Indo-Gangetic plain. (Kumarrao JVDK, Johansen C and Rego TJ eds.). pp. 3-13. ICRISAT, Patancheru, Andhra Pradesh.

Nigam RK, Chaturvedi SK, Ali M, Vishwadhar and Ahmad, R. 2004. Training manual on pulses production. Indian Institute of Pulses research, Kanpur. pp. 71.

Ramakrishna A, Gowda CLL and Johansen C. 2000. Management factors affecting legumes production in the Indo-Gangetic Plain. In: Legumes in rice and wheat cropping systems of the Indo-Gangetic Plain - constraints and opportunities. (Johansen C, Duxbury JM, Virmani SM, Gowda CLL., eds.). pp. 156-165. ICRISAT, Patancheru, Andhra Pradesh.

Ravichandran M. 2009. Pulse Production technology. *In*: Legumes: Their Production, Improvement and Protection. (M. Prakash and S. Murugan, eds). Satish Serial Publishing House, New Delhi. pp. 135-148.

Singh AK and Bhatt BP. 2013. Effects of foliar application of zinc on growth and seed yield of late-sown lentil. *Indian Journal Agricultural Science* 83(6): 622-626.

Singh AK and Singh NP. 2002. Performance of bold seeded lentil varieties under varying seed rates in normal and late sown conditions. *Indian Journal of Agronomy* 47(2): 227-230.

Singh AK, Bhat BP, Sundaram PK, Chndra N, Bharati RC, Patel SK.2012. Faba bean (*Vicia faba* L.) phenology and performance in response to its seed size class and planting depth. *Int. J. of Agril. & Stat. Sci.* 8(1): 97-109.

Singh AK, Manibhushan, Bhatt BP, Singh KM and Upadhyaya A.2013. An Analysis of Oilseeds and Pulses Scenario in Eastern India during 2050-51. *Journal of Agricultural Science* 5(1): 241-249.

Singh C, Singh P and Singh, R 2008. Lentil. In: Modern techniques of raising field crops. Oxford and IBH publication house, New Delhi, pp. 209-219.

Singh KM and Singh RKP. 1995. An Economic Analysis of Lentil Cultivation in N-E Alluvial Plains of Bihar. *Economic Affairs* 40(3): 157-163.

Singh SK, Varma SC and Singh RP. 2002. Integrated nutrient management in rice and its residual effect on lentil. *Indian Journal of Agricultural Research* 36(4): 286·289.

Tiwari AK. 2008. Lentil. *In*: Pulses in India- retrospect's and prospects (Tiwari, AK, ed.). Directorate of Pulses Development, Bhopal. Pp. 75-86.

Vankteswarlu U and Ahlawat IPS. 1993. Effcet of soil moisture regimes, seed rate and phosphorus fertilization on growth and yield attributes of late sown lentil (*Lens culinaris* L.). *Indian Journal of Agronomy* 38(2): 236-243.

Zohary D and Hopf M. 2000. Domestication of plants in the old world: the origin and spread of cultivated plants in West Africa, Europe and the Nile valley. Oxford University Press, New York, USA.

Citation

Singh AK. Singh AK Ray IP Lala, Ram V and NJ Singh. 2018. Agronomic Management of Lentil for Getting Higher Production. In: Scientific Lentil Production. Singh AK, Bhakta N, Sangale UR, Manibhushan, Sundaram PK, Kumar S and Yasin JK. (eds.) Society for Upliftment of Rural Economy Varanasi, India. Pp. 165-189.

❑❑❑

Chapter 10

Scientific Production Techniques of Lentil (*Lence culinaris* Madik)

Lal Singh, Parmeet Singh, Amar Jeet Gupta and Anil Kumar Singh

SYNOPSIS

A leguminous plant native to Southwest Asia, having flat pods containing lens-shaped, round, flattened, edible seeds is called lentil (*Lens culinaris*). The short, flat and oblong pods contain 1 or 2 seeds. Germination is hypogeal. Lentils are divided into two groups, according to their size: the large lentil (macrospermae) and the small lentil (microspermae). One of the most well-known varieties in the West is the round European lentil that has a biconvex disk shape and is green or brown in color, still with its skin. The Egyptian or red lentil has no skin and is smaller and more round. Lentils can vary in shape, color, texture and flavor, depending on the species. Grains are used as *dal* (whole or dehulled) and are also used in preparations, viz. *dal-moth*. Taking into consideration different ecologies, regions, countries and continents lentils are adapted throughout world. Its cultivation has been taken up by large, medium and small farmers mainly under rainfed but also in irrigated ecosystems. The lentil growers in different countries face the challenges of biotic and abiotic stresses more or less of the same magnitude which are responsible for the low productivity and stagnation in the production.

Keywords: Climate, Lentil, Production, Soil, Plant protection.

Introduction

Lentil (*Lens culinaris*) is leguminous annual plant with whitish or pale blue flowers. Its pods contain two greenish-brown or dark-colored seeds, also called lentils, which when fully ripe are ground into meal or used in soups and stews. It is one of the oldest crops that originated in near East and Mediterranean region. Probably indigenous to South west Asia and known to have been used as early as the Bronze Age, the lentil was introduced to Greece and Egypt before biblical times and was one of the first food plants cultivated in Europe. It had spread to Europe, India and China, and now it is introduced and cultivated in most sub-tropical and warm temperate region. It is grown throughout Northern and Central India (Singh *et al.*, 2011). It occupies 1.34 million ha and contributes 0.88 million tonnes to pulse production. It is mainly cultivated in Madhya Pradesh, Uttar Pradesh, Bihar and West Bengal, with 85 % of total production. Lentils are unusually high in protein content and are much used for food in Europe, especially by the poor, and increasingly in the United States. Many varieties are cultivated, for the seeds as well as for forage. Lentil seeds, from their shape, gave their name to the magnifying lens. The gulfweed (seaweed) is sometimes called sea lentil. Lentils are classified in the division Magnoliophyta, class Magnoliopsida, order Rosales, family Leguminosae. Lentils are relatively tolerant to drought, and are grown throughout the world (Singh *et al.*, 2015a). The FAO reported that the world production of lentils for calendar year 2009 was 3.917 million metric tonnes, primarily coming from Canada, India, Turkey and the United States. About a quarter of the worldwide production of lentils is from India, most of which is consumed in the domestic market. Canada is the largest export producer of lentils in the world.

Table 10.1 Top ten lentil producers of the world – 2009

Country	Production (tonnes)	Country	Production (tonnes)
Canada	1,510,200	Ethiopia	123,777
India	950,000	China	120,000
Turkey	302,181	Syria	102,461
United States	265,760	Iran	83,985
Australia	143,000	Bangladesh	60,537
		World	3,917,923

Fig. 10.1 Healthy optimum plant standoff lentil promising good reap at the end

Soil and climate

Lentil is grown on a wide range of soils ranging from light loamy sand to heavy clay soil in northern parts, and in moderately deep, light black soils in Madhya Pradesh and Maharastra. But clay loam soils are ideal for its cultivation. Its range of cultivation extends to as altitude of 3,500 m in North West hills. The optimum temperature for its growth and development ranges from 15 to 25^0 C. This crop is very sensitive to high temperature and high humidity.

Recommended varieties

The small seeded varieties resistant to rust are PL 406, PL 639, Asha (B 77), LL 147, Pusa Vaibhav, PL 4 and L 4147, and bold seeded varieties resistant to rust and tolerant to wilt are L 4076, LH 84-8, DPL 15 and DPL 62. Bold seeded variety JL 3 released from central zone is tolerant to wilt. Small seeded, tolerant to rust and wilt are KLS 218, HUL 57 VL masoor 103 and VL masoor 4. Bold seeded rust resistant varieties are Vipasha, Priya (DPL 15), Sheri (DPL 15) and PL 5. Bold seeded tolerant to rust are Sapna (LH 84-8), Arun, Malika (K 75). Shalimar Masoor -1. (Released for SKUAST-K). Recommended for pulse growing belts upto 1650 m amsl. Matures within 200-205 days. Resistant to frost and winter chilling. Resistant to root rot and leaf spot, grains are bold, macro-sperma type. Protein content in seed is 20-21 % and seed index is 4-5 gram. Yield potential is 7-9 q/ha (Singh *et al.*, 2008).

Cropping system

Lentil is generally grown as rainfed crop during *rabi* after rice maize and pearl millet. In intercropping it is grown with barley and mustard. It is also grown as an intercrop in autumn-planted sugarcane. In north eastern plains it is also grown as *utera* crop after rice. The seeds are broadcast in standing crop of rice just before harvest (Singh *et al.,* 2011).

Land preparation

For preparation of land, after harvest of *kharif* crop, field should be ploughed once with soil-turning plough. The soil should then be worked with cultivator cross harrowing and planking so that it is well pulverized.

Sowing time

Seed can be sown in first fortnight of November in temperate condition. In other parts on India the seed is sown in second fortnight of October under rainfed conditions. However it can be sown in first week of November in irrigated areas. Under late sown conditions the seed can be sown up to 15 December (Singh *et al.,* 2013b).

Seed rate

The seed rate of 30-40 kg /ha is recommended for small seeded varieties and 50-60 kg/ha for bold seeded varieties.

Fig. 10.2 Light brown coloured lentil **Fig. 10.3** Dark green coloured lentil

Seed inoculation

Make 10% gur/jagary solution and mix *Rhizobium* species @ 5-10 gm / kg seed in the solution. Seed should be dipped in the solution for 10 minutes followed by drying under the shade. Do not treat seeds with fungicides in case seeds are being inoculated.

Sowing method

Seed should be sown in lines at a spacing of 25 cm x 10 cm inter and intra row (Fig. 10.4), (Singh *et al.*, 2013a).

Nutrient management

Fertilizers should preferably be applied on the basis of soil test (Fig. 10.5). However, in absence of soil test following fertilizer schedule should be adopted in addition to inoculation of *Rhizobium* and PSB same as given below. All the fertilizers are applied as basal (Singh *et al.*, 2012b).

Fig. 10.4 Line sown lentil

Table 10.2 Effects of nutrient management on lentil production

Fertilizers	Recommended nutrient dose / ha	Fertilizer quantity/ha
FYM	10 tonnes/ha at the time of first ploughing	
VAM	-	4-5 kg
Urea	30 kg N	15 kg as basal
DAP	60 kg P_2O_5	130 kg
MOP	30 Kg K_2O	50 kg

Fig. 10.5 Managing nutrient for anticipating bumper harvest at the end

Weed management

Pre-emergence spray of pendimethalin @ 0.75 kg a.i./ha within 2-3 DAS followed by two manual weeding (Fig. 10.6), in spring, first at last week of February to first week of March and second at 20-25 days after first weeding if required under temperate conditions. However, under tropical and sub-tropical condition pre-emergence spray of pendimethalin @ 0.75 kg a.i./ha within 2-3 DAS followed by one manual weeding at 40-50 DAS should be done (Singh *et al.*, 2015b).

Fig. 10.6 Weed free lentil crop, promise for minimum use of pesticide

Irrigation

If soil is not in proper moisture condition, pre-sowing irrigation should be given to ensure proper germination of seeds. After sowing, irrigation should be given as & when required (Fig.10.7). Pre-flowering and pod formation stage is most critical to moisture stress. So avoid moisture deficit during this stage.

Fig. 10.7 Managing water efficiently, always tricky for lentil production

Harvesting

Harvesting should be done by uprooting the plants or cut with sickle (Fig.10.8), when more than 80 % crop matures and care should be taken to avoid shattering of the pods while harvesting bundling and transporting to threshing floor (Fig.10.9). Harvested crop dried on the threshing floor and threshed by beating with sticks or trampling with bullocks.

Fig. 10.8 Lentil harvesting at full swing

Fig. 10.9 Bundled lentil crop for further processing

Yield

The average yield is 6-8 q/ha under rainfed and 10-12 q/ha under irrigated conditions may be produced.

Storage

Before storage, seed should be thoroughly dried in the sun. This is essential to reduce infestation of store grain pests.

Nutritional Values

Lentil or masoor is recognized as one of the most nutricious pulse crops, ranking next to chickpea (gram) amongst *rabi* pulses. The Nutritional Value for: lentils, dry, cooked/ boiled (per 3.5 oz/100 g). Water 69.6%, protein 9.0 g, fat, 0.4 g, carbohydrates 20 g, fiber 3.9 g, calories 116. Excellent source: folic acid and potassium. Good source: iron and phosphorus. Also contains: magnesium, zinc, thiamine, copper, niacin, vitamin B6 and pantothenic acid.

Table 10.3 Nutritional value (raw dry weight basis)

Nutrition	Value per 100 g (3.5 oz)	Nutrition	Value per 100 g (3.5 oz)
Energy	1,477 kJ (353 kcal)	Folate (vit. B_9)	479 μg (120%)
Carbohydrates	60 g	Calcium	56 mg (6%)
Sugars	2 g	Iron	7.54 mg (58%)
Dietary fiber	31 g	Magnesium	122 mg (34%)
Fat	1 g	Phosphorus	451 mg (64%)
Protein	26 g	Potassium	955 mg (20%)
Water	10.4 g	Sodium	6 mg (0%)
Thiamine (vit. B_1)	0.87 mg (76%)	Zinc	4.78 mg (50%)

Source: USDA Nutrient Database

With about 30% of their calories from protein, lentils have the third-highest level of protein, by weight, of any legume or nut, after soybeans and hemp. Proteins include the essential amino acids isoleucine and lysine, and lentils are an essential source of inexpensive protein in many parts of the world, especially in West Asia and the Indian subcontinent, which have large vegetarian populations (Callaway, 2004). Lentils are deficient in two essential amino acids, methionine and cysteine (Randy, 2011). However, sprouted lentils contain sufficient levels of all essential amino acids, including methionine and cysteine. Lentils also contain dietary fiber, folate, vitamin B_1, and minerals. Red (or pink) lentils contain a lower concentration of fiber than green lentils (11% rather than 31%) (USDA). *Health* magazine has selected lentils as one of the five healthiest foods (Raymond, 2006)· Lentils are often mixed with grains, such as rice or small pasta, which results in a complete protein dish. Lentils also have some anti-nutritional factors, such as trypsin inhibitors and relatively high phytate content. Trypsin is an enzyme involved in digestion, and phytates reduce the bio-availability of dietary minerals. The phytates can be reduced by soaking the lentils in warm water overnight. Lentils are a good source of iron.

Diseases and pests

The major diseases of lentil are rust in northern plains and wilt in central zone. Use of resistant varieties is helpful in controlling the disease (Singh *et al.*, 2012a). Pests are not a major problem in lentil. In some years aphid population rises due to rise in temperature during January or February in sub-tropical areas and in March or April in temperate areas. However, some of the diseases and pests are described below:

Diseases and their management

1. Ascochyta blight
2. **Causal organism:** *Ascochyta pisi*

Symptoms

The disease develops on seedlings during cool, wet weather, resulting in the blackening of stem at the soil line which extends up-words and ultimately may cause plant death. Leaf infection results in the formation of small purple to black spots. Under dry conditions these lesions remain small but under moist conditions the lesions enlarge. Stem lesions also enlarge and often coalesce to completely girdle the stems giving the entire lower plant portion a black appearance.

Management

Maintain field sanitation by removing or ploughing pea refuge immediately after harvest before wind and rain may disperse the pathogens. Seed treatment with carbendazim 50WP or mancozeb 75WP or captan 50WP @ 2-3g/kg of seed effectively reduces the disease. When disease is anticipated, spray the crop with carbendazim 50WP @ 0.05% or mancozeb 75WP @ 0.3% or chlorothalonil 75WP @ 0.3%

3. Powdery mildew

Causal organism: *Erysiphe pisi*

Symptoms

The first symptoms appear as white, off-coloured spots on the upper surface of the lower and older leaves. These spots increase in size and appear as white, powdery areas. The disease progresses in susceptible cultivars and covers entire plant with white, powdery growth. Tissue beneath infected areas may turn purplish in colour and cleistothecia are formed in the mature lesions. Severe infection results in early crop senescence and reduces the seed quality and quantity.

Management

Collect and destroy the crop debris by burning or deep ploughing. Seed treatment with triadimefon 25 WP or mancozeb 75 WP or captan 50 WP @ 2-3 g/kg of seed is effective in reducing the initial disease level. When disease is anticipated, spray the crop with triadimefon 25 WP @ 0.05% or hexaconazole 5 EC @ 0.03% or dinocap 48 EC @ 0.05% or difeniconazole 25 EC @ 0.03%.

4. Pythium seed and seedling rot

Causal organism: *Pythium ultimum, P. aphanidermatum* and *P. debaryanum.*

Symptom

The disease causes rot in seed. The rotted seeds become soft and soil surrounded become heavily infiltrated by fungus hyphae. If radicles or plumules are produced, they may become soft, watery and translucent, with or without cotyledon rot. The pathogen generally attacks the juvenile tissues such as root tips. The damage is more severe when the soil moisture is high.

Management

Seed treatment with metalaxyl MZ 72 WP or captan 50 WP or mancozeb 75 WP @ 2-3g/kg of seed. Cultural practices that promote rapid seed germination, emergence and uniform plant stand minimize this problem. Factors such as planting too deep, soil crusting and poor seed vigour increase Pythium seed and seedling damage.

5. Rhizoctonia seedling blight

Causal organism: *Rhizoctonia solani*

Symptoms

The initial symptoms appear as water-soaked lesions on hypocotyls and epicotyl, and as reddish brown lesions above - and below - radicals as they emerge. The growing points of seedlings may die as these emerge from the soil and secondary shoots may arise from basal nodes near the seed. This may continue as several successive shoots are damped off by the pathogen. In general, if primary shoot is killed the plant never recovers to produce a normal crop. On older plants, reddish brown, sunken lesions may occur on epicotyls and may occasionally cause girdling.

Management

Rotation with cereals seed treatment with carbendazim 50WP or captan 50WP or mancozeb 75WP @ 2-3g/kg.

6. Fusarium wilt

Causal organism: *Fusarium oxysporum* f.sp. *pisi*

Symptoms

The symptoms incited by the pathogen include downward curling of leaves and stipules. The basal inter-node may be thickened, and the leaves and stem become more brittle and rigid than leaves of healthy plants. The root system appears normal, however, a longitudinal section may show yellow to orange colouration often in the vascular tissue of the root and stem. Often this vascular discoloration extends to the upper stem. As the disease develops, yellowing progresses from the lower leaves to the stem apex.

Management

Pea seeds should not be planted more frequently than once in five years in heavily infected fields.The disease can be significantly reduced by tillage practices to reduce soil compaction, better fertility and adopting the practices which favour soil moisture and root penetration. Seed treatment with carbendazim 50WP or captan 50WP or mancozeb 75WP @ 2-3 g/kg of seed. Avoid application of ammonical fertilizers as these tend to increase the susceptibility of peas to this disease. Soil amendment with pine needles or crucifers crop trash.

Insect pests and their management

1. **Pea pod borer** {Etilla zincknella (Hubn.)}

   The newly emerged tiny greenish caterpillars feed on floral parts and subsequently they bore into the pods to feed on the seeds. Young larvae also feed on tender portion of leaves and shoots. The pest remains active from 3rd week of May to 2nd week of June (pod filling stage).

Management

For peas, spray endosulfan 35 EC or malathion 50 EC@ 140 ml/ 100 lit of water at the onset of first flush of flowering and repeat after 15 days interval, if necessary.

2. **Leaf miner** *(Phytomyza horticola)*

   The maggots mine the leaves and form zig zag white galleries resulting in the loss of green matter and also affecting flowering and pod formation. The pest remains active from month of April (pre-flowering stage) upto June

Management

Spray dimethoate 30 EC@ 100 ml/ 100 lit of water. Soil application of carbofuran 3G @32.5 kg/ha will take care of leaf miners and aphids.

3. **Gram pod borer** {*Helicoverpa armigera* (Hubn.)}

   Young larva feeds on tender leaves, buds, flowers and subsequently it bores into pods and feeds on seeds with its head and part of body inside the pod while rest of body part outside. The infestation continued from 3rd week of May to 2nd week of June (pod filling stage)

Management

Install Biotraps @ 10 traps/ha from April to June. Spray endosulfan 35 EC@ 200 ml/100 lit of water+ NSKE@5% at flowering stage.

4. Aphid

Suck sap from tender shoots of the plants resulting in stunted growth. Leaves may sometimes turn yellow and dry up. The formation of flowers and pods is adversely affected.Pest remains active from April to July.

Management

Spray of dichlorvos 76 EC @ 70 ml/ 100 lit of water or monocrotophos @ 0.04% solution is effective for aphid control.

References

Callaway JC. 2004. Hempseed as a nutritional resource: an overview. *Euphytica* 140: 65-72.

http://www.bitterpoison.com/protein/11248

http://www.glisonline.com/aminoacids.php

Randy Sell. 2011. "Lentil". North Dakota State University Department of Agricultural Economics. Archived from the original on 2009-06-21. Retrieved 2011-12-14.

Raymond Joan. 2006. "World's Healthiest Foods: Lentils (India)". *Health Magazine*.

Singh AK, Bhat BP, Sunram PK, Kumar S, Bharati RC, Chandra N and Rai M.2012b.Study of Site Specific Nutrients Management of Cowpea Seed Production and Their Effect on Soil Nutrient Status. *Journal of Agril.* Sci. 4(10): 191-198.

Singh AK, Bhatt BP, Sundaram PK, Gupta AK and Singh Deepak. 2013a. Planting geometry to optimize growth and productivity faba bean (*Vicia faba* L.) and soil fertility. *J. Environ. Biol.* 34(1): 117-122.

Singh AK, Kumar P and N Chandra. 2013b. Studies on yield production of mung bean (*Vigna radiate*) sown at different dates. J. *Environ. Biol.* 34: 1007-1011.

Singh AK, Manibhushan, Chandra Naresh and Bharati RC. 2008. Suitable crop varieties for limited irrigated conditions in different agro climatic zones of India. *Int. J. Trop. Agri.* 26(3-4): 491-496.

Singh AK, Singh D, Singh AK, Gade RM and Sangle UR. 2012a. Good Agronomic Practices (GAP) - An efficient and eco-friendly tool for sustainable management of plant diseases under changing climate scenario. *J. Plant Disease Sci.* 7(1):1-8.

Singh AK, Singh SS, Prakash V, Kumar S and Dwivedi SK. 2015a. Pulses Production in India: Present Status, Bottleneck and Way Forward. *Journal of Agrisearch* 2(2): 75-83.

Singh P, Singh L, Singh P, Lone BA, Qayoom S, Ahmad L, Ganai MA, Kanth RA and Fayaz A. 2015b. Response of lentil (*Lens culinaris* Medik) and weeds to different weed management practices under temperate conditions. *Journal of Agrisearch* 2(1): 72-74.

Singh, AK, Meena, MK, Bharati RC. 2011. Sulphur and Zinc Nutrient Management In Rice Lentil Cropping System. International Conference on "Life Science Research for Rural and Agricultural Development. 27-29 December, 2011, CPRS Patna (Bihar). pp 66-67.

USDA nutrient database: http://www.nal.usda.gov/fnic/foodcomp/search/

Citation

Singh L, Singh P, Gupta AJ and Singh AK. 2018. Scientific Production Techniques of Lentil (*Lence culinaris* Madik) In: Scientific Lentil Production. Singh AK, Bhakta N, Sangale UR, Manibhushan, Sundaram PK, Kumar S and Yasin JK. (eds.) Society for Upliftment of Rural Economy Varanasi, India. Pp. 191-204.

❑❑❑

Chapter 11

Enhancing Lentil Productivity Under Rainfed Ecosystem

SK Singh

SYNOPSIS

Lentil is 1.48 m ha with production of 1.30 MT. Approximately 11.65 million ha area is rice fallows in the country providing ample scope for expansion of lentil crop. Lentil is also relay cropped with rice in which the lentil seeds are broadcast in the standing rice crop 10-15 days before rice harvest under the relay or utera system or immediately after the harvest of rice with minimum tillage under rice-lentil sequential system. We are still using the traditional varieties available with them or available in their vicinity and thus there is a wide gap in average and potential yield that can be bridged by the use of newly developed improved varieties associated with adoption of scientific cultivation. There are two critical stages in life span of lentil for irrigation i.e. pre-flowering 50 days after sowing) and pod filling (80 days after sowing). Water requirement varies from 180 to 200 mm under different agro-climatic regions. With the availability of improved varieties of lentil, it can be grown even after the harvest of long duration lowland rice. The best time of sowing is from second fortnight of October to the first week of November. Twice foliar spray of 2 % urea just before flowering and after 20-25 days are recommended specially when there is lack of atmospheric & soil moisture under rainfed situations. Lentil being a slow growing crop in the initial stages, suffers adversely from weed competition during initial period of its growth up to about 60 days after sowing. The most important disease is seedling mortality caused by a fungus, *Fusarium oxysporum* f. sp. Hairy caterpillar, Semi-looper do some damage. Lentil crop should be harvested when the pods are ripe, but the plants have not become dead ripe to avoid shattering of grains. A well managed crop yields about 20-25 q grains/ha.

Keywords: Cropping system, Lentil, Rainfed ecosystem, Utera.

Introduction

In India, lentil is cultivated in four agro-climatic regions viz., North hill zone, North West plain zone, North east plain zone and central zone. It occupies 1.51 million ha area with production of 0.96 million tones. In India it was grown in 1.48 m ha with production of 1.30 MT during 2010-11 (Kumar *et al.*, 2013). Among all the pulse crops, lentil has experienced a positive growth in area and production from 0.91 million ha and 0.43 million tones in 1980-82 to 1.46 million ha and 0.90 m tones in 2004-2006. This has mainly been possible due to cultivation of improved varieties and adoption of appropriate production technologies. Approximately 11.65 million ha area is rice fallows in the country. Uttar Pradesh (0.35 m ha), Bihar & Jharkhand (2.20 m ha), Madhya Pradesh (1.44 m ha), Chhattisgarh (2.94 m ha), West Bengal (1.72 m ha), Odisha (1.22 m ha), and Assam (0.54 m ha) are the major states with substantial area as rice fallows, providing ample scope for expansion of lentil crop in these states. Lentil is grown as a post-rainy season crop mostly under rainfed farming which is not only thirsty but hungry too. Substantial area under lentil in northeastern states is under monoculture in rotation with rice, maize, sorghum, cotton, jute, etc. However, it is also grown as intercrop with wheat, barley, mustard and linseed. In these states, lentil is also relay cropped with rice in which the lentil seeds are broadcast in the standing rice crop 10-15 days before rice harvest under the relay or utera system or immediately after the harvest of rice with minimum tillage under rice-lentil sequential system. In general, lentil is grown on residue of rice crop and has to survive on the leftover moisture and nutrients by the previous crop (Ali *et al.*, 2012).

Fig. 11.1 Field view of lentil crop at different growth stage due to different sowing dated

In both cropping systems, lentil depends predominantly on residual soil moisture. In rice-lentil sequential system, sowing of lentil is delayed due to the late harvesting of long duration rice varieties hence it experiences forced maturity as a result of increasing temperature, resulting in poor yields (Fig.11.1). In utera system, lentil plants are exposed to waterlogged conditions in the early stages, resulting in sub-optimal plant populations and poor crop establishment, though the system allows early sowing of lentil. The average yield of lentil in India (629 kg/ha) is 25% lower than the world average (1053 kg/ha). The reasons for low yield are occurrence of various biotic, abiotic (early water logging, terminal drought, cultivation of lentil in nutrient-poor marginal and acidic soils with minimal inputs) and edaphic factors at different growth stages. Plant stand and seed yield are dependent on the moisture status of soils at the time of sowing the lentil. Therefore, combinations of technology in the field of crop improvement, natural resource management and crop protection have to be deployed together for enhancing lentil productivity under rainfed ecosystem.

Selection of variety

The farmers of Uttar Pradesh, Madhya Pradesh, Bihar, West Bengal and Assam, in general are cultivating lentil in a large area but they are not able to get the higher yields because they are still using the traditional varieties available with them or available in their vicinity and thus there is a wide gap in average and potential yield that can be bridged by the use of newly developed improved varieties associated with adoption of scientific cultivation. More effort should be devoted to management interventions to realize the full potential of germplasms. New varieties i.e. Arun (Fig. 10.2) and Mallika, rust resistant varieties DPL-15, DPL-62, rust resistant varieties with different plant types–Lens–4076, and Pusa Baisakhi. are bigger in size while Pant L-639, Subrata, HUL-57, IPL-81(Noori), KLS 218, Moitree, NDL-1, PL-5, PL-6 and PL-8, S-173, Pusa-8, L 9–12, Pant L–234, Pant L–209, BR–25, Pusa–6, B-17, S-177, C-31, S-256, Pant L-406 are small sized variety (Singh *et al.*, 2008).

Soil Requirements

Lentil crop is being grown on a variety of soils ranging from light loam textured alluvial soils to heavy black cotton soil. It can be grown in poorer soils, low lying paddy soils and even in moderately saline and alkali soils. Acid soils however, are not suitable for lentil unless acidity is corrected by liming. Well drained, loam soils having neutral reaction are best suited for lentil cultivation.

Water management

There are two critical stages in life span of lentil for irrigation i.e. pre-flowering 50 days after sowing) and pod filling (80 days after sowing). Water requirement varies from 180 to 200 mm under different agro-climatic regions. Out of which, pod filling stage is the most critical. Therefore, efforts should be made to avoid moisture stress at this stage by providing a protective irrigation. Generally, in lentil crop, irrigation should be applied at 50% available soil moisture depletion. Singh *et al.* (2006) found significantly superior (1.65 t/ha) response when irrigation was provided by Low Energy Water Application (LEWA) over sprinkler (1.59 t/ha) and surface irrigation (1.37 t/ha) methods at Patna. For quality seed production at dry site, water may be added in varying amount by line source sprinkler system or at a wet site rain out shelter may be used to exclude the moisture. A comparative study on growth parameters under irrigated and non irrigated conditions revealed that in non irrigated conditions there is low biomass build up and availability of dry matter is low for partitioning. The rate of dry matter partitioning is predominantly the major determination of yield (Bhattacharya and Chandr, 1997). Lentil crop requires an optimum amount of moisture throughout the crop season (Fig. 11.3). It has a shallow root system and any moisture deficit affects the growth, development and productivity. In case of severe deficit, morphological and physiological changes like reduction in leaf area, changes in leaf orientation, reduction in plant height, development of extensive root system, stomata closure, accumulation of free proline etc. are evident. As the crop growth progress, the soil moisture depletes and the crop invariably faces terminal stress. Situation improves if there is some amount of rainfall during winter season. For severely drought prone areas selection for early flowering is required. The major sources of moisture supplement in rainfed areas are rainfall, which is quite erratic and uncertain so that the harvesting of rainwater is essential. The following practices are adopted for moisture conservation under rainfed conditions:

Fig.11.2 Field view of lentil crop at the onset of (podding) reproductive phase

Fig. 11.3 Water management in Lentil : Key to boost production

Timely Sowing: Crop should be sown just after withdrawal of monsoon (Mid October) to utilize conserved soil moisture.

Use of organic manure: The organic manure increase the water holding capacity of soil and retain soil moisture for long time and it also check soil losses through evaporation.

Use of mulches: In wide row of lentil crop, mulching increases 25 to 30% crop yield. It reduces the moisture losses through evaporation, controls soil temperature, reduces weed emergence and improves soil microbial activity.

Use of anti-transparent: Water is a major constraint under rainfed areas where crop suffers due to moisture stress. Crop requires water for transpiration and evaporation and these may be minimized by spraying of anti-transpirant on standing crop such as Kaoline, asphalt, cycocel etc. Asphalt checks the transpiration of water by scaling leaf stomata and cycocel also reduces vegetative growth of plant and enhances reproductive phase of crops, resulting in higher grain production.

Rotations and Intercropping

Lentil is generally grown after the harvest of kharif crops. With the availability of improved varieties of lentil, it can be grown even after the harvest of long duration lowland rice. The major sequences are:

Rice – (both upland and lowland)/maize/sorghum/pearl millet/pulses/oilseeds – lentil.

Lentil is grown as an intercrop in association with faba bean (Fig. 11.4), barley, toria, linseed, rapeseed and mustard in various row ratio. It is also grown as an intercrop in autumn sugarcane. Two rows of lentil may be fitted 30 cm apart between two rows of sugarcane.

Fig. 11.4 Close-up view of intercropping of Lentil with faba bean

Land Preparation

The land should be prepared by giving one deep ploughing followed by criss-cross harrowing and planking to make the soil friable and weed free so that seeding could be done at a uniform depth. There should be proper moisture at the time of sowing to ensure uniform germination.

Seed and Sowing

Timely sowing is the key factor in the realization of full yield potential of high yielding varieties of lentil (Singh *et al.*, 2013b). The best time of sowing is from second fortnight of October to the first week of November depending upon the soil moisture condition in rainfed areas. Delay in planting causes reduction in yield (Singh *et al.*, 2012). If sowing is delayed for some reason, the reduction in yield could be minimized by closer spacings. Sowing should be done in rows behind a country plough or with the help of a seed drill at row to row spacing of 30 cm and plant to plant spacing of 15-20 cm at a depth of 3-4 cm in moist soil layer. If broadcast sown, the seeds should be mixed thoroughly in the soil to reach to the moist soil layer. The seed rate of normal sown crop is 30-40 kg/ha while for late and broadcast sown crop, the seed rate should be increased to 50-60 kg/ha. Before sowing, the seeds should be treated with Bavastin @ 1 g/kg seed.

Rhizobium Culture

The seeds should be treated with specific rhizobium culture before sowing for higher production.

Manures and Fertilizers

Application of well rotten FYM / compost @ 10t/ha is always advantageous for improving the physical conditions and water retention capacity of the soils. It is also necessary to apply a small starter dose of nitrogen to lentil crop in the initial stage. The N, P, K are applied in the ratio of 15-20 kg N, 40-50 kg $P\text{-}_2\ O_5$ and 20-25 kg K_2O /ha for increased yield of lentil. The entire amounts of N, P, K fertilizers should be applied in furrows 3-4 cm below the seeds at the time of sowing. Twice foliar spray of 2 % urea just before flowering and after 20-25 days are recommended specially when there is lack of atmospheric & soil moisture under rainfed situations. Sulphur is one of the essential nutrient, supplement sufficiently through SSP or any other source (Fig. 11.5). Lentil is generally grown in sequence with rice in late sown conditions. In general paddy fields are deficient in zinc. Therefore, lentil crop may also show Zn deficiency. In the initial stage of Zn deficiency, the leaflets start falling. The deficiency can be rectified by spraying a solution of 0.5% $ZnSo_4$ + 0.25% $CaCO_3$ at the appearance of first deficiency symptoms (Singh *et al.*, 2011). The Zn deficient soils should be supplied with Zn@ 3-5 kg/ha (Fig. 11.5), at the time of land preparation every alternate year (Singh and Bhatt, 2013)

Fig. 11.5 Sulphur and Zinc management for optimizing lentil production

Weed Management

Lentil being a slow growing crop in the initial stages, suffers adversely from weed competition during initial period of its growth up to about 60 days after sowing. Therefore, the crop should be kept weed free up to this period. One or two hand or mechanical weedings followed by hoeing on 25-30 DAS and 45 to 50 DAS will keep the crop weed free (Fig.11.6). Weedicides such as pendimethalin 30 EC @ 1.5 kg a.i./ha (4-5 litre in 800-1000 liters of water) as pre-emergence spray may be used as an effective herbicide. It should be well incorporated in the soil before sowing (Singh *et al.*, 2015).

Fig. 11.6 Weed free lentil field: An assurance for pest free good crop

Plant Protection

Diseases

Several fungal diseases caused extensive damage to lentil crop (Singh *et al.*, 2013a). The most important one include seedling mortality caused by a fungi, *Fusarium oxysporum* f. sp. *lentil* and *rolfsii* and is detected when seedlings start drying up in two ways. In one case, the seedlings turn yellow and then dry up which is known as seedling wilt and in the second type which is called collar rot, the seedlings collapse while still green and then dry up. The next is wilt caused by a fungus *Fusarium oxysporum* f. sp. *lentil* and characterized by reduced or checked growth of plants, yellowing of leaves, dying of plants and under development of roots. The other one is Rust caused by the fungus *Uromyces fabae* and characterized by the appearance of pink or brown pustules on leaves and stems later turning to black and dying up of plants. The other two diseases are powdery mildew caused by a fungus *Erysiphe polygoni* and characterized initially by small whitish patches on the lower surface of the leaves and further spreading to leaves, stems, and pods as whitish powdery mass, and Downy Mildew caused by *Peronospora lentis,* initially observed in the form of light greenish to yellow spots on the upper surface of the leaves. Later the whole affected leaves are covered by cottony growth of the fungus and the plants remain stunted. Recently a new killer disease known as *stemphylium* blight like late blight of potato is very commonly occurring. Lentil variety Maitry is having resistance to this disease. Spraying of Rubral or Bavistin should be sprayed as preventive measure.

Control Measures

Control measures of these fungal diseases include – use of healthy and disease free seeds of resistant varieties like Pant L-406, L-9-12, etc., collection and burning of affected debris; treating the seeds with Agrosan G.N. (0.25%) before sowing; spraying the crop with 0.2% Dithane M-45 two times at 15 days interval to control rust, and spraying the crop with wettable sulphur like Sulfex or Elosal (0.3%) two times at 15 days interval to control Powdery Mildew.

Insect pest

Not many insect pests damage the lentil crop. However, some of the insects like hairy caterpillar, Semi-looper do some damage. They may be controlled by spraying Thiodan 35 EC (Endosulfan) @ 1.25 liters in 800 liters of water per hectare. For control of Pod borer, spray monocrotophos (Nuvacron 40 EC) @ 1 ml/liter of water. About 600-800 liters of this solution is required for spraying one hectare.

Harvesting and Threshing

Lentil crop should be harvested when the pods are ripe, but the plants have not become dead ripe to avoid shattering of grains (Fig. 11.7a to 11.7d). The harvesting is generally done by hand with the help of a sickle. After proper drying, the plants are threshed by beating with sticks or under bullocks feet or tractor wheels to separate the grains. After threshing, clean the grains and dry them under sun to bring the moisture content at 12% level for safe storage.

Fig. 11.7(a) Lentil crop ready to harvest

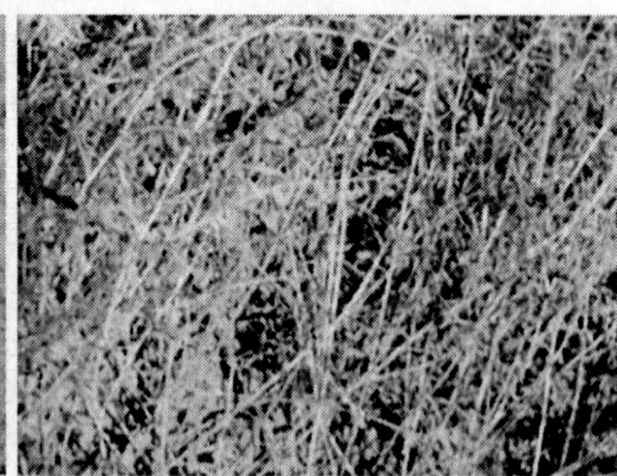

Fig. 11.7(b) Over mature lentil crop waiting for harvest

Fig. 11.7(c) Harvesting of lentil is under progress

Fig. 11.7(d) Harvested lentil crop

Yield

A well managed crop may yields about 20-25 q lentil seed /ha. A well managed rainfed crop with one supplemental irrigation can produces 8-12 q lentil seed /ha. A rainfed crop may produced 6-10 q lentil seed/ha. Intermittent winter rain during may increase grain yield by 20-30 %.

Conclusion

In India approximately 11.65 million ha area is rice fallows which providing sufficient scope for expansion of lentil crop. Lentil cultivation is also done by practicing utera system or immediately after the harvest of rice with minimum tillage under rice-lentil sequential system. There is a wide gap in average and potential yield that can be bridged by the use of newly developed improved varieties associated with adoption of scientific cultivation. Water requirement varies from 180 to 200 mm under different agro-climatic regions. Lentil can be grown even after the harvest of long duration lowland rice. Lentil sowing can be done during second fortnight of October to the first week of November. Lentil suffers adversely from weed competition during initial period of its growth up to about 60 days. The most important disease is wilt caused by a fungus, *Fusarium oxysporum* f. sp. and among insects pest hairy caterpillar, Semi-looper do some damage. Lentil crop should be harvested when the pods are ripe, but the plants have not become dead ripe to avoid shattering of grains. A well managed crop yields about 20-25 q grains/ha.

References

Ali RI, Awan TH, Ahmad MM, Saleem U and Akhtar M. 2012. Diversification of Rice-Based Cropping Systems to Improve Soil fertility, Sustainable Productivity and Economics. *J. Animal & Plant Sciences* 22(1): 108-112.

Bhattacharya A and Chandra, S. 1997. Physiological analysis of lentil seed yield under soil moisture stress and non stress conditions. *Legume Research* 20(1): 52-56.

Kumar H, Dikshit HK, Singh AM, Sharma DK, Kumari J and Singh A. 2013. Analysis of stability for seed iron and zinc concentration in elite lines of lentil. *Krishika Shodh Patrika* 2(1): 33-38.

Singh AK and Bhatt BP. 2013. Effects of foliar application of zinc on growth and seed yield of late-sown lentil. *Indian J. Agril. Sci.* 83(6): 622-626.

Singh AK, Bhatt BP, Singh KM, Kumar Abhay, Manibhushan, Kumar Ujjawal, Chandra Naresh and Bharati RC 2013a. Dynamics of powdery mildew (*Erysiphe trifolii*) disease of lentil influenced by sulphur and zinc nutrition. *Plant Pathology Journal* 12(2):71-77.

Singh AK, Bhatt BP, Sundaram PK, Chandra N, Bharati RC and Patel SK. 2012. Faba bean (*Vicia faba* L.) phenology and performance in response to its seed size class and planting depth. *Int. J. of Agril. & Stat. Sci.* 8(1): 97-109.

Singh AK, Kumar P and N Chandra. 2013b. Studies on yield production of mung bean (*Vigna radiate*) sown at different dates. J. *Environ. Biol.* 34: 1007-1011.

Singh AK, Manibhushan, Chandra Naresh and Bharati RC. 2008. Suitable crop varieties for limited irrigated conditions in different agro climatic zones of India. *Int. J. Trop. Agri.* 26(3-4): 491-496.

Singh P, Singh L, Singh P, Lone BA, Qayoom S, Ahmad L, Ganai MA, Kanth RA and Fayaz A. 2015. Response of lentil (*Lens culinaris* Medik) and weeds to different weed management practices under temperate conditions. *Journal of Agrisearch* 2(1): 72-74.

Singh SK, Singh AK and Batta RK. 2006. Yield performance of different irrigation methods and sulphur sources on lentil, chickpea and mustard. *Int. J. of Tropical Agriculture* 24(1-2):141-145.

Singh, AK, Meena, MK and Bharati RC. 2011. Sulphur and Zinc Nutrient Management in Rice Lentil Cropping System. International Conference on "Life Science Research for Rural and Agricultural Development. 27-29 December, 2011, CPRS Patna (Bihar). pp. 66-67.

Citation

Singh SK. 2018. Enhancing lentil productivity under rainfed ecosystem. In: Scientific Lentil Production. Singh AK, Bhakta N, Sangale UR, Manibhushan, Sundaram PK, Kumar S and Yasin JK. (eds.) Society for Upliftment of Rural Economy Varanasi, India. Pp. 205-216.

□□□

Chapter 12

Integrated Nutrient Management in Lentil

Jagannath Pathak, Saurabh Sharma, Anil Kumar Singh and Sushil Dimree

SYNOPSIS

Nutrient imbalance is one of the major abiotic constraints limiting productivity of pulses including lentil. The natural gift of the biological N_2 fixation, enable lentil as well as all pulse crops to meet 80-90 per cent of their nitrogen requirements, hence a small dose of 15 - 25 kg N/ha is sufficient to meet out the requirements of most of the legume crops including lentil. However, in emerging cropping systems which includes rice and other high nutrient demanding crops a higher dose of N (30-40 N/ha) had shown beneficial effect. Phosphorus deficiency in soils is wide spread and most of the pulse crops have shown good response to 20 -60 kg P_20_5 N/ha depending upon nutrient status of soil, cropping system and moisture availability. Response to potassium application is location specific. In the recent years, use of sulphur (20-30 N/ha) and some of the micronutrient such as Zn, B, Mo and Fe have improved productivity of pulse crops considerably in many pockets. Band placement of phosphatic fertilizers and use of bio-fertilizers enhances the efficiency of applied as well as native P. Foliar nutrition of some micro nutrients proved quite effective. The amount and mode of application is determined by indigenous nutrient supply, moisture availability and genotypes. Balanced nutrition is indispensable for achieving higher productivity. At the same time, in view of increasing nutrients demand, there is immense need to exploit the alternate source of nutrients viz., organic materials and bio-fertilizers to sustain the productivity with more environment friendly nutrient management systems. The environmental issues and other hazards emerging out of the imbalanced use of nutrients should also be addressed properly.

Keywords: Nutrients, lentil, INM, Micronutrients.

Introduction

Lentil (*Lens esculenta Moench*) may have been one of the first agricultural crops grown more than 8,500 years ago. Production of this cool season annual crop spread from the Near East to the Mediterranean area, Asia, Europe and finally the Western Hemisphere. It may have been introduced to the United States in the early 1900s. The crop has received little research attention to improve its yield and quality. It grows well in limited rainfall areas of the world. Lentil is a pulse (grain legume) crop. In North America much of the acreage is in eastern Washington, northern Idaho, and western Canada where drier growing season conditions prevail. It has been grown in that area since the 1930s as a rotation crop with wheat. Most of the lentil production in the United States and Canada is exported, but domestic consumption is increasing. Lentil is a protein/calorie crop. Protein content ranges from 22 to 35%, but the nutritional value is low because lentil is deficient in the amino acids methionine and cystine. Lentil is an excellent supplement to cereal grain diets because of its good protein/carbohydrate content. It is used in soups, stews, casseroles and salad dishes (Singh *et al.*, 2011). Sometimes they are difficult to cook because of the hard seed coats that result from excessively dry production conditions. Lentil plants are slender, semi-erect annuals with compound leaves (4 to 7 pairs of leaflets) with a tendril at the tips. Plants normally range from 12 to 20 inch tall, the taller plants resulting from cool growing season temperatures, good moisture and good fertility. Plants can have single stems or many branches depending upon the population in the field. Flowering begins on the lowest branches, gradually moving up the plant and continuing until harvest. Each flower produces a short pod containing one or two lens-shaped seeds. Flowers can be white, lilac or pale blue in color and are self-pollinated. At maturity plants tend to lodge because of their weak stems. Lentil produced in North America has larger seeds than that from India and the Near East. The seeds (2 to 7 mm in diameter) come in colors of tan, brown, or black and some varieties produce purple or black mottled seeds.

Integrated Nutrient Management

The basic concept of integrated nutrient management (INM) or integrated plant nutrition management (IPNM) is the adjustment of plant nutrient supply to an optimum level for sustaining the desired crop productivity (Deokaran *et al.*, 2014 and Singh *et al.*, 2011). It involves proper combination of chemical fertilizers, organic manure, crop residues, N_2 fixing crops like pulses and oilseeds such as soybean and bio-fertilizers suitable to the system of land use and ecological, social and economic conditions. The cropping system rather than an individual crop, and farming system rather than an individual field, is the focus of attention in this approach for the development of INM practices for various categories.

INM use five major sub-concepts

1. Plant nutrients stored in the soil.
2. Plant nutrients, those present in the crop residues, organic manure and domestic wastes.
3. Plant nutrients purchased or obtained from outside the farm.
4. Plant nutrients losses e.g. those removed from the field in crop harvest and lost from the soil through volatilization (ammonia and nitrogen oxide gases and leaching (nitrate, sulphate etc.)
5. Plant nutrients outputs e.g. nutrient uptake by the crops at harvest time.

The integrated plant nutrient management or integrated nutrient management thus firstly operates at plot level, optimizing the utilization of plant nutrients from diverse sources which are locally available, in order to improve the agronomic efficiency of such nutrients and at the same time reducing the losses of nutrients (Fig. 12.1). Organic recycling is especially promoted through INM for facilitating nutrient recycling to improve soil fertility and productivity at low cost.

Fig. 12.1 Effect of sulphur and zinc nutrition on lentil performance at early and later growth stage

Goals of INMO

- To maintain soil productivity.
- To ensure productive and sustainable agriculture.
- To reduce expenditure on costs of purchased inputs by using farm manure and crop residue etc.
- To utilize the potential benefits of green manures, leguminous crops and bio-fertilizers.
- To prevent degradation of the environment.
- To meet the social and economic aspirations of the farmers without harming the natural resource base of the agricultural production.

Components of the INM

The main objective of the INM is to efficiently utilize all the sources of the

- Plant nutrients.
- Soil nutrients.
- Chemical fertilizers.
- Organic manure, soil amendments and crop residues.
- Green manure.
- Biological nitrogen fixation.
- Bio-fertilizers.

Implementation of INM activities

Different stages of implementation of INM are as follows:

1. Diagnosis phase: collection of background information.
2. Analysis of constraints.
3. Preparing potentiality and feasibility summary.
4. On-farm demonstrations.
5. Evaluation of INM activities.

In the first stage of diagnostic phase information with regard to the following is collected and analyzed:

- Farming/cropping systems,
- Crop varieties grown,
- Awareness about soil fertility problems,
- Use of chemical fertilizers,
- Use of organic manure,
- Availability of fertilizer and other inputs,
- Irrigation sources and practices,
- Soil testing service facility.

Constraints in the adoption of INM technologies

The INM technologies must be compatible with the local farming system if they are to find acceptance and adoption. Therefore, attention must be paid to examine the interactions among different components of INM and the management of crops and animals that form the farming system. Some of the agronomic parameters which need attention are cropping pattern, intercropping practices, biological condition of the field (weeds, diseases, and insects), soil conditions, irrigation facilities and climatic conditions.

Common constraints encountered by the farmers in adoption of INM technology are as follows:

- Non-availability of FYM
- Difficulties in growing green manure crops
- Non-availability of quality bio-fertilizers
- Non-availability of soil testing facilities
- High cost of chemical fertilizers
- Non-availability of water
- Lack of knowledge and poor advisory services
- Non-availability of improved seeds
- Poor soil conditions
- Non-availability of credit facilities

Mineral nutrition

Since centuries, it is known that roots of terrestrial plants obtain nourishment from the soil. During the first half of the nineteenth century, it was found that plants need certain chemical elements referred to as essential elements and that elements are absorbed by roots principally as inorganic ions. The term mineral nutrient is generally used to refer as inorganic ion obtained from the soil and required for plant growth. The process of absorption, translocation and assimilation of nutrients by the plants is known as mineral nutrition.

Essential Elements

Plant needs 16 elements for their growth and completion of life cycle. They are: carbon, hydrogen, oxygen, nitrogen, phosphorus, potassium, calcium, magnesium, sulphur, iron, manganese, zinc copper, boron, molybdenum, and chlorine. In addition, four more elements viz., sodium, cobalt, vanadium, and silicon are absorbed by some plants for special purposes. All these elements are not required by all plants. Among these, all carbon atoms and most of the oxygen atoms are derived from the carbon dioxide which is assimilated principally in photosynthesis.

Criteria of Essentiality

Plant analysis using modern techniques reveals that plant body contains about 30 elements and in some cases as many as 60 elements. An element is considered as essential, when plants cannot complete vegetative or reproductive stage of life cycle due to its deficiency; when this deficiency can be corrected or prevented only by supplying this element; and when the element is directly involved in the metabolism of the plant.

Forms of mineral elements absorbed by the plants

According to this criterion, sodium is considered as non-essential. However, sodium is known to increase yields of some crops like sugar beets, turnips and celery. Therefore, farmer considers sodium as essential element. Nicholas (1961) proposed the term 'functional nutrient' for any mineral element that functions in plant metabolism.

Classification of Essential Elements

The essential elements can be classified based on the amount required, their mobility in the plant and soil, their chemical nature and their functions inside the plant.

Amount of Nutrients

Depending on the quantity of nutrients present in plants, they can be classified into three:

Basic Nutrients

The basic nutrients viz. carbon, hydrogen and oxygen, constitute 96% of total dry matter of plants. Among them, carbon and oxygen constitute 45% each. The total dry matter produced by rice crop in one season is about 12 t ha^{-1}.

Macronutrients

The nutrients required in large quantities are known as macronutrients. They are N, P, K, Ca, Mg, and S. Among these, N, P, and K are called primary nutrients and Ca, Mg and S are known as secondary nutrients. The later are known as secondary nutrients as they are inadvertently applied to the soils when N, P, and K fertilizers which contain these nutrients are used.

Micronutrients

The elements which are required in small quantities are known as micronutrients or trace elements. They are Fe, Zn, Cu, B, Mo, and Cl. These elements are very efficient and minute quantities produce optimum effects (Singh *et al.,* 2011). On the other hand, even a slight deficiency or excess is harmful to the plants.

Chemical Nature

The nutrients can be classified into cations and anions and metals and non-metals based on their chemical nature (Table 12.1).

Table 12.1 Classification of plant nutrient on the basis of cations and anions

Cations	K, Ca, Mg, Fe, Mn, Zn,Cu
Anions	NO_3, H_2PO_4, SO_4
Metals	K, Ca, Mg, Fe, Mn, Zn,Cu
Non- metals	N, P, S, B, Mo, Cl

Functions in plants

Based on the functions, nutrients are grouped into four

1. Elements that provide basic structure to the plant - C, H and O.
2. Elements useful in energy storage, transfer and bonding - N, S, and P.
3. Elements necessary for charge balance - K, Ca, and Mg. These act as regulators and carriers.
4. Elements involved in enzyme activation and electron transport - Fe, Mn, Zn, Cu, B, Mo, and Cl. These elements are catalysers and activators.

Mobility of nutrients in the soil

Mobility of nutrients in the soil has considerable influence on availability of nutrients to plants and method of manure application. For plants to take up these nutrients, two processes are important: movement of nutrient ions to the absorbing root surface, and roots reaching the area where nutrients are available. In the case of immobile nutrients, the roots have to reach the area of nutrient availability and forage volume is limited to root surface area. For highly mobile nutrients, the entire volume of the root zone is forage area. Based on the mobility in the soil, the nutrient ions can be grouped as mobile, less mobile and immobile. The mobile nutrients are highly soluble and are not adsorbed on clay complex; e.g.: NO_3^-, $SO_4^=$, $BO_3^=$, Cl^- and Mn^{++}. Less mobile nutrients are also soluble, but they are adsorbed on clay complex and so their mobility is reduced; e.g.: NH_4^+, K^+, Ca^+, Mg^{++} and Cu^{++}. Immobile nutrient ions are highly reactive and got fixed in the soil; e.g.: $H_2PO_4^-$, $HPO_4^=$ and Zn^{++}.

Mobility of nutrients in plants

Knowledge of the mobility of nutrients in the plant helps in finding what nutrient is deficient. A mobile nutrient in the plant moves to the growing points in case of deficiency. Deficiency symptoms, therefore, appear on the lower leaves (Table 12.2).

Table 12.2 Classification o f nutrients based on its Mobility in plants

Nutrients	Category
N, P and K	highly mobile
Zn	moderately mobile
S, Fe, Mn, Cu, Mo and Cl	less mobile
Ca and B	Immobile

Primary (macro) Nutrients

Primary (macro) nutrients are nitrogen, phosphorus, and potassium. They are the most frequently required in a crop fertilization program. Also, they are need in the greatest total quantity by plants as fertilizer.

Nitrogen

- Necessary for formation of amino acids, the building blocks of protein
- Essential for plant cell division, vital for plant growth
- Directly involved in photosynthesis
- Necessary component of vitamins
- Affects energy reactions in the plant

Phosphorus

- Involved in photosynthesis, respiration, energy storage and transfer, cell division, and enlargement
- Promotes early root formation and growth
- Improves quality of fruits, vegetables, and grains
- Vital to seed formation
- Increases water-use efficiency
- Hastens maturity

Potassium

- Carbohydrate metabolism and the break down and translocation of starches
- Increases photosynthesis
- Increases water-use efficiency
- Essential to protein synthesis
- Important in fruit formation
- Activates enzymes and controls their reaction rates
- Improves quality of seeds and fruit
- Improves winter hardiness
- Increases disease resistance

Secondary Nutrients

The secondary nutrients are calcium, magnesium, and sulphur. For most crops, these three are needed in lesser amounts than the primary nutrients. They are growing in importance in crop fertilization programs due to more stringent clean air standards and efforts to improve the environment.

Calcium

- Utilized for continuous cell division and formation
- Involved in nitrogen metabolism
- Reduces plant respiration
- Aids translocation of photosynthesis from leaves to fruiting organs
- Increases fruit set
- Essential for nut development in peanuts
- Stimulates microbial activity

Magnesium

- Key element of chlorophyll production

- Improves utilization and mobility of phosphorus
- Activator and component of many plant enzymes
- Increases iron utilization in plants
- Influences earliness and uniformity of maturity

Sulphur

- Integral part of amino acids
- Helps develop enzymes and vitamins
- Promotes nodule formation on legumes
- Aids in seed production
- Necessary in chlorophyll formation (though it isn't one of the constituents)

Micronutrients

The micronutrients are boron, chlorine, copper, iron, manganese, molybdenum, and zinc. These plant food elements are used in very small amounts, but they are just as important to plant development and profitable crop production as the major nutrients. Especially, they work "behind the scene" as activators of many plant functions.

Boron

- Essential for germination of pollen grains and growth of pollen tubes
- Essential for seed and cell wall formation
- Promotes maturity
- Necessary for sugar translocation

Chlorine

- Interferes with P uptake
- Enhances maturity of small grains on some soils

Copper

- Catalyzes several plant processes

- Major function in photosynthesis
- Major function in reproductive stages
- Indirect role in chlorophyll production
- Increases sugar content
- Intensifies color
- Improves flavor of fruits and vegetables

Iron

- Promotes formation of chlorophyll
- Acts as an oxygen carrier
- Reactions involving cell division and growth

Manganese

- Functions as a part of certain enzyme systems
- Aids in chlorophyll synthesis
- Increases the availability of P and Ca

Molybdenum

- Required to form the enzyme "nitrate reductase" which reduces nitrates to ammonium in plant
- Aids in the formation of legume nodules
- Needed to convert inorganic phosphates to organic forms in the plant

Zinc

- Aids plant growth hormones and enzyme system
- Necessary for chlorophyll production
- Necessary for carbohydrate formation
- Necessary for starch formation
- Aids in seed formation

In addition to the 13 nutrients listed above, plants require carbon, hydrogen, and oxygen, which are extracted from air and water to make up the bulk of plant weight.

Forms of nutrient elements absorbed by the plants

Plants absorb nutrients from the soil as i) single nutrient element, un-combined with other nutrients and ii) essential nutrient element, combined with other elements to form nitrate, phosphate, sulphate etc (Table 12.3).

Table 12.3 Forms of nutrient elements absorbed by the plants

Nutrient elements	Forms absorbed by plants
Absorb as a single nutrient element	
Potassium	K^+
Calcium	Ca^{2+}
Magnesium	Mg^{2+}
Iron	Fe^{2+} (Ferrous) Fe^{3+} (Ferric)
Manganese	Mn^{2+} (Manganous) Mn^{4+} (Manganic)
Copper	Cu^+ (Cuprous) Cu^{2+} (Cupric)
Zinc	Zn^{2+}
Chlorine	Cl^-
Silicon	$Si(OH)_4$
Cobalt	Co^{2-}
Sodium	Na^+
Absorb as a combined form	
Nitrogen	Ammonium (NH_4^-), Nitrate (NO_3^-)
Phosphorus	HPO_4^{2-} , $H_2PO_4^-$
Sulphur	SO_4^{2-}
Boron	H_3BO_3 (Boric acid), $H_2BO_3^-$ (Borate)
Molybdenum	MoO_4^{2-} (Molybdate)
Carbon	CO_2 (Mostly through leaves)
Oxygen	CO_2 (Mostly through leaves), O^{2-},OH^-, CO_3^{2-}
Hydrogen	HOH (Hydrogen from water), H^+
water	H^+, OH^-

Factors influencing nutrient availability

1. Natural supply of nutrients in the soil which is closely tied up to parent material of that soil and vegetation under which it is develop.
2. Soil P^H as it affects nutrient release.
3. Relative activity of micro-organism which plays a vital role in nutrient release and may in the case of mycorrhiza directly function in nutrient uptake.
4. Fertility addition in the form of commercial fertilizers, animal manure and green manure.
5. Soil temperature, moisture and aeration.

Deficiency Symptoms

When nutrient is not present in sufficient quantity, plant growth is affected. Plants may not show visual symptoms up to a certain level of nutrient content, but growth is affected and this situation is known as *"hidden hunger"*. When a nutrient level still falls, plants show characteristic symptoms of deficiency. These symptoms though vary with crop, have a general pattern. These are generally marked by diseases and other stresses and so need careful and patient observation on more number of plants for typical symptoms. The deficiency symptoms appear clearly in crops with larger leaves.

Identification of Deficiency Symptoms

The deficiency symptoms can be distinguished based on the region of occurrence, presence or absence of dead spots and chlorosis of entire leaf or inter-veinal chlorosis. The region of appearance of deficiency symptoms depends on mobility of nutrient in plants. The nutrient deficiency symptoms of N, P, K, Mg, and Mo appear in lower leaves because of their mobility inside the plants. These nutrients move from lower leaves to growing leaves thus causing deficiency symptoms in lower leaves. Zinc is moderately mobile in plants and deficiency symptoms, therefore, appear in middle leaves. The deficiency symptoms of less mobile elements (Fe, S, Mn and Cu) appear on new leaves. Since Ca and B are immobile in plants, deficiency symptoms appear on terminal buds. Chlorine deficiency is less common in crops.

Toxicity Symptoms

When a nutrient is present in the soil in excess of plant's requirement, the nutrient is absorbed in higher amounts which causes imbalance of nutrients or disorder in physiological processes. Unlike deficiency symptoms, toxicity symptoms are less common.

Nitrogen

Excess nitrogen causes delay in maturity and increases succulency. The adverse effects of excess nitrogen are lodging and abortion of flowers. Crops become susceptible to pest and diseases.

Phosphorus

Excess P causes deficiency of iron and zinc. In some plants like maize, leaves develop purple colouration and plant growth is stunted.

Iron

Tiny brown spots appear on lower leaves of rice starting from tips and spreading towards bases. Leaves usually remain green. In extreme case, the entire leaf turns purplish brown in colour.

Manganese

The plant is stunted and tillering is often limited. Brown spots develop on the veins of the leaf blade and leaf sheath, especially on lower leaves.

Boron

Chlorosis occurs at the tips of the older leaves, especially along the margins. Large, dark brown, elliptical spots appear subsequently. The leaves ultimately turn brown and dry up.

Soil

Lentil is adapted to all soil types, from sand to clay loam, if there is good internal drainage. Lentil does not tolerate flooded or waterlogged soils, and does best on deep, sandy loam soils high in phosphorus and potassium. Good drainage is required, because even short periods of exposure to waterlogged or flooded field conditions kill plants. A soil pH near 7.0 is best for lentil production.

Fertility and Lime Requirements

Inoculation with the proper *Rhizobium* will provide the nitrogen requirements of lentil. However, if available nitrogen is low (organic carbon less than 0.5%), an early nitrogen supplement of about 20 kg/ha is required to sustain the young plants until root nodulation develops. If applied as starter, this material should be applied adjacent to but not in contact with the seed. Smaller amounts of N (20 kg/ha) may provide some benefit on higher organic carbon soils, especially where drainage is somewhat restricted. Care should be taken not to provide too much nitrogen, which could inhibit nodulation and produce excessive vegetative growth at the expense of seed yield. Phosphorus and potassium are recommended for maximum yields on soils testing medium or low. Band applications of P may be of some benefit on very low P or high pH (above 7.6) soils. Lime needs for lentil have not been well established. However, it is likely similar to that of other leguminous vegetables. As for most legumes, sulphur needs are medium to high, and responses may be seen on light colored sandy soils where manure has not been applied.

Effect of organic manures and bio-fertilizers on soil and plant

Nutrients held in organic combinations become available to plants only slowly in many cropping season as a result of natural control over nitrification affected by the heterotrophic organisms. Prevent conditions leading to the deficiency of secondary and micronutrients in the soil. An organic residue, grasses and others when incorporated into the soil not only improve the soil physical properties but also improve the soil microbial life. The addition of organic residues like straw not only retards humus depletion but also enriches the nitrogen status of soil by fixing atmospheric nitrogen through the process of photochemical fixation. When organic residues are blended with different calcium phosphates such as basic slag and rock phosphates, the fixation of atmospheric nitrogen by humus is accelerated. As soil organic matter is a steady supplier of available nitrogen, phosphorus, potash, calcium and other plant nutrients, lands rich in human population can support increased production without the use of chemical fertilizer.

Nutrient requirement for lentil

Out of 16 essential elements, pulses specially need adequate amount of P, Ca, Mg, S and Mo. Phosphorus is required for proper root growth and growth of rhizobia. Calcium and magnesium are required to stimulate growth and to increase the size of the nodules, pod formation and grain setting. Sulphur is required for nodulation and protein synthesis. Molybdenum for nitrogen fixation and assimilation and boron for reproduction are the requirement. Nutrient removal and uptake for producing one tonne of biomass, pulse crops in general remove about 30-50 kg N, 2-7 kg P_20_5, 12-30 kg K_2O, 3-10 kg Ca, 1-5 kg Mg, 1-3 kg S, 200-500 g Mn, 5 g B, 19 g Cu and 0.5 g Mo from soil.

Nitrogen: Mahon and Child (1979) found that small doses of N fertilizers applied to an annual pulse are beneficial if nodule initiation is delayed. In dry pea, N application at 20 to 60 kg ha^{-1} increased seed yield by an average of 9% in one quarter of 58 trials conducted in Alberta (McKenzie *et al.*, 2001). Application of fertilizer N increased dry bean (*Phaseolus vulgaris* L.) seed yield proportionally in southern Manitoba (McAndrew and Mills, 2000).

Phosphorus: Phosphorus is the second most critical plant nutrient overall, but for pulses it assumes primary importance owing to its important role in root proliferation and there by atmospheric nitrogen assimilation. Havlin *et al.* (2005) reported that phosphorous is relatively immobile (moves very little) in the soil. Most crops recover only 10 to 30% of the P fertilizer in the first year following application. Recovery varies widely depending on soil type and conditions, the crop grown and application method. McVicar *et al.* (2010) found that lentil has a relatively high requirement for phosphorus to promote development of its extensive root systems and vigorous seedlings, disease, and drought tolerance because of P application.

Potassium: Potassium application generally neglected, resulting in imbalance nutrient supply and lower crop yields. Under intensive cropping systems, large amounts of K are removed leading to serious depletion of soil K reserves. Pulses such as pigeon pea and lentil remove about 60 and 52 kg K_2O/t grains, respectively. The application of potassium regulates the utilization of other nutrients in the plant system. The range of lentil responses to K was between 3 to 16 kg grains per kg K_2O. Average pigeon pea and pea responses to 20 kg K_2O/ha were 14 and 7 kg grain per kg K_2O, respectively. In a separate study, an average increase of 5 kg grain per kg K_2O was recorded for chickpea in northern states of India. It should be noted that economic response to 40 kg P_2O_5 was apparent in lentil, pigeonpea, and pea, while in chickpea and black gram, economic responses were recorded at 30 and 20 kg P_2O_5/ha (Srinivasanrao *et al.,* 2013).

Sulphur: Shukla and Behera (2012) found that grain yield of gram increased with the S application @20 kg and 40 kg/ ha over control and maximum yield was obtained when S was applied @ 40 kg/ha through gypsum. Shukla *et al.* (2012) reported that among pulses, chickpea, field pea, pigeon pea require 40 kg S/ha while lentil, green gram, black gram and cluster bean need 20 kg S/ha to produce optimum crop yields (Singh *et al.,*2011). (Singh, 1999) reported that average removal of sulphur by one tonne of pulses ranges between 4-8 kg as compared to 3-5 kg by cereal crops. Similarly pulses from one hectare remove sulphur between 5-10 kg annually which depends upon nature of crop, soil and environmental factors. He also reported that application of 60-120 kg S/ha through pyrites (22 % sulphur) increased the mean seed yield of lentil by 178-322 kg/ ha in 3-4 trials on red soils of Bundelkhand of Uttar Pradesh over control (Singh and Bhatt, 2015).

Integrated nutrient management: Arya *et al.* (2007) conducted an experiment on chickpea + mustard intercropping system and reported that integrated nutrient application (INA) [50% recommended dose of fertilizer (RDF) + FYM at 5 t/ha+ biofertilizers (Rhizobium + PSB)] significantly enhanced average plant height, crop biomass, leaf area index, number and dry weight of root nodules of chickpea and plant height and crop biomass in mustard as compared to control. INA also produced significantly more grain and biological yield of chickpea and mustard and more chickpea equivalent yield (2664 kg/ ha) and net return (US $617/ha) as compared to control grown under intercropping system.

Fertilizers doses: Fertilizers should be tailored on soil test basis if soil test is not possible then NPK may be applied @ 20:60:20 kg ha^{-1}. Entire dose of NPK should be applied at the time of sowing. Nitrogen topdressing not required at all. Starter dose of nitrogen is essential during the initial growth phase of lentil for proper nodulation.

Conclusion

Nutrient imbalance is one of the major abiotic constraints limiting productivity of pulses including lentil. It required less fertilizer in comparison to cereals and oilseeds but time and method of application is important. The inbuilt mechanism of biological N_2 fixation enable lentil crops to meet 80-90 per cent of their nitrogen requirements, hence a small dose of 15-25 kg N ha^{-1} is sufficient to meet out the requirements of this crops. In the recent years, use of sulphur (20-30 kg ha^{-1}) and some of the micronutrient such as Zn, B, Mo and Fe have improved productivity of pulse crops considerably in many pockets. Band placement of phosphatic fertilizers and use of bio-fertilizers enhances the efficiency of applied as well as native P. There is urgent need to address environmental issues and other hazards emerging out of the imbalanced use of nutrients.

References

Arya RL, Varshney JG and Kumar Lalit. 2007. Effect of integrated nutrient application in chickpea + mustard intercropping system in the semi-arid tropics of North India. *Communications-in-Soil-Science-and-Plant-Analysis*. 38(1/2): 229-240.

Deokaran, Singh SB and Ram Kewal. 2014. Effect of Zinc and Boron application on Yield of Lentil and Nutrient Balance in the Soil under Indo-Gangetic Plain Zones. *Journal of AgriSearch* 1(4): 206-209.

Havlin JL, Beaton JD, Tisdale SL and Nelson WL. 2005. Soil Fertility and Nutrient Management: *An Introduction to Nutrient Management*. 7th Edition, Pearson/Prentice Hall, Upper Saddle River, NJ. p.515.

Mahon JD and Child JJ. 1979. Growth response of inoculated pea (*Pisum sativum*) to combined nitrogen. *Can. J. Bot.* 57: 1687–1693.

McAndrew DW and Mills K. 2000. Nitrogen fertilizer in dry bean in Manitoba. *In Proc. Third Pulse Crop Research Workshop.* Winnipeg, MB. Nov. 19–21. pp. 72-75.

McKenzie RH, Middleton AB, Solberg ED, DeMulder J, Flore N, Clayton GW and Bremer E. 2001. Response of pea to rhizobia inoculation and starter nitrogen in Alberta. *Can. J. Plant Sci.* 81: 637–643.

McVicar R, McCall P, Brenzil C, Hartley S, Panchuk K, Mooleki P, Vandenberg A and Banniza S. 2010. *Lentils in Saskatchewan*. Fact Sheet. Saskatchewan Ministry of Agriculture.

Shukla AK and Behera SK.2012. Progress report for years 2007-2010 of the AICRP of Micro-and secondary-nutrients and pollutant elements in soils and plants. IISS Bhopal, pp. 42.

Shukla AK, Behera SK, Rao AS and Singh AK.2012. State - wise micro and secondary nutrients recommendations for different crops and cropping systems. *IISS Research Bulletin* 1/2012, pp. 22.

Singh AK and Bhatt BP. 2015. Late sown lentil performance in response to foliar application of zinc. *Bangladesh Journal of Botany* 44(1): 125-128.

Singh AK, Bhat BP, Sunram PK, Kumar S, Bharati RC, Chandra N and Rai M. 2012. Study of site specific nutrients management of Cowpea seed production and their effect on soil nutrient status. *Journal of Agril.* Sci. 4(10): 191-198.

Singh M V 1999. Sulphur management for oilseed and pulse crops. *IISS Bulletin 03,* Bhopal, p. 5.

Singh, AK, Meena, MK Bharati RC. 2011. Sulphur and Zinc Nutrient Management in Rice Lentil Cropping System. International Conference on "Life Science Research for Rural and Agricultural Development. 27-29 December, 2011, CPRS Patna (Bihar).pp 66-67.

Srinivasanrao Ch, Masood Ali, Ganeshamurthy AN, and Singh KK. 2003. Potassium requirement in pulse crops. *Better Crops Internationals.* 17 (1): 8-11.

Citation

Pathak J, Sharma S and Singh AK. 2018. Integrated nutrient management in lentil In: Scientific Lentil Production. Singh AK, Bhakta N, Sangale UR, Manibhushan, Sundaram PK, Kumar S and Yasin JK. (eds.) Society for Upliftment of Rural Economy Varanasi, India. Pp. 217-236.

❑❑❑

Chapter 13

Managing Sulphur and Zinc for Sustainable Rice-Lentil Production in Indo-Gangetic Plain

Anil Kumar Singh, MK Meena and RC Bharati

SYNOPSIS

Sulphur and zinc are one of the important secondary and micro nutrients showing deficiency symptom in the rice-lentil growing tract. Hence both the nutrient was applied to rice and their direct and residual response was ascertained to rice and lentil in sequence. Pooled results indicate that maximum rice yield (7.72 t/ha) was recorded with combined application of 30 kg sulphur and 6 kg zinc, whereas corresponding minimum rice yield (7.09 t/ha) was recorded with control where no application of zinc and sulphur was done. Similarly maximum lentil seed yield (1243 kg/ha) was recorded with combined application of 30 kg sulphur and 6kg zinc whereas corresponding minimum lentil seed yield of 960 kg/ha was recorded with control. Highest N uptake (281.2 kg/ha) was recorded with 6 kg zinc application, whereas P and K uptake was highest with 40 kg sulphur application. Highest extra net return (₹ 6611/ha) was recorded in the plots which received combined application of 30 kg sulphur and 6kg Zn. Maximum B: C ratio was noticed in with 20 kg sulphur application.

Keywords: Cropping System, Indo Gangetic Plains, Lentil, Rice, Sulphur, Zinc, Content and uptake B: C Ratio.

Introduction

Rice and lentil are very important in the Indo-Gangetic Plain (IGP) for their contribution to human and animal nutrition, as components of indigenous cropping systems, and as restorers of soil fertility (Ali *et al.*, 2012; Joshi, 1998; Reddy, 2009 and Ramakrishna *et al.,* 2000). Cereals and legumes play important role in the quality of life on human kind and form a major crop for human food, animal feed and forage. Rice is staple food of not only Bihar but also for India and South Asia (Deokaran, *et al.,* 2014). Lentil is the important legume crop mainly grown in residual soil moisture in Eastern part of India and prominent source of vegetable protein (Singh *et al.,* 1997; Singh *et al.,* 2012 and Singh *et al.,* 2011). India is the largest producer, importer and consumer of pulses in the world. However, pulses production has been stagnant at between 11 and 14 million tonnes over the last two decades. Per capita pulses consumption over the years has come down from 61g /day in 1951 to 30 g/day in 2008 (Reddy, 2009). Substantial increases in the production of cereals and grain legumes in the IGP can be achieved by fine tuning management aspects and thereby increasing total system productivity and sustainability (Ramakrishna *et al.,* 2000). Rice - lentil cropping system is practiced with no application of chemical fertilizers or only high analysis chemical fertilizers only to rice crops.

There is stagnation and deterioration in productivity of both the crops in a cropping sequence as well as on individual basis because zinc and sulphur deficiency is very common in track of Eastern Indo-Gangetic alluvial plains, where high intensive cropping of rice–wheat is practiced and oilseeds and pulses are raised (Singh and Bhatt, 2015 and Singh and Singh, 2008). Rice –lentil cropping system is important cropping system not only for Bihar but also for Eastern UP, West Bengal and Odish. In Bihar it is second to rice –wheat. Some part of country and in pockets of Bihar especially in Tal and Diara region, which occupies 3.00 lakh ha where either rice- lentil cropping or mono cropping of lentil is possible (Singh *et al.,* 2011). In Indo Gangetic plains (IGP) of Bihar, there is tendency to use indiscriminate amount of nitrogenous fertilizers and very limited amount of other nutrients' containing high analysis chemical fertilizers. In general in this part of country lentil is grown on residue of rice crop and have to survive on the leftover nutrients and moisture by the previous crop (Ali *et al.,* 2012).On an average to produce one tonne of grain of high-yielding varieties of rice, remove about 22 kg N, 7 kg P_2O_5, 32 kg K_2O, 5 kg MgO, 4 kg CaO, 1 kg S and 40g Zn from the soil. Similarly for producing one tonne of biomass, lentil crop in general remove about 40 kg N, 3.9 kg P_2O_5, 85 kg K_2O, 8.0 kg S 7.0 kg Ca, 9.0 kg Mg, 35 g Zn from soil (Ahlawat and Ali, 1993). There is a need to ascertain and promote the uses of types of fertilizers required to correct the deficiency of all these nutrients especially zinc and sulphur. Zn deficiency is the most widespread micronutrient disorder in lowland rice and application of Zn along with NPK fertilizer increases the grain yield dramatically in most cases (Singh and Bhatt, 2015 and Singh and Singh, 2008). Deficiency symptoms of Zn in rice crop and sulphur in lentil are frequent and reduction in yield of rice and lentil is often blamed to zinc sulphur deficiency (Deokaran, *et al.,* 2014). There is need to investigate role of zinc and sulphur in the sustainable production of rice and lentil and their effects on fertility and productivity status of soil (Ahlawat and Ali, 1993 and Sahaa *et al.,* 2007). Keeping in the view of the importance of rice - lentil cropping system in the IGP and role of sulphur and zinc nutrient in their crop physiology and ultimately in the economic yield this experiment was undertaken.

Fig. 13.1 Filed view of lentil crop at early growth stage; showing residual effects of sulphur and zinc on lentil in rice lentil cropping system

Materials and Methods

To evolve suitable nutrient management system with respect to one secondary nutrient (sulphur) and one micro nutrient (zinc) under rice - lentil cropping system for Indo- Gangetic plains of Bihar, field experiment consisting total 16 treatment combination was tested i.e. four level of sulphurS_1(0kg), S_2(20kg), S_3(30kg), and S_4(40kg) and zinc Zn_1(0kg), Zn_2(4kg), Zn_3(5kg) and Zn_4(6kg). These treatments were applied in combination at ICAR Research Complex for Eastern Region Patna in randomized block design (RBD) replicated with thrice. Both sulphur and zinc applied to rice crop and their direct effects on rice and residual effect on lentil was investigated. Different sources were used for sulphur and zinc. Other nutrients especially NPK and agronomic management practices was as per recommended practices and was same for all treatment. No major incidence of pests and disease were noticed during the course of experimentation. The texture of soil of experimental field was silty clay loam with mean pH value of 6.8, electrical conductivity 0.16 dS/m in 1:2 soils: water solution, organic carbon 0.68 per cent, with available nitrogen 244.7 kg/ha, available phosphorus 28.6 kg/ha, available potash 185.8 kg/ha, sulphur 8.3 kg/ha and zinc 0.8kg/ha. The plot size was kept 10.0 m X 5.0 m. Seed yield (kg/ ha) was estimated based on seed weight per plot adjusted to 12 % moisture. Though only seed yield recorded has been given and discussed in this paper. Nutrient content on whole plant basis for both the crop was analysed at harvest and uptake was also calculated accordingly. Physical and chemical parameter of soils were recorded on pH, EC (dS/m), OC (%), available P (kg/ha) and available K (kg /ha) before sowing and after harvest crop. The soil samples were analyses following the procedure described by AOAC (1980). Regular analysis of variance was performed for each trait for all three seasons and the combined (Pooled) analysis over seasons after testing error variance homogeneity was carried out according to the procedure outlined by Gomez and Gomez (1984), using the MSTATC version 2.1 (Michigan State University, USA) statistical package design. Significant differences between the treatments were compared with the critical difference at (± 5%) probability by LSD.

Results and Discussion

Effects of sulphur and zinc on performance of rice and lentil

Grain yield of rice was also affected significantly with the levels of both the factors/nutrients (S and Zn). Maximum (7.57 t/ha) and minimum (7.20 t/ha) grain yield was obtained with 6 and 0 kg/ha (Mondal *et al.*, 2004; Sahaa *et al.*, 2007; Singh and Gupta, 1986 and Singh *et al.*, 1997). To get clear-cut response of both the factor on rice yield, combined two way data is presented in table 12.1. Perusal of data revealed that minimum rice yield (7.09 t/ha) was recorded with absolute control plots where no application of zinc and sulphur was done during entire experimentation period. Whereas corresponding maximum (7.72 t/ha) rice yield was recorded with combined application of 30 kg sulphur and 6 kg zinc (Mondal *et al.*, 2004 and

Singh *et al.*, 2011). This combined analysis suggests that for better output and for balanced nutrition combined application is advocated.

Similarly, highest lentil seed yield (1147 kg/ha) was recorded with 30 kg sulphur treatment whereas, lowest yield (1015 Kg/ha) was noticed with no application of sulphur (Singh *et al.*, 2002; Singh and Singh, 2008 and Thiyagarajan *et al.*, 2003). To get comprehensive and combined residual

Table 13.1 Effect of sulphur and zinc nutrition on rice seed yield (kg/ha) during 2010

Treatments	S_1(0kg)	S_2(20kg)	S_3(30kg)	S_4(40kg)	Mean of Zn
Zn_1(0kg)	7.09	7.17	7.23	7.27	7.19
Zn_2(4kg)	7.22	7.31	7.36	7.39	7.32
Zn_3(5kg)	7.29	7.46	7.52	7.53	7.45
Zn_4(6kg)	7.31	7.58	7.63	7.53	7.51
Mean of S	7.23	7.38	7.43	7.43	
CD (±5%)	Mean effects 0.12			Interaction 0.27	

response of both the nutrients on lentil seed yield two data is presented in table 13.2. Results of interaction of both the nutrient was synergistic in manner up to extant possible except at highest level. Results revealed that minimum lentil seed yield of 960 kg/ha was recorded with control, where nothing has been applied, whereas corresponding maximum lentil seed yield (1243 kg/ha)was recorded with combined application of 30 kg sulphur and 6kg zinc (Singh and Bhatt, 2015; Singh and Gupta, 1986; Singh *et al.*, 2011 and Thiyagarajan *et al.*, 2003).

Table 13.2 Effect of sulphur and zinc nutrition on seed yield (kg/ha) yield of lentil

Treatments	S_1(0kg)	S_2(20kg)	S_3(30kg)	S_4(40kg)	Mean of Zn
Zn_1(0kg)	960	1003	1073	1103	1035
Zn_2(4kg)	1038	1138	1118	1125	1105
Zn_3(5kg)	1070	1125	1153	1131	1120
Zn_4(6kg)	992	1143	1183	1112	1108
Mean of S	1015	1102	1132	1118	
CD (±5%)	Mean effects 34.7			Interaction (69.3)	

Fig. 13.2 Filed view of lentil crop at onset of reproductive growth stage; showing residual effects of sulphur and zinc on lentil in rice lentil cropping system

Effects of sulphur and zinc nutrition on content and uptake of NPK in rice and lentil

To know the effect of sulphur and zinc on NPK concentration (%) and uptake (kg/ha) in rice (grain + straw) and lentil (grain + straw) under rice–lentil cropping systems, samples were analyses for above purpose. Data presented in table 13.3 and table 13.4, indicates that concentration and uptake of nitrogen, phosphorus and potassium due to application of sulphur and zinc, influenced significantly. Highest concentration (1.25 and 1.96 percent) of nitrogen in rice and lentil was recorded in plots fertilized with zinc 6kg. The lowest concentration (1.10 and 1.83 percent)of nitrogen was noticed in case of rice and lentil plots treated with 0 kg sulphur and 0 kg zinc. Likewise response was also obtained in case of phosphorus and potassium concentrations in both the crops. Highest (0.57 %) and lowest (0.42 %) P concentration in rice recorded with 0 kg zinc application and 0 kg sulphur application. This is clear-cut indication of sulphur induce P deficiency and antagonistic relation of P and Zn (Islam *et al.*, 2009; Singh and Singh, 2008 and Thiyagarajan *et al.*, 2003). K concentration in rice and lentil had been influenced by both the tested nutrients. Sulphur influenced rice greatly whereas, in case of lentil K concentration gets improved with increasing doses of Zn (Table 13.3). It was noticed that the concentration of all the three major nutrients (NPK) across the treatment was higher side in case of lentil in comparison to rice (Singh *et al.*, 2001; Sahaa *et al.*, 2007).

Table 13.3 Effect of sulphur and zinc nutrition on NPK concentration (%) in rice (grain + straw) and lentil (grain + straw) under rice –lentil cropping systems

Treatments	N (%)		P (%)		K (%)	
	Rice	Lentil	Rice	Lentil	Rice	Lentil
S_1(0kg)	1.10	1.85	0.42	0.74	0.73	0.91
S_2(20kg)	1.17	1.89	0.48	0.76	0.75	0.93
S_3(30kg)	1.21	1.91	0.46	0.79	0.74	0.87
S_4(40kg)	1.23	1.98	0.52	0.85	0.83	0.98
Zn_1(0kg)	1.09	1.83	0.57	0.71	0.78	0.93
Zn_2(4kg)	1.19	1.84	0.51	0.73	0.76	0.95
Zn_3(5kg)	1.23	1.92	0.48	0.66	0.78	0.76
Zn_4(6kg)	1.25	1.96	0.44	0.57	0.76	0.94
CD (±5%)	0.08	0.11	0.05	0.07	0.03	0.06

Uptake of NPK by rice and lentil is accordingly to the content and their above ground biomass they produced. Since biomass produced by the respective crops and contents of NPK are influenced significantly, similar trend was also repeated in case of uptake. In general maximum uptake of all three nutrients was recorded in case of rice (Table 13.4). Highest N uptake 281.2kg/ha was recorded with 6 kg zinc application, corresponding maximum uptake of P (119.2 kg/ha) and K (179.2 kg/ha) was recorded in the plot supplemented with sulphur @40 kg/ha in the cropping system (Deokaran, *et al.*, 2014; Islam *et al.*, 2009; Sahaa *et al.*, 2007; Singh and Gupta, 1986 and Singh and Singh, 2008).

Effects of sulphur and zinc nutrition on fertility status of soil

Perusal of data presented in table 13.5 vindicate that the initial value of pH and EC was recorded normal, however organic carbon (%) and K was in medium in category. Phosphorus is the only mineral which content was recorded and classified as high (28.6 kg/ha). Sulphur and zinc content was recorded below bench mark level. After the end of three years crop cycle, pH and EC value were recorded normal. There are gradual build-up of organic carbon (%), N, P and K due to application of sulphur and zinc; however significant build-up was noticed in case of sulphur and zinc content in the soil of experimental plots and it cross the critical limits and categorized as normal. The data shows that every parameter was improved with the application of the minerals, this not only improving overall productivity of cropping system but also improving fertility status of soils as well (Deokaran, *et al.*, 2014).

Fig. 13.3 Close-up view of lentil crop at onset of reproductive growth stage; showing residual effects of sulphur and zinc on lentil in rice lentil cropping system

Table 13.4 Effect of sulphur and zinc nutrition on NPK uptake (kg/ha) in rice (grain + straw) and lentil (grain + straw) under rice–lentil cropping systems

Treatments	Nitrogen (kg/ha)			Phosphorus (kg/ha)			Potash (kg/ha)		
	Rice	Lentil	Total	Rice	Lentil	Total	Rice	Lentil	Total
S_1(0kg)	194.0	46.9	240.9	74.1	18.8	92.8	128.7	23.1	151.8
S_2(20kg)	205.6	53.4	259.0	84.3	21.5	105.8	131.8	26.3	158.1
S_3(30kg)	214.3	55.4	269.8	81.5	22.9	104.4	131.1	25.3	156.3
S_4(40kg)	222.9	58.3	281.2	94.2	25.0	119.2	150.4	28.8	179.2
Zn_1(0kg)	190.0	50.1	236.7	97.6	19.3	116.9	133.5	25.3	158.9
Zn_2(4kg)	212.5	51.9	264.3	91.1	20.7	111.7	135.7	26.9	162.6
Zn_3(5kg)	218.2	53.8	271.9	85.1	18.5	103.6	138.4	21.3	159.6
Zn_4(6kg)	223.5	55.7	279.2	78.7	16.2	94.9	135.9	26.7	162.6

Table 13.5 Effect of sulphur and zinc nutrition on soil fertility status

Treatments	pH	EC ds/m	OC (%)	N(kg/ ha)	P(kg/ ha)	K(kg/ ha)	S (ppm)	Zn (ppm)
Initial Value	6.80	0.19	0.71	244.7	28.6	185.8	0.81	8.3
Status (I)	Normal	Normal	Medium	Low	High	Medium	Low	Low
S_2(20kg)	6.85	0.19	0.72	257.3	33.1	190.6	0.94	9.2
S_3(30kg)	6.91	0.20	0.73	268.9	30.1	193.7	0.99	9.7
S_4(40kg)	6.95	0.20	0.74	286.5	27.3	189.9	1.07	9.8
Zn_1(0kg)	6.86	0.18	0.70	242.9	28.5	183.4	0.83	8.1
Zn_2(4kg)	6.84	0.19	0.71	254.3	30.7	185.9	0.91	8.7
Zn_3(5kg)	6.84	0.19	0.74	271.4	32.9	187.3	0.96	9.7
Zn_4(6kg)	6.83	0.20	0.76	289.4	34.2	193.2	0.98	1.1
Status (F)	Noral	Normal	Medium	Low	High	Medium	Normal	Normal

Feasibility of developed technology

To test the feasibility of developed technology i.e. sulphur and zinc nutrition in rice-lentil cropping system in the Indo-Gangetic plains of Bihar, input and output relation was worked out based on extra investment incurred for added inputs as other cost of cultivation is remain same and also to stay away from detailed explanation. Extra expenditure, extra incomes and net extra returns were calculated and presented in table 13.6. Highest extra net returned (.6611/ha) was noticed in case of combined application of 30 kg sulphur and 6 kg zinc

Fig. 13.4 Close-up view of lentil crop clearly showing effects of sulphur and zinc on lentil maturity

Table 13.6 Effect of sulphur and zinc nutrition on economics of developed technology

Treatments	S_1(0kg)			S_2(20kg)			S_3(30kg) (./ha)			S_4(40kg)			Mean of Zn		
	Expenditure	Extra incomes	Net extra returns	Expenditure	Extra incomes	Net extra returns	Expenditure	Extra incomes	Net extra returns	Expenditure	Extra incomes	Net extra returns	Expenditure	Extra incomes	Net extra returns
Zn_1(0kg)	0	0	0	200	1451	1251	300	3426	3126	400	4351	3951	225	2307	2082
Zn_2(4kg)	1000	2548	2548	1200	5393	4193	1300	5198	3898	1400	5528	4128	1225	4667	3691
Zn_3(5kg)	1200	3685	3685	1400	5913	4513	1500	6901	5401	1600	6439	4839	1425	5734	4609
Zn_4(6kg)	1400	1962	1962	1600	6996	5396	1600	8211	6611	1800	5992	4192	1600	5790	4540
Mean of S	900	2049	2049	1100	4938	3838	1175	5934	4759	1300	5577	4277			

Table 13.7 Effect of sulphur and zinc nutrition on economics of extra cost of cultivation v/s excess return

Treatments	Extra Gross Income (. /ha)		Gross Extra Income (.ha)	Extra cost of Cultivation (./ha)	Net Profit (./ha)	BC ratio*
	Rice	Lentil				
S_1(0kg)	0	0	0	0	0	0
S_2(20kg)	753	2446	3199	200	2999	15.9
S_3(30kg)	1075	3293	4368	300	4068	14.6
S_4(40kg)	1230	2994	4224	400	3824	10.6
Zn_1(0kg)	0	0	0	0	0	0
Zn_2(4kg)	1032	1715	2747	1000	1747	2.7
Zn_3(5kg)	1603	1968	3571	1200	2371	3.0
Zn_4(6kg)	2015	1842	3857	1400	2457	2.8

***Note:** Benefits Cost Ratio (B:C Ratio) was calculated based on gross extra income and extra cost of cultivation involved.

Likewise extra Benefits Cost Ratio (B: C Ratio) of added inputs and corresponding output is presented in the table 13.7. It is interesting to note that B: C ratio has follows different fashion for different added nutrients. With increasing dose sulphur the response ratio in terms of per rupee investment is decreasing in order with maximum B: C ratio (15.9) in case of plot treated with 20 kg sulphur and minimum (10.6) with 40 kg sulphur. In case of zinc application, the response ratio recorded increasing in trend up to 5kg (B: C ratio of 3.0) and then starts decreasing (Bastia *et al.*, 2008; Singh *et al.*, 2001 and Singh and Singh, 2008).

References

Ahlawat IPS and Ali M. 1993. Fertilizer Management in Food Crops. In: Nutrient Management for Pulses - A Review (Tandon, H.L.S. Eds.). pp. 114-138. FDCO, New Delhi.

Ali R I, Awan T H, Ahmad M M, Saleem U and Akhtar M. 2012. Diversification of Rice-Based Cropping Systems to Improve Soil fertility, Sustainable Productivity and Economics. *J. Animal & Plant Sciences* 22(1): 108-112.

Bastia D K, Garnayak, L M and Barik T. 2008. Impact of rice - based cropping systems for higher productivity and income per unit area. *Indian J. Agron* 53(1): 63-66.

Deokaran, Singh SB and Ram Kewal. 2014. Effect of Zinc and Boron application on Yield of Lentil and Nutrient Balance in the Soil under Indo-Gangetic Plain Zones. *Journal of AgriSearch* 1(4): 206-209.

Islam M, Ali S and Hayat R. 2009. Effect of integrated application of phosphorus and sulphur on yield and microutrient uptake by chickpea (*Cicer arietinum* L.). *Int. J. Agric. Biol.* 11: 33–38.

Joshi P K 1998. Performance of Grain Legumes in the Indo-Gangetic Plain in Residual Effects of Legumes in Rice and Wheat Cropping Systems of the Indo-Gangetic Plain. Legumes in Rice and Wheat Cropping Systems of the Indo-Gangetic Plain, pp 3-13. Kumarrao, J.V.D.K. Johansen, C. and Rego, T. J. (Eds). ICRISAT, Patancheru, Andhra Pradesh.

Mondal S S, Ghosh A, Acharya D and Maiti D. 2004. Production potential and economics of different rainfed rice (*Oryza sativa*) - based utera cropping systems and its effect on fertility build-up of soil. *Indian J. Agron* 49(1): 6-9.

Ramakrishna A, Gowda C L L and Johansen C. 2000. Management factors affecting legumes production in the Indo-Gangetic Plain. Legumes in rice and wheat cropping systems of the Indo-Gangetic Plain - constraints and opportunities, pp 156-165. Johansen, C. Duxbury, J.M. Virmani, S.M. Gowda, C.L.L. (Eds.) ICRISAT, Patancheru, Andhra Pradesh.

Reddy A A. 2009. Pulses Production Technology: Status and Way Forward. *Economic & Political Weekly*. 44(52): 73-80.

Sahaa P K, Ishaqueb M, Salequeb M A, Miahb, M A M, Panaullahb G M and Bhuiyanb NI. 2007. Long Term Integrated Nutrient Management for Rice Based Cropping Pattern: Effect on Growth, Yield, Nutrient Uptake, Nutrient Balance Sheet, and Soil Fertility. *Communications in Soil Science and Plant Analysis* 38(5-6): 579-610.

Singh A K, Meena M K and Bharati R C. 2011. Sulphur and Zinc Nutrient Management in Rice Lentil Cropping System. 27-29 December, 2011(in) *Proceedings of International Conference on Life Science Research for Rural and Agricultural Development*, pp. 66-67. CPRS Patna Bihar.

Singh A K, and Singh N P. 2008. Yield and uptake of primary nutrients by large seeded varieties of lentil under varying seed rates in normal and late sown conditions. *Journal of Food Legumes* 20(2): 187-189.

Singh AK and Bhatt BP 2015. Late sown lentil performance in response to foliar application of zinc. *Bangladesh Journal of Botany* 44(1): 125-128.

Singh AK, Bhat BP, Sundaram PK, Kumar S, Bharati RC, Chandra N and Rai M.2012.Study of Site Specific Nutrients Management of Cowpea Seed Production and their Effect on Soil Nutrient Status. *Journal of Agril.* Sci. 4(10): 191-198.

Singh G, Singh V, Singh O P and Singh R K.1997. Production potential of various cropping systems in flood prone areas of Eastern Uttar Pradesh. *Indian J. Agron.* 42(1): 9-12.

Singh K, and Gupta VK. 1986. Response of chickpea to zinc fertilization and its critical level in soils of semi-arid tropics. *Fertilizer Research* 8: 213-218.

Singh S K, Verma S C, Singh R P. 2001. Effect of integrated nutrient management on yield, nutrient uptake and changes in soil fertility under rice (*Oryza sativa*)-lentil (*Lens culinaris)* cropping system *Indian J. Agron.* 46(2): 191-197.

Singh S K, Verma S C. and Singh R P. 2002. Integrated nutrient management in rice and its residual effect on lentil. *Indian J. Agric. Res.* 36(4): 286-289.

Thiyagarajan T M, Backiyavathyand M R and Savithri P. 2003. Nutrient Management for Pulses - A Review. *Agric. Res* 24(1): 40-48.

Citation

Singh AK, Meena MK and Bharati RC. 2018. Managing sulphur and zinc for sustainable rice -lentil production in Indo Gangetic Plain In: Scientific Lentil Production. Singh AK, Bhakta N, Sangale UR, Manibhushan, Sundaram PK, Kumar S and Yasin JK. (eds.) Society for Upliftment of Rural Economy Varanasi, India. Pp. 237-249.

❏❏❏

Chapter 14

Efficient Tillage Management Options for Optimizing Lentil Production under Drought-Prone Rainfed Rice–Lentil Cropping System

SS Singh, AK Singh, S Kumar, JS Mishra, AA Haris, UR Sangle, BP Bhatt, Shishir K Singh, A Yadav, US Singh and S Singh

SYNOPSIS

To improve land productivity, food and nutritional security in this region, rice- lentil is one of the major cropping systems sustained under residual moisture condition. The study was aimed at to identify suitable lentil genotypes which can perform better under varying field conditions and management practices prevalent in the region. Five lentil genotypes and two tillage practices were evaluated under rice-lentil cropping system. The performance of different lentil genotypes with zero tilled and conventional tillage was variable under puddle transplanted rice-lentil cropping system. Among the tested genotypes, Mallika was better (1.46 t/ha) under zero tillage whereas Pusa Vaibhav (1.52t/ha) was superior under conventional tillage. In case of lentil succeeding unpuddled transplanted rice, the performance of Pusa Vaibhav was best (1.53 t/ha) under zero as well conventional tillage. Overall, the performance of lentil crop was better under unpuddled transplanted rice condition. Lentil variety Pusa Vaibhav performed equally well with zero as well as conventional tillage under unpuddled transplanted rice, whereas Mallika performed well under zero tillage in sequence with puddle transplanted rice. Both the genotypes being of medium duration are suitable for rice- fallow areas of eastern region of India.

Keywords: Cropping System, Lentil, Tillage options, genotypes.

INTRODUCTION

Rainfed rice system in eastern India has high scope to increase productivity, water and input use efficiency (Singh *et al.*, 2014b). Of the rice area of 28 million hectares in eastern region, only 47 percent is under irrigation and rest is rainfed which has enormous scope to increase productivity, water and input use efficiency. In Bihar 1.63 lac ha of land are affected by drought, which has potential to produce additional tonnes of food grains. Frequent occurrence of drought creates water scarcity problem at the time of sowing of different crops (Singh *et al.*, 2012). There is need of development of effective cropping system for rainfed area, so that farmers may get maximum output. To improve land productivity, food and nutritional security in this region, rice- fallow areas must be brought under cultivation with pulse crops that can survive in residual moisture. Lentil is a major legume crop and main source of dietary protein in eastern region (Ali *et al.*, 2012 and Singh *et al.*, 2013a). Rice - lentil is a second prominent cropping system of Bihar under rainfed agriculture or water scarce condition (Singh *et al.*, 2014a and Singh *et al.*, 2013b). Many lentil genotypes are evaluated fitting into this situation. These genotypes would respond differently when preceding rice crop was taken under puddle and unpuddled conditions. The puddling in rice has many adverse affects on soil structure that may influence the productivity of subsequent crops. The lentil genotypes are also tried under Zero tillage and Conventional tillage. The effort is to save energy and resources using ZT following suitable varieties. ZT also provide adequate time for planting in generally late available fields after long duration rice crop in preceding season. Shortage of irrigation water and labour during peak periods make transplanting and manual weeding costly, invariably causing delays in farm operations (Laik *et al.*, 2014). As we knows that puddling is prerequisite for rice transplanting deteriorates the physical properties of soil, which forms a compacted layer (plough pan) that restricts percolation of water causing water logging and resists root penetration and growth of the following crops such as lentils in rainfed systems where land preparation is done quickly to restrict evaporation of soil moisture (Singh *et al.*, 2014b). This study was aimed at to identify suitable lentil genotypes which can perform better under varying field conditions and management practices prevalent in the region. Due to puddling land preparation becomes difficult and requires more energy to achieve proper soil tilth for sowing of succeeding crops in dry season and it may cause some negative impact also on succeeding dry season crops, especially on pulses which require a good root growth for better nitrogen fixation (Laik *et al.*, 2014). As an alternative practise, rice seedlings can be planted after rainfall without puddling at saturation level of soil on clean field which save the soil from harmful effect of puddling and encourages good growth of succeeding dry season crop. Farmers normally go for surface seeding of lentil to avoid delay in planting and to harness maximum benefits of residual soil moisture. However, the same benefits can be achieved through use of conservation agriculture option of zero tillage (ZT) with significantly enhanced grain harvest of lentil over what is achievable through surface seeding. ZT reduces the cost of cultivation by approximately Rs 2,500-3,000/ha, facilitates early seeding (7-10 days), provides better scope for utilization of underutilized / unutilized or rice-fallow lands and enhances system sustainability and profitability. In contrast, conventional system increases the cost of cultivation, delay the planting, reduces the yield and responsible for less utilization of underutilized/ fallow lands. Therefore in this experiment we will test the effect of zero tillage and conventional tillage on various genotypes of lentil and the overall effect of puddled and unpuddled rice on the productivity of succeeding lentil crop.

Keeping the above facts in mind, an experiment was planned and executed to test the overall effect of puddled and unpuddled rice on the productivity of succeeding lentil crop. Five lentil genotypes viz. Arun, PL406, Mallika, Pusa Vaibhav and Shivalik and two tillage practices i.e., zero tillage (ZT) and conventional tillage were evaluated under puddled and un-puddled transplanted rice field at research farm of ICAR-RCER, Patna, Bihar, India during 2012-13 to 2013-14 dry season. The experiment was laid out in split plot design, with three replications under both puddled and unpuddled transplanted rice field condition. Observations of yield and yield attributes such as germination percentage, biomass, number of pods per plant, number of seed per pod, number of nodules per plant, harvest index and were recorded.

The treatments were lentil genotypes and tillage methods.

Lentil varieties: 5

1. Arun,
2. PL 406,

3. Mallika,
4. Pusa Vaibhav
5. Shivalik

Tillage: 2

Zero Tillage (ZT)

Conventional Tillage (CT)

Treatments: 10

Design: RBD

Replications: 3

Plot Size: 5mx5m= 25 m^2

Seed rate and spacing

30-35 kg/ha

- Line to line distance: 25 cm
- Plant to plant distance : 10 cm

2. Field preparation (As per the treatment)
3. Fertilizer use 20 kg N and 40 kg P_2O_5 per ha
4. Irrigation (Only supplemental irrigation was considered)
5. Weeding

Fields were kept weed free as much as possible especially early in the season. First hand weeding was done after 25-30 DAS and the second after 45-50 DAS. The lentil crop was established following BMP but the land preparation was as per the treatment. Seed Treatment: The seeds were treated with 2.5 g Difoltan, Thiram or Capton per kg of seed 24 hours before sowing. Treat the seed with 2.5 g Chlorpyriphos 20 EC insecticide @ 3 ml/kg seed for control of cut worm. Treated the seed with Rhizobium and PSB culture; @ 500 g of *Rhizobium* and 1.5 kg of PSB culture for one hectare seed.

Fig.14.1 Field view of lentil crop under different tillage management practices

RESULTS AND DISCUSSION

This field trial was conducted for comparative evaluation of lentil genotypes under different soil configuration for rainfed drought- prone ecosystem. Observations were recorded and were analysed and presented in tables. Data presented in table 14.1 reveal the performance of different lentil genotypes under rice (PT) - lentil cropping system differ under zero tilled as well as conventional tillage conditions .There was difference in response by tested lentil genotypes. In case of pods/ plant, maximum (35.1) was recorded with genotype Mallika under conventional tilled situations. Mallika was better (1.46 t/ha) under zero tillage condition as compare to other tested genotypes. Maximum grain yield (1.52t/ha) was recorded with genotype Pusa Vaibhav under conventional tilled situation.

Fig. 14.2 Field view of lentil crop under different tillage management practices at different maturity stages

Table 14.1 Lentil performance under rice (Puddled)- lentil cropping system

Varieties	Tillage	Pods / plant	Grain yield (t/ha)
Arun	ZT	16.8	1.10
	CT	20.2	1.19
Pl-406	ZT	23.3	1.34
	CT	25.3	1.46
Mallika	ZT	33.3	1.46
	CT	35.1	1.51
Pusa Vaibhav	ZT	29.3	1.45
	CT	31.3	1.52
Shivalik	ZT	31.4	1.36
	CT	34.7	1.44
CD± (5%)	3.1	0.04	
V	1.88	0.14	
T	1.19	0.09	
V*T	2.66	0.19	

Performance of lentil genotypes under rice (UPT)-lentil cropping system differ under ZT and CT shown in table 14.2 In case of pods/ plant, maximum (33.1) was recorded with genotype Shivalik under conventional tilled situations. The performance of cultivars like and Pusa Vaibhav was better (1.53 t/ha) under zero tillage as well conventional tillage.

Table 14.2 Lentil performance grain yield (t/ha) under rice (UPT)-lentil cropping system

Varieties	Tillage	Pods / plant	Grain yield (t/ha)
Arun	ZT	27.2	1.26
	CT	29.8	1.12
Pl-406	ZT	26.8	1.48
	CT	29.8	1.42
Mallika	ZT	30.7	1.50
	CT	29.2	1.51
Pusa Vaibhav	ZT	27.4	1.53
	CT	29.8	1.52
Shivalik	ZT	32.3	1.53
	CT	33.1	1.47
CD± (5%)			
V	1.56	0.11	
T	0.99	0.07	
V*T	2.21	0.15	

Performance of tested lentil cultivar under puddled transplanted and unpuddle transplanted were summarized in table 14.3. Results indicate that in case of lentil grown under puddled transplanted rice field, conventional tilled field produced significant maximum lentil yield (1.43 t/ha) over zero tilled field performance (1.34 t/ha). Similarly in case of under unpuddled transplanted rice field where lentil was grown, significant maximum lentil yield (1.46t/ha) was recorded in case zero tilled as compare to conventional (1.41 t/ha).The overall lentil productivity (1.44 t/ha) in UPT system is higher over PT system (1.39 t/ha).

Table 14.3 Lentil grain yield (t/ha) under various tillage combinations under rice based cropping system

	Puddled Transplanted		**Un-puddled Transplanted**	
	Con. Tilled	**Zero Tillage**	**Con. Tilled**	**Zero Tillage**
Arun	1.19	1.11	1.18	1.26
PL 406	1.46	1.34	1.48	1.48
Mallika	1.52	1.47	1.51	1.51
Pusa Vaibhav	1.53	1.45	1.52	1.53
Shivalik	1.47	1.36	1.47	1.53
Average	1.43	1.34	1.41	1.46
Average of PT system= 1.39				
Average of UPT system=1.44				
CD± (5%)				
V	0.14	0.11		
T	0.09	0.07		
V*T	0.19	0.15		

Data presented in table 14.4 showed that cultivars like Mallika, Pusa Vaibhav and Shivalik perform better under conventional tillage as compared to zero tilled across the previous crop grown condition be it puddled transplant or un-puddle transplant of rice. Over all Pusa Vaibhav and Mallika produced highest (1.49t/ha) under zero tillage conditions, corresponding (1.53t/ha) seed yield was obtained by cultivar Pusa Vaibhav under conventional tillage conditions. B: C ratio was highest in un-puddled transplanted rice field condition (2.98) followed by puddled conventional tillage (2.78).

Table 14.4 Lentil varietal performance grain yield (t/ha) under various tillage combinations in rice based cropping system

Genotypes	Con. Tilled	Zero Tillage
Arun	1.23	1.11
PL 406	1.47	1.38
Mallika	1.51	1.49
Pusa Vaibhav	1.53	1.49
Shivalik	1.50	1.41
	CD± (5%)= 0.04	

Conclusion

Performance of lentil variety Pusa Vaibhav (1.53t/ha) was found best under conventional tillage, whereas Mallika and Pusa Vaibhav Performed equally good (1.49 and 1.49 t/ha) under zero tillage.The performance of different lentil genotypes with zero tilled and conventional tillage was variable under puddle transplanted rice-lentil cropping system. Among the tested varieties, Mallika was better (1.46 t/ha) under zero tillage whereas Pusa Vaibhav (1.52 t/ha) was superior under conventional tillage. In case of lentil succeeding un-puddled transplanted rice, the performance of Pusa Vaibhav was best (1.53 t/ha) under zero as well conventional tillage

References

Ali RI, Awan TH, Ahmad MM, Saleem U and Akhtar M. 2012. Diversification of rice-based cropping system to improve soil fertility, sustainable productivity and economics. *Journal of Animal and plant sciences* 22(1): 108-12.

Laik Ranjan, Sharma Sheetal, Idris M, Singh AK, Singh SS, Bhatt BP, Saharawat Yashpal, Humphreys E and Ladha JK. 2014. Integration of conservation agriculture with best management practices for improving system performance of the rice –wheat rotation in the Eastern Indo-Gangetic Plains of India. *Agriculture, Ecosystems & Environment* 195(6): 68-82.

Singh AK, Manibhushan, Bhatt BP, Singh KM and Upadhyaya A. 2013a. An Analysis of Oilseeds and Pulses Scenario in Eastern India during 2050-51. *Journal of Agril. Sci.* 5(1): 241-249.

Singh AK, Meena MK, Bharti RC and Gade RM 2013b. Effect of Sulphur and zinc management on yield, nutrient uptake, change in soil fertility and economics in rice (*Oryza sativa* L.) -lentil (*Lens Cularis*) cropping system. *Indian journal of Agricultural Sciences* 83(3): 344-348.

Singh AK, Singh KM, Bharati RC, Chandra N, Bhatt BP and Pedapati A. 2014a. Potential of Residual Sulphur and Zinc Nutrition in Improving Powdery Mildew (Erysiphe trifolii) Disease Tolerance of Lentil (*Lens culunaris* L.).*Communication in Soil Science and plant analysis* 45: 2807-2818.

Singh SS, Singh AK, Sundaram PK. 2014b. Agrotechnological options for upscaling agricultural productivity in eastern indo gangetic plains under impending climate change situations: A review. *Journal Agrisearch* 1(2): 55-65.

Singh AK, Bhat BP, Sundaram PK, Chndra N, Bharati RC, Patel SK.2012. Faba bean (*Vicia faba* L.) phenology and performance in response to its seed size class and planting depth. *Int. J. Agril. & Stat. Sci.* 8(1): 97-109.

Citation

Singh SS, Singh AK, Kumar S, Haris AA, Sangle UR, Bhatt BP, Yadav A, Singh US, Singh S and Singh Shishir K. 2018. Efficient tillage management options for optimizing lentil production under drought-prone rainfed rice–lentil cropping system. In: Scientific Lentil Production. Singh AK, Bhakta N, Sangale UR, Manibhushan, Sundaram PK, Kumar S and Yasin JK. (eds.) Society for Upliftment of Rural Economy Varanasi, India. Pp. 251-259.

□□□

Chapter 15

Water Management for Profitable Lentil Production

AA Saad, Parmeet Singh, Zahida Rashid, FA Pir, Amjad Masood and Wasim Raja

SYNOPSIS

Lentil is an important winter legume, a rich source of protein. Lentil plants are herbaceous with quadrangular stem with 4-6 primary branches sprout out directly from the main stem, emerging from the cotyledon node below ground or from nodes above ground. The plant is hardy and can tolerate frost and severe winter to a great extent. It is also moderately drought tolerant crop. Due to its indeterminate growth habit, the plant continues to flower as long as growing conditions like moderate air temperature and soil moisture remain favourable for vegetative growth. The GDD was higher with irrigated condition with more during vegetative to flowering and flowering to podding, while, it was more only during podding to maturity with non-irrigated condition. Excess soil moisture before the plant is in full bloom can delay and reduce seed set. Excess moisture near the time of harvest may encourage the spread of fungal diseases. Planting dates of lentil experiments showed that end of October was the optimal period in north western plain zone (NWPZ) whereas in north eastern plain zone (NEPZ), second fortnight of November was optimal date subject to adequacy of moisture at sowing time. Lentil crop is rarely grown as irrigated crop. Lentil grown as rainfed crop requires 1-2 supplemental irrigations in case of failure of winter rains during the growing season. First irrigation is scheduled at 6 weeks after sowing and the second at flowering or pod formation stage. It requires at least moderate moisture of 150-250 mm during its growing period of 150-170 days to produce a full seed set. Research indicated that productivity of lentil increased with increase in irrigation frequency and total water use increasing the available water to 300 mm will double lentil yield to 3.52 t /ha. In other experiment on water management, seed and biological yield of lentil increased by supplementary irrigation as the average seed yield of 1.04 t/ ha and biological yield of 4.27 t /ha under dry farming condition reached to 1.81 and 6.21 t /ha, respectively.

Keywords: Lentil, Resource conservation technologies, soil moisture, Water requirement, Water-use efficiency.

Introduction

The lentil [*Lens culinaris* L. (Medik)] is a food legume with typically lens-shaped seed; one or two grow in a short-beaked pod. The plant originated and was among the first crops to be domesticated in the Near East. It has been a part of the human diet since Neolithic times. Lentil seeds have high levels of protein (24 – 26%) which is an essential nutrient for human beings. In Indian sub-continents, lentils are mainly consumed as *dhal* (seeds boiled and mashed into soup) served with rice and *roti*. The important lentil-growing countries of the world are India, Canada, Turkey, Bangladesh, Iran, China, Nepal and Syria. India is the second largest producer of lentil after Canada in the world. It is mostly grown in northern plains, central and eastern parts of India covering the states of Madhya Pradesh, Uttar Pradesh, Bihar and West Bengal. Lentil is a cool season pulse crop and therefore it is not an economical crop in hot tropics. The expansion of irrigated agriculture in northern India has led to displacement of pulses by rice, wheat and maize in large area. Areas under pulses are declined by 20% in north India, while increased by 32% in central and southern India. Most of pulses are grown under moisture stress conditions in rainfed or with limited irrigation. The strongest determinant of growth and yield of lentil under rainfed condition is rainfall and its distribution. Large inter-seasonal fluctuations in weather resulted in larger inter-seasonal fluctuations in water use and yield of lentil. Therefore, water management is one of the most important parts of management practices in lentil production (Singh *et al.*, 2013a).

Lentil crop and its agro-ecosystem

Lentil plants are herbaceous with quadrangular stem with 4-6 primary branches sprout out directly from the main stem, emerging from the cotyledon node below ground or from nodes above ground. The leaves are small, compound and pinnate and sometimes form tendrils at the end of leaflets. Generally, the root system is shallow (60 cm) characterized with slender taproot and fibrous lateral roots. However, rooting pattern ranges from a many-branched shallow root system to less-branched deep root system. The taproot and lateral roots bear round to elongate N_2- fixing nodules in the surface layer of soil. Lentil plants have an indeterminate growth habit. The growth of lentil crop is adversely affected at temperatures above 27 °C. Lentil is a long day plant hence grown as a winter season crop with conserved soil moisture after monsoon season in the northern states with annual rainfall of 350–500 mm. Unlike chickpea, it remains unaffected by rain at any stage of its growth including flowering and fruiting. The plant is hardy and can tolerate frost and severe winter to a great extent. It is also moderately drought tolerant crop. Cold temperature favours vegetative growth while warm temperature enhances maturity of the crop. The optimum temperature for its growth is 18–30 °C. Low temperature less than 10 °C delays germination and reduces vegetative growth. Lentil is raised on light loams and alluvial soils in northern India, and on well-drained, moderately-deep, light-black soils in central India (Singh *et al.*, 2013b). It is also grown on poor soils in low-lands in the northern parts of India. Lentil does best on leveled or slightly rolling land with soil pH ranges from 6.0 to 8.0. The crop can withstand moderate alkalinity but cannot tolerate high salinity and water logging.

Due to its indeterminate growth habit, the plant continues to flower as long as growing conditions like moderate air temperature and soil moisture remain favourable for vegetative growth. Therefore, stress of low moisture and low nitrogen are required to encourage seed set and hasten maturity. The flowering continues also until they encounter heat, frost, mechanical damage or chemical desiccation. This indeterminate growth habit is most predominant in late maturity group varieties, but all current lentil varieties have indeterminate growth habit. The concept of growing degree-days (GDD) has been used to estimate growth stages and maturity dates. The GDD was higher with irrigated condition with more during vegetative to flowering and flowering to podding, while, it was more only during podding to maturity with non-irrigated condition (Vijay Laxmi, 2012). Excess soil moisture before the plant is in full bloom can delay and reduce seed set. Excess moisture near the time of harvest may encourage the spread of fungal diseases (Singh *et al.*, 2014a).

Large-scale on-farm trials conducted by several State Agriculture Universities in five states of India (Chhattisgarh, Jharkhand, Orissa,

West Bengal and eastern Madhya Pradesh) have clearly shown that lentil can be successfully grown after rice harvest. The crop are also grown under *utera* system (relay cropping) in low lands of the eastern Indo-Gangetic plains, which involves broadcasting of seedsin standing matured rice crop without tillage operation (surface seeding) but at huge investment of seeds, as the germination is usually very poor due to lack of better contact of seeds with moist soil (Ali *et al.*, 2012).

Multi-location studies under All India Coordinated for Pulse Improvement Project on planting dates of lentil showed that end of October was the optimal period in north western plain zone (NWPZ) whereas in north eastern plain zone (NEPZ), second fortnight of November was optimal date subject to adequacy of moisture at sowing time. In the Central zone where moisture is the major constraint, an early sowing by mid of October is ideal to ensure proper germination. The yield of lentil declined at the rate of 14.8 kg/ha/day in NWPZand 30 kg/ha/day in NEPZ when planted beyond end of December. In *utera* system, planting time largely depends upon cessation of monsoon rains and maturity of rice crop. In low lands, water from rice field is drained by the end of September or early October and small seeded lentil are broadcasted 4-7 days before harvest of rice (Ali and Mishra, 2000). The optimum seed rate of microsperma lentil (small seeded; 15-25 g/1000-seed) is 40-45 kg/ha, whereas for macrosperma lentil (large seeded; >30 g/1000-seed), it is 55-60 kg/ha. Under *utera* system microsperma lentil are generally grown with increased seed rate of 60 kg/ha (Ali *et al.*, 2012).

Water requirement of lentil

The grain legumes are grown under water-limited conditions. Yields are necessarily limited by the amount of water available to support growth. To quantify this relationship, the following mechanistic expression of maximum grain yield (Y_{max}) was proposed by Sinclair and Vadez (2012).

Y_{max}d" W * HI * k/VPD

Where,

W= Total amount of water available for transpiration expressed in t /ha.

HI = Harvest index,

k = mechanistically defined transpiration coefficient (5–6 Pa for grain legumes).

VPD = mean vapour pressure deficit weighted (Pa) for daily transpiration rate distribution.

The weighted daily VPD is ~75% of the difference between the maximum and minimum vapour pressure.

In water-limited environments, a grain legume crop may not have any more than 150mm (1500 t /ha) of water available to it. Also, the season of growth is arid and the value of VPD may be 2.5 kPa (2500 Pa) or more. Assuming a grain legume crops with HI equal to 0.35 and a k value of 6 Pa, Y_{max} can be calculated as follow:

Y_{max}d" 1500* 0.35 * 6/2500 = 1.26 t/ha.

Clearly, high grain legume yields cannot be expected in such water-limited conditions. Doubling the available water to 300 mm will double yield to 3.52 t /ha. It shows irrigation is needed to increase the productivity of crops growing under water-limited environments like, dry farming, dryland and rainfed conditions. Sinclair and Vadez, (2012) also indicated that aside from irrigation requirement under arid or semi-arid condition, a possible plant trait that could be selected for genetic modification is plant behaviour to decrease the effective-VPD. While VPD is seemingly an environmental variable, the crop could have suppressed leaf gas exchange under high VPD condition, usually occurs in midday, so that the daily effective-VPD for transpiration is shifted to a lower value. A search for this trait within the germplasm of all grain legumes selected for limited watered conditions is needed. In environments of intermittent drought, not only would this trait result in a lower effective-VPD, but it would conserve soil water to extend the period that a crop can remain productive between rain events and avoid the consequences of severe drought.

Crop type is a chief factor influencing irrigation water needs. Crops differ both in terms of their daily water needs and the duration of their total growing period. Crops with high daily needs and a long total growing season require much more water than those with relatively lower daily needs and shorter growing seasons. In general, water requirement in grain legumes is less as compared to cereals. Water requirement of a crop also depends on stored water of soil, amount of precipitations in previous years and soil depth and texture

(Tiwari and Vyas, 1994). Thus availability of soil moisture has been reported to be the pivotal point for realization of yield potential. There are strong linear relationships between yield and moisture supply in lentil (Silim *et al.*, 1993). Lentil requires at least moderate moisture of 150-250 mm during its growing period of 150-170 days to produce a full seed set. Research indicated that productivity of lentil increased with increase in irrigation frequency and total water use (Singh *et al.*, 2014b and Salehi *et al.*, 2006).The biological and seed yields were higher under irrigated condition compared to non-irrigated condition (Vijay Laxmi, 2012).

Seeding of lentil with adequate soil moisture

Lentil has been adapted to grow during the cool season when evapotranspiration is minimal. Planking after seeding is also very important to minimize moisture loss. Their growth usually depends on stored soil moisture for a large part of their crop duration. Poor emergence with deeper seeding (10 cm) and poor germination with shallow seeding (5 cm)might occur on account of low soil moisture in the surface layer. Most of the cool-season pulses exhibit hypogeal germination (Ali *et al.*, 2012). Availability of adequate soil moisture is essential for planting relatively at deeper depth (Fig. 15.1). Deeper planting reduces incidence of *Fusarium* wilt as compared to shallow planting and also promote better root development (Kaul*et al.*, 1974 and Singh *et al.*, 2014a). Seeding depth of 2.5 to 7.5 cm is advised. Efforts should be made to conserve late monsoon rains through appropriate water-conservation measures like contour bunding, deep ploughing, mulching and water harvesting ponds. In general, a pre-sowing irrigation helps in uniform germination and ensures optimum plant population and growth when soil moisture is inadequate at planting (Ali *et al.*, 2012). As long as moisture is available, the crop will continue to flower and set pods. If growers wait for the crop to dry naturally under high-moisture conditions, they risk compromising the integrity and value of the crop. Generally, irrigation scheduling is widely adopted on the basis of moisture sensitive periods and climatological approach using IW: CPE ratio.

Fig. 15.1 Line sown lentil crop under residual soil moisture condition

Moisture sensitive period

The crop requires 150-250 mm of water depending on soil and climate. Lentil crops do not need any irrigation if there is plenty of soil moisture at sowing time and followed by one or two winter showers. In areas of high moisture stress, irrigation may considerably improve the crop productivity. In NWPZ and NEPZ, response to irrigation is generally low due to winter rains and high relative humidity during its growing season. In general legumes need not be irrigated in their early vegetative stages as it may induce excessive vegetative growth as well as poor aeration in heavy textured soil, thus do more harm than good by interfering with the nodulation and oxygen requirement of roots. There are certain growth and developmental stages/period at or during which a crop is affected severely due to moisture stress and the yield loss cannot be compensated by adequate moisture supply in other stages or periods, known as *moisture sensitive growth stages* or critical growth stages for moisture stress. The most critical growth stage for moisture stress in lentil is *pod formation* followed by *flowering initiation* (Ali and Mishra, 2000).It was also reported that the *seedling* and *flowering* stages of lentil were the most sensitive to water availability and drought stress (Salehi, *et al.*, 2006).

Supplemental and deficit irrigation

The term 'supplemental irrigation' generally refers to a *rainfed crop* receiving additional irrigation during the whole season or during *sensitive growth stages.* In contrast to the supplemental irrigation, 'deficit irrigation' generally refers to a *fully irrigated crop* from which water is withheld during certain *tolerant growth stages*. Deficit irrigation has been investigated as a valuable strategy for dry regions (Pereira *et al.*, 2002). Lentil crop is rarely grown as irrigated crop. The availability of improved yielding varieties during recent decade's encouraged Indian farmers for growing pulses as an irrigated crop (Laxmipathi Gowda *et al.*, 2013).Lentil grown as rainfed crop requires 1-2 supplemental irrigations in case of failure of winter rains during the growing season. First irrigation is scheduled at 6 weeks after sowing and the second at flowering or pod formation stage. Maghooli *et al.* (2012) reported that supplementary irrigation in dry farming of lentil especially at 50% flowering stage 'I3' increased mean seed yield of lentil and its components significantly in comparison to no supplementary irrigation 'I1'(Fig. 15.2). In the same experiment, lentil seed yield increased significantly with plant density at 300 plants m^2. The interaction effect of supplementary irrigation and plant density was not significant for seed yield. The relative reduction of water stress during flowering stage prolonged the growth and seed filling periods and improvement of 100 seeds weight. Increase in plant density increased canopy closure rate, leaf area index, environmental factors use efficiency, yield components and ultimately yield.

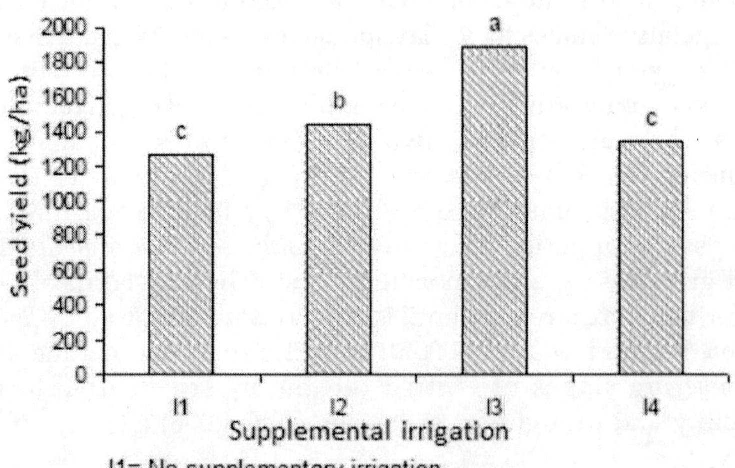

I1= No supplementary irrigation
I2= Supplementary irrigation at planting
I3= Supplementary irrigation at 50% flwering
I4= Supplementary irrigation during pod filling

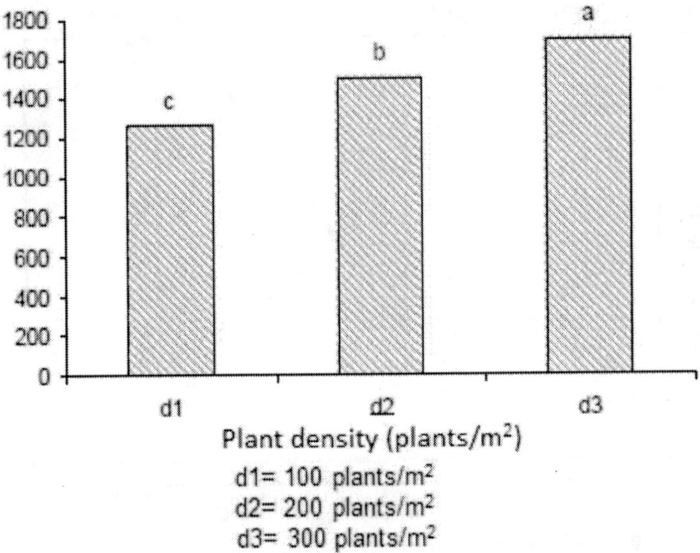

Fig. 15.2 Effect of supplementary irrigation at different growth stages and plant density

The densities lower than optimum did not cover the soil in late season having high temperature and evaporation rate from the soil surface and subsequently the soil water decreased which increased the effect of water deficit.

Water-use efficiency by supplemental and deficit irrigation

Traditionally, agricultural research focused primarily on maximizing total production. In recent years, focus of agriculture production has shifted to the limiting factors in production systems, notably the availability of either land or water. Supplemental irrigation with a limited amount of water, if applied to rainfed pulses during critical stages can result in substantial improvement in yield and crop water productivity (WP) or water-use efficiency (WUE).Under dry land or semi-arid condition, supplementary irrigation could significantly prevent from undesirable effects of water deficit (Pathak *et al.*, 2009; Rockström *et al.*, 2010). WP or WUE is defined as the ratio of the mass of marketable yield (Y_a) to the volume of water consumed by the crop as evapotranspiration (ET_a), i.e. $WP = Y_a / ET_a$. The relationship between Y_a and ET_a can be expressed and explained by crop water production function (CWP-function) as shown by Zhang, 2003; Geerts and Raes, 2009 in Fig.15.3.

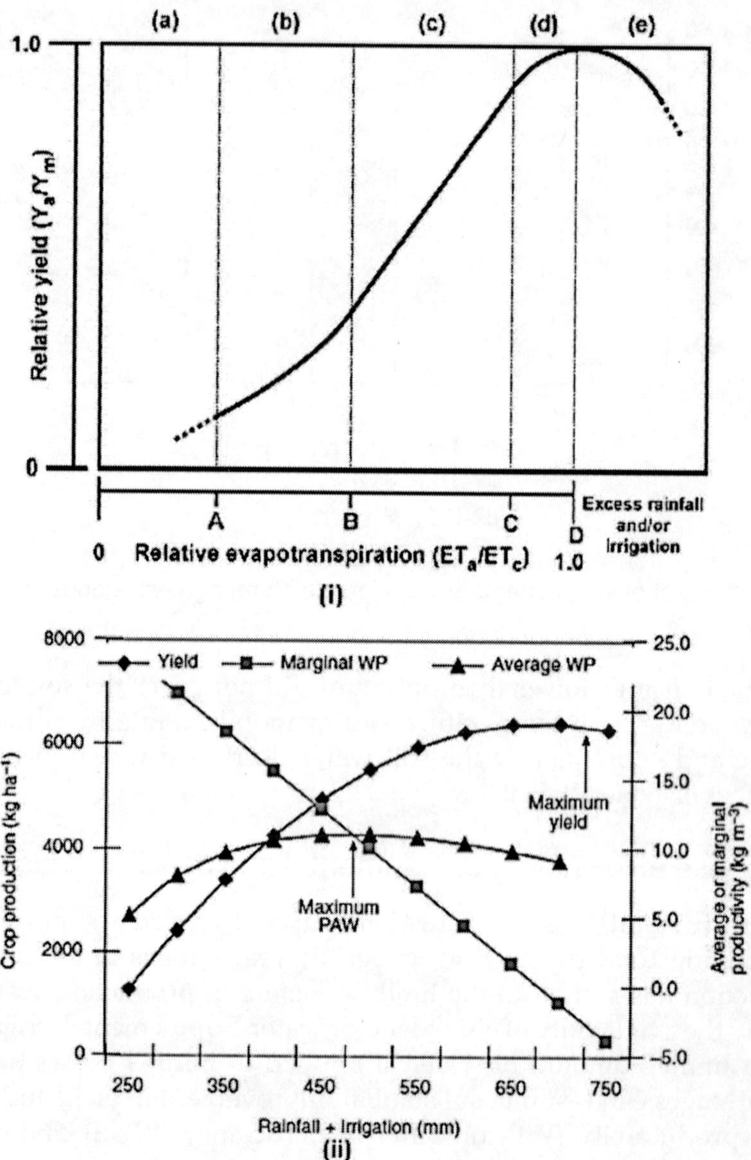

Fig. 15.3 CWP- function explained by (i) Geerts and Raes (2009) and (ii) Zhang (2003)

The CWP-function has a logistic or sigmoid shape (Fig. 14.3i) by plotting relative yield (ratio of actual yield, Y_a, to maximum possible yield under given agronomic conditions, Y_m) versus relative evapotranspiration (ratio of actual evapotranspiration 'ET_a' or seasonal amount of water that is evapotranspired to crop ET under non-stressed standard conditions 'ET_c' also called as seasonal crop water requirement). Within the CWP-function, different sections can be distinguished that may vary in width or that may even be absent. *Section-a* showed low yield or total yield loss happened with insufficient water supply, resulting very low WP. The yield and WP can only be increased if a considerable amount of water is added until *section-b* is reached. Once a minimum amount of water (A) is guaranteed by residual moisture, rainfall or irrigation, yield and WP start to increase with increase in water level. This section has *concave* shape. With additional water supply the production function can become nearly *linear* upto point B in *section-c*. Evapotranspiration increases with water supply, so as the relative ET is reached close to 1 and the increased CWP-function with decrease rate often observed (*section-d*). This section may be quite large for some crops (drought-tolerant), while it may be almost absent in some crops (drought-sensitive). If irrigation/rainfall is excess, the CWP-function has more pronounced S-shape, creating an additional *section-e*. In this section, applying more water than required by ET_c will not increase yield as the water is lost through evaporation and deep percolation, yield might even decline as a result of water logging or nutrient leaching from root zone. On the basis of possible shape of CWP-function in different sections, Geerts and Raes (2009) differentiated drought-tolerant (showing convex shape of CWP function in *section-d*without *section-c*) and drought-sensitive crops (showing linear shape of CWP-function in*section-c*). On this basis, a drought-tolerant crop exhibits maximum WP reached for ET_a lower than ETc, whereas drought sensitive crop displays WP increases until full water requirements are met. Oweis *et al.* (2004) reported a convex quadratic CWPs function for lentil grown as rainfed crop and found that seed and biological yield of lentil increased by supplementary irrigation as the average seed yield of 1.04 t/ ha and biological yield of 4.27 t /ha under dry farming condition reached to 1.81 and 6.21 t/ha, respectively. While supplemental irrigation is not always available, where it is available, it is a means of significantly increasing grain legume yields and WUE (Siddique *et al.*, 2012). Similarly, Zhang (2003) demonstrated a simple model of crop water production function (Fig 14.5 ii) in irrigated condition to achieve maximum productivity of applied water (PAW) from rain and irrigation in relation to *crop productivity* (Y) which is a function of the amount of water received by the crop in terms of rainfall (P) and irrigation (I) i.e. Y= f (P, I); *average productivity* (é) which is output divided by input i.e. é = Y /(P + I) and marginal productivity

(v) which is change in production associated with the addition of one unit input i.e. v= "Y/" (P + I). The maximum yield is achieved when the marginal yield is equal to zero. The arrows in the figure indicate that the maximum PAW value occurs at a lower value of applied water than maximum yield does. Zhang (2003) reported that the yield of lentil increased linearly with the amount of water supplied both from rainfall and irrigation (Fig. 15.3). The most productive use of water was reached with about 440–500 mm of water supply (140–180 mm irrigation) in northern Syria, where PAW in lentil gradually increased with increased in water supply and reached a plateau at a maximum PAW of 0.4 kg/ m³

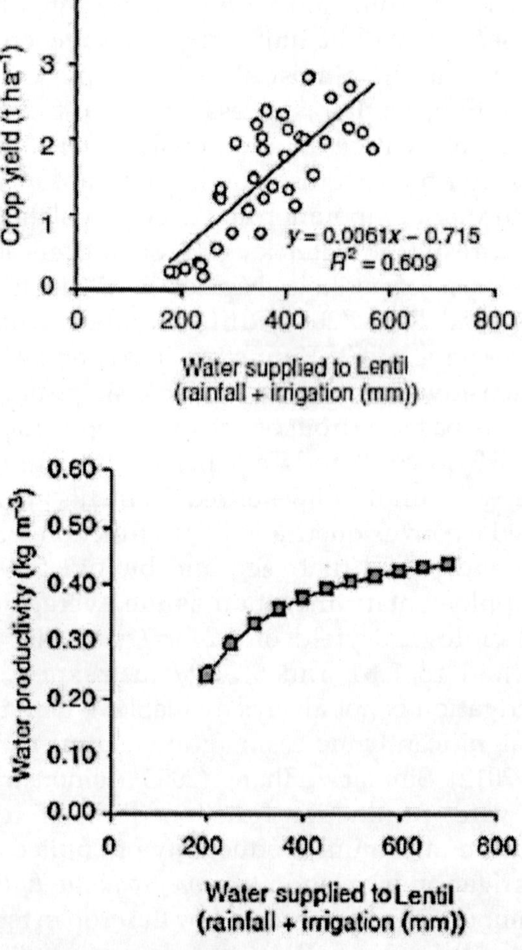

Fig. 15 4 Crop-water production functions in lentil presented by Zhang, 2003.

Lentil can be grown under late sown condition but with assured irrigation facility.The importance of irrigation increased under late sown condition due to poor root development. In irrigated condition, *deficit irrigation* should be practice. The deficit irrigation leads to higher productivity of the water (rainfall and irrigation) than can be attained with fully irrigation and can, therefore, be used for improving the productivity of water in semi-arid areas (Zhang, 2003). Deficit irrigation requires more control over the amount and timing of water application than full irrigation practice. However, information on when and how much to irrigate and sensitive stage to water stress on the basis of climate and soilare needed in order to reduce unwanted effects of water stress on production. Venkateswarlu and Ahlawat (1993) observed significant yield increase owing to irrigation in late sown lentil, higher yield was obtained under wet moisture regime (0.6 IW: CPE) as compared todry moisture regime (0.35 IW: CPE) under Delhi conditions on sandy loam soils.

Fig. 15.5 Managing water efficiently is a key to maximize lentil production, being a rainfed crop

Resource conservation technologies for higher crop productivity and WUE

1. Laser land levelling

Laser land levelling is the process of smoothening within + 2 cm from the average elevation of the field using laser equipped bucket which scraps from higher places and spread onto the low lying areas. It facilitates uniform water distribution, resulting proper crop stand, uniform growth and maturity and yield enhanced by 10 – 25%. In addition to this water can be saved by 35 – 45 % and increased the cultivable area by 3 – 6 % due to reduction in area required for bunds and channels (Jat *et al.*, 2004; and Singh *et al.*, 2014b).

2. Zero tillage practice

Crops grown without or with minimum tillage operation reduces the cost of cultivation, thereby increasing the profit margin of the farmers. In some cases comparable or higher yield is also recorded with zero tillage practice. Cultivation of lentil under *utera* system in the eastern Indo-Gangetic plains is an example of zero tillageas surface seeding which involves broadcasting of seeds in standing matured rice crop without tillage operation. But the requirements of seeds were high, as the germination was usually very poor due to lack of better contact of seeds with moist soil as well as mortality due to Fusarium wilt. Farmers used local land races of lentil at relatively high seed rate (100-150kg/ha) to offset seed mortality due to soil-borne pathogens as well as to ensure proper plant population. Results of farmers participatory trials conducted in this region revealed that zero-till lentil with reduced seed rate (30kg/ha) sown at 5-6 cm deep helps in reducing wilt incidence besides improving grain yield (1.53t/ha) over conventional tillage (1.22t/ha) and surface seeding (0.91 t/ha). Further, in an adaptive trials conducted in the region on evaluation of cultivars and economizing seed rate for surface seeding of lentils, it was observed that reducing seeds rate up to 25 kg/ha did not affected yields significantly. Cultivars like HUL57 and Arun yielded 40% higher than farmers' grown variety PL639 (Kumar *et al.*, 2011, Laik *et al.*, 2014 and Singh *et al.*, 2014b).

3. Raised Bed planting

This is water-saving technology, which saves 30-40% water depending upon soil type. In this technique, after preparation of land all three activities, viz. bed formation, placement of fertilizer and sowing are done in single operation. In situations where sowing can be delayed due to pre-sowing irrigation, dry seeding can be done on raised beds followed by irrigation immediately after seeding. Irrigation can be efficiently given at pod filling stage, without fearing of pod damages and crop lodging.

4. Crop residue management

Mulching is an efficient way to reduce evaporation, soil water depletion, moderating soil temperature under rainfed condition to improve WUE. Effects of irrigation and straw mulch on WUE and grain yield of lentil under well and limited irrigation conditions were evaluated, significantly increased in lentil yield was reported by application of 2 t/ha straw mulch (Amini *et al.*, 2013).

Conclusion

Lentil is a long day herbaceous shallow rooted plant and is well adapted to cool climate. This crop is mostly restricted to the northern states of India with annual rainfall of 350–500 mm. The crop is generally grown in winter season on conserved soil moisture after south-west monsoon. Lentil requires at least moderate moisture of 150-250 mm during the growing season to produce a full seed set. It is moderately drought tolerant crop but does not tolerate water-logging condition. Due to its indeterminate growth habit, lentil will often continue to flower as long as growing conditions remain favourable for vegetative growth. Therefore, low moisture stress is required to encourage seed set and hasten maturity. The most critical growth stage for moisture stress is pod formation followed by flowering initiation. 1-2 supplemental irrigations are required for higher productivity and water-use efficiency. First irrigation is scheduled at 6 weeks after sowing and the second at flowering or pod formation stage. Resource conservation technologies like laser levelling, zero-tillage, crop residue mulching and raised bed planting increased crop productivity and water productivity of lentil.

References

Ali M and Mishra JP. 2000. *Technology for production of winter pulses*. IIPR, Kanpur, Uttar Pradesh, India. pp 24.

Ali M, Mishra JP, Ghosh PK and Naimuddin. 2012. Rabi (winter) Pulses (Chickpea, lentil, fieldpea, lathyrus and French beans) (In) *Textbook of Field Crops Production-Food grain Crops Volume I.* (Tech. ed.) Rajendra Prasad. Directorate of Knowledge Management in Agriculture, Indian Council of Agricultural Research, KAB I, Pusa, New Delhi, India, pp. 320–378.

Amini Rouhollah Alami-Milani Morteza and Dabbagh Mohammad Inasab Adel. 2013. Effect of different irrigation treatments and mulch on water use efficiency of lentil (*Lens culinaris* Medick.). *International Journal of Biosciences* 3(4): 44–49.

Geerts Sam and Raes Dirk. 2009. Deficit irrigation as an on-farm strategy to maximize crop water productivity in dry areas. *Agricultural Water Management* 96: 1275–1284.

Jat ML, Pall S, Subba Rao AVM, Kuldeep Sirohi, Sharma SK and Gupta RK. 2004. Laser Levelling – the precursor technology for resource conservation in irrigated ecosystem of India, pp. 9-10. National Conference on *Conservation Agriculture: Conserving Resources-Enhancing Productivity*, September 22-23, 2004, New Delhi.

Kaul JN, Sekhon HS and Dhingra KK. 1974. Plant mortality in gram G-130 as influenced by irrigation, depth of sowing, seedrate and potash level (In) *Proceedings of the All India Coordinated Rabi Pulses Workshop*, UPIAS, Kanpur, India.

Kumar S, Singh RV, Piggin C, Haddad, A, Ahmed S. and Kumar Raj. 2011. No-till lentil: An option for profitable harvest in dry areas. *Grain legumes* 57: 39-42.

Laik Ranjan, Sharma Sheetal, Idris M, Singh AK, Singh SS, Bhatt BP, Saharawat Yashpal, Humphreys E and Ladha JK. 2014. Integration of conservation agriculture with best management practices for improving system performance of the rice –wheat rotation in the Eastern Indo-Gangetic Plains of India. *Agriculture, Ecosystems & Environment* 195(6): 68-82.

Laxmipathi Gowda CL, Srinivasan S, Gaur PM and Saxena KB. 2013. Enhancing the productivity and production of pulses in India (In) *Climate change and sustainable food security*, (eds) P.K. Shetty, S. Ayyappan and M.S. Swaminathan. National Institute of Advanced Studies, Bangalore and Indian Council of Agricultural Research, New Delhi. pp. 145–159.

Maghooli E, Khourgami A and Bitarafan Z. 2012. Optimal management of single irrigation and plant density effects on yield and yield Components of lentil in autumn planting at Kermanshah region of Iran.*International Journal of Science and Advanced Technology* 2(2): 78–83.

Oweis T, Hachum A and Pala M. 2004.Lentil production under supplemental irrigation in a mediterranean environment.*Agricultural Water Management* 68(3): 251–265.

Pathak P, Sahrawat KL, Wanin SP, Sachanb RC and Sudi R. 2009. Opportunities for water harvesting and supplemental irrigation for improving rainfed Agriculture in semi- arid areas (In) *Rainfed agriculture: unlocking the potential*. CAB International Publishing, Wallingford Oxfordshire, UK, pp. 197–221.

Pereira LS, Oweis T and Zairi A. 2002. Irrigation management under water scarcity. *Agricultural Water Management* 57: 175–206.

Rockström J, Karlberg L, Wani SP, Barron J, Hatibu N, Oweis T, Bruggeman A, Farahani J. And Qiang Z.2010. Managing water in rainfed agriculture -The need for a paradigm shift. *Agricultural Water Management* 97(4): 543–550.

Salehi M, Hagnazari A and Shekari F. 2006. The study of morpho-physiological traits of lentil (*Lens culinaris* Medik) in relation with grain yield under normal and drought conditions. *The 9th Iranian Crop Sciences Congress*, 27–28 August, pp. 534.Abourehan Campus, University of Tehran.

Siddique KHM, Johansen C, Turner NC, Jeuffroy MH, Hashem A, Sakar D, Gan Y and Alghamdi SS. 2012. Innovations in agronomy for food legumes.A review. *Agronomy for Sustainable Development* 32: 45–64.

Silim SN, Saxena, MC and Erskine W. 1993. Adaptation of lentil to the Mediterranean environment. II. Response to moisture supply. *Experimental Agriculture* 29: 21–28.

Sinclair TR and Vadez V. 2012.The future of grain legumes in cropping systems.*Crop and Pasture Science* 63(6):501-512. http://dx.doi.org/10.1071/CP12128.

Singh AK, Manibhushan, Bhatt BP, Singh KM and Upadhyaya A. 2013a. An Analysis of Oilseeds and Pulses Scenario in Eastern India during 2050-51. *Journal of Agril. Sci.* 5(1): 241-249.

Singh AK, Meena MK, Bharti RC and Gade RM 2013b. Effect of Sulphur and zinc management on yield, nutrient uptake, change in soil fertility and economics in rice (*Oryza sativa* L.) –lentil (*Lens Cularis*) cropping system. *Indian journal of Agricultural Sciences* 83(3): 344-348.

Singh AK, Singh KM, Bharati RC, Chandra N, Bhatt BP and Pedapati A. 2014a. Potential of Residual Sulphur and Zinc Nutrition in Improving Powdery Mildew (Erysiphe trifolii) Disease Tolerance of Lentil (*Lens culunaris* L.).*Communication in Soil Science and plant analysis* 45: 2807-2818.

Singh SS, Singh AK, Sundaram PK. 2014b. Agrotechnological options for upscaling agricultural productivity in eastern indo gangetic plains under impending climate change situations: A review. *Journal of Agrisearch* 1(2): 55-65.

Tiwari RJ and Vyas M D. 1994.Effect of soil moisture content on the field emergence and yield of Lentil. *Lens Newsletter* 21(1): 20–21.

Venkteswarlu U and Ahlawat IPS. 1993. Effects of soil moisture regimes, seed rates and phosphorus fertilization on growth and yield attributes and yield of late sown lentil (*Lens culinaris* L.). *Indian Journal of Agronomy* 38(2): 236–243.

Vijay Laxmi. 2012. Phenology, dry matter distribution and yield attributes under normal and drought stress conditions in lentil (*Lens culinaris* Medik). *Journal of Food Legumes* 25(4): 300–305.

Zhang H. 2003. Improving water productivity through deficit irrigation: examples from Syria, the north China Plain and Oregon, USA (In) *Water productivity in agriculture: limits and opportunities for improvement* (eds) W. Kijne, R. Barker and D. Molden. CAB International Publishing, Wallingford Oxfordshire, UK, pp. 301–309.

Citation

Saad, AA, Singh P, Rashid Z, Pir FA, Masood A and Wasim R. 2018. Water management for profitable lentil production. In: Scientific Lentil Production. Singh AK, Bhakta N, Sangale UR, Manibhushan, Sundaram PK, Kumar S and Yasin JK. (eds.) Society for Upliftment of Rural Economy Varanasi, India. Pp. 261-278.

□□□

Chapter 16

Boosting Pulse Production Through Efficient Water Management in the Era of Eminent Climate Changes

Anil Kumar Singh, Ashutosh Upadhyaya, KM Singh and Sushil Dimree

SYNOPSIS

India has travelled a lot on rough path to reach highway in respect to agricultural growth and development to match or even exceed to population growth. The gap between demand and supplies of pulses are partially attributed by change in food habits, population growth, being produced on poor & marginal soils, low productivity and poor adoption of production technology as whole. Presently we are consuming more vegetable oils than required for balanced diet. Indian is world largest homeland of vegetarian population and world leader in pulses production and imports large quantities to provide protein supplements to it population. Indian pulse production has been stuck between 14 and 15 MT since mid-nineties, resulting in poor intake (33 g/capita/day) India imports 2-3 Mt of grain legume on regular basis presently India is in the mid-way of self-sustaining in grain legume production. Indian will produced plenty with respect to grain legume by the 2050, probably able to sustain their legume production under stress environments. The most important way to increase grain yield of food legumes per unit area under stress environments will be proper technological backup with matching infrastructure, timely availability of quality inputs along with policy support are major issues which needs to be address properly sooner or later. Efficient water management is one of the critical inputs as in general perception is that legumes need no supplementary water, whereas research finding revealed that need based watering at critical stages are capable to improved production by 15-25 % depending up how much stress has already being faced by the standing crop till now.

Keywords: Climate Change, Pulses, Water Management.

Introduction

India is no longer begging bowl, with respect to food grains, thanks goes to policy planner researcher and the most to our farmers who make this dream to happen in our life time. India has travelled a lot on rough path to reach highway in respect to agricultural growth and development to match or even exceed to population growth. Legumes are the basic ingredient in the diets of a vast majority of the Indian population, as they provide a perfect mix of vegetarian protein component of high biological value when supplemented with cereals (Andrews and Hodge, 2010; Ali and Gupta, 2012). India recorded an all-time high production of food grains in 2011-12, despite increase in vulnerability to intensifying volatilities of the climate change. The country has also succeeded in achieving an annual agricultural growth rate of about 3.5% during the past five years, which is higher than that observed since the past decades (Singh, 2013). The National Food Security Bill (NFSB) is on the Parliament's anvil for the last about two years. On several counts, the Government is keen to have the Bill passed soon and an ordinance has been passed to the implement with immediate effects. The proposed Bill will give 67 per cent of India's population (numbering nearly 840 million people based on current total population) five kilograms of rice, wheat and coarse cereals per month at heavily subsidized prices, while those in the poorest category (numbering around 250 million) will receive 35 kilograms of food grains. Enactment of the Bill stipulates that the Rights will be met essentially from home-grown grains. The implications are obvious - accelerated, enhanced and sustained (climate-resilient) national food and agricultural production. Grain legumes are not only important sources of proteins but also offer vitamins and minerals, popularly known as "Poor man's meat" and "rich men vegetable". For an active normal body grain legume requirement is about 40 g per day or 14.6 kg per person per year (Narasinga Rao, 2010). Data presented in table 16.1 clearly indicated that availability and intake decrease with the period of time, which was 60.5 g/day during 1950-51 and 31.6 g during 2010-11 (GoI, 2012).

Table 16.1 Per capita consumption of vegetable oil and grain legume

Year	Grain legume	
	kg/year	g/day
1950-51	22.1	60.5
1960-61	25.2	69.0
1970-71	18.9	51.9
1980-81	11.3	30.9
1990-91	15.2	41.6
2000-01	11.6	31.8
2010-11*	11.5	31.6

Sources: Economic Survey 2011-12 and www.Indiastat.com/Ministry of Agriculture, Govt. of India.

Since republic (1950) productivity of legume has been increased by 0.56 times respectively. The compound rate of grain legume has been presented in table 15.2, since 1950-51, leading to India as a largest producer, importer and consumer of grain legume (Chand and Singh, 2012; GoI, 2012).

Table 16.2 Compound growth rate achieved in grain legume and oilseeds production in India during the period 1950-2011

Particular	Grain legume
Productivity (kg/ha)	441
Productivity (kg/ha)	689
Growth (%)	0.56
Overall compound growth rate (%)	1.011
Compound growth rate (%) Area	0.52
Compound growth rate (%) Production	1.27
Compound growth rate (%) Productivity	0.73

Source: Department of Agriculture and Cooperation, Ministry of Agriculture, Govt. of India

Time series data at all India is presented in table 16.3 showing decadal growth dynamics in respect to grain legume, since 1950-51, also provides glimpses of success story of grain legume production (Reddy 2004; GoI, 2012). Area production and productivity were registers incremental growth with the time at all India levels, though fluctuations were noticed in case of all the parameters. At All Indian levels remarkable, 30 per cent, growth in area under grain legume (26.28 Mha) was notice during 2010-11.

Table 16.3 All-India area, production and productivity of grain legume

Year	Area (Mha)	Production (Mt)	Productivity (Kg/ha)
1950-51	19.09	8.41	441
1960-61	23.56	12.70	539
1970-71	22.02	11.69	531
1980-81	22.46	10.63	473
1990-91	24.66	14.26	578
2000-01	20.35	11.08	544
2010-11*	26.28	18.09	689

* Fourth Advance Estimates;

Source: Department of Agriculture and Cooperation, Ministry of Agriculture, Govt. of India

Important grain legume crops of India

As rainfed crops, legumes are mostly mixed or intercropped with cereals or long-season and widely-spaced crops. Cropping systems research has identified - compatible crops for intercropping and the optimum proportions and spatial arrangements that yield maximum intercropping advantage.

Table 16.4 Important legume crops of India

Legume crop		
Kharif	**Rabi**	**Summer**
Pigeon pea	Chickpea	Mung bean
Mung bean	Pea	Urd bean
Urd bean	Winter legumes	Cowpea
Cowpea	Khesari (Lathyrus)	
Moth bean	Cowpea	
Horse gram	Kidney bean	

The statistics of importance of grain legumes in India are shown in table 16.3. Chickpea or gram (*Cicer arietinum* L.) and pigeonpea or "tur" (*Cajanus cajan*) which together account for 61 per cent of the total pulse production, are the principal grain legumes; chickpeas are grown in the post-rainy season (October – March) and pigeonpeas are planted in the rainy season (June-October). Other legumes grown exclusively in the post-rainy season are lentil (*Lens culinaris* Medic), "khesari" or grasspea (*Lathyrus sativus* L.) and peas (*Pisum sativum* L). Mungbean or green gram (*Vigna radiata*) "urd" or black gram (*Vigna mungo*) and cowpea (*Vigna sinensis*) are grown in both seasons but the post-rainy season crop is possible only in the warmer parts of the country. Where irrigation is available these can also be grown in the summer. Other grain legumes grown in the rainy season with limited regional importance are field beans (Lablab purpureus L.) Sweet moth bean (*Vigna acantifolia*), cluster beans (*Cyamopsis tetragonolobus*) and soybean (*Glycine max*). Groundnut (*Arachis hypogaea*) is the major oil seed legume grown in both seasons but in the post-rainy season the crop is mostly confined to the irrigated areas of peninsular India (Table 16.4). Improved varieties of major legume crops, released for its cultivation in India has been listed in Table 16.5.

Agricultural productivity

Irrigated area produce 2.5 to 3.5 t/ ha, whereas in case of rainfed it is still ranged in between 0.8 to 1.0 /ha. Is it not a self-explanatory to focus on the important of efficient water management in agriculture production? Foregone narration is enough to indicate the priorities and possibilities for water management in legumes (Table 16.6).

Table 16.5 Improved varieties of Legume Crops, India

Grain legumes	Suitable crop varieties for different regions			
	North & North Western	North Eastern	Central	Peninsular
Pigeon pea Extra early maturing and early maturing (120-140 days)	Prabhat, UPAS-120, ICPL81, 4-84, Pant A-l	T28, T64, C-501. J-11, Punjabi, M-13, GAUG-10	HY5, J9-19, CO-3, DL74-1	Sehore l97, HY1, HY5, 4-84, CO-1
Pigeon pea Medium maturing (160-180 days)	Mukta, T21, Pusa ageti	No. 148, Khargone 2, GS-1, LRG30, C-11, R60, BDN-2	No. 148, Khargone 2, GS-1, RG30, C-11, R60, BDN-2	BDN1, JA3, GS-1, HY2
Pigeon pea Late maturing (200 days)	B517, T7, T-17, NP (WR)15	Gwaliro 3, HAU2, Bahar, Basant	NP (WR) 15, T7, Gwaliro 3	-
Urdbean	T9, T27, Pusa 1, Mash 48, Mash 1-1, U-l9, U36	Type 77, T9, Type 65, Pusa 1	Pusa 1, No. 55, Khargone 3, T9, Mash 48, Navin, B-102, B76	T9, No. 55, D6-7, CO-1, CO-3, ADT-1, PDM 71-1, PDM 71-2
Mungbean	ML-l,Type 1, G6S, Typc-2, Type 51, PS-10, Jawahar 45, ML 5	Shela, D45-6, Guj-l,Guj-2, K851	Jawahar 45, Khargone 1, Kopargaon, Pusa Baisakhi, Krishna-11	CO-1, ADT-1, Jawahar 45, Pusa Baisakhi
Cowpea	Pusa, Phalguni, No. 455, 779, FS68, EL48	Same as Western Zone	Pusa ophasal; Pusa phalguni	New era, C-152, U-16, U37, 779, K16, PTbl
Chickpea (Desi)	C-235, Type-3, H208, F61, BG203, K468, Pant G 114	G-130, C-235, Type-3, K850, H208, BG209, JG74, Pant G 110	JG315, BG200, K468, BG203	Annegiri, Chaffa, JG221, Jyothi, BDN 9-3
Chickpea (Kabuli)	K4, K5, L-550, C-104, Hima	L-550, Hima, K4	L550, C-104	L550, C-104
Pea	T-163, K9, PS-110, PS60, EC33866, Bonnville	T-163, PS409, PS49, BR-12	PS-110, K9, L-116	PS49.PG2, L-116
Lathyrus	PS-1, P393, S3	PS-1, P396, S3	PS24, LSD6, PS6,P396	PS-1, P24, P396
Lentil	Type 36, L9-12, Pant L209, Pant L406, Pusa 4, Pant 838	LG7, Pusa 8, T36, BR25, No. 26, B77	Pant 838, JLs-1, L9-12, P43, Pusa 1	-
Groundnut	T28, T64, C-501. J-11, Punjabi, M-13, GAUG-10	AK-12-24	AJ-12-24, Gangapuri, Jyoti, Kopergoan	TMV2, Kadiri 3, Spanish improved, TMV-10, S206, Pol-1, Pol-2

Table 16.6 Productivity of different crop group under irrigated and rainfed ecosystem

Crop group	Productivity (t/ha)	
	Irrigated	Rainfed
Cereals	2.5 to 3.5	0.8 to 1.0
Legumes	1.4-2.1	0.9-1.1

Importance of climate and legume options

India as a whole, simply enjoy the twenty one agro-ecological zones eight phyto-geographical regions and fifteen ago-climatic zone, all these classification numerated is just to know that Indian environment is result of rich and complex ecological, environmental diversity which enable to evolve several crops and its various species, due diversity at greater length and same time one can find similarity too leads to accommodate all most all type of plant in its environment successfully (Andrews and Hodge, 2010 and Abraham, 2004). Only a limited number of crops are adapted to the climatic conditions and the farmer must sow the crop best suited to the moisture conditions encountered at that time. Moisture is so dominantly limiting, that "soil improving" crops are much less effective than in more humid areas. Success with rigid or complex sequences is difficult in the face of widely varying rainfall. To achieve the optimum production potential under different abiotic and biotic stress condition, selecting crops and theirs varieties based on criteria described in this presentation would certainly boost crop production efficiently in India (Singh and Kumar, 2009). Studies on climate change have underscored two points; first that atmospheric commons, namely the earth's carbon absorbing capacity, is finite and depletable and that growth of GHG emissions, even at their present level pose a threat to humankind (Singh *et al.*, 2013). Carbon pollution is causing the world's climate to change not only on the magnitude of the change but also on the potential for irreversibility, resulting in extreme weather, higher temperatures and more droughts. Our earth is undoubtedly warming. This warming is largely the result of emissions of carbon dioxide and other Greenhouse Gases (GHG's) from human activities including industrial processes, fossil fuel combustion, and changes in land use, such as deforestation etc. Day by day the cycle of climate on earth is changing. Global warming has led to season shifting, changing landscapes, rising sea levels, increased risk of drought and floods, stronger storms, increase in heat related illness and diseases all over the world. This has resulted due to emissions of Green House Gases (GHG's) from various anthropogenic activities (IPCC, 2007). An increase of temperature by 1°C it would be equivalent to a 150 km Northward shift of isotherms (lines joining places with similar temperature) or about 150 m lower altitude. There is a 5 per cent decrease in rice yield of every °C rise in temperature above 32 °C (IPCC, 2007 and Salih and Hardallou, 1986). According to recent report by Inter Governmental Panel on Climate Change (IPCC) by 2100 AD, due to global warming the average global surface temperature is projected to increase by 1.1 to 4.0 °C above 1990 levels for low emission scenario of greenhouse gas (GHG). The reduce length of growing seasons as a result of climatic change is causing detrimental effects on agriculture, In south Asia, crop yield could be decreases up to 50 per cent by 2050 if suitable measures not be take. The most significant impact of climatic change is expected in respect of availability of water. Why is there a need for agriculture to aid in the mitigation of climate change and concurrently better adapt itself for change, despite of occupying 40-50 percent of the total land surface agriculture contribution is meagre to the extent of 10-12 percent of total global anthropogenic greenhouse gas (GHG) emissions (Singh *et al.*, 2012a). The forecasted deviations in agro-climatic conditions especially temperature and their connected impacts on rainfall and resulting accessibility of water to crops and extreme weather circumstances are all probable to interrupt substantially the potential of pulse production. The furthermost upsetting portion of the forecast is the estimated increase in winter time and summertime temperatures by 3.2°C and 2.2°C respectively, by 2050 (Wani *et al.*, 2003). Such uncharacteristic rises drives surely have an adverse impact on pulse production in the form of a reduction in total crop-cycle duration. Grain legumes like mung bean and urd bean are short-duration crops (65-75 days). Further reduction in crop duration will amount to a lower yield per unit area (Singh *et al.*, 2012a). Since region is inhabitant of poor's among the poor, food grain production is still a main task to feed the ever-increasing population and literally less area has been left for production of both non-food crop and very little scope exists for its horizontal expansion, prominent to vertical expansion with improved technology. However inclusion of summer pulse/legumes particularly short duration mung bean under irrigated condition (Singh *et al.*, 2012a). Inclusion of grain legume in cropping system as intercrop particularly crop with wide space like sorghum, maize sugarcane etc. will certainly instruments stability in respect to oilseeds and especially grain legume production and will bound to prosperity in this region (Singh and Kumar, 2009).

Water resources Scenario of India

Indian location in world map is unique in several contexts; it is located in Asian Region particularly in South Eastern Region, which is again much diversified in all respect having all most all kind of climate right from tropical to temperate high land to below sea level cultivated area in Kerala. India is having 7.0 percent of total geographical area of world with 16 percentage of whole universe population. India having 329 M ha totals geographical area out of which 142 M ha area under cultivation with less than 150 % cropping intensity due to lack of assured irrigation system. Apart from having more area in rain fed cultivation total area under non suitable category, relatively less or unsuitable for general cultivation is listed in table 16.7 which is ultimate produce of agro ecological evolution process, which needs suitable intervention to brought back gradually under partial or full cultivation (Singh, 2012; Singh, 2013 and Wani *et al.*, 2003).

Table 16.7 Type and extent of land degradation in India

Type of degradation	Area (Million ha)
Water erosion	57.16
Wind erosion	10.46
Salt Affected and water logging	9.52
Shifting cultivation	2.38
Ravines	2.68
Degraded forest	24.90
Others	1.34
Total	107.43

Water management and grain legume production

Poor soil and agro-climatic conditions not only compel late sowing of winter legumes, which leads to reduced length of growing period but also necessitate to sustain cold injuries at early vegetative phase which freeze all biological activities for prolonged period. A sudden rises in temperature after that, not only induces forced maturity but simultaneously invites several biotic stress viz., diseases and insects pests (Ali *et al.*, 2012; Reddy, 2009 and Singh and Singh, 2008). Traditionally winter legumes sowing is delayed up to last week of November and some time under extreme circumstances it goes up to the first fortnight of December, obviously due to reasons already explained. However, optimum sowing time of winter legumes is first fortnight of October (Singh *et al.*, 2013 and Ramakrishna *et al.*, 2000). Consequent upon delayed planting, early encounter with severe cold, growth and development of winter legumes crop gets hampered for a considerable period. Subsequently plants get comparatively less time to complete their lifecycle which, by and large forces maturity (Ramakrishna *et al.*, 2000). In Indian IGP, normal sown winter legumes is a medium duration (130-150 days) crop, while under late sown conditions it is forced to complete its life cycle in 105±5 days (Joshi, 1998; Ramakrishna *et al.*, 2000; Reddy, 2009; Singh and Singh, 2008 and Singh *et al.*, 2012). Typically, a late sown winter legume undergoes three distinct phases and considerable degrees of phenological modifications are bound to happen. Eventually, winter legumes crop during its early seedling phase grows slowly due to its energy invested in the initial establishment (Singh *et al.*, 2002 and Singh *et al.*, 2012). However, in mid-phase, insignificant growth and development is observed. This poses serious threat to realization of full yield potential due to cold injuries. This phase is very important for creating source of channelizing the energy at later stage. In the last and most important phase winter legumes faces heat injury, resulting in early onset of reproductive phase, causing imbalance in resources and inputs, biotic stress and forced maturity (Joshi, 1998; Dixit *et al.* 2009; Reid *et al.*, 2011 and Singh *et al.*, 2012). To improve the winter legumes production under late sown conditions of Indian IGP, critical examination of situation revealed that, interventions to boost vegetative growth during early and mid-phase of life, to create base / source is the basic necessity that can be achieved by accelerated vegetative growth, and finally unilateral translocation of photosynthate to sink during reproductive stage (Guilfoyle and Hagen, 2001and Pandey and Gautam, 2009) are therefore essential.

Cropping pattern

Moisture is basic necessity for existence of any farm of life including agriculture. Crops and cropping patter is interdependence and interchangeable largely influenced by growing period of particular region which is again an outcome of soil and climatical considerations. An attempt has been made to quantify the agricultural/cropping activities based on length of growing period basically depends on availability of water/moisture to support successful crop production (Table 16.8).

Table 16.8 Choice of cropping pattern for different growing periods

Length of growing period	Efficient cropping system
<75 days	Perennial vegetation Monocropping of short duration pulses
75-140 days	Monocropping
140-180 days	Intercropping
> 180 days	Double cropping

Criteria for selecting crop

Selection of crop/species based on their tolerance to dry spell/ condition, temperature salinity etc. Fallowing crops mentioned in table 15.8 and 15.9 may be selected as per their tolerance power (degree of tolerance) to environmental abiotic stress and off course requirement. Crop succeeds under series of event, interaction with surrounding environment particularly with soil, water and weather conditions upon which biotic stress is buildup (Andrew and Hodge, 2010). One should select crops for their agroclimatic situation based on consideration mentioned elsewhere in this article. Survival of fittest and adoption to the extreme condition is two widely accepted theories in this modern biological system but in real situation both are seen in combination because nature is great leveler in one or other respects for coexistence of above said theories partially in agricultural production system some crops requires more water where as some need comparatively less than others (Loss, 1997). Economic plants which are more concern in agricultural systems are annual, biennial or some perennial herbs, based on agro-ecological conditions and availability of water crops are simply divided in to two category water loving plants (hydrophytes, cryophytes etc.) and some of them having less affinity to water are (sandophytes). Based on experience gained due to experimentations and evolves from centuries lists of crop are given in the table 16.9 which needs less water than others. Crops are also categorized based on their relative degree of tolerance to the limited water. Generally grasses and beans are hardy in nature and considered more tolerant than others.

Table 16.9 Legume crops for limited water supply

Scientific Name	Common Name	Degree of Tolerance*
Leucaena leucacephala	Leucaena	2.0
Phaseolus vulgaris	Common Bean	1
Vigna unguiculata	Cowpea	1.5
Cajanus cajan	Pigeon Pea	2.0
Dolichos lablab	Lablab Bean	2.5
Vigna radiata	Mung Bean	2.0
Phaseolus acutifolius	Tepary Bean	2.5
Vigna aconitifolius	Mat Bean	2.5
Tylosema esculentum	Marama Bean	3.0

Rated from 0 (no tolerance) to 3 (high tolerance)

Growing season and rainfall requirements

Under limited irrigation condition, duration of crops and their total water requirement and breakup of requirement is very much essential. According to availability of scanty water one should select crops as per in table 16.10 and 16.11. List may be endless but some of most widely grown and used only are listed with respect to days taken to complete vegetative phase and post vegetative(reproductive) phase for long duration (late maturing) and se well as short duration (early maturing) genotypes. Requirement for rainfall under sandy soils and clay soils from sowing to full vegetative growth, at flowering and at grain formation but not at harvesting time (Andrew and Hodge, 2010).

Soil Consideration

Before selecting particular crops and their specific variety we must considered the type of soil, depth of soil, fertility status, salt tolerant capacity, pH of the soil and minimum depth of ground water along with the capacity of tolerance to short periods of water logging and moisture stress. According to above consideration, requirements for different principle crops are laid down in Table 16.10.

Table 16.10 Growing season and total water requirements for selected drought tolerant crops

Crops Name	Number of Days				Rainfall after Sowing				Rainfall in Month of Flowering		Rainfall at Harvest
	From sowing to 50% flowering		From sowing Toharvest		Early Maturing		Late Maturing		No-moisture storage (Sand)	Maximum Moisture Storage (Clay)	
	Early Maturing	Late Maturing	Early Maturing	Late Maturing	Nomoisture storage (Sand)	Maximum Moisture Storage (Clay)	Nomoisture storage (Sand)	Maximum Moisture Storage (Clay)			
Ground Nut	60	90	95-110	110-140	300 mm/ 3 months	125 mm/ 3 months	500 mm/ 4 months	300 mm/4 months	60 mm/ 10 days	60 mm/ 10 days	30 mm/ 10 days
Chick Pea	---	----	120	180	200-300 mm/3 months	0 mm/3 months	300-400 mm/month	100 mm/ 4 months	25 mm/ 10 days	25 mm/ 10 days	25 mm/ month
Pigeon Pea	100	210-300	150-180	270-360	500-1000 mm/6 months	300 mm/ 6 months	700-1300 mm/12 months	700-1300 mm/12 months	40 mm/ 10 days	40 mm/ 10 days	50 mm/ month

Table 16.11 Soil requirements for selected drought tolerant crops

Soil Characteristics		Legumes			
		Ground Nut	Pea	Gram	Cowpea
Texture	Heavy			"	
	Medium	"	"	"	"
	Light	"	"		"
Minimum rooting Depth (cm)	Deep (90+)				
	Med. (60-90)	"	"	"	"
	Shallow (30-60)				
Fertility	High				
	Medium	"	"	"	"
SaltTolerance	Good				
	Moderate				
	Poor	"		"	"
pH Range	6.0-8.0	5.5-7.5	5.5-7.5	5.5-7.5	"
Tolerance to short periods of water logging		Low	Medium to Low	Medium to Low	Medium to Low
Minimum depth of ground water (cm)		60	30-50	30-50	40

Water management in grain legumes

It is well known fact that three basic resources, viz. climate, soil and water, determine the nature of crops that can be grown successfully in a particular region (Singh and Kumar 2009).An efficient utilization of these resources is essential for optimum production of food for human consumption. Under a given set of environmental conditions production of crop is limited by the availability of nutrition and water. Soil provides anchorage for the plants and also serves as a reservoir of water and nutrients required by them. While chemical fertilizers supplement the low nutrient supplying capacity of the soil, there is no substitute of water for production of crops. Efficient management of water, a limited resource as it is, is of utmost importance for sustaining and increasing the production. Water, crucial to life and existence, is an important aspect in the cultivation (Singh, 2012). Drought stress has become the major limiting factor on plant growth and yield. Water deficit during the reproductive growth is considered to have the most adverse effect on crop productivity.

Fig. 16.1 Surface wetting is efficient way to manage water in legumes including lentil crop

Water use efficiency in grain legumes

Water use efficiency is defined here as yield of plant product (tonnes of wheat grain, Y) per unit of crop water use (mega litres of water lost by evapotranspiration, ET), and is important in all areas of plant production. Water use efficiency (Y/ET) is the outcome of an entire suite of plant and environmental processes operating over the life of a crop to determine both Y and ET. Consequently, biomass production per unit ET, has been used extensively as an interim measure of water use efficiency. ET comprises non-productive evaporation (E) of water from the soil surface and productive transpiration (T) of soil-stored water by the plant. Evaporation of free water from leaf surfaces adds to non-productive evaporation (interception evaporation) (Ali, 2008). Crop grown in a sunny and hot climate needs more water per day than the same crop grown in a cloudy and cooler climate. There are, however, apart from sunshine and temperature, other climatic factors which influence the crop water need. These factors are humidity and wind speed. When it is dry, the crop water needs are higher than when it is humid. In windy climates, the crops will use more water than in calm climates. The highest crop water needs are thus found in areas which are hot, dry, windy and sunny. The lowest values are found when it is cool, humid and cloudy with little or no wind. From the above, it is clear that the crop grown in different climatic zones will have different water needs and thus water use efficiency also varies accordingly (Ali, *et al.*, 2012). A generalized view has been presented in the table 16.11 and table 16.12 which represent that legumes have genetically low water use efficiency as compare to food crop especially belongs to family Poaceae also called Gramineae. It was noticed by several researcher that dicot plants are more efficient than monocots,similarly C_3 plan are performed better than C4 plants (Table 16.13), (Keller *et al.*, 2000).

Table 16.12 Water use efficiency (WUE) of different crops

High	Medium	Low
Maize, Sorghum, and Sugarcane	Wheat, Barley and Oats	Green gram, Pigeon pea, Soybean, Peas and Rice

Table 16.13 Water use efficiency (WUE) in C3 and C4 species

Class	Water Use Efficiency (mg dry weight/g water)	
	C3 species	C4 species
Dicotyledons	3.44	1.59
Monocotyledons	3.14	1.49

Irrigation Requirement

The crop requires total 4-5 irrigations i.e. two irrigation at an interval of 7 days and three irrigations at an interval of 15 days. Legume plants with relatively high water requirements are very sensitive to soil drought as lesser rainfall during the vegetation season is one of the most important environmental factors limiting the crop yield. Irrigation is not recommended about 2 weeks after planting. Excess water at seeding slows growth and may increase root rots. During the rest of the season, moisture levels in the topsoil should be kept at or above 50% of available water. Critical stages at which water stress should be avoided to the crop are two critical times during bloom and pod set. Soil type does not affect the amount of total water needed, but affect frequency of water application. Growth and development of legume crops in general is very sensitive to water stress. That sensitivity is a result of its maximum depth of rooting is relatively shallow, approximately 0.9 m (Hebblethwaite, 1977). An experiment conducted in Eastern Sudan reported more economic yields were when the crop was irrigated at 21-day intervals up to flowering and at 7-day intervals after the onset of flowering. The highest grain and total biological yields were obtained when irrigation was at intervals of 14/7 days (Farah *et al.*, 1990). In a field experiment at loamy-sand soil in Saudi Arabia, it is reported that considerable grain yield of most of legumes seed can be achieved if irrigated at 15% soil moisture depletion from field capacity, which was not significantly different from the yield at field capacity (Naeem, 2008).

Fig. 16.2 Close-up and field view of view lentil irrigation in progress

Critical growth stage for grain legumes

Drought of different intensity is experienced in rainfed areas at various growth stages of legumes. Response to limited irrigation has been observed in most of the grain legumes. Among various crops, common bean *(Phaseolus vulgaris L.; French bean)* was found to be more responsive to irrigation followed by pea. The success of mung bean as a catch crop (during summer months) in the rice-wheat system is solely dependent upon adequate supply of irrigation plants (Keller *et al.*, 2003). Latesown chickpea in sequence with rice also needs irrigation compared to the normal sown crop, probably due to restricted root growth with late sowing. Pre flowering or flower initiation and post podding has been found to be the most critical stage in most of the legumes. However, the initial soil profile moisture and soil types largely determine the requirement of subsequent irrigation. Similarly, excess moisture or waterlogging reduces oxygen concentration in the rhizosphere and thus affects BNF activity and nutrient availability with consequent yield reduction. Therefore, it is imperative to provide good drainage, especially in low-lying areas.

Fig. 16.3 Lentil crop at flowering stage under efficient water management condition

Grain legume performance in response to water management

Most of legumes is best adapted to the more moist agriculture areas and does best under relatively cool growing conditions. Hot, dry spells result in wilting of the plants and may reduce seed set. The crop should be grown with caution in the brown soil zones and on droughty, light-textured soils unless irrigation is available, as most of legumes respond very well to irrigation. Agronomy of irrigated most of legumes is similar to dry land production. Yields can be much higher than dry land production; however, special attention must be paid to prevent losses due to diseases, such as *botrytis* and *ascochyta*.

Fig. 16.4 Lentil crop proceeding toward physiological maturity under efficient water management condition

Grain yield

Since, there is a certain correlation between dry matter production and grain yield in most of legumes crop; it appears that the maintenance of adequate levels of water throughout the vegetative growth of most of legumes is essential for high yields (Golezani *et al.*, 2009). The most of legumes is regarded as a drought-sensitive crop and the major factor restricting its cultivation is the high year-to-year yield variability usually due to drought stress. Water stress decreases the final leaf area, net photosynthesis, light use efficiency, pod retention and filling by reducing the availability of assimilates and distorting hormonal balance. Water limitations considerably reduces grain yield of cultivars, due to large reductions in growth, grain filling duration, grain weight and grains per plant. Superiority of well-watered (I1:70 mm evaporation from Class A pan) plants in growth and grain filling duration resulted in production of comparatively more and larger grains and consequently higher grain yield per unit area (Table 16.14)

Fig. 16.5(a) Harvesting of Lentil under efficient water management condition

Fig. 16.5(b) Quality harvest of lentil seed under efficient water management condition

Table 16.14 Comparison of means of maximum grain weight, grain filling rate, grain filling duration, grains per plant and grain yield of three faba bean cultivars under different irrigation conditions

	Treatment	Grain filling rate (mg d-1)	Grain filling duration (day)	Maximum grain weight (mg)	Grains per plant	Grain yield (gm-2)
Irrigation	I1	38.45a	43.16a	1400.0a	9.156a	493.4a
	I2	33.45a	38.69a	1214.5b	5.378b	264.4b
	I3	35.34a	30.00b	1116.9c	3.267c	157.0c
Cultivar	Aquodolce	35.46a	37.72a	1233.0b	5.322b	266.9b
	Barakat	35.41a	40.42a	1356.2a	7.411a	412.5a
	Saraziri	34.26a	33.88b	1141.7c	5.067b	235.4b

Source: Goelezani *et al.*, 2009

Note: different letters in each column indicating significant difference at P<0.05. I1, I2 and I3 denotes irrigation after 70,100 and 130 mm evaporation from Class A pan respectively Therefore, sufficient water supply during plant growth and development is necessary to ensure a satisfactory supply of assimilate to the grains via an extensive and long-lived foliage. Water shortage exerted a large adverse influence on dry matter accumulation, crop growth rate and relative growth rate of most of legumes cultivars (Golezani *et al.*, 2009). Ageeb *et al.* (1989) reported that irrigation to most of legumes at 7-day interval increased seed yield and the number of plants/m2, while the number of pods/plant and 100-seed weight were decreased. Many studies have reported very substantial increases in seed yields as a result of proper irrigation; including studies in regions where rainfall is abundant throughout the growing season Water requirements have not been determined, however, when growing most of legumes as a cover crop. Proper irrigation scheduling is expected to differ when the crop is grown as a cover crop compared with when it is grown for seed production. Stock and El-Naggar (1980) concluded that the optimum soil water content during flowering was at 40–60% of the available water and that either higher or lower water content resulted in sub optimal seed yields. Shuaibani, 2009 reported that water stress leads to significant decrease in number of days to flowering and maturity (Table16.15). The highest reduction in seed yield was detected in T1 and T2 treatments, which may be due to unfavorable conditions of plant growth as a result of lower water supply.

Table 16.15 Influence of water deficit on some growth parameters of faba bean two growing seasons (combined analyses of two seasons)

Treatments	Parameters					
	Plant height (cm)	No. of leaves/ plant	Leaf area per plant (cm2)	Plant weight (g)	No. of days to 50% flowe-ring	No. of days to 50% of matur-ing
T1-Total applied water 2000m3	49.11	54.67	1222.67	978.3	44.63	123.25
T2-Total applied water 3000m3	51.75	44.33	1205.67	1132.3	46.13	124.38
T3-Total applied water 4000m3	62.58	105.33	2336.67	1325.3	46.38	125.13
T4-Total applied water 6000m3	76.10	112.67	2493.33	2176.10	45.50	130.38
T5-Total applied water 7000m3	82.16	112.98	2542.22	2706.9	45.25	137.5
T6-Total applied water 7500m3	85.25	111.67	2431.56	281.3	45.88	140.63
LSD at 0.05	7.00	5.42	44.53	343.35	0.98	2.25

Source: AL-Suhaibani N.A., 2009.

Husain *et al.* (1988b) found that irrigation had little effect on faba bean growth in New Zealand when precipitation was adequate early in the growing season. Study indicated that farmers can reduce the need for irrigation by planting the crop later in the year when temperatures are cooler and winter rains are more reliable. In fact, with sufficient soil moisture available at the beginning of the season, it appears that the water requirements for growing faba bean can be adequately met by winter precipitation in central California, particularly when rainfall is above normal. Ouda *et al.* (2010) found that irrigating most of legumes with 80% of full fresh or drainage water could reduce yield by 7%. However, reschedule irrigation and applying 3156 and 3366 m3/ha of fresh and drainage water, respectively reduced yield losses to less than 1%. These findings implied that avoiding sensitive growth stages to water stress in faba bean could help in saving an ample amount of irrigation water.

Table 16.16 Influence of water deficit on some yield parameters of faba bean in two growing seasons, (combined analyses of two seasons)

Treatments	Parameters				
	No. of tillers/ plant	No. of pods/ plant	Seed weight per plant (g)	100-seed weight (g)	Biological yield (t/ha)
T1-Total applied water 2000m3	5.11	9.41	508.3	72.66	3.23
T2-Total applied water 3000m3	5.94	8.95	553.3	82.99	3.74
T3-Total applied water 4000m3	6.05	11.79	612.1	87.44	4.37
T4-Total applied water 6000m3	6.61	13.60	902.5	86.10	7.18
T5-Total applied water 7000m3	7.19	15.29	1285.5	88.63	8.93
T6-Total applied water 7500m3	7.03	17.14	1210.1	93.21	8.52
LSD at 0.05	0.98	2.43	139.4	2.96	1.13

Source: AL-Suhaibani N.A., 2009.

Biochemical composition

Kelly (1973) attributed differences in the protein contents of legumes to environmental conditions, genotypes, and agricultural practices. Fiber content values showed significant variation among genotypes at both locations ranging from 7.76% (Aquadulce) under rainfed conditions to 11.04% (S82408-1-2-3) under irrigation conditions. The lower values under rainfed than under irrigated conditions could be attributed to environmental: the non-favorable temperature during seed filling stage and other factors, Thus, Ash, protein and fiber were higher under irrigation conditions, whereas the opposite was true for fat and carbohydrate (Saxena *et al.*, 1991). Grain legumes are in general susceptible to water logging. Laser leveled bays are suitable if well drained. If drainage is not good, then beds should be considered. Irrigation times should be kept as short as possible as water logging will cause temporary growth reductions and will affect yield. Drainage is a combination of surface and internal drainage. Surface drainage can be improved with large capacity drains with good outfall, laser-leveled and smoothed bays and beds or spinner cuts. Internal drainage is related to soil structure and can be improved with gypsum where appropriate, and with pasture rotations. Minimizing cultivation, particularly of dry soil, can help to preserve soil structure and internal drainage. If drainage is less than ideal, rains following irrigation can lead to prolonged water logging and subsequently reduce yield. Pre-irrigation and sowing into moisture is a strategy successfully employed on many farms. The alternative is to rely on rainfall. Dry sowing can be successful by ensuring that the earliest rain is used to germinate the crop. The irrigation during flowering should not be too late as this may hasten the end of flowering and severely affect yield. If water is available it should be planned in such a way that plant remains free from water stress (Tewati and Virk 1996).

Water productivity

The term water productivity is used to denote the amount of value of product over volume or value of water depleted or diverted or used. The value of product may be expressed in terms of biomass, grain or money. Crop water productivity may be computed during crop period considering the production or value of production from crops and water used during the period (either on the basis of irrigation water or total water used including rainfall). At the field level agronomists evaluate the productivity of water through water use efficiency, the ratio of yield to water consumed (kg/m_3) by the crop through evapotranspiration (Doorenbos and Kassam 1979; Kinje *et al.*, 2003) or as the yield per unit depth of water per area kg/ha/mm (Gregory, 1989). Loss *et al.*, reported that under Mediterranean-type environments, WUE for dry matter production and seed yield from early sown most of legumes (up to 36 and 14 kg ha"1 mm "1, respectively) were equivalent to cereals and greater than those for other grain legumes. Irrigation water-use efficiency was highest when plants were irrigated at 50% Etc as observed by Bryla *et al.* (2003) in subsurface drip irrigated most of legumes in California. Under deficit irrigation in Egypt, water productivity value was 1.14 and 1.21 kg/ m3 under the application of 3156 and 3366 m3/ha of fresh and drainage water, respectively as reported by Ouda *et al.*, 2010. WUE generally increased with supplemental irrigation (SI), up to the 2/3 SI level, and then decreased (except for early sowing) beyond that.

Under rainfed conditions, early sowing resulted in the highest water use efficiency as reported from Syria by Oweis *et al.*, 2005. The normal sowing date (Mid December) resulted in the maximum water productivity. Late sowing steadily resulted in the lowest water use efficiency under all levels of water availability.

Conclusion

We have achieved impressive economical and agricultural growth, but still incidences of hunger, malnutrition and poverty are unacceptably high, thanks our policy planner to implementation of The National Food Security Bill (NFSB) though by ordinance to provide food to poor's. Expectations of impending climates specify immense modifications in temperature, rainfall pattern, humidity and soil moisture regimes. Changes in climate not only influence the entire cropping system but also affect the performance of cultivars of different field crops including legumes. There is pressure on food legume to maintain or even to be higher yielding under the expected climate changes. Apart from other technologies and inputs like improved genotype with respect to improved resistance to multiple biotic stresses, water management will certainly going to be very crucial due multiple abiotic stresses.

References

Abraham Blum. 2004. The Mitigation of drought stress, plant stress.com;- Henry T. Nguyen, Abraham Blum (Ed.), Physiology and Biotechnology Integration for Plant Breeding.

Ageeb, OAA, Salih FA and Ali MA. 1989. The effect of sowing date, watering interval and intercropping with sorghum and maize on the yield of faba bean. *FABIS-Newsletter*. 24: 8-10.

Ali M. 2008. Technological Approaches for increasing pulse production in India. In Proceedings of National Symposium on Technological Innovation for Resource Starved Farmers in Global Perspective, CSAUA&T Kanpur 28-30 April.

Ali M and Gupta S. 2012. Carrying capacity of Indian agriculture: pulse crops. *Cur. Sci.* 102: 874-881.

Ali R I, Awan T H, Ahmad M M, Saleem U and Akhtar M. 2012. Diversification of rice-based cropping systems to improve soil fertility, sustainable productivity and economics. *J. Animal and Plant Sciences.* 22: 108-112.

Ali, AE, Ahmed GEE and El-Handallou EB. 1982. Faba bean and their role in diets in Sudan. *In*: Hawtin, G. and C. Webb (eds.), *Faba Bean Improvement,* pp: 317–8. Martinus Nijhoff Publishers, The Hague AOAC (Association of Official Analytical Chemists), 1984. *Official Methods of Analysis,* 14th Ed. Washington, DC, USA.

Al-Suhaibani NA. 2009. Influence of Early Water Deficit on Seed Yield and Quality of Most of legumes under Arid Environment of Saudi Arabia. *American-Eurasian J. Agric. & Environ. Sci.* 5: 649-654, 2009.

Anonymous. 2011. Agricultural Statistics at a Glance 2011, Department of economics and statistics, department of agriculture and cooperation, Ministry of Agriculture, Govt. of India.

Arshad M and Frankenberger Jr WT.1998. Plant growth regulating substances in the rhizosphere: Microbial Production and Functions. *Adv. Agron,* 62: 46-151.

Ashton FM and Monaca TJ.1991. Weed Science: Principle and Practices.3rd Edition. John Wiley & Sons. Inc., New York, 342pp.

Audus LJ .1977. Plant Growth Substances: Chemistry and Physiology. Leonard Hill, London, Vol. I, 240pp.

Bartel B .1997. Auxin Biosynthesis. *Ann. Rev. Plant Physiol. Plant Mol. Biol.* 48: 49-64.

Cerning JA, Saposnik and Guilbot A.1975. Carbohydrate composition of horsbeans (*Vicia faba*) of different origins. *In*: Hebblethwaite, P.D. (ed.): 159–80. *The Most of legums* . Butterworths, London.

Chand R. 2009. Demand for food grains during the 11 Plan and towards 2020. Policy Brief No. 28.National Centre for Agricultural Economics and Policy Research (pp. 1-4). New Delhi.

Chand R and Singh RB. 2012. Major issues in Agriculture Sector for the Twelfth Plan. NAAS News 12(1): 6-11.

Cooper PJM. 1991. Fertilizer use, crop growth, water use and WUE in Mediterranean rainfed farming systems. Pp 135-152 in: Harris HC, Cooper PJM and Pala M (Eds.) Soil and crop management for improved water use efficiency in rainfed areas. Proceedings of an International Workshop, Ankara, Turkey, May 1989. Aleppo, Syria: ICARDA.

David R, Bryla Æ, Gary S, Banuelos Æ and Jeff P. Mitchell. 2003. Water requirements of subsurface drip-irrigated most of legums in California. *Irrigation Sciences.* 22: 31–37.

Doorenbos J and Kassam AH.1979. Yield response to water. FAO Irrigation and Drainage Paper 33, FAO, Rome, Italy.

Farah SM, Fakki El, Gorashi H and Ali AM. 1990. Faba bean water management at New Halfa [Sudan]. *Agricultural water management*. ISSN 0378-3774.

Farah SM, Fakki El, Gorashi H, and Ali AM. 1990. Most of legumes water management at New Halfa [Sudan]. *Agricultural water management*. ISSN 0378-3774.

GoI. 2012. Data Base on Indian Agriculture, Department of Agriculture and Cooperation. Ministry of Agriculture, Retrieved from http://dacnet.nic.in.

Golezani KG, Ghanehpoor S, Mohammadi-Nasab AD. 2009. Effects of water limitation on growth and grain filling of most of legumes cultivars. *Journal of Food, Agriculture and Environment* 7: 442–447.

Gregory PJ. 1989. Water-use efficiency of crops in the semiarid tropics. In: *Soil, Crop and Water Management in the Sudano-Sahelian Zone*. ICRISAT, Patancheru, India, pp. 85–98.

Hebblethwaite PD, Ingram J, Scott RK and Elliot J. 1977. Some factors influencing yield variation of field beans *(Vicia faba* L.). Proceedings of the Symposium on the Production, Processing and Utilization of the Field Beans *(Vicia faba* L.) (Thompson, R., Ed.). Bulletin No. 15, Scottish Horticultural Research Institute, Invergrowie, U.K. pp. 20-27.

Husain MM, Hill GD and Gallagher JN. 1988. The response of field beans (*Vicia faba* L.) to irrigation and sowing date. 2. Growth and development in relation to yield. *Journal of Agriculture Sciences* 111: 233–254.

IPCC, 2007. Climate Change 2007: Mitigation. Contribution of Working Group III to the Fourth Assessment Report of the Intergovernmental Panel on Climate Change. Metz B, Davidson OR, Bosch PR, Dave R and Meyer LA. (eds), Cambridge University Press, Cambridge, United Kingdom and New York, NY, USA.

Joshi PK. 1998. Performance of grain legumes in residual effects of legumes in rice and wheat cropping systems of the Indo-Gangetic plain. In Legumes in rice and wheat cropping systems of the Indo-Gangetic plain. Eds Kumarrao J V D K, Johansen C and Rego TJ. ICRISAT, Patancheru, Andhra Pradesh pp. 3-13.

Joshi PK. 2009. Prices and Market Intervention in Pulses., paper presented in Brainstorming workshop on Issues and Strategies for Increasing Productivity and Production of Pulses in India, organized by the Indian Council of Agricultural Research and Ministry of Agriculture, Government of India, New Delhi, 9-10 June.

Keller A, Sakthivadivel R and Seckler D. 2000. 'Water scarcity and the role of storage in development' IWMI research report 39, IWMI: Colombo.

Kelly JF. 1973. Increasing Protein Quantity and Quality. *In*: Milner, M. (ed.), Nutritional Improvement of Food Legumes by Breeding, p. 179. The Protein Advisory Group of the United Nations System, New York, USA.

Kinje Jacob, Barker R, and Molden D. 2003. Improving Water Productivity in Agriculture: Editors' Overview, in Jacob Kinje and others (Eds.) Water Productivity in Agriculture: Limits and Opportunities for improvement, Comprehensive Assessment of Water Management in Agriculture. UK: CABI Publishing in Association with International Water Management Institute.

Loss SP, Siddique KHM and Tennanta D. 1997. Adaptation of most of legumes (*Vicia faba* L.) to dryland Mediterranean-type environments III. *Field Crops Research*. 52: 17-28.

Mitchell Andrews and Simon Hodge. 2010. Legumes Cultivars for Stress Environments: In Climate Change and Management of Cool Season Grain Legume Crops edts. SS Yadav and Robert Redden. pp. 243-281.

Muhammad AI, Muhammad K, Sher Muhammad, Shahzad M A, Nawaf S and Naeem A. 2012 Integrated use of *Rhizobium leguminosarum*, plant growth promoting rhizobacteria and enriched compost for improving growth, nodulation and yield of lentil (*Lens culinaris* Medik.) *Chilean Journal of Agricultural Research*. 72: 104-110.

Musallam IW, AL-Karaki GN, Ereifej K and Tawaha M. 2004. Chemical Composition of Most of legumes Genotypes under Rainfed and Irrigation Conditions. *International journal of Agriculture Biology*. 6: 2-6.

Mushari A and Al-Naeem. 2008. Effect of Irrigation Scheduling on Growth Parameters and Water Use Efficiency of Barely and Most of legumes Crops in Al-Ahsa, Saudi Arabia. *American Journal of Plant Physiology*. 3: 111-120.

Nadarajan N and Gupta S. 2010. Scenario of pulses production in India. Agriwatch. pp. 26-27. Narasinga Rao, B. S. 2010. Nutrient requirement and safe dietary intake for Indians. NFI Bull., 31: 1-8.

Pala M and C Studer. 1999. Cropping systems management for improved water use efficiency in dryland agriculture. Paper presented at the International Conference on: Water Resources Conservation and Management in Dry Areas. 3-6, December 1999. Amman, Jordan.

Ramakrishna A, Gowda CLL and Johansen C. 2000. Management factors affecting legumes production in the Indo-Gangetic Plain. In Legumes in rice and wheat cropping systems of the Indo-Gangetic plain - constraints and opportunities. Eds Johansen C, Duxbury J M, Virmani S M, Gowda C L L. pp. 156-165. ICRISAT, Patancheru, Andhra Pradesh.

Reddy AA. 2009. Pulses Production Technology: Status and way forward. *Economic and Political Weekly*. 44: 73-80.

Reddy AA and Reddy G P. 2010. Supply side constrains in production of pulses in India: A case study of lentil. *Agricultural Economics Research Revie*. 23: 129-136.

Reddy, A. A.2004. Consumption Pattern, Trade and Production Potential of Pulses. Economic & Political Weekly, XL (48): 54-60.

Reddy AA. 2009. Pulses Production Technology: Status and Way Forward. *Economic & Political Weekly* XLIV (52): 73-80.

Reid DE, Ferguson BJ, Hayashi S, Lin YH and Gresshoff PM. 2011 Molecular mechanisms controlling legume autoregulation of nodulation. *Annals of Botany* 108: 789-795.

Salih, FA and El Hardallou SB. 1986. The yield performance and proximate analysis of twelve cultivars of most of legums grown at five sites in the northern region of the Sudan. *FABIS*, 15: 52–55.

Samiha A. Ouda, Mouhamed A Shreif and Rashad Abou Elenin. 2010. Increasing water productivity of most of legums grown under deficit irrigation at middle Egypt.14th International Water Technology Conference, IWTC 14, 2010, Cairo, Egypt.

Saxena MC, Silim SN and Matar A. 1991.Agronomic management of most of legumes for high yields. *Options Mediterraneans* 10: 91–96.

Singh AK and Kumar Pravesh.2009. Nutrient management in rainfed dryland agro ecosystem in the impending climate change scenario. Agril. Situ. India. LXVI (5): 265-270.

Singh AK, Bhatt BP and Singh D. 2013. Efficient cultural practices – curbing plant disease in the climate change era. *Indian Farming* 63(1): 6-11.

Singh AK, Manibhushan, Bhatt BP, Singh KM and Upadhyaya A. 2013.An Analysis of Oilseeds and Pulses Scenario in Eastern India during 2050-51. *Journal of Agril. Sci.* 5(1): 241-249.

Singh AK, Manibhushan, Chandra Naresh and Bharati RC. 2008. Suitable crop varieties for limited irrigated conditions in different agro climatic zones of India. *Int. J. Trop. Agri.* 26(3-4): 491-496.

Singh AK, PK Sundaram and Verma Nidhi. 2012. Harnessing production potential of dry land agriculture through efficient nutrient management (ENM) under climate change condition. *Prog Agri.* 12(2): 229-240.

Singh AK. 2012. Water management : Priorities and Possibilities of Investment for Accelerated and Stabilizing Agricultural Growth in Bihar, Patna. National Conference on Priorities and Possibilities of Investment for Accelerated and Stabilizing Agricultural Growth in Bihar, Saturday, 17 March 2012, Patna17th March, 2012. pp. 12-14.

Singh AK, Sangle UR and Bhatt BP. 2012a. Mitigation of imminent climate change and enhancement of agricultural system productivity through efficient carbon sequestration and improved production technologies. *Indian Farming* 61(10): 5-9.

Singh AK, Singh D, Singh AK, Gade RM and Sangle UR. 2012b. Good Agronomic Practices (GAP)-An efficient and eco-friendly tool for sustainable management of plant diseases under changing climate scenario. *J. of Plant Disease Sci.* 7(1): 1-8.

Singh RB. 2013. Climate Smart Agriculture towards an Ever-Green Economy NAAS, India 34p.

Stock HG, El-Naggar ES.1980.Untersuchungen zur Ermittlung des optimalen Beregnungsregimes fur Ackerbohnen. *Arch Acker-Pflanzenbau Bodenkd.* 24: 665–672.

Tewatia, BS and Virk AS.1996. Nutritional potential of faba bean for improved productivity in ruminants- A Review. *FABIS.* 38/39: 2–11.

Theib Oweis, Ahmed Hachum and Mustafa Pala. 2005. Faba bean productivity under rainfed and supplemental irrigation in northern Syria. *Agricultural Water Management.* 73: 57-72.

Wani S, Pathak P, Sreedevi T, Singh H and Singh P. 2003. Efficient management of rainwater for increased crop productivity and groundwater recharge in Asia', in Water productivity in agriculture: Limits and opportunities for improvement, Kijne JW, Barker R and Molden D (Eds). Wallingford: CAB International: 199-215.

Citation

Singh AK, Upadhyaya A, KM Singh and Dimree S. 2018. Boosting pulse production through efficient water management in the era of eminent climate change. In: Scientific Lentil Production. Singh AK, Bhakta N, Sangale UR, Manibhushan, Sundaram PK, Kumar S and Yasin JK. (eds.) Society for Upliftment of Rural Economy Varanasi, India. Pp. 279-308.

❑❑❑

Chapter 17

Weed Management for Optimizing Lentil Production

Vishram Ram, AK Singh and Anil Kumar Singh

SYNOPSIS

Lentil is grown in rotation rice and other *kharif* crops in India. In comparison with pea and chickpea, it is short in stature, shallow rooted and slow to emerge and has an open growth habit that often allows emergence and establishment of weeds. As a result, lentil is often the least competitive crop grown in rotation in India. Early establishing weeds, particularly those emerging prior to or with the crop, have a competitive advantage over lentil and severely reduce yield if not controlled. Later emerging weeds compete with lentil during pod and seed development and directly reduce seed size and quality.

Keywords: Crop, competitiveness, critical period, lentil, weed.

Introduction

Agriculture is highly vulnerable to climate change. Higher temperature eventually reduces yields of desirable crops while encouraging weed and pest proliferation. Changes in precipitation patterns increase the likelihood of short-run crop failures and long run production declines. Although there will be gain in some crops in some regions of the world, the overall impacts of climate change on agriculture are expected to be negative, threatening global food security. In developing countries like India, climate change will cause yield declines for the most important crops and results in price increases (Dubey *et al.*, 2011).

Fig. 17.1 Weed free lentil crop during early stage of growth, ensuring better management for

Weed Management and Crop Productivity

Can manipulation of interactions among weeds, crops and other biota produce adequate weed suppression to reduce or eliminate herbicide use? The answer is a qualified yes (Fig.17.1). There is widespread agreement that effective long-term ecological weed management strategies involve the use of multiple tactics, sometimes referred to as the 'many little hammers' approach. In this way, multiple interaction points among weeds, their micro-environment and other organisms are targeted to reduce weed biomass, seed production and seed survival. In developing weed management strategies, the following must be taken into account: i) not only do crops and weeds compete directly for resources (light, water and nutrients), but indirect effects mediated through changes in pest or disease dynamics can also be important; ii) the presence of weed seed banks in the soil requires long-term management strategies to reduce both the annual input of new seeds and pre-existing seed bank numbers by increased seed predation or loss of viability and iii) if selection pressures are sufficiently strong over extended periods of time, weed populations can evolve into more competitive populations better adapted to agricultural field conditions than populations from non-agricultural areas (Weinig, 2005 and Singh *et al.,* 2015). The strongest selection pressure acting upon weed populations in recent years is herbicide use, and not surprisingly herbicide-resistant weeds now pose major challenges for growers, especially in reduced or no-till systems (Murphy and Lemerle, 2006).

A holistic approach to weed management should include both short- and long-term strategies. Tactics fall into two main categories: those that reduce weed growth and fecundity during the growth cycle and those that reduce weed seed survival (Singh *et al.*, 2015). Examples of the first group include the use of competitive crop varieties, manipulating crop seeding density and spatial arrangement, tillage, intercropping, use of allelopathic residues and suppressive mulches, and targeted use of bio-control agents. The second group includes various soil management techniques (e.g. reduced tillage, residue management and organic matter inputs) and the use of weed-suppressive crop rotations that reduce seed survival due to enhanced seed predation or infection by pathogens (Singh *et al.*, 2012).

Increasing Crop Competitiveness

Crop competitiveness can be increased by the use of weed-suppressive cultivars, manipulation of seeding density and spatial arrangement, and fertility or water management. Relatively little work has been done to breed more competitive crop varieties (Fig.17.2). It is time consuming to do large screenings in the presence/absence of weed competition, but comparative competition studies done with a wide range of varieties can help identify easily measurable traits associated with higher competitive ability and enable indirect screening for competitiveness, a much less time consuming option (Gibson *et al.*, 2003). Although there is still debate, competitive traits include early plant vigor and canopy development, allelopathy, the ability to deplete a given resource and/or plasticity in root and shoot growth to respond rapidly to temporal and spatial patchiness in resource availability. Early plant vigor is particularly important in annual crop production. Extensive research has identified the stage(s) in which crops are most susceptible to weed competition, the so-called critical weed-free period, and this is often during the crop establishment and early growth stages (Seeme *et al.*, 2003).

Fig. 17.2 Hand weeding is one of the best known old age practices for obtaining maximum weed free filed

Critical period for weed management

The critical period of weed control (CPWC) is defined as the period of crop growth when the crop must be kept weed free to prevent yield loss due to weed interference (Erman *et al.*,2008). The CPWC is determined from two components: (1) the critical timing of weed removal (CTWR), the maximum period of time a crop can be exposed to early season weed competition before a yield loss threshold is reached, and (2) the critical weed-free period (CWFP), the minimum duration of a weed-free period required following planting to prevent yield losses above a threshold, generally defined as 5%. The CTWR and the CWFP delineate the beginning and end of the CPWC, respectively. Results of CPWC studies in lentil have been quite variable. Moreover, the work defining the CPWC in lentil has been conducted primarily in Asia, where lentils are more commonly produced. Only a single previous work on the CPWC has been conducted on lentil in North America, and that study focused strictly on interference by wild oat (*Avena fatua L.*). There is a need to determine the CPWC for lentils in the cropping systems, as the weed spectra and the environment are unique. The CPWC in lentil has been estimated to be between 30 and 60 dars after emergence (DAE) in India. In Syria, the CPWC has been estimated to be between 60 and 90 DAE. It was estimated a CPWC between 49 and 56 DAE in Jordan. No yield loss thresholds were reported in these studies. Its CPWC between 38 and 92 DAE in Jordan based on a 10% yield loss threshold. Based on 5%

yield loss threshold in Iran estimated a CPWC between 14 and 28 DAE and 28 and 42 DAE depending on location. In India estimated a CPWC beginning prior to crop emergence and lasting until 56 DAE based on 5% yield loss threshold. The average CPWC for lentil reported over the last two decades is between 25.7 and 54.8 DAE. Generally, the CPWC for a given crop species varies greatly and depends on the density of weed infestation, weed species characteristics, crop species characteristics, climatic conditions, and environment. The CPWC can also differ due to planting densities, and the duration of the CPWC can be affected by weed densities (Ahmad *et al.*, 2009). Because most weed species germinate discontinuously, weed emergence is not as predictable as crop emergence. Typically, the CPWC concept typically assumes that weed emergence, species composition, and density are spatially homogenous across an environment. Variability in environment, crop characteristics, and other factors causes CPWC estimates to be specific to a crop within a given region. The CPWC is unique for each crop, and it varies for each crop in differing environmental conditions. Seed size often differs among lentil cultivars. Yield loss due to weed competition could potentially differ between large and small-seeded lentil cultivars (seeded at the same density) due to differing amounts of carbohydrates and nutrients available to the seedling from the cotyledons of the seed, which influences early seedling vigour and competitiveness (Gulluoglu and Arioglu, 2009).

Cultural practices of weed management

Preventing the movement of weed from infested into un-infested areas is a crucial component of control. Both sanitation and quarantine are required in order to prevent the dispersal of seeds. The minute seeds can easily be transferred from one field to another by cultivation, and also by water, wind, and animals. However, the most significant seed transfer agents are people, transportation vehicles, and farming machines, which easily transfer seeds and contaminated soil. Once a field is infested with parasitic weeds, controlling its seed production is very difficult. Extermination of seeds before their spread to new fields and regions is a crucial component in weed prevention program (Panetta and Lawes, 2005). Quaternary ammonium salts have been found effective in seed eradication with complete seed kill achieved under commercial conditions at 0.5% a.i. of the disinfectant didecyldimethyl ammonium bromide. Early plantings of lentil are more severely infected by weeds. Delayed sowing is the best documented traditional method for. It also reduces S. *gesnerioides* and *Cynodon dactylon* infection. However, delaying sowing date implies shortening grain filling, which is detrimental for yield. Thus, options are needed that combine the yield benefit of early sowing with a decreased weed infestation. Hand weeding can only be recommended in cases of limited infestation to prevent any further increase in the weed population and to reduce the seed bank in the soil. However, even when hand weeding is still commonly used in some countries where no other feasible means of control are available and the wages for labor are cheap, it is only practical in preventing build-up of weed seeds in slightly infested soils.

Fig. 17.3 Ideal plant stand in the weed free plots of lentil

Solarization utilizes the sunlight in the summer to produce high temperatures under clear polyethylene mulch that covers the soil for several weeks. Mulching wet soil with transparent polyethylene sheets for a period of 4-8 weeks during the warmest season proved to be effective for a number of vegetables but also lentil (Mauromicale *et al.*, 2001). Cultivation of the soil can strongly affect the weed seed bank. Minimum tillage can contribute to weed control by reducing the amount of viable seeds incorporated into the soil. No tillage has been reported to considerably reduce *O. crenata* infection on faba bean and lentil (López-Bellido *et al.*, 2009). On the other hand, deep-ploughing has been recommended to bring the seeds into a depth, where they cannot germinate due to the lack of oxygen, but expectations have not been met sufficiently perhaps due to problems with the costs and practicability. Nitrogen compounds and manure fertilization has potential for control of broomrape species. Nitrogen in ammonium form affects negatively root parasitic weed germination and/or elongation of the seedling radical. In addition, manure fertilization augments the killing effect of solarization on *O. crenata* seeds. Intercropping is already used in India as a low-cost method of controlling weeds. It has recently been shown that intercrops with oat (*Avena sativa L.*) or with fenugreek (*Trigonella foenum-graecum L.*) or berseem clover (*Trifolium alexandrinum L.*) can reduce *O. crenata* infection on legumes being allelopathy a major component for the reduction. The finding that germination of seeds exposed to synthetic germination stimulants is inhibited in presence of oat or fenugreek roots suggest that oat roots might be exuding substances that inhibit *O. crenata* seed germination. This has been confirmed in a subsequent work, and trigoxazonane identified from fenugreek or benzoxazolinones from oat root exudates that might be responsible for inhibition of *O. crenata* seed germination. Considerable genetic variation in allelopathic activity has been found within crops which may allow for selection of more allelopathic cultivars.

Rotation may have direct and indirect impacts on weed in infested areas. While trap- and catch-crops in rotation may reduce to some extent the seed bank in soil, as discussed above, other rotation crops may have allelopathic effects on weed seeds. Decreasing host cropping frequency cannot, by itself, solve the weed problem. A nine-course rotation would be required to prevent the seed bank increases in lentil based cropping system. The use of trap crops offers the advantage of stimulating weed seed germination without being parasitized, contributing to seed bank depletion. However, given the high amount of seeds in infested soils, it cannot be expected that the application of trap crops or germination stimulants will eliminate the seed bank in the soil immediately. It is a feature of weeds that their numerous tiny seeds stay viable for a long time and never do germinate all at the same time. The application of trap crops therefore should be included in the regularly rotation and fallow management for the infested fields. The use of cereals could be a suitable solution due to the high frequency in which they are cultivated and its importance for human consumption. For that reason, breeding on cereal germplasm for the maximum inductor effect on broomrape seeds could be of great interest. In fact clear genotypic effects have been reported in sorghum in induction of *S. gesnerioides* germination. Growing the selected varieties with high inductor potential for suicidal broomrape germination in rotations introduced in infected soils could shorten the time necessary between host crops. Catch crops in contrast to trap crops are plants which are heavily parasitized. Harvesting or destroying them after the appearance of the weed would dramatically reduce the seed potential in the soil. The ideal solution would be a catch crop of agronomical interest by itself.

Fig. 17.4 Ideal plant stand in the weed free plots of lentil at podding stage

Chemical Weed Management

Successful results from herbicide application depend heavily on numerous interacting factors. Many of the biological factors involved are not fully understood, and are out of your control so give careful attention to the factors that you can control. Annual weeds compete with lentil mainly when the crop is in its earlier stages of growth 5 leaf stages. Weeds should be removed no later than 6 weeks after sowing to minimize losses. However, only rarely are selective herbicides completely non-toxic to the crop. Early post-emergence control nearly always results in higher yields than the chemical applied after branching.

Points to remember for the successful use of herbicides

Plan the operation- check paddock sizes, tank capacities, water availability and supply. Carefully check crop and weed growth stages before deciding upon a specific post-emergent herbicide. Read the label-check to make sure the chemical will do the job. Note any mixing instructions, especially when tank mixing two chemicals. Follow the recommendations on the label. Conditions inhibiting plant cell growth, like stress from drought, water logging, poor nutrition, high or low temperatures, low light intensity and disease or insect attack are not conducive to good herbicide uptake and movement. Use good quality water, preferably from a rainwater tank. Water quality is very important. Bore, hard, dirty or muddy water needs hotup to improve results with certain herbicides.Use good equipment checked frequently for performance and output. Check boom height with spray pattern operation for full coverage of the target. Check accuracy of boom width marking equipment. Check wind speed, a light breeze helps herbicide penetration into crops. Do not spray when wind is strong. Do not spray if rain is imminent or when heavy dew or frost is present. Calculate the amount of herbicide required for each paddock and tank load. Add surfactant where recommended. Select the appropriate nozzle type for the application. Beware of compromising nozzle-types when tank mixing herbicides with fungicides or insecticides. Be aware of spraying conditions to avoid potential spray drift onto sensitive crops and pastures, roadways, dams, trees, watercourses or public places. Keep a record of each spray operation.

Grass and broadleaf weeds are a very serious problem in lentil production.Typical troublesome weeds include wild oats, various mustards, nightshades; *Amaranthus palmeri, Chinopodium album, Lactuca serriola, Matricaria discoidea, Thlaspi arvense* and *Matricaria spp.* Lentil is poor competitors against weeds and require an intensive weed control program. Weeds also contribute exudates at harvest that stains lentil seed. Stained seed is considered inferior in quality. Weeds also interfere mechanically with lentil harvest. Yield loss and reduced seed quality from weed competition, if left uncontrolled, would range between 50-100%.

Chemical desiccant use varies on a yearly basis, and depends on the extent of the weed infestation and the natural dry down of the lentil crop at maturity. Desiccant herbicides are important in years of warm wet, springs and cool, wet, summers that promote luxuriant plant growth. Under such conditions lentils will continue to flower and set pod and weeds will continue to grow as long as moisture is available. If growers must wait for natural dry down to occur under such high moisture conditions, they will risk, pod shattering, sprouting, seed coat slough, and seed bleaching. Weeds that remain luxuriant under available moisture will also mechanically impair the harvest of the crop. In years when weeds are less threatening, lentils are mechanically swathed or direct combined. The following herbicides are effective for weed management in lentil

Paraquat dichloride (Gramoxone Extra): Use varies on a yearly basis. In years of little weed problems, no paraquat is used. In years of high weed infestation, paraquat is applied to 10-15% of the lentils in late season as a harvest-aid to desiccate lentils and weeds. It's typically applied once per year @ 0.35 kg *a.i.*/ ha.

Glyphosate (Roundup Ultra, Roundup Ultra RT): It is applied to 5-10% of the acreage as a pre-harvest aid or a spot treatment to control escaped weeds such as *Cirsium arvense, Matricaria spp.* and grassy weeds. It's usually applied once per year @ 0.85 kg *a.i.*/ha. As a spot treatment, glyphosate is especially useful in the low, wet areas of fields to control persistent weeds. As a pre-harvest aid, its use is similar to the paraquat use as a desiccant. This is considered an important herbicide because of its lower residual fate.

Imazethapyr (Pursuit) : Most widely used herbicide to control broadleaf weeds. It is used in 75% lentil grown in the world. Most often applied pre-plant incorporated in combination with triallate to control wild oats. The chemical should apply once per year @ of 0.053 kg *a.i.*/ha.

Pendimethalin (Stomp 33EC): Apply pendimethalin 1.0 kg *a.i.*/ ha. as pre-emergence for weed control in lentil, when crop is grown as intercrop with other winter crops.

Metribuzin (Sencor DF 75%): Metribuzin is the most common post-emergence herbicide used to control *Matricaria spp.* and *Chinopodiu album*. It is used @ 0.22 kg *a. i.*/ha. Once in a year.

Integrated Weed Management

Clearly, there is no single ecological silver bullet to replace herbicide use, but an increasing numbers of studies are testing suites of tactics together; including modified crop rotations, planting arrangements, residue management and tillage (Sosnoskie *et al.*, 2006). In one case, a combination of crop rotation, crop sequence, no-till, residue management and competitive crop canopies has enabled wheat growers in the world to reduce pesticide use by 50%. Some researchers used an empirical life cycle simulation based on demographic knowledge of major weed species to identify the best rotations and crop sequences for reducing weed growth and seed production. Others have shown significant reductions in weed seed banks under different management systems. For example, over a

6-year period weed seed density in the soil increased in conventional and no-till systems, but declined in reduced input and organic systems; however, despite the seed bank declines weeds were not effectively controlled in the reduced input and organic systems over a 12-year period. Elsewhere after six years reduced tillage in combination with crop rotation had increased weed diversity but reduced seed density by 80% (Murphy *et al.,* 2006).

There are many options for further improvement in developing weed-suppressive cropping systems, especially, if more competitive crop varieties become commercially available, we increase our knowledge of weed and seed bank ecology and better utilize modeling tools such as life cycle simulations, and further refine cover crop and residue management strategies. Weed suppression goals need to be linked to the management of the whole pest complex; to do this will require stronger collaborations among weed scientists, entomologists, pathologists, nematologists, plant breeders and soil scientists. Other pests and diseases are impacted by changes in tillage, crop rotation, fertility, etc., and weeds themselves can be hosts for crop pests or provide refugia for natural enemies that can enhance biological control. Unraveling the effects of different components of the pest complex will require innovative research designs and statistical analysis (Kranz, 2005).

Conclusion

This section has focused on how to reduce weed competition with crops; however, there are also concerns about losing positive services that weeds can provide. There is evidence that weeds can be good hosts of arbuscular mycorrhizal fungi (AMF) that can play important roles in improving nutrient availability, suppressing pathogens and perhaps in the bio-control of weeds (Vatovec *et al.,* 2005). Others have demonstrated the role of weeds in sustaining biodiversity. For example, in Europe weeds found in stubble after grain harvest greatly affected the number of birds using the field, notably linnets (*Carduelis cannabina*), and loss of weeds is thought be one of a number of causes in the overall decline of birds in farmland. Work in the UK also revealed that arable weeds support high insect diversity reinforcing the need to balance weed control and biodiversity conservation (Marshall *et al.,* 2003).

References

Ahmadv and G, Mondani F and Golzardi F. 2009. Effect of crop plant density on critical period of weed competition in potato. *Scientia Horticulturae* Amsterdam 121: 249-254.

Dubey SK, Uma Sah and Singh SK. 2011. Impact of climate change on pulse productivity and adaptation strategies as practiced by the pulse growers of Bundelkhand region of Uttar Pradesh. *Journal of Food Legume* 24(3): 230-234.

Erman M, Tepe IK, Bükün B, Yergin R and Takesen M. 2008. Critical period of weed control in winter lentil under non-irrigated conditions in Turkey. *African Journal of Agricultural Research* 3: 523-30.

Gibson KD, Fischer AJ, Foin TC and Hill JE. 2003 Crop traits related to weed suppression in water-seeded rice (*Oryza sativa* L.). *Weed Science* 51: 87-93.

Gulluoglu L and Arioglu H. 2009. Effects of seed size and in-row spacing on growth and yield of early potato in a Mediterranean environment in Turkey. *African Journal of Agricultural Research* 4: 535-541.

López-Bellido RJ, Benítez-Vega J, López-Bellido L. 2009. No-tillage improves broomrape control with glyphosate in faba-bean. *Agronomy Journal* 101:1394-1399

Marshall EJP, Brown VK, Boatman ND, Lutman PJW, Squire GR and Ward LK. 2003. The role of weeds in supporting biological diversity within crop fields. *Weed Research* 43: 77-89.

Mauromicale G, Restuccia G and Marchese A. 2001. Soil solarization, a non-chemical technique for controlling *Orobanche crenata* and improving yield of fababean. *Agronomie* 21: 757-765.

Murphy CE and Lemerle D. 2006. Continuous cropping systems and weed selection. *Euphytica* 148: 61-73.

Murphy SD, Clements DR, Belaoussoff S, Kevan PG and Swanton CJ. 2006. Promotion of weed species diversity and reduction of weed seedbanks with conservation tillage and crop rotation. *Weed Science* 54: 69-77.

Panetta FD, Lawes R. 2005. Evaluation of weed eradication programs: the delimitation of extent. *Diversity and Distribution* 11: 435-442.

Seem JE, Creamer NG and Monks DW. 2003. Critical weed-free period for 'Beauregard' sweetpotato (*Ipomoea batatas*). *Weed Technology* 17: 686-695.

Singh AK, Singh D, Singh AK, Gade RM and Sangle UR. 2012. Good Agronomic Practices (GAP) - An efficient and eco-friendly tool for sustainable management of plant diseases under changing climate scenario. *J. Plant Disease Sci.* 7(1): 1-8.

Singh P, Singh L, Singh P, Lone BA, Qayoom S, Ahmad L, Ganai MA, Kanth RA and Fayaz A. 2015. Response of lentil (*Lens culinaris* Medik) and weeds to different weed management practices under temperate conditions. *Journal of Agrisearch* 2(1): 72-74.

Vatovec C, Jordan N and Huerd S. 2005. Responsiveness of certain agronomic weed species to arbuscular mycorrhizal fungi. *Renewable Agriculture and Food System* 20: 181-189.

Weinig C. 2005. Rapid evolutionary responses to selection in heterogeneous environments among agricultural and nonagricultural weeds. International Journal of Plant Science 166: 641-647.

Yenish J. 2009. Lentils. Pages 202-204 in E. Peachey, D. Ball, R. Parker, J. P. Yenish, D.W. Moroshita, and P. Hutchinson, eds. Pacific Northwest Weed Management Handbook. Moscow, ID: Educational Publications.

Citation

Vishram Ram, Singh AK and Singh AK. 2018. Weed management for optimizing lentil production. In: Scientific Lentil Production. Singh AK, Bhakta N, Sangale UR, Manibhushan, Sundaram PK, Kumar S and Yasin JK. (eds.) Society for Upliftment of Rural Economy Varanasi, India. Pp. 309-321.

❑❑❑

Chapter 18

Weed Constraints and their Management in Lentil Crop

Parmeet Singh, AA Saad, FA Pir, Purshotam Singh, Lal Singh, Waseem Raja, Amjad Masood and Zahida Rashid

SYNOPSIS

Lentil is one of the oldest legumes for mankind. Weed communities continuously change both quantitatively and qualitatively as they react to the biotic and edaphic components of the environment. Most of lentil fields were dominated by *A. Tenuifloius*, probably due to its wide ecological amplitude as it can grow both on light and heavy soils. However the major weeds are *Chenopodium album, Fumaria parviflora, Melilotus alba, Chenopodium murale, Anagalis arrvensis, , Asphodelus teunifolius, C. oxycantha, S. Halapence, F. indica, M. Denticulata and L. sativus & Vicia sativa*. So all the factors discussed should be applied in holistic manner to manage the biggest constraint of weeds for profitable production.

In general the period from 30 to 60 days after sowing is critical period for lentil crop and is most crucial for competition with weeds. Good weed control is essential because lentils grow slowly during winter and are poor competitors with weeds. Avoid sowing lentils in fields with a history of severe broad-leafed weed problems. Pre herbicides are often ineffective on late emerging weeds and available post broad leaved weeds can result in crop damage. Integrated weed management (IWM) combines the use of biological, cultural, mechanical, and chemical practices to manage weeds, so that reliance on any one weed management technique is reduced.

Keywords: Crop rotation, IWM, Lentil, Weed management.

Introduction

Lentil (*Lens culinaris* medic) is one of the earliest domesticated plants known to humanity (Sarkar and Erskine, 2006). This crop is important for human nutrition and sustains large population in South

Asia and the Mediterranean region. Lentil is classified according to seed size, seed coat and cotyledon colour. Chilean lentil is large seeded with a thousand seed weight of 50 gms or greater while Persian lentil is small seeded with a thousand seed weight of 40 gms or less (Mavikar *et al.,* 2006). The major seed coat classes of lentil are green and red (Mcvicar *et al.,* 2006). Green lentil generally has a pale seed coat and yellow cotyledon where as red lentil has dark seed coats and red or orange cotyledons. Green lentil is typically marketed as whole seed, while red lentil is marketed either whole or dehulled and split.

Over the last 30 years, many agronomic improvements to lentil have been made such as improved disease resistance, height, lodging tolerance and yield potential (Sarkar and Erskine, 2006). However, weed control is a major concern and one of the greatest limiting factors in lentil production (Erman *et al.,* 2004). Weeds are major concern and weed control is foremost issue in lentil production. Infect Erman *et al.,* 2004 consider weeds as the most important factor affecting lentil yield. Weed competition in lentil has resulted in yield losses of 14% upto 100 % (Swanton *et al.,* 1993; Boerboom and Young, 1995; Singh *et al.,* 1996). Weeds can also cause problem for mechanical harvest (Brand *et al,* 2007). Lentil is a poor competitor with weeds due to its short structure, slow canopy closure and slow rate of development especially early in the growing season (Elkoka *et al,* 2004; Kirkland et al, 2000; Black Shaw *et al.,* 2002). As a result lentil is ranked as least competitive crop growing. The physiological and morphological characters, generally increase the competitive ability of the crop with weeds like early emergence, rapid leaf expansion, the formation of a dense canopy, increased plant height, early vigorous growth and increased root size (Black Shaw *et al.,* 2002). Unfortunately lentil does not possess many of these aforementioned properties. Lentil develops relatively slowly, especially early in the growing season which creates the opportunity for the growth and development of weeds (Erman *et al.,* 2004; Elkuca *et al.,* 2004; McDonald *et al.,* 2007). Lentil canopy closes late in comparison to other crops and is often sparse with open space available for weeds to establish (Elkuca *et al.,* 2004; Kirkland *et al.,* 2000) lentil is also a short plant (Elkuca *et al.,* 2004) reaching heights of 30 to 40 cm and combined with open canopy structures provides little shades and increased opportunity for weed emergence. Lentil roots are relatively shallow which allows moderate drought tolerance. However, this limits the uptake of water and nutrients when in competition with weeds (Mc vicar *et al.,* 2006). Lentil's low competitive ability is compounded when growing season temperatures are low or when moisture is scarce. Furthermore lentils does not tolerate water logged conditions or saline soils.

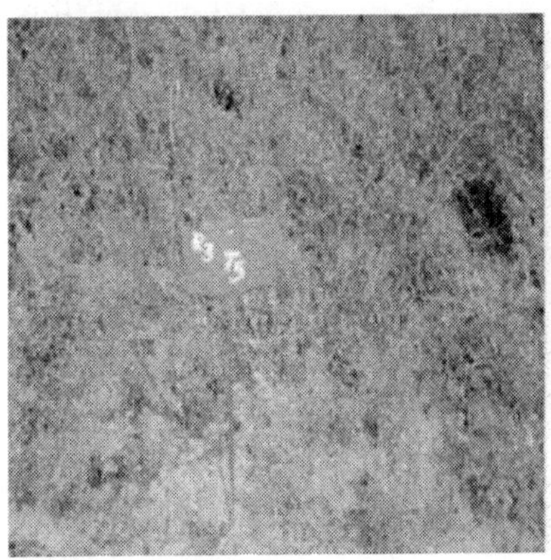

Weed infestation of lentil crop in temperate Kashmir

Herbicide Injury of lentil crop in temperature Kashmir

Fig. 18.1 Weed infestation and herbicides injury in lentil

Therefore the combined physiological and morphological characteristics of lentil result in relatively low competitiveness with weeds. Consequently, simultaneous growth of weeds and lentil plants resulted in the loss of plant height, auxiliary branch number, pod number per plant; seed yield and shoot dry weight of lentil plants as compared to the treatment of full absence of weeds during vegetative season as tabulated below

Table 18.1 Influence of weed management practices on the lentil performance

Treatments	Dry Weed Biomass (q/ha)	Pulse yield of lentil (q/ha)	% increase
Unweeded check	48.0	3.6	–
Hand-weeded	–	13.5	2750
Pendimethaline (PE, 1.0 kg/ha)	15.21	13.2	266.7
CD (P=0.05)	1.5	1.4	–

(*Source:* OP Gupta, 2012)

Weed communities of Lentil fields

Weed communities are never static nor are they necessarily at equilibrium as they change in response to external and internal forces. Weed communities continuously change both quantitatively and qualitatively as they react to the biotic and edaphic components of the environment. Weed identification and record keeping was essential in strategizing successful weed management practice. Thus the information regarding their distribution, ecological importance is a pre-requisite to identify the most noxious weeds in any agricultural system. Moreover comparison of weed communities helps to understand the patterns of interspecific population shift during the growth season of the crops. Most common weeds observed in lentil field are presented in Table 18.2.

Table 18.2 Most common weeds observed in lentil field

1. *Anagallis arvensis* L.
2. *Cenchrus ciliaris* L.
3. *Cousinia thomsonii* C. B. Clarke
4. *Artemesia Scoparia* Waldst & Kit
5. *Centaurea iberica* Trev, Ex, Spreng
6. *Cynodon dactylon* (L.) Pers.
7. *Asphodelus tenuifolius* Cavan
8. *Chenopodium album* L
9. *Dactyloctenium aegyptium* (L.) P. Beaur
10. *Avena fatua* L.
11. *Chenopodium murale* L
12. *Dicanthum annulatum* (Frossk) Stapf
13. *Buglossoides arvensis* L
14. *Cirsium arvensis* (L.) Scop.
15. *Emex australis* Steinch
16. *Calendula arvensis* L.
17. *Citrullus colocynthis* (L.) Schard
18. *Eragrostis poaeoides* P. Beauv.
19. *Carthamus oxycantha* M. B
20. *Convolvulus arvensis* L.
21. *Euphorbia dracunculoides* Lam

22. *Euphorbia helioscopia* L.
23. *Fumaria indica* (Hausskn), K. N
24. *Galium aparine* L.
25. *Heliotropium europeaeum* L
26. *Lathyrus aphaca* L.
27. *Lathyrus sativus* L.
28. *Launaea nudicaulis* N. K. N (nonless)
29. *Malva parviflora* L.
30. *Medicago denticulate* Willd
31. *Orobanche aegyptiaca Pers*
32. *Otostegia Limbata* (Btr.) Bioss
33. *Pentanema vestitum* (Wall, ex D. C) Ling
34. *Silene arenisa* C Koch
35. *Sonchus asper* (L.) Pers.
36. *Sorghum halapense* (L.) Pers
37. *Spergula fallax* (Lowe) E. H, L Krause.
38. *Rifolium polycerata* L
39. *Veronica agrestis* Auct non L.
40. *Vicia monantha* L.
41. *Withania somnifera* (L.) Dunal

Most of lentil fields were dominated by *A. Tenuifloius*, probably due to its wide ecological amplitude as it can grow both on light and heavy soils (Gupta, 1987), has low water requirement (Ashiq *et al.*, 1996), but develop think stands with adequate moisture (Gupta, 1987). Further Sultan and Nasir, 2007 reported that *A. Tenuifolius, C. oxycantha, S. Halapence, F. indica, C. Album, M. Denticulata and L. sativus* are most problematic weeds in lentil fields and will be the problematic weeds in future. A noxious weed *Parthenium hysterophorus* is recorded, particularly on the road sides and field margins in the form of thick stands. There is a strong likelihood that it will transfer from field margins into fields in near future (Sultan and Nasir, 2007). It is prerequisite to enquire bioecological features, which include dormancy, germination, seedling development, emergence, vegetative growth, flowering, seed setting, maturity and seed dispersal of prevalent weed species for designing an optimal control method. Further the dominant weeds in lentil fields were broadleaved so, herbicide should be used accordingly.

Critical Period of Weed Control (CPWC)

Critical period of weed control (CPWC) was first documented by Nieto *et al*, 1968. It is defined the period in a crop life cycle in which weeds must be controlled to prevent yield losses and is a useful tool to determine the optimal timing of the weed control (Knezevic *et al.*, 2002). In practical terms, the CPWC is used to determine the length of time that weeds must be controlled in a crop to prevent significant yield losses. It can be used to optimize herbicide application, tillage or other means of weed control. Therefore utilizing the CPWC can reduce prophylactic herbicide usage, environmental and ecological degradation (Kenzevic *et al.*, 2002), as well as the cost of weed control (Mohammad *et al.*, 2005). The CPWC has been determined in many crops and has helped producers to identify the optimal weed control timing and method. The CPWC has been determined for lentil (Mohmed *et al.*, 1997; Singh *et al.*, 1996), However results are region specific. In general the period from 30 to 60 days after sowing is critical period for lentil crop and is most crucial for competition with weeds.

Weed Management

Good weed control is essential because lentils grow slowly during winter and are poor competitors with weeds. Avoid sowing lentils in fields with a history of severe broad-leafed weed problems. Vetches, tares, medics, clovers, Wild Radish, Bed straw, Bifora and self-sown pulses are particularly hard to control. Preventing seed set of broad-leafed weeds in the year prior to growing lentils is important. It is essential to plan your weed control strategy before sowing. Delaying sowing is an option in most areas except the lower rainfall areas. This can enable several weed kills before sowing. It is not advisable to keep the lentil fields weed free always. Because it may not be economical. So to avoid any significant loss to our economic lead it is recommended that the weed infestation particularly at critical stage should be managed scientifically. This can be achieved by maintaining weed intensity and dry weight below threshold level. In weed control we aim to limit the growth of unwanted plants, both in space and time, without any attempt to eliminate them from the scene. Following approaches are generally being adopted, each having its own benefits and limitations.

1. Preventive Approach
2. Physical or Mechanical Approach

3. Cultural or Agronomic Approach
4. Chemical measures Approach
5. Biological measures Approach
6. Integrated Approach.

Physical or Mechanical Approach

It involves use of manual or mechanical energy for destroying weeds. It is the oldest but most common method of weed control. For weeding the wide row crops, a variety of implements run with animal and tractor power have been developed during the last 200 years or so. Practices in lentils, such as harrowing or rotary hoeing fields after emergence, represent a viable alternative for organic producers. Producers should be aware that mechanical weed control practices should be applied with caution due to the high sensitivity of lentil shoots and roots to damage. If harrowing or hoeing is planned, be sure to increase seeding rates because mechanical damage may reduce stands.

Cultural or Agronomic Approach

In practical weed management strategy cultural practices is of pivotal role. It includes manipulation of soil conditions to meet optimum requirement of lentil crop, selective placement of nutrients, maintaining high crop populations and proper insect and disease management for vigorous crop. Eventually it can be achieved by proper seedbed preparation, variety selection, proper sowing and crop establishment, insect and disease precautions and management, nutrition management, and irrigation scheduling. For example, Baird *et al.*, 2009, observed that increasing the seeding density of lentils translated into increased yield and decreased weed biomass. However, while cultural practices are at the backbone of an integrated weed management plan, they alone may not be enough to secure adequate weed control in lentil fields. Increasing crop plant density seemed to be an important mean to increase lentil competitive ability and enhance the efficacy of partial weed suppressive means like mechanical control and chemical control at moderately reduced rate. This also seemed to be possible due to the wide range of optimum densities in this crop.

Crop rotation

It is one of the most powerful tools to manage weed problems through agronomic ways. The help that crop rotation offer in weed control is not always been recognized before their effectiveness is proved. It is stated that with each crop appear certain typical weeds (Crop associated weeds) which are much less serious in some crops. These weeds increase in proportions rapidly, if their favourable crop is grown simultaneously, but can be sabotaged by growing different crop. It is very important to carefully consider the crop rotation history prior to growing lentils, as this crop is very sensitive to many herbicides. The high persistence SU-herbicides commonly used in small grain crops such as Ally (Metsulfuron), Glean (Chlorsulfuron), can damage subsequent annual legumes, including lentils. Due to their persistence, it may be necessary to wait up to four years before seeding lentils, depending on the product, application rate, and weather conditions.

Stale seedbed

By this agronomic way we can reduce the weed seed bank in lentil fields. These are the beds where one to two flushes of weeds are destroyed before planting of any crop. Here the weed seeds which are mostly concentrated in upper 4-5 cm depth of soil are slightly irrigated to stimulate their germination. These germinated flush of weeds are destroyed wither by using any contact herbicide or by a shallow non inverting type tillage implement. Depending upon the time available one or two flushes of weeds can be destroyed in this manner. Then lentil crop seeds are sown on these plots where weed seed bank is already exhausted due to stale seedbed treatment.

Chemical Approach

Weed management in lentil production is difficult due to poor competitiveness of the crop and the availability of only a few efficacious herbicides for control of broad leaf weeds (Wall and Mc Millan, 1994). Pre herbicides are often ineffective on late emerging weeds and available post broad leaved weeds can result in crop damage (McDonald *et al.*, 2007). There are only three herbicides registered for control in lentil i.e *Ethalflurolin* [{N-ethyle-N-(2-methyle-2-propanyle-2 4 dinitro-4 (trifluromethyle) benzenamine[; trade name Edge™, *Trifluralin* [{2,6 dinitro-N, N-dipropyl-4-(trifluoromethyl) benzenamine)} trade name Treflan™] and *Metribuzin* [{4-amino-6-(1,1- dimethylethyl) -3-

(methylthio)-1,2,4 trizin-S-(4H)-One]; trade name Sencor™] (Sensman, 2007). Triflurolina and Ethalfluralin are classified as dinitroanalines (DNA) and are mitotic inhibitors (Group three) with activity on both grass and broad leaf weeds. DNA are pre herbicides and affect roots and shoots of emerging seedlings resulting in the appearance of swollen root tips and shoots. These herbicides have limited activity on cruciferous weeds (Wall and McMullen, 1994) and their activity and efficacy can be influenced by soil and environmental conditions (Backie *et al.,* 2006). Furthermore the application of these herbicides can result in recropping restriction for some crops.

Metribuzin is classified as photosynthetic inhibitor (Group 5 herbicide) and affects only broadleaf weeds. The herbicides are foliar applied and has limited soil activity with a half life of only 2.6 weeks (Bouchard *et al.,* 1982). The resulting injury includes chlorosis and necrosis of oldest leaves within 1-3 days (Senseman, 1997). Metribuzin provides limited control of broadleaved weed species (Malik and Town ley-smith, 1990), suppressing wild mustard (*Sinapsis arvensis* L), Lambs quarters (*Chenopodium album* L) and Sink weed (*Thalpsis arvensis* L . Metribuzin can be phytotoxic to lentil with symptoms including leaf drop, reduced yield and plant death. Metribuzin can also result in reduced lentil seed size (with possibly low germination) (Malik and Townley-smith, 1990). To prevent such injury Metribuzin must be applied early in the growing season (McVicar *et al.,* 2006) which may not correspond to optimal weed control timing. Further, Metribuzin can be applied in combination with graminicides due to antagonism of grassy weeds (Kirkland *et al.,* 1989). Graminicides must be applied approximately five to seven days after Metribuzin. This delays grass weed control and is generally inconvenient and expensive for the producer to make two herbicide applications to control both grass and broadleaf weeds.

Avoiding herbicide damage

Lentil is susceptible to the soil residues of some herbicides used in previous years. Lentil is also sensitive to herbicide drift. Inadvertent drift from broadleaf weed herbicides can be detrimental and cause slow crop growth, reduce yield, or both. Sprayer tanks should be thoroughly cleaned before applying any crop protection product on lentil, and producers should inform their neighbours about the location of their lentil crop. During periods of crop stress (heat, drought, frost, or after land rolling) the ability of the lentil crop to tolerate herbicide application may be reduced. Crop injury can be reduced by waiting approximately four days after the crop stress before applying the herbicide, by maintaining water volumes at label recommendations and by applying the product during the evening. Under adverse conditions, most post sowing pre-emergent herbicides are capable of causing damage to lentil. Even post emergent application can also cause crop damage in some circumstances. In most cases damage can be attributed to the products solubility and:-

- Sowing too shallow
- Applying the herbicides to the dry soil followed by heavy rainfall
- Rate too high for soil type
- Variety susceptible differences
- Uneven soil surface
- Washing of herbicides into press wheel furrows.

Crop topping

Lentil can be successfully crop topped with Gramoxone (R) to control seed set of annual rye grass, however care is needed as timing is important. Ideally, lentil should be crop topped as late as practical to minimize yield reductions, but the growth stage of the grass often determines the timing. They cannot be crop topped too ealy because grain quality can be affected with green kernels arising from premature ripening. Lentil variety choice based on maturity time can therefore be important. The lentils ideally should have finished flowering and have well formed seeds in the pods before commencing spraying. The lentils should be starting to turn yellow/brown at this time. Beware of coloured foam marker dyes staining seed through the pods.

Weed Wiping

Weed wiping is being successfully used in lentils to prevent weed seed set of herbicide resistant rye grass and other tall weeds in the crop. Crop height needs to be consideration in lentil variety choice.

Preventing herbicide resistance

Herbicide resistance is the innate ability of a weed biotype to survive and reproduce after treatment with an herbicide dose that would normally be lethal. To reduce the risk of developing herbicide resistant biotypes, producers should rotate among herbicides with different modes of action applied either as tank mixes, premix formulations, or sequential applications. Also, producers should rotate management practices. Finally, crop rotation is an excellent tool to reduce the selective pressure on herbicide resistant weeds.

Biological Approach

It involves use of living organisms. A successful bio-agent should be host specific, adjustable to the new environment, rapid destroyer of the target weed, easy to multiply and effective on several texa of the weed in question. It is unfortunate biological control has given the last resort in weed management. However, with more rapid development of bio-herbicides and other fast acting bioagents, this thinking is bound to change. There is scarcity of literature about use of bioagents and alleochemicals for weed management in lentil crop.

Integrated Approach

There are four major principles of integrated weed management. First manipulation of crop habits with the objective to make it more completive, second is to break the survival mechanism of the weeds in the soil. Third, adoption of crop husbandry practices which discourage establishment of perennial and parasitic weeds e.g. crop rotations and fourth, practice should be flexible for accommodation of other approaches and ecofriendly too. Integrated weed management (IWM) combines the use of biological, cultural, mechanical, and chemical practices to manage weeds, so that reliance on any one weed management technique is reduced. The main goals of an IWM program are to 1. use preventive tools to maintain weed density at a level that does not harm the crop, 2. prevent shifts toward more difficult to control weeds, and 3. develop agricultural systems that maintain or improve crop productivity, farm revenues, and environmental quality. Designing a successful IWM program requires understand-ing the different biological and ecological factors that influence the short-, mid-, and long-term dynamics of weeds in agricultural settings.

Weed control methods such as hand weeding (Elkoca *et al.*, 2004) and physical control methods such as crop tillage (Boerboom and Young, 1995) are effective and common practice in some lentil growing regions. However they are labour intensive, time demanding and expensive. Weed management is integral to ensure profitable lentil production due to crops poor competitiveness with weeds. Cultural, physical and chemical methods are often employed to manage weeds in lentil crops (Brand *et al.*, 2007).

Breeders Concern for Weed Control

To address weed control and crop selectivity concerns, plant breeders at the university of Saskatchewan crop development centre developed herbicide tolerant lentil varieties (Chant, 2004). Imidazolinone (IMI) tolerant varieties were first commercially introduced in 2006. The varieties has increased tolerance to IMI herbicide until the II node stage. Imazethapyr/Imazamox are the IMI herbicides registered for use in IMI- tolerant lentil. IMI tolerance represents a new technology for weed control in lentil and information is required to optimize weed control and subsequently maximize yields.

Conclusion

Lentil is an important pulse crop and grows almost all the climatic conditions. It can withstand to a large extent against frost. Due to its morphology and growing habit, it is succumbs from competition of weeds especially under stress conditions. Initially it is slow growing crop, hense infested badly with weeds. It takes 45-60 days to spread on ground. First 30-60 days are considered most critical. Weed infestation is one of the limiting factors in achieving optimum yield of lentil. The extent of yield reduction depends upon time, duration and intensity of weed infestation and weed competition with crops for growth resources. Major weeds are *Chenopodium album, Fumaria parviflora, Melilotus alba, Chenopodium murale, Anagalis arrvensis, , Asphodelus teunifolius, C. oxycantha, S. Halapence, F. indica, , M. Denticulata and L. sativus & Vicia sativa*. So all the factors discussed should be applied in holistic manner to manage the biggest constraint of weeds for profitable production.

References

Ashiq M, Nayyar MM and Shaw ML. 1996. Weeds of crops and their control. Agronomy Deptt. Ayub Agriculture Research Institute. Faisalabad. pp. 82.

Baird JM, Shirtliffe SJ and Walley FL. 2009. Optimal seeding rate for organic production of lentil in the northern Great Plains. *Canadian Journal of Plant Science* 89: 1089-1097.

Beckie HJ, Harker, KN, Hall LM, Warwick SI, Légère A, Sikkema PH, Clayton GW, Thomas AG, Leeson JY, Séguin-Swartz G and Simard MJ. 2006. A decade of herbicide-resistant crops in Canada. *Can. J. Plant Sci.* 8: 1243-1264.

Blackshaw RE, JT O'Donovan, KN Harker and X Li. 2002. Beyond herbicides: new approaches to managing weeds. p.305-312. Proceedings of the International Conference on Environmentally Sustainable Agriculture for Dry Areas; Shijiazhuang, Hebei, China. Lethbridge, AB, Canada, Dobing Enterprises.

Boerboom, CM and Young FL. 1995. Effect of postplant tillage and crop density on broadleaf weed control in dry pea (*Pisum sativum*) and lentil (*Lens culinaris*). *Weed Technol*ogy. 9: 99-106.

Bouchard DC, Lavy TL and Marx DB. 1982. Fate of metribuzin, metolachlor and fluometuron in soil. *Weed Sci.* 30: 629-632.

Brand J, Yaduraju NT, Shivakumar BG and McMurray L. 2007. Weed management. p 159-172. *In* S.S. Yadav, D.L. McNeil, and P.C. Stevenson (ed.). Lentil: An ancient crop for modern times. Springer, Dordrecht, The Netherlands.

Chant SR. 2004. Imidazolinone tolerance in lentil (*Lens culinaris* Medik.). M.Sc. Thesis, University of Saskatchewan, Saskatoon, SK. pp 95.

Elkoca E, Kantar F and Zengin H. 2004. Effects of chemical and agronomical weed control treatments on weed density, yield and yield parameters of lentil (*Lens culinaris* L. Cv. Erzurum-89). *Asian J. Plant Sci.* 3: 187-192.

Erman M, Tepe I, Yazlik A, Levent R and Ipek K. 2004. Effect of weed control treatments on weeds, seed yield, yield components and nodulation in winter lentil. *Weed Res.* 44: 305-312.

Gupta OP. 1987. Predominent weeds of Indian Agriculture. Advances in Weed Science, A case study of Indo-Pak Subcontinent. Proc. Pak-Indo-US Weed Control workshop. NARC, Islamabad, March 11-14. Pp. 3309-317.

Gupta OP. 2014. Modern Weed Management. 3rd edition. Agrobios (India). Pp. 373.

Kirkland KJ, Holm FA and Stevenson FC. 2000. Appropriate crop seeding rate when herbicide rate is reduced. *Weed Technology* 14: 692-698.

Kirkland KJ, Harker KN and O'Sullivan PA. 1989. Influence of metribuzin and cyanazine on the phytotoxicity of graminicides on wild oats and barley. *Can. J. Plant Sci.* 68: 195-203.

Malik N and Townley-Smith L. 1990. Performance of imazethapyr on pulse crops. *Weed Technol.* 4: 791-798.

McDonald GK, Hollaway KL and McMurray L. 2007. Increasing plant density improves weed competition in lentil (*Lens culinaris*). *Aust. J. Exp. Ag.* 47: 48-56.

McVicar R, Pearse P, Brenzil C, Hartley S, Panchuk K, Mooleki P, Vandenberg A and S Banniza. 2006. Lentil in Saskatchewan. [Online] Availableat www.agriculture.gov.sk.ca/.Default.aspx? DN=ce32b 7bd-c7ad-4b9e-a560 (verified 18 December 2008). Saskatchewan Ministry of Agriculture.

Mohammadi G, Javanshir A, Khooie FR, Mohammadi SA and Zehtab S. 2005. Critical period of weed interference in chickpea. *Weed Res.* 45: 57-63.

Nezevic SZ, Evans SP, Blakenship EE, Van Acker RC and Lindquist JL. 2002. Critical period for weed control: the concept and data analysis. *Weed Sci.* 50: 773-786.

Nieto J, Brondo MA and Gonzalez JT. 1968. Critical periods of the crop growth cycle for competition from weeds. *PANS.* 14: 159-166.

Nikam, SR, Moerkerk M and Cousens R. 2010 Weed seed contamination in cereals and pulse crops. *Thirteen Australian Weed Conference.* Pp. 59-6.

Senseman SA (ed.). 2007. Herbicide Handbook. 9th ed. WSSA, Lawrence, KS.

Sultan S and Nasir ZA. 2007. Intra-annual variation in weed communities of Lentil fields in CHakwal Pakistan. *Pak Journal of Botany* 38(5): 1471-1479.

Swanton CJ, Harker KN and Anderson RL. 1993. Crop losses due to weeds in Canada. *Weed Technol*ogy. 7: 537–542.

Wall DA and McMullan PM. 1994. Effectiveness of several new selective herbicides in lentils (*Lens culinaris). Crop Prot.* 13: 553-557.

Citation

Singh P, Saad AA, Pir FA, Singh P, Singh L, Raja W, Masood A and Rashid Z. 2018. Weed Constraints and their Management in lentil Crop. In: Scientific Lentil Production. Singh AK, Bhakta N, Sangale UR, Manibhushan, Sundaram PK, Kumar S and Yasin JK. (eds.) Society for Upliftment of Rural Economy Varanasi, India. Pp. 323-336.

□□□

Chapter 19

Insect Pests of Lentils (*Lens culinaris* L.) and their Management

Ajay Kumar Pandey, RP Maurya and Promod Mall

SYNOPSIS

The most serious and frequent occurring insect pest of the stored grain is *Callosobruchus chinensis* which cause severe damage in storage, bean pod borer (*Etiella zinckenella*) and aphid. The management of insect pest of the crop is crucial to optimizing production. To achieve target of additional production of pulses, it is necessary to concentrate efforts on management of insect pests. Identification of these insect pests with their biology and management strategies is of great significance to design the management strategies most effectively and economically.

Keywords: Insect pests, lentils, management, pea aphid, lentil pod borer.

Introduction

Lentil, probably originated in the Near East and rapidly spread to Egypt, Central and South Europe, the Mediterranean basin, Ethiopia, Afghanistan, India, Pakistan, China and later to the new World including Latin America (Ladizinsky, 1979; Duke, 1981; Cubero, 1981). Lentil is a nutritious food legume having relatively high content of protein (22%), carbohydrate and calories (3250 kcal/kg) with fast cooking characteristics (Muehlbauer *et al.*, 1985; Saskatchewan Agriculture and Food, 2005). Furthermore, the residues including husks, dried leaves, stems, fruit walls and bran (residues) can be fed to livestock. In India, Lentil (*Lens culinaris*) is one of the world's most important winter season pulse crop after chickpea and pea. India is producing 14.76 million tons of pulses from an area of 23.63 million hectare. Among the major pulse crop like chickpea, pigeon pea, mung bean, urdbean and lentil, lentil is grown on 15 lakh hectare area (7.88% of total area under the major pulse crops) with the production of 9.5 lakh tonnes (7.79% of total pulse production) with the productivity of 633.3 kg per ha. Among all the state of India, Madhya Pradesh, Uttar Pradesh and Bihar are the major states which produce lentil. Out of total area under lentil production, 40.53 percent (6.08 lakh ha) land is under the cultivation of lentil in Uttar Pradesh while in Madhya Pradesh and Bihar, it is 37.6 percent (5.64 lakh ha) and 10.8 percent (1.62 lakh ha), respectively. These states contribute 45.76, 30.21 and 12 percent share in lentil production with the productivity of 715.46, 703.7 and 508.87 kg/ ha, respectively. The magnitude of production and productivity of lentil depends on various abiotic and biotic factors. Out of many factors, this crop is exposed to a broad spectrum of insect species to attack at different stage of crop. The insect pests like aphid, leaf weevils, lygus bug, cutworm, thrips, bud weevil, pod borers, bean pod borer (*Etiella zinckenella*) and leaf minors (*Liriomyza* spp, *Phytomyza* spp.) have been recorded to cause damage to lentil crop (Van der Maesen, LJG. 1972). The most serious and frequent occurring insect pest of the stored grain is *Callosobruchus chinensis* which cause severe damage in storage, bean pod borer (*Etiella zinckenella*) and aphid. The management of insect pest of the crop is crucial to optimizing production. To achieve target of additional production of pulses, it is necessary to concentrate efforts on management of insect pests. Identification of these insect pests with their biology and management strategies is of great significance to design the management strategies most effectively and economically. However, major problems of these insects have rearely been recorded in the field. The details of major insect pests of lentil are as follows:

1. Pea aphid (*Acyrthosiphon pisum*)

Family: Aphididae

Order: Homoptera

Pea aphid, *Acyrthosiphon pisum* (Harris) is a widespread insect pest of leguminous crops including lentils (*Lens culinaris*). Beside the field peas as main hosts, pea aphid (*Acyrthosiphon pisum* also attacks on other legumes crops including lentil, faba beans, sweet peas, etc. Adult pea aphids are soft-bodied, slow-moving and light to dark green in colour. They are pear shaped, about 3 mm long and 1.5 mm wide with long legs. The antennae of the pea aphid have narrow dark bands at the tip of each segment. The nymphs are closely resemble the adults but smaller than the adult. They feed and produce wingless females that give rise to winged and wingless females. In late September or October, females lay eggs or sometime directly nymph on leaves and stems. The nymph start sucking the plant sap and reach to maturity in 5 to 20 days depending on the weather. The optimum temperatures for rapid development of pea aphids range from about 23 - 28 °C. Both, adults and nymphs suck the sap form the soft parts of plant. The drainage of plant sap results in symptoms like deficiency of water. The severely infested leaves become curled which eventually affect the yield.

Management

Early sowing of lentil crop may reduce yield loss. Among the various predator, lady bird beetle, syrphid flies, flower bugs and damsel bugs reduce the aphid population as these predator predate on aphid. At least five species of parasitoid belonging to Aphidiidae are known to attack or parasitized the pea aphids. The most common parasitoid is known as *Aphidius ervi*. Spraying of systemic insecticides like Malathion 50 EC @2 ml/1 have been found to be effective.

2. Lentil pod borer (*Etiella zinckenella)*

Family: Pyralidae

Order: Lepidoptera

The lima bean pod borer *Etiella zinckenella* (Lepidoptera: Pyralidae) is a major agricultural pest which attacks various leguminous crops (Tabata and Yasuda, 2011; Naito and Masaki, 1962; Oatman, 1967; Whalley, 1973) and has been reported from Japan, Australia, Indonesia, U.S.A., Mexico, West Indies, South America, Europe, Sri Lanka and India. This insect feed on lentil, pigeon pea, lima bean, green peas, cow pea, horse gram etc. Adult is brownish grey moth with a white stripe along the anterior margin of each forewing and transverse yellow band and pale hind wings and the wing expanse is 25 cm. Larva is greenish initially, having 5 black spots on the prothorax, turns pink before pupation. The pest breeds throughout the year. The moths are nocturnal in nature and emerge in February-March. The mated female lays whitish, elliptical eggs singly or in small groups on various parts of the plant including developing pods. At 25°C temperature, eggs hatch within five days and first instar larvae come out. The larva feed on floral part and developing pods and becomes full grown in 10-27 days. Pupation takes place in the soil at the depth of 2-4 cm and its development is completed in 10-15 days. There are five generations in a year. The damage is caused by the larvae which web together top foliage with flowers and feed upon them. They also bore into pods to feed on developing grains. After feeding, larva moves to another pods. Owing to this type of feeding on grains, the reduction in yield may be up to 5 percent.

Management

- Deep summer ploughing eliminates immature stages of this pest.
- Spray NSKE 5% twice followed by triazophos 0.05% has been found effective.
- Apply any one of insecticides at 25 kg/ha. chlorpyriphos 1.5 % DP, quinalphos 4D, carbaryl 5D or spray insecticides like quinalphos 25 EC @ 1000 ml/ha.
- Among the newer pesticides larvin 75WP @ 562.5 g a.i./ha, flubendiamide 480 SC @ 48 g a. i. /ha, flubendiamide 480 SC @ 36g a. i. /ha have been found effective to manage this pest (Kumar and Shivaraju, 2009).

3. Leaf Miner (*Phytomyza atricornis*)

Family: Agromyzidae

Order: Diptera

This pest is widely distributed in northern India and feeds on Mustard, cauliflower, cabbage, lentil, potato etc. The adults are two winged flies having greyish black meso-notum and yellowing frons. The pest is active from December to April or May. Pupation is inside leaf mine. Both winter and summer are passed in pupal stage. Only the larvae are destructive and they make prominent whitish tunnels in the leaves. The adult emerge out at the beginning of December and after mating, start laying eggs in leaf tissues. The eggs hatched in 2-3 days and neonate maggots mine the leaf and continue to feed between the lower and upper epidermis of leaf by making zig-zag tunnels. The mining of mesophyll of leaf by maggot leads to development of irregular blotches on leaves. Maggots feed inside the mine and become full grown in about 5 days and pupate within the galleries. The adult emerge from the pupae in 6 days and the life cycle is completed in 13-14 days. When the attacked leaves are held against bright light, the minute slender larvae can be seen feeding within the tunnels. The larvae of this pest mine the leaves to feed on the leaf tissue resulting in series of triangular blotch symptom in which eggs are deposited or irregular mines are the visible symptoms of attacked by this pest. The large number of tunnels made by the larvae interferes with photosynthesis and the proper growth of the plants making them look unattractive.

Management

Several pesticides administered separately as well as in a combination of sprays, dust and granular formulations have been found effective in controlling the damaged by leaf miner (Atwal *et al.*, 1969). The application of phosphamidon, dimehtoate and monocrotophos effectively controlled this pest. Some newer insecticides like imidacloprid, acetamiprid, acephate, profenofos, thiamethoxam, NSKE have been found to be effective for the management of this pest (Harendra and Saravanan, 2008). Maggots are parasitized by some braconids, eulophids. Foliar spray with dimethoate 2 ml/l is found to be effective.

4. Lygus bug *(Lygus pratensis)*

Family: Miridae

Order: Hemiptera

Lygus bugs are considered the most important polyphagous insect of pest of both herbaceous crops and fruit trees (Eric *et al.*, 2012). In lentil crop, several species like *Lygus lineolaris* (the tarnished plant bug), *L. borealis, L. elisus* and *L. keltoni* have been observed in different countries which are destroying flower buds and seeds of various crops including lentil. The tarnished plant bug has more than 300 recorded host species. Lygus bugs are small, oval-shaped insects that feed on a variety of crops and weeds. Adult is about 3 mm wide and 6 mm long with pale green to reddish brown or black colour. Lygus bugs share characteristics with all true bug which include the distinctive, triangular or 'V' shaped marking on the dorsal surface of body and membranous wingtips. Adults are active and fly readily when approached. Immature lygus bugs (nymphs) are light green and wingless. Young nymphs are often mistaken for aphids, which are similar in size and shape. However, lygus bug nymphs are much more active, are harder bodied. Several black spots, usually five, occur on the backs of nymphs. There are five moults after which immature stage become adult. Wing buds are evident in the fourth and fifth instar. Lygus bugs survive the winter protected in the ground litter and crop residues. During the spring adults lay eggs and feed on various plants. The eggs hatch into nymphs in about 10 days and reach maturity in a month. Damage is caused by the piercing–sucking mouthparts to suck plant sap from the developing flower buds and seeds., which punctures the pods and seed coats injecting a toxic substance into the plant parts (Pulse News, 2003) resulting in visible lesions to surfaces of stems, buds, flowers and pods. Seeds that have been fed on will collapse or shrink, as well as darken, and will lose their quality and viability. They also create small, dark, circular patches on the pod surface and on the underlying seeds. The spots result in some economic losses due to low marketing prices offered by grain buyers (Summerfield *et al.*, 1982 and Summerfield *et al.*, 1992; Crop Profiles, 2001). Feeding lygus bugs produces depressed, chalk-coloured lesions (chalky spot) on the lentil seeds. Host plants of lygus include alfalfa, canola, lentils, potato, strawberries, vegetable crops, fababean, tree fruits and weeds such as redroot pigweed, stinkweed and wild mustard.

Management

Two applications of acephate (0.2%) reduce the population of Lygus bug.

5. Bud thrips *(Frankliniella* spp.)

Family: Thripidae

Order: Thysanoptera

About 18 thrips species on legumes have been recorded out of which *Thrips tabaci* Lind. *Frankliniella intonsa* (Trybom) and *Thrips fuscipennis* Haliday, are most common. About 90 % of the total number of thrips occurred on peas (*Pisum sativum* L) and lentil (*Lens culinaris* (Medic.), bean (*Phaseolus vulgaris* L.) and soybean (*Glycine max* (L.) Merr. Significantly more *T. tabaci* infested lentil and peas than beans and soya bean (Pobozniak. 2011). It is found in all kinds of vegetation and is a major pest of onion from November to May. However, it also causes damage to leguminous crop also. The female live about 15 to 30 days and lay eggs singly in leaf tissue with the help of sharp ovipositor. The eggs hatch after 4th to 9th day of egg lying and first instar nymphs come out which start feeding on different pant parts. Nymphs are full grown in 5 to 10 days. The pupation occurs in ground. Both nymphs and adults cause damage by sucking the sap from leaves, stems and floral parts resulting in spotted appearance. The infestations also lead to dropping of flowers. The infested leaf express white patches due to drainage of sap and finally attacked leaf become curled and wrinkled.

Management

Spraying of cypermenthrin 0.1 % reduces the population by inhibiting the thrips to lay the eggs (Ting *et al.*, 2012).

6. Cutworm (*Agriotis* spp.)

Family: Noctuidae

Order: Lepidoptera

This pest occurs thorough out India including Punjab, U. P., Orissa, Bihar, Bengal, Assam, M.P. Maharashtra, Karnataka, Rajasthan, Tamil Nadu etc and cause severe damage to wheat, barley, maize, gram, potato and tobacco crops including lentil crop. Ever through they are called gram cutworms; they are polyphagous attacking besides gram, potato, tobacco, wheat, cucurbits, cotton, pea, lady's finger, other vegetables etc. Adult is a dark brown or whitish coloured moth with reddish tinge. Forewings are dark coloured with patches while the hind wings are light coloured. Caterpillars are dark grey or dirty black in coloured with grey appearance and have the habit of coiling up at the slightest touch. The moths are great fliers, found on the hills in the summer and in the plains in the winter and vice versa. They are active during night and migrate in large number of swarm to reach the plains in October. After mating they lay eggs on the under surface of the leaves of the most plant or even in the soil. The larvae remain hidden in the soil during the day and feed at night on young shoots or underground tubers. Pupation occurs in earthen coon in the soil. The life cycle takes longer in the winter. The damage is done by the caterpillars which cut the stems at ground level during night and drag it below the soil for feeding. The larvae cause more damage than what they eat. In old plant, they cut shoots and feed upon leaves which result in ultimately reduction in the yield. In severe infestation, whole field is covered with cut plants. However, in lentil, it is a sporadic and does not severe attack every ear. It remains active from November to January.

Management

1. Soil application of chlorpyriphos (1.5%) dust @ 25 kg per ha at the time of land preparation or seed treatment with chlorpyriphos 20 EC @ 500 ml dissolved in 5 liter water for one quintal of seed.
2. Collection and destruction of larvae can be done if possible.

7. Pulse beetle (*Callosobruchus chinensis*)

Family: Bruchidae

Order: Coleoptera

It is a serious pest of pulses crops which has been reported from all over India and many other countries also. This pest attacks all kinds of pulses like gram, pea, soybean, pigeon pea moong, urd, lentil etc. Adult is small, chocolate coloured about 3 to 5 mm long with serrate antennae. The indentifying characteristic is that the wings do not cover abdomen completely. The eggs are laid singly glued to a single seed. Single seed may contain 5 to 8 eggs. After 4-8 days, neonate larvae come out form the eggs. The first instar grub bore the seed and grows by eating the seed content. The adult come out from the pupa by cutting round holes in the seed coat. The damage is caused by grubs by eating out the entire content of the grains leaving only the shell. Infestation starts from the field. Maximum activity is seen from June to August. The adult do not cause damage except egg laying. The grubs alone eat away the entire seed content. Beetles start attacking the pods in the field itself from where the infestation is carried to godown. Holes cut by the escaping adults are easily seen. If the container is poorly aerated, fungus and foul smell could also develop.

Management

1. The ginger rhizome powder and neem leaf powder have been reported to be effective in reducing the egg lying of the beetle.
2. The lentil varieties 79-1 have been found resistant against *Callosobruchus analis*.

3. Clean and whitewash the store before storing the grains. Cracks, cervices and holes present in the floor, walls and ceiling of the store should be filled up.
4. The bags should be disinfected before storing the grains. The bags should be dipped in a solution of Malathion prepared by taking 2 ml of Malathion 50 EC in one liter of water for 10 minutes and dried in shade before use.
5. Only clear and dried grain having moisture less than 10 percent should be stored.
6. Reduction of the survival of the beetle population has been recorded in seed treated with Neem leaf powder.

8. Pod borer (*Helicoverpa armigera*)

Family: Noctuidae

Order: Lepidoptera

It is polyphagous insect pest which feeds on all most pulse crop incluiding vegetable like potato, okra, tomato, vegetable pea and chillies. Moths are pale brown in colour with 'V' shaped specks on forewings while larvae are greenish to brownish in colour with dark brown linings along the sides of the body. A fertilized female lay yellowish eggs about 250 to 600, singly on various plant parts. The incubation period is about 5-8 days. The young larvae feed on the foliage and pods. The full grown larvae pupate in soil by making earthen cells which ends with in 8-15 days. There may be 4-6 generations in a year. Though this pest prefer to feed on gram and red gram, but larva is polyphagous in nature so it can feed most of the pulse crop including lentil. Larvae of this pest first feed on tender plant parts and later it starts feeding on developing grains. In severe condition, larvae consumed the whole pod.

Management

Timely sowing or early maturing cultivars reduce the possibility of heavy infestation of *H. armigera* to pulse crops. Use of *Helicoverpa* tolerant variety is recommended. Mixed intercropping with non-preferred host like linseed is effective to reduce the pest attack. From flowering stage adopt following IPM practices-

1. Erect pheromone traps @ 10/ha to monitor the pest.

2. Erect bird perches @ 50/ha to attract predatory birds.
3. When eggs and early instar larvae are noticed spray NSKE 5 % or neem based insecticides
4. Use of microbial insecticides:
    - NPV 200 to 500 LE per ha is effective to reduce the pest population.
    - *Bt* formulation 400g or 400 ml/ac thrice at weekly interval in evenings is effective to manage the pest palpation.
5. Collection and destruction of four or fifth instar larvae is effective method to reduce the population in crop grown on small area.
6. Avoid indiscriminate use of insecticides, synthetic pyrethroids and mixtures.
7. On need basis spray
    - Chlorpyriphos 20 EC @ 2.5 ml/l at initiation of flowers
    - Quinalphos 2 ml/l or acephate 1.5 g/l at flowering and fruiting using 750-1000 l of spray fluid with High Volume sprayer.
    - In severe incidence, indoxacarb 1 ml/l or spinosad 0.3 ml/l
8. Adopt community approach.

Among the various insect pest of pulse crop, very few are the major pest of lentil crop. These insect pests do not cause severe damage regularly. Seldom, any one may become serious problem in 4 - 5 years.

Conclusion

Lentil is a nutritious food legume having relatively high content of protein. Lentil suffer attacks of several insect pests, however, the most serious and frequent occurring insect pest of the stored grain is *Callosobruchus chinensis* which cause severe damage in storage, bean pod borer (*Etiella zinckenella*) and aphid. The management of insect pest of the crop is crucial to optimizing production. To achieve target of additional production of pulses, it is necessary to concentrate efforts on management of insect pests. Identification of these insect pests with their biology and management strategies is of great significance to design the management strategies most effectively and economically.

References

Atwal AS, Chaudhary JP and Ramzan M. 1969. Studies on the biology and control of pea leaf miner, *Phytomyza atricornis* (Meigen). *Journal of Res. Punjab*. 6: 163-169.

Crop Profiles 2001. Crop profile for lentil in, Montana. http:www.scarab.msu.montana.du

Cubero JI. 1981. Origin, taxonomy and domestication. p. 15-38. In: C. Webb and G. Hawtin (eds.), Lentils. C.A.B., London, UK.

Duke JA. 1981. Handbook of legumes of world economic importance. Plenum Press, New York. p. 52-57.

Eric Conti, Francesca Frati, and Gianandrea Salerno. 2012. Oviposition Behaviour of *Lygus rugulipennis* and its Preferences for Plant Wounds. *Journal of Insect Behavior* 25(4): 339-351.

Kumar Ashok CT and Shivaraju C. 2009. Evaluation of newer insecticide molecules against pod borers of black gram. *Karnataka J. Agric. Sci.* 22(3): 521-523.

Ladizinsky G. 1979. The origin of lentil and wild gene pool. *Euphytica* 28: 179-187.

Naito A and Masaki J. 1962. Studies on the distribution of lima bean pod borer, *Etiella zinckenella* Treitschke and soybean pod borer, *Grapholitha glycinivorella* Matsumura. *J. Cent. Agric. Exp. Stn.* 2: 145–228.

Oatman ER. 1967. An ecological study of the lima-bean borer (*Etiella zinckenella* Tr.) in Southern California. *Ann. Entomol. Soc. Am.* 60: 552–555.

Singh Harendra and Saravanan L. 2008. Seasonal incidence and management of pea leaf miner *Phytomyza horticola* Goureau infesting pea. *International Journal of Plant Protection* 1: 33-37.

Summerfield RJ, Muehlbauer FJ, Short RW. 1982. Lygus bugs and seed quality in lentils (*Lens culinaris* Medic.). Agricultural Reviews and manuals, Agriculture Research Service, USDA, 43 p.

Summerfield RJ, Short RW, Muehlbauer FJ. 1992. Lygus bug on lentil in the United States. *In:* Proceeding of the Second International Food Legume Research Conference on Pea, Lentil, Faba bean, Check pea and Grass pea, 12–14 April, Cairo, Eygpt, pp. 859–876.

Tabata Jun and Yasuda Koji. 2011. The Sex Pheromone of the Lima Bean Pod Borer *Etiella zinckenella* (*Treitschke*) JARQ. 45(1): 47-50.

Ting Yang, Geert Stoopen, Gerrie Wiegers, Jing Mao, Caiyun Wang, Marcel Dicke and Maarten A. Jongsma 2012. Pyrethrins Protect Pyrethrum Leaves Against Attack by Western Flower Thrips. *Frankliniella occidentalis. Journal of Chemical Ecology* 38(4): 370-377.

Van der Maesen, LJG. 1972. *Cicer* L., a monograph of the Genus, with special reference to the chickpea (*Cicer arietinum* L.), its ecology and cultivation, Commun. Agric.University. Wageningen

Whalley PES. 1973. The genus Etiella Zeller (Lepidoptera: Pyralidae): A zoogeographic and taxonomic study. *Bull. Br. Mus. (Natl. Hist.) Entomol.* 28: 1–21.

Citation

Pandey AK, Maurya RP and Mall P. 2018. Insect pests of lentils (*Lens culinaris* L.) and their management. In: Scientific Lentil Production. Singh AK, Bhakta N, Sangale UR, Manibhushan, Sundaram PK, Kumar S and Yasin JK. (eds.) Society for Upliftment of Rural Economy Varanasi, India. Pp. 337-348.

❑❑❑

Chapter 20

Integrated Disease Management of Lentil (*Lens culinaris* Medikus subsp. *culinaris*)

RC Shakywar, Shyam Singh and HK Singh

SYNOPSIS

Lentil (*Lens culinaris* Medikus subsp. *culinaris*) is grown world over in cool (*Rabi*) season. Biotic stresses such as diseases, insect-pests, nematodes and weeds substantially reduce the yield of lentil in farmers' fields. Among these, fungi and viruses are the largest and most important groups affecting all parts of the plant at all stages of growth of cool (*Rabi*) season crop but prokaryotes (bacteria) are less effective against the crops. According, to estimates made in India nearly 10-15% of food legumes production is lost due to diseases alone. Among144 fungi, diseases caused by species of Ascochyta and Botrytis are of great importance to lentil and other members of family Fabaceae. The viruses of major economic importance on lentil belong to the, carlavirus, furoviruses, luteovirus, nanovirus and potyviruses. Scrutiny of the literature on diseases of lentil and their management reveals new records of diseases, loss estimations, symptoms, etiology of causal agents, mode of spread, favorable conditions, identification of host plant resistance and integrated disease management strategy. Integrated disease management (IDM) modules for important fungal, bacterial and viral diseases of lentil (ascochyta blight, anthracnose, botrytis grey mould, fusarium wilt, powdery mildew, rhizoctonia root rot, rust, bacterial blight, alfalfa *mosaic virus, bean yellow mosaic virus, cucumber mosaic virus, pea enation mosaic virus and pea seed borne mosaic virus*) are discussed. The IDM involves the individual component of disease management such as host plant resistance (HPR), agronomic practices, judicious use of fungicides, pesticides for vector management, bio-pesticides for pathogen, risk forecasting that operate on different aspects of the disease etiology, such that they complement each other and can be applied together in farmers' fields collectively to provide farmers with maximum economic return.

Keywords: Integrated Disease Management, Lentil, *Ascochyta* Blight, *Fusarium* wilt.

Introduction

Pulse play a significant role in improving source of revenue, nutritional security of farmers and populations in less developed countries as well as sustainability of agriculture in dry areas worldwide. Lentil (*Lens culinaris* Medikus subsp. *culinaris*) is one of the important and most nutritious *rabi* pulses. It has the potential to cover the risk of rainfed farming. It is also used as a cover crop to check the soil erosion in problem areas. The plants are ploughed back into the soil as green manure also. It derives the name *Lens* from the lens shaped seeds. It is mostly eaten as '*dal*'. The *dal* is made by splitting the grain in 2 cotyledons, which are deep orange red or orange yellow in colour. The whole grain is also used in some of the dishes. It is also rich in calcium (560 ppm), iron and niacin. It has the lowest content of lectins and trypsin inhibitors among legumes. Since it is a leguminous crop, it improves the fertility of soil biological nitrogen fixation. Lentil seeds also provide a source of starch for textiles and printing. Lentil residues form important livestock feed. Lentil floor is used for thickening of soups. It is mixed with wheat flour in bread and cake production (Lazou and Krokida, 2010). Proteins include the essential amino acids like as isoleucine and lysine and are an essential source of economical protein in many parts of the world. Lentils are deficient in two essential amino acids methionine and cysteine. Lentils also have some anti-nutritional factors such as trypsin inhibitors and relatively high phytate content. Trypsin is an enzyme involved in digestion, and phytates reduce the bio-availability of dietary minerals. The phytates can be reduced by soaking the lentils in warm water overnight (Callaway, 2004). Lentil grown universal is prone to attack by a large number of plant pathogens *i.e.* fungi, bacteria, viruses, phytoplasmas, nematodes and phanerogamic plant parasitic, which result in severe economic losses globally. Among these, fungi and viruses are the prevalent and perhaps the most important groups affecting all parts of the plant at all stages of growth. Diseases of lentil such as ascochyta blight, anthracnose, botrytis grey mould, fusarium wilt, powdery mildew, rhizoctonia root rot, rust, bacterial blight, alfalfa *mosaic virus, bean yellow mosaic virus, cucumber mosaic virus, pea enation mosaic virus and pea seed borne mosaic virus (Table 19.1)*. Around 45 viruses are reported to infect legumes (Bos *et al.*, 1988; Makkouk *et al.*, 2003) worldwide.

Table 20.1 Lentil diseases and their causal organism / agents

S.No.	Diseases	Causal Organism
A.	**Fungal diseases**	
1.	Ascochyta Blight	*Ascochyta lentis*
2.	Anthracnose	*Colletotrichum truncatum*
3.	Botrytis Grey Mould	*Botrytis cinerea* and *Botrytis fabae*
4.	Fusarium wilt	*Fusarium oxysporum* f.sp. *lentis*
5.	Powdery mildew	*Erysiphe pisi* DC., and *E. polygoni*
6.	Rhizoctonia root rot	*Rhizoctonia solani*
7.	Rust	*Uromyces vicia-fabae* (Pers.) de Bary.
B.	**Bacterial diseases**	
8.	Blight	*Pseudomonas syringae* pv. *syringae*
C.	**Viral diseases**	
9.	Alfalfa *mosaic virus*	*Alfalfa mosaic virus*
10.	*Bean Yellow Mosaic Virus*	*Bean Yellow Mosaic Virus*
11.	*Cucumber mosaic virus*	*Cucumber mosaic virus*
12.	*Pea enation mosaic virus*	*Pea enation mosaic virus*
13.	*Pea seed borne mosaic virus*	*Pea seed borne mosaic virus*

However, only few are of major economic concern with respect to specific regions. Among the more important groups are the carlavirus, furoviruses, luteovirus, nanovirus and potyviruses. The Potyviruses are the most important overall causing economically important diseases in lentil as well as legume crops. Many of the viruses are seed borne in their legume hosts; some are sufficient to have enabled worldwide distribution. There are reviews dealing with the individual foliar diseases of cool season and warm season legumes. In this chapter, attempts have been made to address the management of major fungal, bacterial and viral diseases of lentil on issues related to importance and disease management strategies for these diseases globally. Integrated disease management strategies for economically important diseases of lentil have been discussed in details with reference to the available research results on the symptoms produced on the plant, causal organism/agents, etiology/disease cycle and epiphytology, mode of spread, favorable conditions and integrated disease management. A successful integrated disease management (IDM) strategy is one under which lentil have been protected from the yield-reducing effects of the pathogen description the later to economic in significance. The IDM involves a total system approach to the suppression of pathogen populations to a level where higher yields can be obtained and enables the farmers to achieve maximum economic return. In the IDM system, the individual component of disease management such as host plant resistance, agronomic practices, judicious use of fungicides, pesticides for vector management, bio-pesticides etc., need to be compatible or complimentary.

Management of lentil diseases

Integrated disease management (IDM) of lentil is an approach that uses all available management strategies to maintain disease pressures below an economic injury threshold. It does not advocate a routine chemical application program to prevent disease but promotes the integration of cultural, physical, biological and chemical management strategies. The routine application of fungicides for cover purposes is not proper, as it does not hub the proper awareness on the actual difficulty and can go ahead to struggle and prospective environmental issues.

The crucial objectives of any IDM plan should be to attain at least the following:

Decrease the opportunity of introducing diseases into the crop and avoid creating circumstances appropriate for disease concern and increase. A sound disease management program integrates a few universal principles and concepts into the overall production system for that particular crop. It is important to understand the concepts of disease management and the strategies developed from them and to make modifications appropriate for a particular crop situation.

The disease triangle concept

The disease triangle (Fig. 20.1) shows the importance of the three factors necessary for disease development:

- Pathogen (virulence) or disease-causing organism (eg. fungus, bacterium or virus)
- Plant (susceptible) for the pathogen to infect
- Environment (favorable) in which the first two will interact

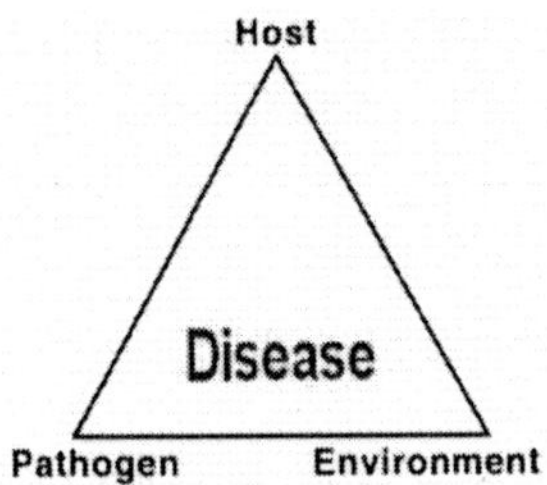

Fig. 20.1 Classical disease triangle

The main emphasis in research and development to combat lentil diseases is on host resistance and chemical management where ever applicable and quite often these components of disease management were practiced in isolation to each other. Recently, a shift in scientific thinking and practice in the management of grain lentil diseases has been seen and greater emphasis was on identifying, evaluating and integrating location specific components of integrated disease management (IDM). In general IDM has followed the principles of IPM (Jeger, 2000). The location specific IDM of food legumes is primarily based on host plant resistance (HPR) or genetic resistance; additionally other components of diseases management. In some environments, IDM may require a single component used alone (usually HPR) or in combination with one other component (such as fungicide seed treatment) to adequately combat diseases of lentil.

In this chapter, lentil disease management practices to economically important have been given (Table 20. 1). Management of viral diseases has been discussed separately under the virus disease management section.

A. FUNGAL DISEASES

1. *Ascochyta* Blight

Ascochyta blight occurs in all lentil growing areas of world, but is more likely to be conspicuous in the cool, southern high rainfall areas. Current commercial varieties are unlikely to suffer significant yield losses due to infection but pod infection is likely to result in seed discolouration that may reduce the market value of the grain.

Symptoms

The symptoms of disease are spot-like lesions develop on leaves, stems, pods and seeds. Small black pycnidia become visible with maturity. Premature leaflet drop, stem breakage and stem die back may occur. Affected crops under severe infection may be blighted and seed may become shriveled, reduced in size and discolored. Flowers and pods could abort leading to yield loss.

Causal organism: *Ascochyta lentis*

Etiology: The fungus survives for two seasons on infected stubble. Fruiting bodies (pseudothecia) form on the previous season's stubble in response to autumn rain and cold temperatures. Windborne ascospores are released from these fruiting bodies following rain. These ascospores can spread considerable distances, however the majority fall within the first few hundred meters.

Mode of spread: The disease is spread primarily by seed infection.

Favorable conditions: The fungus is generally survived in cool and wet weather. If crop rotation is not used, the severity of infection may increase. Non-resistant cultivars grown and planting low quality seed will increase the likelihood of infection. Residue from previous crops provides high inoculum.

Integrated disease management

Fungicide spray may be needed at flowering to control the infection of pods that causes grain discolouration. Use disease free seed as the first step in controlling this disease. Growers should also isolate new crops from the previous season's stubble to minimize the spread of infection into the crop by winds blown trash and spores. Wide rotations should also be maintained to minimize the disease carryover within a paddock.

2. *Anthracnose*

Anthracnose is a relatively new disease of lentils. The disease is causing yield losses in excess of 50%. It is much more destructive than ascochyta blight. This is significant since ascochyta completely destroyed some lentil crops north of Devils Lake in 1986. Anthracnose is widespread in Manitoba. It was detected in commercial lentils near Makoti in 1992 and was destructive in 1993. It destroyed or severely damaged the lentil variety trials at Carrington in 1993 and several years since.

Symptoms

Anthracnose can be first noticed as brown lens-shaped lesions appear on the leaves. They are less distinct than those of ascochyta blight. Small bristle-like spore bearing structures appear on affected stems. The crown and lower stem areas become bleached and then turn brown. The disease can cause plant defoliation and premature death. Small black fruiting bodies are produced on dead tissue. Crown infections reduce plant vigour and cause premature ripening.

Causal organism: *Colletotrichum truncatum*

Etiology: Spores overwinter on seed and infected lentil stubble. The spores are released during the spring and early summer. Soil borne spores infect and kill the lower leaves and can cause premature plant death. Upper plant parts are usually infected later in the season from spores that are spread by splashing rain.

Mode of spread: Wind plays little role in the spread of this disease other than the fact that the fungus can be spread by dust during harvest.

Favorable conditions: Humid to moist weather conditions, rain splash, wind and sprinkler irrigation. Infected fields nearby, leaf wetness for 18-24 hours and moderate temperature is ranging from 20-25 °C.

Integrated disease management

Use a crop rotation of at least 3 years. Avoid peas and faba beans in the rotation since there is evidence that they are susceptible to at least some of the same strains of the anthracnose fungus that attack lentil. Grow cultivars that have shown some resistance to anthracnose. Foliar fungicides have also been effective against this disease.

3. *Botrytis Grey Mould* (BGM)

Botrytis Grey Mould tends to occur on lentils in the cooler growing regions south of the Great Eastern Highway and in the higher rainfall coastal areas north of Perth. No serious Botrytis Grey Mould infections have as yet occurred in commercial lentil crops in country.

Symptoms

The disease becomes apparent under wet, spring conditions. All above ground parts of the plant can be affected. Grey to dark brown lesions develops on stems, leaves, flowers and pods. Under humid conditions the lesions become covered with a fluffy grey mould. Pod infection result in the production of shriveled and discolored seed, and result in a reduction in yield and grain quality.

Causal organism: *Botrytis cinerea* and *Botrytis fabae.*

Etiology: The fungus can survive as several forms these include in infected seed, sclerotia in the soil in old infected residues and on alternate host plants. Infected residues are an important source of fungal inoculum. Spores are produced on old residues are carried by the wind into new crops where infection can occur.

Mode of spread: *BGM* often *spreads* plant to plant within a row.

Favorable conditions: High relative humidity, moderate temperatures with high moisture, rain and sprinkler irrigation is favorable conditions for development of disease.

Integrated disease management

Digger, Cassab and Nugget have moderate resistance to botrytis grey mould. Use seed with the lowest level of Botrytis infection available.

Ensure crops are at least 500 m away from last year's lentil stubble and if possible up-wind. Always apply a fungicide dressing to the seed before sowing. Apply a foliar fungicide at early flowering if disease is present.

4. *Fusarium* wilt

This disease causes serious yield losses in those countries where the pathogen is known to occur. Yield losses of up to 60% may occur under favorable conditions (Singh *et al.*, 2007).

Symptoms: Disease appears in patches at seedling and adult stages. Seedling wilt is shown by sudden drooping followed by drying of leaves and the whole seedling. Symptoms are seen from flowering to pod filling stag. Stunting of the plant wilting of top leaves yellowing of foliage and death of the plant are the apparent symptoms on the infected plants. A slight reduction of lateral root growth is seen in a healthy root system.

Causal organism: *Fusarium oxysporum* f.sp. *lentis*

Etiology: Chlamydospores are formed abundantly and rapidly (2-4 weeks), but formation may be slow (4-6 weeks). Chlamydospores are usually formed singly or in pairs, but may be found in clusters or small chains. They may be either terminal or intercalary and are most obvious in hyphae on the agar surface, although they may appear in submerged hyphae. On Potato Dextrose Agar (PDA), colony morphology varies widely. *Fusarium oxysporum* usually produces a pale to dark violet or dark magenta pigment in the agar, but some isolates produce no pigment at all. Mycelia may be floccose; sparse or abundant; and range in colour from white to pale violet. Abundant pale orange or pale violet macro conidia are produced in a central spore mass in some isolates. Small pale brown, blue to blue-black or violet sclerotia may be produced abundantly in some isolates.

Mode of spread: The fungus is soil borne and may survive in the soil in the form of chlamydospores for long time.

Favorable condition: Warm spring and hot dry summer. Optimum temperature is around 25°C.

Integrated disease management

Crop rotation is best for eliminating soil borne infection. Removal of infected plant and deep summer ploughing. Growing of resistance varieties. Seed treatment with Carbendazim reduce early incidence of wilt. Seed treatment with Thiram + *Trichoderma harzianum* also reduce wilt incidence.

5. *Rhizoctonia* root rot

Symptoms

Reddish to dark brown root lesions can develop on epicotyls and hypocotyls. Brown discoloration occurring near the soil line on the epicotyls could girdle the stem. Seeds may rot, or when growing, develop rot or pre and post-emergence damping-off. Lesions developing on roots that eventually progress enough to pinch them off can be observed. Stem lesions can develop at or below the soil line that may expand above the soil line to the lower branches in older plants. Leaves can turn yellow from the base of the plant up. Circular patches of stunted plants may be observed in the field. A plant that can be easily pulled out from the soil is indicative of rot.

Causal organism: *Rhizoctonia solani*

Etiology: The fungus lacks fruiting bodies and spores. The mycelium is light-brown, thick in which black sclerotia are formed. Sclerotia are variable in form, small and loosely connected by mycelial threads. The fungus has a wide host range and can infect the roots of several.

Mode of spread: The fungus is soil borne and may survive in the soil in the form of sclerotia for long time.

Favorable conditions: Sandy soil favors infection, particularly when organic matter is high. Cool temperatures (11-18^{0}C) and root knot nematode presence make plants more susceptible to infection, although the pathogen tolerates a wide range of soil temperatures. Infection is more likely when paired with diseased seed, compacted soil, herbicide carryover and non-rotation of crops.

Integrated disease management

Use of crop rotation for 2-3 years with non-leguminous crops. *Seed treatments are a cheap and effective method for suppressing the disease.*

Seed treatment with bio-agents (*Trichoderma viride* @ 10g/Kg seed) / Indofil M-45 @ 3 g/Kg of seed. Always avoid infected seed. Use of foliar application of fungicides like as Indofil M-45, Blitox-50 etc. @ 3g/1 of water.

6. *Powdery* mildew

Powdery mildew another important disease of lentil has been reported from many countries such as Cyprus, Ethiopia, India, Siberia, Sudan, Syria, Tanzania and the former USSR. In North America, powdery mildew is often observed after lentil started flowering. The disease poses a serious problem on breeding material in plastic or glass houses in both India and Syria and in India it is also recorded in off-season nurseries in Trans Himalayan regions, like Lahaul Spiti and Sangla etc., but it is rarely seen in the field during the cropping season.

Symptoms

A fine powdery, white growth of conidia and mycelium initiates as small spots and rapidly spreads to cover the entire surface of leaves, stems and pods. Later, the leaflets become dry and curled, and are shed prematurely, causing considerable reduction in yield and seed quality. The seeds from infected plants remain small and shrivelled.

Causal organism: *Erysiphe pisi* DC., *E. polygoni* DC. and endoparasite *Leveillula taurica* (Lév.) Arnaud.

Mode of spread: The anamorph stage is responsible for spread of the disease.

Favourable conditions: Moderately high temperatures and moderate relative humidity favour the disease development.

Integrated disease management

Foliar sprays with fungicides benomyl, tridemorph, aqueous sulfur, karathane (dinocap), calixin or sulfex (ferrous bisulfide) as well as certain insecticides (Quinalphos, Tnazophos, Phoxim) applied at 10-15 days interval are effective in suppressing mildew growth (Beniwal *et al.*, 1993). Many lentil genotypes are reported resistant to powdery mildew (Tikoo *et al.*, 2005), and should be planted whenever possible. The teleomorph stage has been reported to occur in India and Sudan (Chitale *et al.*, 1971). Field and crop sanitation, Dithane M-45 or Carbendazim at 2.5 g/lit should be sprayed.

7. *Rust*

It is an autoecious fungus; completing its life cycle on lentil. The disease occurs during the flowering/early podding stage as aecia, which may develop into secondary aecia or uredia. The resulting aeciospores and uredospores lead to a further disease spread in the crop season.

Symptoms

The plants give dark brown or blackish appearance visible as patches in the field. Early in the season the rust produces small creamy, yellow spots on leaves. These are the spore masses of fungus. The spots are soon replaced by the familiar orange / brown pustules (uredia) surrounded by a pale halo (similar in appearance to rusts on cereal crops) on leaves and stems. Uredia leave a rust-like deposit when wiped with your finger; this deposit is the spores, which can be dispersed by wind to create further infections. Later in the season, rust infections on the stem appear as black / brown masses usually surrounded by a pale halo.

Causal organism: *Uromyces vicia - fabae* (Pers.) de Bary.

Mode of spread: The telial or resting stage, of the fungus which allows it to survive over summer.

Favorable conditions: The fungus survives in summer as teliospores. High humidity and cloudy weather with temperatures of 20-22 ^{0}C favor disease development.

Integrated disease management

It is advisable to use clean seeds without rust contaminations and to treat the seed with a suitable fungicide such as diclobutrazole or Bavistin @ 2g/kg of seed. Preventive fungicide sprays of Mancozeb at early disease development stage have been recommended. The use of host plant resistance is the best means of rust management. Field resistance of lentil to rust is governed either by a single dominant gene by two independently inherited dominant genes by two complementary genes or by three independently inherited genes (Basandrai *et al.*, 2005). Use of multiple disease resistance genotypes PL 01 and PL 02. Foliar application of systemic fungicides (tilt or contaf @ 0.2% /l of water) before disease initiation.

B. BACTERIAL DISEASE

8. *Blight*

Bacterial blight is rarely seen in different parts of India. The disease is more common and severe outbreaks occur about once a decade. *Severe epidemics can result in crop failure; however losses are usually less than 20%.*

Symptoms

The disease first becomes evident as small, dark-green, water-soaked lesions on leaves and stipules. The lesions may enlarge and coalesce but are always delimited by the veins and develop a characteristic fan shape. The lesions on the leaflets turn yellowish and later brown and papery; lesions on the pods are sunken and turn olive-brown. Lesions may also develop on stems near ground level. These begin as water-soaked areas, which later turn olive-green to dark brown. Stem lesions may coalesce causing the stem to shrivel and die. Stem infection may spread upwards to the stipules and leaflets. Pre-emergence and post-emergence damping-off may occur and even advanced plants may be killed.

Causal organism: ***Pseudomonas syringae pv. syringae***

Etiology: The bacteria can survive on seed. The disease commonly becomes established within a field by sowing infected seed. During wet weather Infection may occur at any stage of plant growth and is most prevalent following frosts.

Mode of spread: The bacteria spread from infected to healthy plants by rainsplash, wind-borne water droplets and plant to plant contact.

Favourable conditions: Plants damaged by frosts or any other physical damage are more susceptible to infection. Rainfall, heavy dews, strong winds and cold temperatures provide the most favorable conditions for spread of disease within crops.

Integrated disease management

The use of clean seed will minimize the possibility of disease, provided the land has not been cropped to lentil for several years. Do not use seed from crops identified with bacterial blight during field inspections. To obtain a blight-free crop, lentil should not be sown on land sown to lentil in the previous year or adjacent to pea stubble. Early sown crops are more vulnerable to bacterial blight infection than late sown crops; never sow earlier than recommended for your district. In areas prone to bacterial blight avoid early sowing. When bacterial blight is detected, steps should be taken to prevent the spread of disease. There are copper based compounds that are registered for use in lentil against bacterial blight.

C. VIRAL DISEASES

9. *Alfalfa mosaic virus*

Alfalfa mosaic virus (AMV) was first described in alfalfa in 1931 and later in pea in 1933. AMV is a significant pathogen of lentil in Australia, but apparently not elsewhere.

Symptoms

Infected plants are exhibiting a mild chlorosis with leaves reduced in size and frequently malformed. Necrosis often occurs along leaflet margins. Plants are usually stunted and may die prematurely. Flowering and pod formation are reduced and seeds small and shriveled.

Causal Agent

AMV consists of a tripartite genome containing three single-stranded RNA species, and a fourth sub-genomic RNA from which the coat protein is translated. Virions vary in size and shape from spherical to bacilliform measuring 18 nm in diameter and up to 56 nm in length. Transmission requires the presence RNA-1 consisting of 3644 nucleotides (nt), RNA-2 (2593 nt), and RNA-3 (2037 nt) together with the coat protein or RNA-4 (881 nt). The coat protein has a molecular mass of 24.25 kDa. AMV is serologically unrelated to any other known virus species.

Host range: The virus has a natural and experimental host range that includes nearly 700 plant species in 70 families and is currently found worldwide.

Transmission: AMV is seed-transmitted at levels up to 5% in lentil.

Disease cycle and epiphytology

The virus is transmitted in a non-persistent manner by at least 14 different aphid species of which the pea aphid (*Acyrthosiphon pisum*) is particularly important. AMV may be introduced into lentil fields from infected weed or crop sources but this can vary with geographical location and vector species. One other source of infection is infected seed. The secondary loci infections can occur in lentil colonized by aphid.

Management

An effective strategy for reducing incidence of AMV is to avoid plantings near established alfalfa in which virus incidence can reach near 100% after three years. Application of an approved insecticide may be beneficial in lentil when high aphid populations occur. Use of certified virus-free lentil seed eliminates seed-infected plants that can be a source of secondary spread.

10. *Bean Yellow Mosaic Virus*

It was discovered in common bean in 1925 and was the first virus reported to cause disease in chickpea in 1956. Many different isolates of the virus occur in several important crops throughout the world. It causes significant losses in lentil and other pulse crops also.

Symptoms

Symptoms on lentil include chlorosis, mild mosaic or mottling, and stunting. Leaves are often twisted or curled with necrosis along margins as the effects of the virus infection progress. Flowering and pod formation are reduced and consequently little seed is produced.

Causal Agent

BYMV is a flexuous rod-shaped virion 750 nm in length containing single-stranded RNA and a coat protein with 282 amino acids. The genome is monopartite with 10,000 nucleotides. BYMV is serologically related to *Clover yellow vein virus* (CYVV) which also infects many legume species. The coat protein of BYMV is 68-76% identical at the amino acid level with the coat protein of CYVV. BYMV is more distantly related to the legume-infecting poty viruses *Bean common mosaic virus, Bean common mosaic necrosis virus, Pea seed-borne mosaic virus* and *Soybean mosaic virus* (McKirdy *et al.*, 2000).

Host range: BYMV has a broad host range of over 200 plant species, predominantly in pulse crops.

Transmission: BYMV is transmitted by seed at level of up to 1% in lentil.

Disease cycle and epiphytology

The virus is acquired from infected weed and perennial plant sources by more than 20 aphid species and transmitted in a non-persistent manner. The most common aphids associated with transmission of BYMV to legume crops are *Acyrthosiphon pisum, Macrosiphum euphorbiae* (Thomas), *Myzus persicae, Aphis gossypii* (Glover) and *Aphis fabae,* but other species including cereal aphids may be important if present in high numbers. Common sources for the virus in the field include clover (*Trifolium* spp. and *Melilotus albus*), vetch (*Lathyrus* spp.), and alfalfa (*Medicago sativa*). Consistent with other poty viruses, BYMV is mechanically transmissible. A significant reduction in yield can occur in lentil fields when the incidence of BYMV is high (Makkouk *et al.*, 1992).

Management

Use of lentil seed certified to be free of BYMV will aid in reduction of the virus within the crop. Maintaining low levels of the virus in surrounding crops and weed hosts can be difficult because of the wide host range of BYMV but spread of the virus can be managed by eliminating local weed hosts. Application of insecticides to manage spread of the virus by aphids is ineffective except where the vector has built up large populations in lentil. A few sources of tolerance have been identified in lentil.

11. *Cucumber mosaic virus*

Cucumber mosaic virus (CMV) was first described in 1916 on lentil crop. It can cause severe losses in lentil during seasons of high aphid populations when infection incidences exceed 50%.

Symptoms

Symptoms in lentil include a general chlorosis of the plant, leaf malformation and stunting, often followed by premature death. Leaflets may show a mild mosaic or reddening of leaf margins. A reduction in pod number or filling of pods generally occurs resulting in yield loss.

Causal Agent

CMV is a member of the Family *Bromoviridae,* Genus *Cucumovirus.* It has a tripartite genome encapsidated in isometric particles ca. 29 nm in diameter that contain RNA-1 (3389 nucleotides), RNA-2 (3035 nucleotides), or RNA-3 (2197 nucleotides). A sub-genomic RNA-4 is also encapsidated in the particles that contain RNA-3 (Jones *et al.,* 2008). The coat protein consists of 218 amino acids with a molecular weight of 24.2 kDa. The species has been divided into distinct sero groups called subgroups I and II with the later comprising 70% of isolates (Moury, 2004). CMV is distantly related to other *Cucumovirus* species *Peanut stunt virus* and *Tomato aspermy virus.*

Host range: The virus has a large host range that includes over 1000 plant species belongings in 85 families.

Transmission: Transmission by seed is erratic in lentil and chickpea but can average 0.1% - 1%, occasionally reaching 2% or higher within seed lots.

Disease cycle and epiphytology

CMV can survive from year to year in many weed species or in green vegetation and crops in temperate areas. Perennial legumes such as clover and alfalfa are ready sources of inoculums and many of these hosts may remain symptomless. The virus is transmitted in a non-persistent manner by many different aphid species. A number of these species including cereal aphids may be important in the case of lentil. Aphids can acquire the virus within a few seconds and transmission can occur in an equally short time period. CMV is seed transmitted in lentil, forage legumes and in may weed species. There is more secondary spread in lentil and consequently lentil seed infected with CMV can lead to major losses in commercial crops (Fletcher *et al.*, 1999).

Management

CMV in lentil can be controlled by planting certified virus-free seed in localities where other sources of infection are not present. Application of systemic insecticides to lentil may be useful in reducing aphid populations, especially if colonization is observed. Lentil lines L-4, PAK 3598, ILL1983, Mizia, Medovina 1-G, ILL788, ILL7700, ILL7163, and ILL5405 have been identified with tolerance or partial resistance to CMV. Some tolerance in lentil may exist in ILL590 and the red lentil cultivar Cobber.

12. *Pea enation mosaic virus*

Pea enation mosaic virus (PEMV) is one of the most important and destructive virus diseases of lentil in temperate zones of the world (Makkouk *et al.*, 1999). It has been reported in the Pacific Northwest states and California in the U.S, Canada, Europe, West Asia, North Africa, Egypt, and Iran but is not known to occur in New Zealand and Australia. Severe epidemics occurred in the Pacific Northwest region in 1983, 1990, 1996 and 2005.

Symptoms

Infected lentil plants typically show a reduction in growth. Plants infected at an early growth stage may die especially if co infected with another virus such as *Bean leaf roll virus*. Leaves frequently exhibit leaf rolling and mottling, severely malformed leaves and tip growth, vein clearing and translucent flecks. Pods are also malformed or misshapen. The most characteristic symptom of PEMV infection is the development of enations (hyperplastic outgrowths) on leaf and pod surfaces. While, it is common in pea, it rarely occurs on lentil or chickpea.

Causal Agent

The disease is caused by a symbiotic co-infection of two single-stranded RNA viruses: PEMV-1 (genus *Enamovirus*, family *Luteoviridae*) and PEMV-2 (genus *Umbravirus*, not assigned to a family). The monopartite ssRNA genome of PEMV-1 is 5.7 kb, while that of PEMV-2 is 4.35 kb. A third small RNA of 717 nucleotides has been identified as a satellite RNA whose replication is controlled by PEMV-2. The genome of PEMV-1 contributes the encapsidating coat protein gene required for aphid transmissibility and the PEMV-2 genome enables systemic movement in the plant. The genomes are encapsidated separately in icosohedral virion particles 25 and 28 nm in diameter with three structural proteins of 54 kDa, 21 kDa, and 17.5 kDa. The two particle types are transmitted together, naturally by aphids or experimentally by mechanical inoculation.

Host range: The host range of PEMV is primarily limited to members of the family Fabaceae including chickpea, common vetch, crimson clover, faba bean, lentil and pea although it can also be propagated on non-legume hosts such as *Chenopodium* spp., *Gomphrena globosa* and *Nicotiana* spp.

Transmission: PEMV is not transmitted by seed in lentil or chickpea.

Disease cycle and epiphytology

PEMV is transmitted in a persistent manner by aphids, most commonly the pea aphid (*Acythosyphon pisum*), cowpea aphid (*Aphis craccivora*), green peach aphid (*Myzus persicae*), potato aphid (*Macrosiphum euphorbiae*) and foxglove aphid (*Aulacorthum solani*). The pea aphid can acquire the virus after a minimum of 15 min and transmit after a latent period of 10-12 hr. Transmission is very efficient as susceptible hosts can become infected after a 1 min feeding period. Transmission efficiency is enhanced with longer acquisition and inoculation access periods. The virus is circulative in aphids but not propagative and not transmitted transovarially or parthenogenically. Only one strain of the virus has been reported.

Management

No resistant varieties are known for lentil although lentil Plant Introduction (PI) accessions PI 472547 and PI 472609 have been identified with tolerance to PEMV (Aydin *et al.*, 1987). Aphids can be controlled in lentils by carefully timed applications of appropriate insecticides to reduce colonization and resulting population build-up. Due to their transient feeding behaviour on lentil plants, aphids are rarely seen on these plants in the field even though infection may be widespread. Manage of aphids is best done by eliminating them in other nearby crops such as pea or faba bean where unchecked, populations are subject to increase. Although alfalfa is not a known host of PEMV, avoid planting lentil near these fields, as alfalfa also is one of the primary host plants of the pea aphid.

13. *Pea seed borne mosaic virus*

Pea seed borne mosaic virus (PSbMV) was first identified in Europe in 1966 and then in Japan in 1967. It was subsequently discovered in the U.S in 1969 and in New Zealand in 1980. PSbMV is now found throughout the world. Losses in these crops can be severe under conditions of high infection levels (Hampton, 1982). PSbMV continues to be a highly significant disease in many pulse crops because of potentially high rates of transmission by seed and the ease of international dissemination of breeding lines and cultivars.

Symptoms

Symptoms on leaves are narrowed, reduced in size with mosaic and mottling and chlorosis or necrosis of shoot tips may occur. A downward leaf roll is common and plants are usually stunted. Seed pods may be abnormal in shape and size with reduced fewer seed. Necrotic line patterns that are common on infected pea and faba bean seed are rare on seed of lentil.

Causal Agent

PSbMV is a member of the family Potyviridae, Genus *Potyvirus* and consists of flexuous rod-shaped particles with a modal length of ca. 770 nm containing a monopartite genome of single-stranded RNA. The genome size is 9924 nucleotides and the coat protein has a molecular mass of 34 kDa. PSbMV has a distant serological relatedness to *Bean yellow mosaic virus*. Four pathotypes designated P-1, P-2 (also called lentil strain), P-3, and P-4 has been distinguished by susceptibilities of differential lines of pea and differences in sequence in two coding regions of the RNA. All isolates tested so far on lentil including pathotypes P-1, P-2, and P-3 have been found to infect the host plant.

Host range: The virus naturally infects lentil, chickpea, pea, faba bean, and other legumes.

Transmission: Seed transmission up to 44% has been reported in lentil seed lots.

Disease cycle and epiphytology

The infected seeds and lentil fields are the primary source of inoculums. Once present in the field the mechanism of plant to plant spread is by aphids which transmit the virus in a non-persistent manner. PSbMV is also mechanically transmissible. Other significant pulse crops that may serve as inoculums sources for the virus include pea, faba bean and vetch. At least 22 different aphid species are known to transmit PSbMV but the pea aphid (*Acyrthosiphon pisum*), cowpea aphid (*Aphis craccivora*), green peach aphid (*Myzus persicae*), bean aphid (*Aphis fabae*) and foxglove aphid (*Aulacorthum solani*) are probably the most common vectors. Seed production is severely reduced in infected lentil, with losses over 77% when plants are infected at the pre-flower stage and germination of infected seed can be reduced by nearly 50%. A major contributing factor to the persistence of PSbMV is the continuous use of contaminated seed collected at the end of a season and then used in subsequent plantings.

Management

The use of certified virus-free seed is the most effective method for managing of the virus. Chemical control of the aphid vector is generally ineffective because of the transient nature and non-persistent manner by which the virus is transmitted. Application of pesticide can reduce secondary spread of the virus in lentil crops where large aphid vector populations have become established. Avoid planting near established fields where known infection sources may exist. No lentil cultivars have been developed with complete resistance to PSbMV, although some plant introduction (PI) accessions that include PI 212610, PI 251786, PI 207745 and PI 368648 have been identified that were immune to the virus (Iqbal *et al.*, 2005). The cultivars Crimson, Palouse and Red Chief and Ill 6198 are reported to be tolerant to PSbMV. Resistance to PSbMV in lentil is controlled by a single recessive gene for which the symbol *sbv* has been proposed.

Integrated viral diseases management in lentil

The development of an effective management package for virus diseases is dependent on the availability of basic information required to design an appropriate combination of interventions which can slow down virus disease development. The most important of these are:

i) Character of the causal agent

ii) Approach of spread

iii) Natural balance of the virus disease counting that of its vector

iv) Amount and significance of crop losses

v) Accessibility of heritable conflict

vi) Available crop-protection methodologies and their applicability to specific farming systems and socio-economic situations.

In many locations, however, such complete basic information is not yet available. Management is optimized through IDM approaches, which combine all possible measures that operate in different ways such that they complement each other and applicable in farmers' fields. Thus, control measures can be classified as

i) Those that control the virus

ii) Those that are directed towards avoidance of vectors or reducing their incidence

iii) Those that integrate more than one method.

Cultural practices such as healthy seed, rouging, alteration in sowing dates and use of early maturing cultivars are effective in minimizing virus disease incidence. Almost 50% of viruses affecting in lentil crop are seed-borne (Bos *et al.*, 1988). Seed-borne infections permit the introduction of primary virus inoculums into the field which facilitates secondary spread to reach a serious level in locations where the environmental conditions permit high vector activity. A close relationship between sowing date and the extent of subsequent virus spread is well documented for many crops. Host resistance is the most acceptable component in virus management because it is environment-friendly, practical and economically acceptable to farmers. There are many crop cultivars with adequate levels of virus resistance; in lentil crop against Alfalfa mosaic virus (AMV), Bean yellow mosaic virus (BYMV),Cucumber mosaic virus (CMV), Pea enation mosaic virus

(PEMV) and Pea seed-borne mosaic virus (PSbMV). Another area of host resistance which is not well exploited is resistance to vector(s). Success in reducing virus spread by chemical control of vectors is likely with persistently rather than with non-persistently transmitted viruses Each of the management strategy provides only partial control, but combining genetic resistance, cultural practices, and chemical sprays is expected to lead to improvements. The use of host resistance and one or two well-timed sprays coupled with optimal planting date and early rouging of virus-infected plants could offer reasonable and economic management and stabilize lentil production. Each strategy needs to be affordable by farmers and fulfill the requirements of being environmentally and socially responsible. It must also be compatible with management measures already in use against other pests and pathogens.

Conclusion

In this chapter, development on management of important diseases of lentil has been discussed. The major highlighting is on identifying, evaluating and integrating location specific components of IDM. This is due to dynamic agro-eco systems and socio-economic conditions and due to dynamic nature of host-pathogen interactions/ co-evolution that necessitates refinement of IDM strategies. Our future activities will be focused on the expansion of IDM technology by increasing the awareness of farmers and also on the importance of quality of crop residues of lentil used as an essential component of the prevailing mixed crop livestock system in the country. In addition, IDM strategies will be developed /refined for emerging pathogens such as anthracnose (*Colletotrichum truncatum*). Experience over the last few decades has shown that plant viruses cannot be managed by preventive measures only. The complicated ecology of many of the viruses which affect lentil crop and for this matter crop in general, calls for a strategy which is long term, sustainable, economically acceptable to the resource poor farmers and friendly to the environment. Effective virus-disease management has always been and will continue to be by integrated management, that is by crop management or ecosystem management.

References

Aydin H, Muehlbauer FJ and Kaiser WJ. 1987. *Pea enation mosaic virus* resistance in lentil (*Lens culinaris*). *Plant Dis.* 71: 635-638.

Basandrai D, Basandrai A K, Thakur H L and Thakur S K. 2005. Inheritance of field resistance against lentil rust (*Uromyces viciae fabae*) in lentil (*Lens culinaris* L.). International Food Legumes Research Conference-IV, IARI, New Delhi India October, 18-22, 2005. 336 p.

Bayaa B and Erskine W. 1998. Diseases of lentil. Pages 423-471. In: The Pathology of Food and Pasture Legumes (Allen, D. J. and Lenne, J. M. eds.), CAB International, Wallingford, Oxon, UK.

Beniwal SPS, Bayaa B, Weigand S, Makkouk K and Saxena MC. 1993. Field guide to lentil diseases and insect pests. International Centre for Agricultural Research in the Dry Areas (ICARDA), Aleppo, Syria.

Bos L, Hampton RO and Makkouk KM. 1988. Viruses and virus diseases of pea, lentil, faba bean and chickpea. Pages 591-615. In: World Crops: Cool Season Food Legumes (Summerfield R. J. eds.). Kluwer Academic Publishers, Dodrecht, The Netherlands.

Callaway JC. 2004. Hemp seed as a nutritional resource: an overview. *Euphytica* 140: 65-72.

Chitale K, Tyagi RNS and Bhatnagar LG. 1971. Perfect stage of *Erysiphe polygoni* from India. *Indian Phytopathology* 34: 540-541.

Fletcher JD, Russell AC and Butler RC. 1999. Seed-borne cucumber mosaic virus in New Zealand lentil crops: yield affects and disease incidence. *N. Z. J. Crop Hortic. Sci.* 27: 197-204.

Hampton RO. 1982. Incidence of the lentil strain of pea seed borne mosaic virus as a contaminant of *Lens culinaris* germplasm. *Phytopathology.* 72: 695-698.

Iqbal SM, Tahir MHN, Asghar A and Anjum MS. 2005. Evaluation of lentil genotypes for resistance to *Pea seed borne mosaic virus. Int. J. Agric. Res. Gov. Ecol.* 7: 148-149.

Jeger MJ. 2000. Bottlenecks in IPM. *Crop Protection* 19: 787-792.

Jones RAC, Coutts BA, Latham LJ and McKirdy J. 2008. *Cucumber mosaic virus* infection of chickpea stands: temporal and spatial patterns of spread and yield-limiting potential. *Plant Pathology* 57: 842-853.

Lazou A and Krokida M. 2010. Functional properties of corn and corn-lentil extrudates. *Food Research International.* 43: 609-616.

Makkouk KM, Kumari SG and Al-Daoud R. 1992. Survey of viruses affecting lentil (*Lens culinaris* Med.) in Syria. *Phytopathol. Mediterr.* 31: 188-190.

Makkouk KM, Kumari SG and Bayaa B. 1999. First report of *Pea enation mosaic virus* affecting lentil (*Lens culinaris* Medik.) in Syria. *Plant Dis.* 83: 303.

Makkouk KM, Kumari SG, Hughes JA, Muniyappa V and Kulkarni NK. 2003. Other legumes: Faba bean, chickpea, lentil, pigeonpea, mungbean, blackgram, lima bean, horse gram, bambara groundnut and winged bean. Pages 447-476. In: Virus and Virus-like Diseases of Major Crops in Developing Countries (Loebenstein G. and Thottappilly G. eds.). Kluwer Academic Publishers, Dordrecht, The Netherlands.

McKirdy SJ, Jones RAC, Latham LJ and Coutts BA. 2000. Bean yellow mosaic poty virus infection of alternative annual pasture, forage and cool season crop legumes: susceptibility, sensitivity and seed transmission. *Aust. J. Agric. Res.* 51: 325-345.

Moury B. 2004. Differential selection of genes of cucumber mosaic virus subgroups. *Mol. Biol. Evol.* 21: 1602-1611.

Singh G, Chen W, Rubiales D, Moore K, Sharma YR and Gan Y. 2007. Diseases and their management. In *Chickpea breeding and management* (eds. Yadav, Redden, Chen and Sharma). CAB International pp. 497-519.

Tikoo JL, Sharma B, Mishra SK and Dikshit HK. 2005. Lentil (*Lens culinaris*) in India: Present status and future perspectives. *Indian Journal of Agricultural Sciences* 75: 539-562.

Citation

Shakywar RC, Singh Shyam and HK Singh. 2018. Integrated disease management of lentil (*Lens culinaris* Medikus subsp. *culinaris*). In: Scientific Lentil Production. Singh AK, Bhakta N, Sangale UR, Manibhushan, Sundaram PK, Kumar S and Yasin JK. (eds.) Society for Upliftment of Rural Economy Varanasi, India. Pp. 349-372.

□□□

Chapter 21

Epidemiology Based Management of *Ascochyta* Blight of Lentil

HK Singh, RB Singh and Anil Kumar Singh

SYNOPSIS

Ascochyta blights of lentil caused by *Ascochyta fabae* f.sp. *lentis* (teleomorph *Didymella lentis*)is one of the important foliar disease of lentil. The pathogen survives on infected crop residues as anamorphic (asexual) pycnidia containing conidia and/or teleomorphic (sexual) pseudothecia with ascospores. Conidia spread by rain splash; ascospores by both rain and wind following primary infection. Symptoms develop on all aerial parts of the plant and consist of necrotic lesions in the center of which numerous pycnidia are formed. Conidia spread by splashing rain and serve as secondary inoculum. The interaction between pathogen and host plant under fluctuating environmental conditions is complex, as illustrated with the Ascochyta blight pathosystem. Nevertheless, thorough understanding of this complex system and quantitative information on the interactions between the key factors are required for the development of reliable strategies for disease management.

Keywords: Lentil, *Ascochyta fabae* f.sp. *lentis*, epidemiology, management.

Introduction

Epidemiology is the study of factors affecting the outbreak and spread of infectious diseases (Agrios, 2005). The occurrence and severity of any plant disease is determined by the interaction of a causal pathogen, a susceptible host and favorable environmental conditions. It is necessary to understand the interrelationship between the host, the pathogen and the environment for disease management. Epidemiological studies identify the elements of the dynamic interactions to device control strategies. The life cycles of lentil pathogen (Pedersen *et al.*, 1993) comprise several stages that are affected by the environment. Epidemiological studies enable identification of the critical stages and the key environmental factors that affect them, and the information they yield forms the basis for developing reliable and cost-effective management strategies. In case of lentil ascochyta blight, this chapter summarizes the environmental factors that affect the components of the disease cycle, identifies the key factors, and demonstrates the significance of epidemiological studies in developing effective disease management strategies. Ascochyta blights of lentil caused by *Ascochyta fabae* f.sp. *lentis* (teleomorph *Didymella lentis*)is one of the important foliar disease of lentil. The pathogen survives on infected crop residues as anamorphic (asexual) pycnidia containing conidia and/or teleomorphic (sexual) pseudothecia with ascospores. Conidia spread by rain splash; ascospores by both rain and wind following primary infection. Symptoms develop on all aerial parts of the plant and consist of necrotic lesions in the center of which numerous pycnidia are formed. Conidia spread by splashing rain and serve as secondary inoculum (Kaiser, 1997). Suppression of Ascochyta blight is essential to ensure profitable and stable lentil yields. Management practices rely required firstly to minimize the onset of disease epidemics by reducing or excluding primary inoculum, and secondly by suppressing subsequent spread of an epidemic. Cultural practices such as planting disease-free seed, crop rotation with non-host crops and destruction of infected crop residues in the field are important in reducing the likelihood of severe epidemics. Resistant cultivars, drip irrigation (wherever irrigation is used), and application of foliar fungicides may also be used for disease suppression (Davidson and Kimber, 2007). Nevertheless, adequate and cost-effective management of Ascochyta blight still pose a challenge to lentil growers. Lentil is relatively extensive crops, and repeated application of fungicides is costly and often not profitable. Furthermore, the efficacy of fungicides is variable, and since high levels of resistance are not always available, susceptible (or partially resistant) cultivars are widely used. Under these circumstances, disease management should be based on integration of different management technique, rather than application of a single measure. Construction of an integrated management strategy should be based on the factors governing disease development in time and space.

Use of epidemiological knowledge to reduce primary infections

i) Preventing spread of the disease from adjacent fields

Spatial distribution of Ascochyta blight in lentil fields is not uniform; disease was often more severe close to the edge of the fields and leveled off with increasing distance from the edge. Often, crop residues from adjacent fields serve as source of primary inoculum. Splash was found to be an effective mechanism for dispersing conidia of *Ascochyta* pathogens. In still air, less than half of the dispersed *A. lentis* conidia reached a distance of more than 15 cm from the source, and few spores were collected at distances greater than 70 cm. However, with wind speeds of 2.5 to 5 m/s, conidia in splash droplets were dispersed over great distances. Furthermore, infected leaflets were shed from the plants and blown downwind, and the number of blown leaflets increased with increasing wind speed. Crop residues blown into adjacent fields during harvest or between the postharvest and pre-planting periods, may also be an important means of short- to medium-distance dispersal (meters to kilometers) (Gossen and Morrall, 1986; Pedersen *et al.*, 1994). The environmental factors governing Ascochyta spread from adjacent fields are rain and wind (Fig.21.1a&b). Management practices for preventing this spread are: avoid planting new crops in close proximity to previous lentil fields; use fallow or non-host crops and apply fungicide treatments on the edge of the fields to minimize the horizontal spread of the pathogens (Pedersen *et al.*, 1994). This approach is especially valuable when a susceptible crop is to be grown immediately adjacent to residues from the previous year's crop. In fields used for seed production, to minimize seed infection, the crop at the edge of the field can be harvested separately from the rest.

ii) Controlling infections from diseased seeds

Pathogen causing Ascochyta disease is seed borne, in this crop and infected seeds are important source of inoculum for long-distance spread and primary infections originate from the conidia obtained from seeds of Ascochyta blight-affected crops. In such cases, disease is observed on basal parts of the plants, just after the germination. The spread of the pathogen from infected seed to seedlings is non-systemic, and depends on contact between the emerging growing plant and lesions on the seed coat or cotyledons. Primary infections that originate from infected seedlings occasionally result in distinct foci randomly scattered about the field, with plants in the center of each focus being more severely diseased than those at its margins. Contaminated or infected seeds are important sources for spreading the disease into new production areas (Kimber *et al.*, 2007). A simulation model showed that for susceptible cultivars, attempts to reduce the incidence of primary (seed) infections (*e.g.*, by seed dressing) is inadequate, as an 80% reduction in the incidence of primary infection (from 1 to 0.2%) still allows substantial disease development (affecting 32% of the area) and considerable yield reduction (by 20%) by the end of the season. Accordingly, production of seeds from disease-free fields is a critical requirement for using highly susceptible cultivars. On the other hand, for moderately resistant cultivars, reducing the incidence of primary infection from 1 to 0.2% would reduce the diseased area to 10% and the yield loss to 1%. Use of seed treatment for eradication purposes on highly resistant cultivars may be unnecessary or uneconomical because, even if the incidence of primary infection was 1%, only a small proportion (9%) of the area would be affected and yield loss would be negligible (0.7%). It should be noted that these results reflect the specific values assigned to each of the parameters in the simulation experiments; they do not represent actual field results. Nevertheless, they provide approximate estimates of the contribution of seed treatment to reduction of disease severity and yield losses.

iii) Preventing infections from ascospores

The teleomorph stages of *Ascochyta* in lentil (*D. lentis*) have been identified. The epidemiological role of the teleomorph stage of *D. lentis* is not clearly known. But is other as in chick pea *Didymella rabiei*, pseudothecia maturation typically peaks at the beginning of spring in line with the emergence of new crops; ascospores release decreases drastically or stops altogether by the beginning of summer. After exhaustion of the pseudothecia, no more ascospores are produced, and the pseudothecial walls degenerate. The ascospores are dispersed over long distances by the wind and, as a considerable number of pseudothecia may be formed on infested debris, a great number of ascospores may be discharged into the air. Under environmental conditions conducive to the pathogen, a healthy field may become severely diseased over a short period of time following ascospore discharge (Gamliet-Atinsky *et al.*, 2005). Adequate disease management can be achieved by application of prophylactic sprays, or by initiating spraying when the disease is still at a low level. Thus, combating the primary infections originating from airborne ascospores is the key to acceptable, season-long disease suppression wherever the teleomorph stage plays a significant role in *D. rabiei* epidemiology. As the time of disease onset varies from year to year and in different environments, a tool for predicting the time of pseudothecia maturation and ascospore discharge is needed, and development of such a tool requires quantitative knowledge of the factors affecting pseudothecia formation and maturation, and ascospore dicsharge. Models are developed to predict pseudothecial development and ascospore release. Forty empirical models reflecting the influence of temperature and interrupted wetness on maturation of fungus pseudothecia were developed in Israel. As wetness is often provided by rain, this was the factor used in the models. The models were verified by using data recorded in infected fields of Israel in 1998. Seven of the models were then validated and the following model provided the best predictions: starting at the beginning of October to December, the predictor of the model was assigned one severity value unit when there was a rain event (1 day or more) with >10 mm of rain, with a daily average temperature (during the rainy days) of <15?C. According to the model, pseudothecia mature after accumulation of six severity units, and ascospores will be discharged during the following rain event (> 2 mm). The model provided accurate negative predictions (*i.e.*, that pseudothecia would not mature and ascospores would not be discharged) in four out of four tested cases, and accurate positive prediction (*i.e.*, that pseudothecia would mature and ascospores would be discharged at a particular time) in six out of eight tested cases. In light of these results, growers in Israel are advised to initiate fungicidal spraying of susceptible cultivars after the accumulation of six rain events, in accordance with the empirical criteria listed above. The model has been used commercially since 2001, with appreciable success (Shtienberg *et al.*, 2005).

Use of epidemiological knowledge to reduce secondary infections

i) Preventing infections from conidia

One of the most effective means of suppressing polycyclic pathogens is the use of fungicides. Chemical management of Ascochyta diseases was attempted in many places of the world, with generally variable results. Whereas excellent disease suppression was reported by some workers, others achieved intermittent control efficacy, and in some cases disease control was inadequate. In South Australia, a strategic program for timing the fungicide was proposed, including applications at 6 weeks after sowing, at early flowering, and at early podding. This management program was tested in a series of field experiments, and the results varied. In two of the five experiments chlorothalonil was highly effective, with control efficacy ranging from 88.6 to 95.3%. However, in the other three experiments the fungicide was not as effective, and control efficacy was low, ranging from 21 to 32.1%. Determining the reasons for this inconsistency is crucial for development and implementation of reliable disease management programs. It should be noted that the management program mentioned above reflected host developmental stage only, and ignored the pathogen biology. Ascochyta blights are polycyclic, and secondary cycles are generally initiated by the succession of conidia released from plant tissue to tissue, or from plant to plant. Conidia of *A. lentis* are released only from wet pycnidia: in moist weather, conidia of *A. rabiei* ooze from the pycnidia in a gelatinous matrix, and moisture is necessary to dissolve this matrix and free the spores, which then may be washed or splashed from plant to plant, or scattered in water droplets of driving rain. In dry weather the extruded conidia dry into hard masses on the lesions, to be subjected to dispersal by subsequent wetting. Splash dispersal carries conidia over short distances, but the combination of splashing rain with strong wind may spread spores over considerable distances, particularly if the conidia are contained in small droplets. Ascochyta blight infection and disease development occur in a temperature range of 5 to 30 C, with an optimum of 20 C, and 17 h of wetness is essential to promote severe infections. Germination of conidia, and germ-tube penetration increased linearly with increasing duration of wetness, but disease severity approached optimum after about 18 h of wetness. Dry periods (6 to 48 h) immediately after inoculation sometimes increased disease severity; however, dry periods (> 12 h) after an initial wetting period of 6 h usually had an adverse effect on disease development. Little infection developed in the absence of leaf wetness, even when the RH was 98%, and no infection developed when RH was < 95%. Disease severity increased with increasing periods of darkness after inoculation (Jhorar *et al.* 1988; Trapero-Casas and Kaiser, 1992).

It is obvious that once the pathogen appears in the crop the environmental factor that governs the occurrence and severity of the resultant epidemics is rain (Fig. 21.1 A & 21.1 B). Data recorded in the field experiments were reanalyzed, taking into account the time of spraying in relation to the occurrence of rain events. The analysis clarified most of the causes of the inconsistent efficacy of disease suppression. Adequate suppression of Ascochyta blight occurred when the fungicides were applied at times that enabled them to protect the plants from infections that occurred during the rain events. Whenever the plants were not protected during rain, disease suppression was insufficient and control efficacy was low. There is a direct relationship between control efficacy and the amount of rain.

Another factor that governs the epidemic is the level of resistance exhibited by the cultivar used. It is logical to assume that a given number of dispersed spores would result in a more severe disease in a susceptible cultivar than in a highly resistant one. Thus, both the level of host resistance and the quantity of rain should be taken into account in deciding whether to spray. One way to integrate these two control criteria is to define a rain-quantity threshold for each range of host resistance level. A threshold of 5 mm of rain was used for application of sprays to highly susceptible cultivars; similarly, thresholds of 10 and 20 mm were used for moderately susceptible and moderately resistant cultivars, respectively. Spraying of highly resistant cultivars is not needed at all (Shtienberg *et al.*, 2000).

ii) Preventing pathogen establishment

Protectant fungicides are effective only if applied before the occurrence of infection. Prophylactic application of protective fungicides is uneconomical in most lentil production areas, because of the long period during which protection is needed, and the consequently multiple sprays. Some systemic fungicides are also effective against Ascochyta blight. These fungicides penetrate the host tissue and exhibit post-infection properties that enable them to be applied after the onset of infection. Post-infection application of systemic fungicides may facilitate flexibility in management and restrict fungicide applications to real infection events rather than to anticipated events, as occurs with protective fungicides.Temperature and moisture regimes after the infection period affected disease development and symptom expression. At constant temperatures, the optimum temperature for disease development was about 20?C. The lower and upper limits for disease development were < 5°C and about 30?C, respectively. The optimum temperature for mycelial growth, pycnidial formation, and spore germination is about 20°C, with lower and upper limits around 0 to 5°C and 30 to 32°C, respectively. Under optimal temperature and leaf wetness conditions, the incubation period for *A. lentis* and *A. rabiei* was 5 to 7 days. In lentil, temperature had a little effect on lesion size and number of pycnidia per lesion, but infection frequency was higher at 10 and 15°C than at 25°C.The environmental factors mentioned above affect pathogen establishment and lesion formation, but none of them governs these processes (Fig. 21.1 A & 21.1 B). Studies on efficacy of systemic, curative fungicides (*e.g.*, difenoconazole, tebuconazole and azoxystrobin) showed that the duration of the post-infection activity of these fungicides was up to 3 days, and application of these fungicides enabled Ascochyta blight to be suppressed as effectively as by regular application of sprays. Use of these systemic fungicides enabled the number of sprays to be reduced by about 40% compared with the regular schedule treatment (2.7 *vs.* 4.3 sprays). Taking host resistance into account might enable further reduction in fungicide use (Shtienberg *et al.*, 2006).

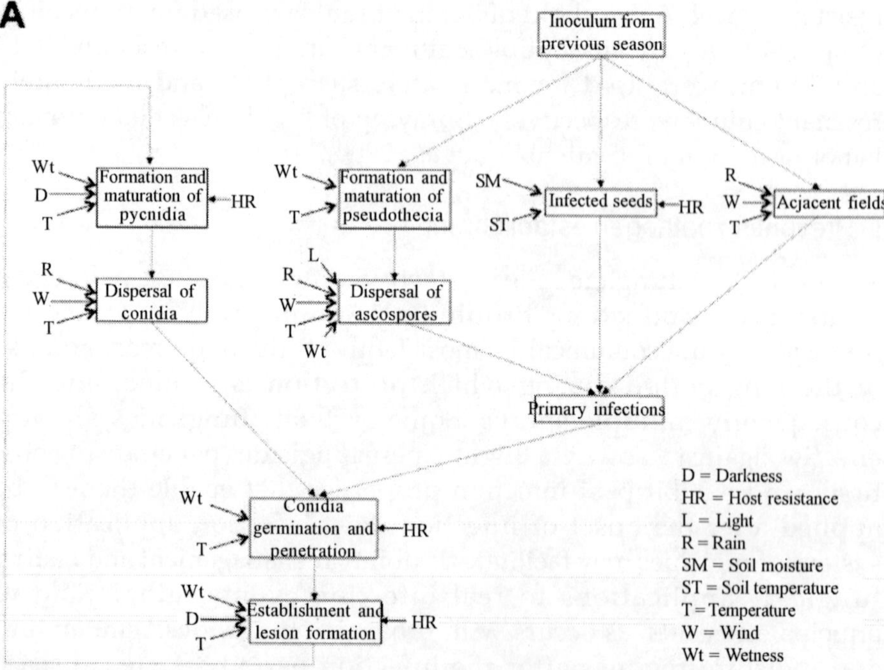

Fig. 21.1 (**a**). Effects of environmental factors on stages in the life cycle of *Didymella lentis*, the causal agents of ascochyta blight of lentil. Factors marked in bold are those governing the specific stage in the life cycle of the pathogens.

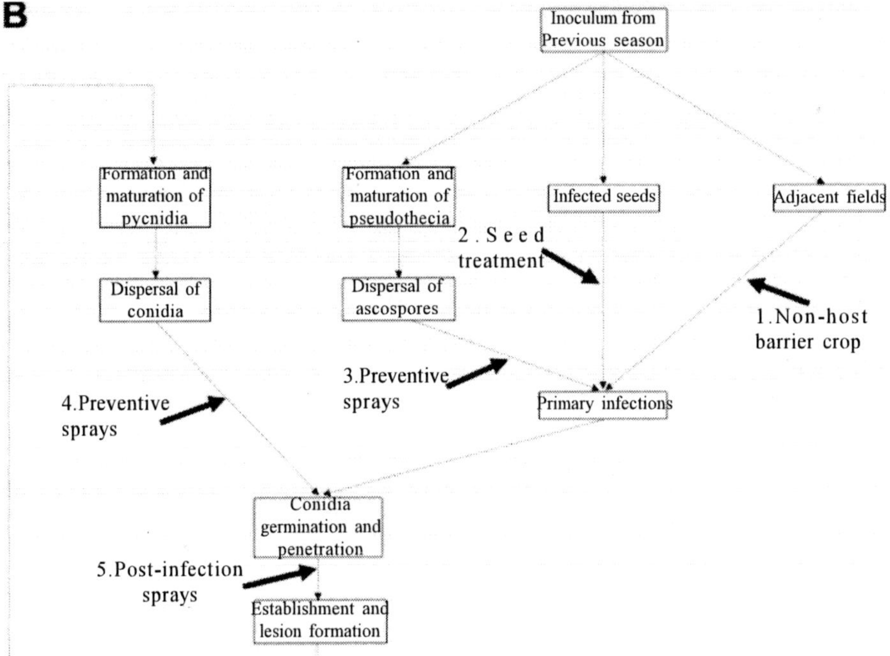

Fig.21.1 (b). The vulnerable stages in the life cycle, and control measures that can be used for disease suppression.

Conclusion

The interaction between pathogen and host plant under fluctuating environmental conditions is complex, as illustrated with the Ascochyta blight pathosystem. Nevertheless, thorough understanding of this complex system and quantitative information on the interactions between the key factors are required for the development of reliable strategies for disease management. In many pathosystems the factors governing the occurrence of epidemics are known or nearly so. However, as the biological systems are complex, interpretation of this knowledge, and development of effective disease management strategies are difficult. Growers need assistance in using existing knowledge in their decision-making procedure. Decision support systems (DSSs) are the tools by which complex knowledge can be formulated in a 'user friendly' way. In this chapter specific examples were given, to demonstrate how epidemiological studies assisted in identifying the vulnerable stages in the disease cycle and in directing the management efforts towards these stages. Taken as a whole, the data presented in Fig. 1 could be the basis for developing a comprehensive DSS for management of Ascochyta blight of lentil, and also of other pulse crops. Components of this system include management of the primary inoculum of the anamorph and teleomorph stages of the pathogens, and preventing the secondary cycles of the anamorph stage (Fig. 21.1A & 21.1B). Not all the indicated components are necessarily applicable in all pathosystems or all countries; nevertheless, they may be used as a general framework. These concepts were implemented over the last 25 years in Israel. Ascochyta blight was the most important yield constraint in the 1980s. Since then, production practices changed markedly, mainly in response to Ascochyta blight. Growers now use cultivars that are moderately susceptible or moderately resistant to *A. fabae* f.sp. *lentis*; the irrigation system has switched from overhead to drip irrigation; seeds are produced in disease-free fields; growers use a forecasting model for timing fungicide applications to protect the crop from primary infections by ascospores, and apply systemic fungicide as a post-infection treatment according to the rain-quantity thresholds mentioned above. Implementation of all these measures enables growers to cope adequately with Ascochyta blight.

References

Agrios GN. 2005. Plant Pathology. Printed in India at Replika Press Pvt. Ltd. Delhi.

Davidson JA and Kimber REB. 2007. Integrated disease management of ascochyta blight in pulse crops. *Eur. J. Plant Pathol.* 119: 99-110.

Gamliel-Atinsky E, Shtienberg D, Vintal H, Nitzni Y, and Dinoor A. 2005. Production of *Didymella rabiei* pseudothecia and dispersal of ascospores in a Mediterranean climate. *Phytopathol.* 95: 1279-1286.

Gossen BD and Morrall RAA. 1986. Transmission of *ascochyta lentis* from infected lentil seed and plant residue. *Can. J. Plant Pathol.* 8: 28-32.

Jhorar OP, Bulter DR and Mathauda SS. 1998. Effects of leaf wetness duration, relative humidity, light and dark on infection and sporulation by *Didymella rabiei* on chickpea. *Plant Pathol.* 47: 586-594.

Kaiser WJ. 1997. Inter- and intranational spread of ascochyta pathogens of chickpea, faba bean and lentil. *Can. J. Plant Pathol.* 19: 215-224.

Kimber RBE, Shtienberg D, Ramsey MD and Scott ES. 2007. The role of chickpea seedling infections in ascochyta blight epidemics. *Eur. J. Plant Pathol.* 117: 141-152.

Pedersen EA, Bedi S and Morrall RAA. 1993. Gradients of *ascochyta* blight in Saskatchewan lentil crops. *Plant Dis*. 77: 143-149.

Pedersen EA and Morrall RAA. 1994. Effect of non-host and fungicide-treated barriers on horizontal spread of ascochyta blight of lentil. *Can. J. Plant Pathol*. 16: 317-325.

Pedersen EA, Morrall RAA, McCartny HA and Fitt BD. 1994. Dispersal of conidia of *Ascochyta fabae* f. sp. *lentis* from infected lentil plants by simulated wind and rain. *Plant Pathol*. 43: 50-55.

Shtienberg D, Gamliel-Atinsky E, Retig B, Brener S and Dinoor A. 2005. Significance of preventing primary infections by *Didymella rabiei* and development of a model to estimate the maturity of pseudothecia. *Plant Dis*. 89: 1027-1034.

Shtienberg D, Kimber RBE, McMurray L and Davidson JA. 2006. Optimisation of the chemical control of ascochyta blight in chickpea. *Aust. Plant Pathol*. 35: 715-724.

Shtienberg D, Vintal H, Brener S and Retig B. 2000. Rational management of *Didymella rabiei* in chickpea by integration of genotype resistance and curative application of fungicides. *Phytopathol*. 90: 834-842.

Trapero-Casas A and Kaiser W J. 1992. Influence of temperature, wetness period, plant age and inoculum concentration on infection and development of Ascochyta blight of chickpea. *Phytopathol*. 82: 589-596.

Citation

Singh HK, Singh RB and Singh AK. 2018. Epidemiology based management of *Ascochyta* blight of lentil. In: Scientific Lentil Production. Singh AK, Bhakta N, Sangale UR, Manibhushan, Sundaram PK, Kumar S and Yasin JK. (eds.) Society for Upliftment of Rural Economy Varanasi, India. Pp. 373-383.

□□□

Chapter 22

Trichoderma: A Potential Bio-Pesticide for Management of Lentil (*Lens Culinaris* Medikus) Diseases

HK Singh, RC Shakywar, Shyam Singh and Anil Kumar Singh

SYNOPSIS

Trichoderma isolates reduced the growth of all the soil borne pathogens significantly and, therefore, can be incorporated for integrated disease management of soil borne plant pathogens. The degree of antagonism varied between and within species of *Trichoderma* against the soil borne plant pathogens. The enzymes and antibiotics produced by *Trichoderma* sp. that appear to be involved in bio-control are strongly influenced by the substrate on which the fungus is grown, and conditions in the laboratory probably occur only rarely in nature or not at all. Temperature also has a profound effect on the production and activities of enzymes and antibiotics associated with bio-control by *Trichoderma* spp. . The presence of other members of the soil micro flora also may influence bio-control activity by inhibiting the growth and development of the bio-control agent or by metabolizing its enzymatic and/or antibiotic products. This may not entirely negate the bio-control activities of *Trichoderma* sp., but it might well limit their efficacy in terms of the length of time that they are effective and the distance from the bio-control focal point that they exhibit influence. *Trichoderma* sp. are the best method for obtaining a potential bio-control agent might be one where the *Trichoderma* sp. is isolated from areas of the plant and soil where, it is expected to function in disease control, and where it is growing under conditions of temperature, moisture, and nutrient availability that approximate those found in nature.

Keywords: Lentil, *Trichoderma* sp., Isolation technique, Identification, Soil borne diseases.

Introduction

Lentil (*Lens culinaris* Medikus) commonly known as "Poor man's meat, as it is one of affordable protein rich legume. It is a rich source of protein, minerals (K, P, Fe, and Zn) and vitamins for human nutrition (Bhatty, 1998). Additionally, because of its high lysine and tryptophane contents, its consumption with wheat or rice provides a balance in essential amino acids in human nutrition. Lentil straw is also a valued animal feed (Erskine *et al.*, 1990). The major fungal diseases affecting lentil grown in the region are Ascochyta Blight (*Ascochyta fabae* f. sp. *Lentis*), Seedling blight (*Rhizoctonia solani*), Stem and Root Rot (*Rhizoctonia solani, Fusarium* spp. and *Botrytis* spp.), Stem Rot. (*Sclerotinia sclerotiorum*), Gray Mold and Pod Rot (*Botrytis cinerea*) and Anthracnose (*Colletotrichum truncatum*). Among the several biotic stresses limiting lentil yield, wilt of lentil is one of serious diseases caused by *Fusarium oxysporum* f.sp. *lentis* and plays a major role in reducing lentil yield (Hamali and Hassanein, 1996). The disease may cause complete crop failure under favorable conditions for disease development and can be the major limiting factor for lentil cultivation in certain areas (Chaudhary and Amerjit, 2002). The disease appears in either the early stage of crop growth (seedling) or during the reproductive stage (adult stage) (Khare, 1981; Stoilova and Chavdarov, 2006). Soil borne pathogens have a broad host range and persist for longer periods in soil by resistant resting structures. These diseases can be effectively controlled to prevent them from causing yield loss. However, management decisions and seasonal conditions play a significant role in disease outbreaks and grain yield. Lentil growers need to implement an integrated approach to disease management in most years to produce a profitable crop. Chemical control of soil borne pathogens provides certain degree of control but at the same time have adverse effects on environment affecting the beneficial soil microorganisms. Besides, its cause human hazardous, phytotoxicity, residual effect and their high cost rate. Therefore, biological control of plant pathogens has been considered as a potential control strategy in recent years and search for these biological agents is increasing. *Trichoderma* is the most commonly used fungal biological control agent and have long been known as effective antagonists against plant pathogenic fungi (Chet, 1987, Kumar and Mukerji, 1996).

Trichoderma spp., are free-living fungi that are common in soil and root ecosystems. They are highly interactive in root, soil and foliar environments. They produce or release a variety of compounds that induce localized or systemic resistance responses in plants. *Trichoderma* strains have long been recognized as biological agents, for the control of plant disease and for their ability to increase root growth and development, crop productivity, resistance to abiotic stresses, and uptake and use of nutrients (Howell, 2003). *Trichoderma* species are considered as promising biological control agents against numerous phytopathogenic fungi including *F. oxysporum* (Sarhan *et al.*, 1999). They are highly interactive in root, soil and foliar environments. They produce or release a variety of compounds that induce localized or systemic resistance responses in plants. *Trichoderma* species have shown efficiency on biocontrol of plant pathogens (Chet and Baker, 1980; Elad *et al.*, 1980; Lifshitz *et al.*, 1986; Mehta *et al.*, 1995; Etebarian 2006). Their efficacy is closely related with local conditions. *Trichoderma* strains have long been recognized as biological agents, for the control of plant disease and for their ability to increase root growth and development, crop productivity, resistance to abiotic stresses, and uptake and use of nutrients (Howell, 2003).

Importance of *Trichoderma*

1. **Disease Control:** *Trichoderma* is a potent biocontrol agent and used extensively for post-harvest disease control. It has been used successfully against various pathogenic fungi belonging to various genera *viz., Fusarium, Phytopthara, Pythium, Rhizoctonia, Sclerotia* and *Verticillium* etc.

2. **Plant Growth Promoter**

   Trichoderma strains solubilize phosphates and micro-nutrients. The application of *Trichoderma* strains with plants such as grasses increases the number of deep roots, thereby increasing the plants ability to resist drought.

3. **Biochemical Elicitors of Disease Resistance**

   Trichoderma strains are known to induce resistance in plants. Three classes of compounds that are produced by *Trichoderma* and induce resistance in plants are now known. These compounds induce ethylene production, hypersensitive responses and other defense related reactions in plant cultivates.

4. **Transgenic Plants**

   Introduction of endo chitinase gene from *Trichoderma* into plants such as tobacco and potato plants has increased their resistance to fungal growth. Selected transgenic lines are highly tolerant to foliar pathogens such as *Alternaria alternata, A. solani*, and *Botrytis cinerea* as well as to the soil-borne pathogen, *Rhizoctonia* spp.

5. **Bioremediation**

   Trichoderma strains play an important role in the bioremediation of soil that are contaminated with pesticides and herbicides. They have the ability to degrade a wide range of insecticides: organochlorines, organophosphates and carbonates (Ranasingh *et al.*, 2006).

Isolation technique

Soil samples should be collected from different agricultural fields, inoculated on various media such as potato dextrose agar, malt extract agar, rose bengal agar and oat flour agar using dilution plate technique and incubated at 25±1°C for 5 days. After an incubation period, colonies determined to be *Trichoderma* spp., according to Watts *et al.* (1988) and Rifai (1969). All *Trichoderma* spp. fungi were purified by hyphal tip/single spore isolation technique and identified on the basis of cultural and microscopic morphological characters.

Identification of *Trichoderma* Species

Trichoderma Selective Medium (TSH) was used for identification of the isolates of *Trichoderma* (Elad and Chet, 1983). 1 ml of soil suspension was taken with the help of 5ml sterilized pipette and poured on the Petri- plate seeded with TSM. The plates were incubated at 28 + 2°C for 5 days. Observation on the appearance of colonies was recorded from 3rd to 5th day. Individual colonies were picked up and maintained in pure culture for further study. *Trichoderma* species were identified and examined under compound microscope on the basis of their cultural and morphological character (Nagamani *et al.*, 2002) and were maintained on PDA slants at 4°C for subsequent studies.

Mass culture

A number of technologies have been developed for mass production of the *Trichoderma* and there is need for quick dissemination of such technologies to the prospective entrepreneurs. Following techniques is given below:

Mass production of *Trochoderma* spp. in solid form /powder form

Mass production of *Trichoderma* spp. in the solid form, sorghum seeds are a suitable substrate for spore production. Appropriate procedure consisted of soaking sorghum seeds in running water for 24 hours, filling in autoclavable bags, sterilizing for 30 minutes. *Trichoderma* suspension should be inoculated on sterilized sorghum seeds and then all bags will be incubated at 25-30 ^{0}C for 10 days. Finally, *Trichoderma* culture will be prepared as powder formulation. Culturing of *Trichoderma* on sorghum seeds contained in autoclavable bags will be suitable for mass production of culture used for greenhouse and field testing. Besides, the sole use of sorghum seeds, spent barley grain mixed with ground sorghum seed at the rate of 15% by weight could be used as a solid substrate which provided high concentration of *Trichoderma* (1.65 x 10^9 colony forming units, cuf/g-product).

Mass production of *Trochoderma* spp. in liquid form

Mass production of *Trichoderma* spp. by liquid substrate fermentation a suitable liquid medium consisted of 15% molasses, 0.3% NH4Cl, 0.1% yeast extract, and 0.05% MgSO4.7H2O and 0.1% KH2PO4 with pH 4.5. The highest fungal biomass was obtained at 13 days of culturing period. Surviving populations of *Trichoderma* spp. in powder formulation prepared from liquid fermentation biomass remained high (108 cfu/g) after storing at 25-30^{0}C for 3 month. Diatomite is the most suitable carrier for production of *Trichoderma* powder form liquid fungal biomass. Viability of *Trichoderma* spp. in powder preparation or on powder coated barley seeds could be extended when they are stored at low temperature (3-5 ^{0}C).

Preparation of *Trichoderma* spp. inoculums

The propagules colony forming unit (cfu) suspension of *Trichoderma* spp. will be prepared in sterile distilled water from 7 days old culture on potato dextrose agar (Rojo *et al.*,2007). The fungal inoculum can be harvested by flooding the culture with in sterilized distilled water and then rubbing the culture surface with a sterile glass rod. The fungal propagules concentration in each suspension will be determined by counting using a haemocytometer slide (Adjusted at 108 cfu/ml). A mixture of milting soybean and talcum powder (w/w) are use as a carrier mixture for antagonistic fungal propagules.

A carrier mixture should be added at rate of 50% to fungal suspension and mix for distribution of fungal propagules.

Evaluation Techniques

The following techniques are used for evaluation of *Trichoderma* spp.

1. **Dual culture technique** (*In vitro*)

   The antagonistic potential of indigenous isolates of *Trichoderma* was evaluated against *F. oxysporum* f.sp. *Lentis* using dual culture technique. (Dhingra and Sinclair, 1995) with slight modifications. Five millimeter diameter mycelial disc of each test antagonist (*Trichoderma* isolates) taken from 7 day old culture was paired against same sized mycelial disc of *Fusarium oxysporum* f.sp. *lentis* at opposite end on PDA (20 ml) contained in 90 mm diameter Petri- plates. The pathogen and antagonist disc were place at equal distances from the periphery of the petriplate. The PDA plates inoculated only with either antagonists and phytopathogen served as control. The plates were incubated at 28+2oC. The experiment was conducted under Completely Randomized Design (CRD). The growth of the pathogen in both test and control experiments were recorded (Fig.22.1). Data were obtained for percent inhibition of radial growth (PIRG) = (R1-R2)/R1 x 100. Where R1 = radial growth of pathogen in control. R2 = radial growth of pathogen in dual culture experiments with antagonists.

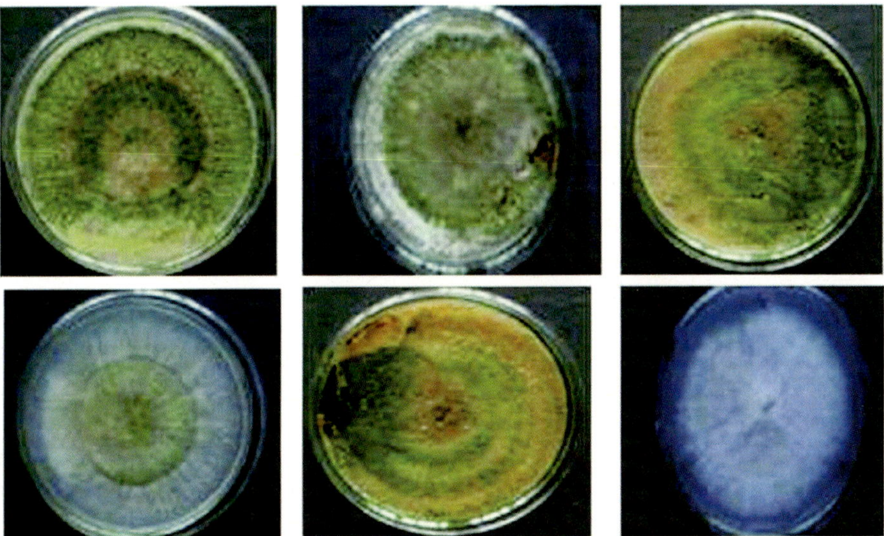

Fig. 22.1 Inhibitory effect of different isolates of *Trichoderma* sp. on mycelia growth of *Fusarium* sp. in dual culture test.
Top: left to right isolate of *T. harzianum* -5, T.h.-7, *T. viride* -2.
Bottom: left to right *T. viride* -18, T *koningii* -9, Control.

2. **Pot culture technique**

   Antifungal activity of *Trichoderma* spp. against fungal pathogens will be evaluated in pots under artificially infestation conditions. The experiment will be designed under greenhouse conditions using pots (40-50 cm diameter) containing 4-5 kg of sterilized loamy clay soil. First, soil will be infested with pathogenic fungus grown on medium @ 5 g/ kg soil in different pots and then the pots will be irrigated for 7 days before *Trichoderma* inoculation. Next, soil will be inoculated with *Trichoderma* fungus @ 5 g/ kg soil, and then pots will be watered up to 7 days before sowing. Ten seeds of faba bean sown in each pot. Five pots as replicates will be used for each treatment as well as the control. The experiment includes non-infested soil (control), soil treated with pathogen and pathogen + *Trichoderma* fungus. Then pots kept under greenhouse conditions till the end of the experiment. Disease incidence and survival (%) of faba bean plants will be recorded at regular intervals.

3. **Field experiment**

   The efficacy of soil treatment with *Trichoderma* spp., in separated block, against the incidence of soil borne diseases, evaluated in a commercial lentil field with a previous history of the disease. The field trial should be designed in Complete Randomized Block with four replications. Each plot should be 3x3 meter in area with four rows of 3m in length and 75 cm in width. First, bio-control agent inoculum will be applied @ 50g/m^2 of soil. The bio-control agent inoculum should be mixed properly with the 20 cm of the soil surface. Next, the soil will be irrigated up to 7 days before sowing. Faba bean seeds of susceptible cultivar planted. The field treatments should be as (1) soil naturally infected (control), (2) soil treated with *Trichoderma* spp.

Mechanisms involve in disease management

One of the most interesting aspects of the science of biological control is the study of the mechanisms employed by bio-control agents to affect disease control. Past research indicates that the mechanisms are many and varied, even within the genus *Trichoderma*. In order to make the most effective use of bio-control agents for the control of plant diseases, we must understand how the agents work and what their limitations are. We can then develop effective means of culturing, storing, applying, and utilizing bio-control agents so that we harness their best effort for disease control. The following mechanisms involved in bio-control with *Trichoderma* species.

Mycoparasitism and antibiotic (toxin) production

Mycoparasitism and antibiotic production as know one of the most salient characteristics of *Trichoderma* spp. and their ability to parasitize on other fungi (Fig.21.2). Therefore, it is not surprising that Weindling (1932) described bio-control by *T. lignorum* of citrus seedling disease, incited by *Rhizoctonia solani*, to mycoparasitism. He described in detail the mycoparasitism of *R. solani* hyphae by the hyphae of the biocontrol agent, including coiling around pathogen hyphae, penetration, and subsequent dissolution of the host cytoplasm. This phenomenon occurred regardless of the supply of external nutrients to the host or mycoparasite. Although he considered the possibility that under certain circumstances *T. lignorum* might act as a competitor for nutrients with *R. solani*, he much favored mycoparasitism as the principal mechanism for biocontrol. Two years later, Weindling (1934) reported that a strain of *T. lignorum* produced a "lethal principle" that was excreted into the surrounding medium, allowing parasitic activity by the biocontrol agent. Weindling (1941), he characterized the "lethal principle", demonstrated that it was toxic to both *R. solani* and *Sclerotinia americana,* and named it gliotoxin (Fig.21.3). Subsequently, it has been demonstrated that the fungus that produced gliotoxin was not *T. lignorum*, but *Gliocladium virens* (Webster and Lomas, 1964) a species that has recently been renamed *Trichoderma virens* (Rehner and Samuels, 1994). In the years following this seminal work, many instances of successful bio-control with *Trichoderma* species have been described to the mechanisms of mycoparasitism and/or antibiosis (Aluko and Hering, 1970, Bliss, 1951 and Chet, 1987). Howell and Stipanovic (1983) isolated and described a new antibiotic, gliovirin, from *Gliocladium* (*Trichoderma*) *virens* (GV-P) that was strongly inhibitory to *Pythium ultimum* and a *Phytophthora* species, but not to *R. solani, Thielaviopsis basicola, Phymatotrichum omnivorum, Rhizopus arrhizus,* or *Verticillium dahliae*. In another instance, Lifshitz *et al*. (1986) showed that control of *Pythium* species on peas by *T. harzianum* (T-12) and *T. koningii* (T-8) was not due to either mycoparasitism or competition. They described bio-control to the production of a toxic factor by the bio-control agent in the spermosphere that inhibited growth of the pathogen. In other studies (Howell *et al.*, 2000 and Howell, 2002) a mutant of *T. virens* deficient for both mycoparasitism and gliotoxin biosynthesis (G6-5) still retained biocontrol efficacy equal to that of the parent strain (G-6) against both *P. ultimum* and *R. solani*. These results indicate that neither mycoparasitism nor antibiosis is required for bio-control of cotton seedling diseases incited by these pathogens.

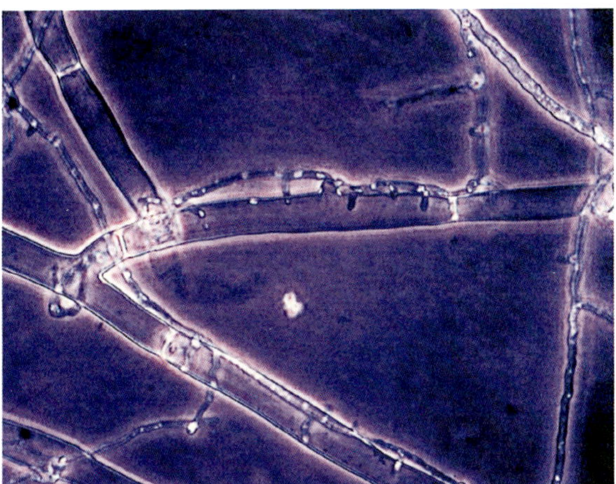

Fig. 22.2 Penetration and haustoria formation within the large hyphae of *Rhizoctonia solani* by the smaller hyphae of *Trichoderma virens*

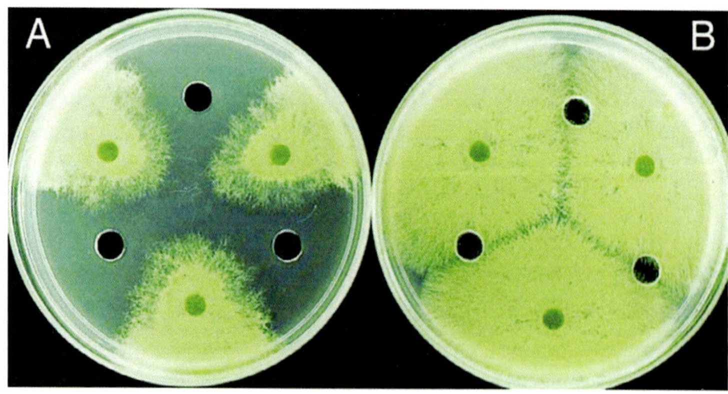

Fig. 22.3 Growth inhibition of *Rhizoctonia solani* by the *Trichoderma virens* produced antibiotic gliotoxin: (A) gliotoxin-amended medium, and (B) non-amended medium

Competition and rhizosphere competence

Trichoderma spp., either added to the soil or applied as seed treatments, grow readily along with the developing root system of the treated plant (Harman, 2000 and Howell *et al.*, 2000). This can be shown easily by simply plating surface sterilized root segments from treated plants on an agar medium (Fig.22.4). After a suitable incubation period, the fungus can be seen growing from virtually all parts of the root. The difficulty in viewing competition through rhizosphere competence as a major mechanism in biological control is that strains of *T. koningii* that are excellent root colonizers exhibit little or no bio-control activity against *R. solani* on cotton seedlings. Although competition through rhizosphere competence may not be among the principal mechanisms that drive biological control, it is certainly a valuable adjunct to those that do. One concept that is associated with competition and rhizosphere competence, the replacement of endogenous fungi on the root surface (Harman, 2001; Anand and Jayarama, 2009) can be difficult to demonstrate. *Trichoderma* spp. is often able to suppress the growth of endogenous fungi on an agar medium and therefore, mask their presence. A case in point is that of *T. virens* treated root segments taken from soil heavily infested with propagules of *Macrophomina phaseolina,* the pathogen that causes charcoal rot in a wide range of crops. When these segments are plated on an agar medium at room temperature, only *T. virens* grows from the root. However, if these same cultures are incubated at 40°C, a temperature at which *T. virens* will not grow, the pathogen grows readily from many parts of the root system (Fig.22.5). This may also occur with other *Trichoderma* species and other pathogens, but it is not easily demonstrated because growth of the bio-control agent can't be suppressed without suppressing the pathogen also. One method of demonstrating pathogen replacement on the roots by bio-control agents might be through the use of fungicide-resistant strains of the pathogen and culture of infested roots on a medium amended with fungicide suppressive only to the bio-control agent.

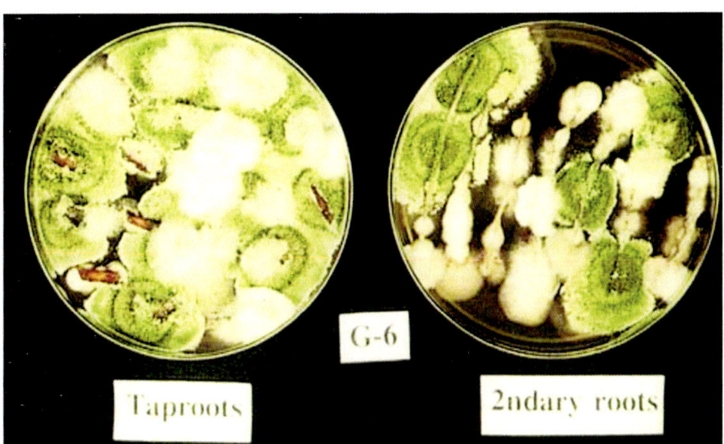

Fig. 22.4 *Trichoderma virens* cultured from tap and secondary roots of cotton plants grown from
T. virens – coated seed in field soil

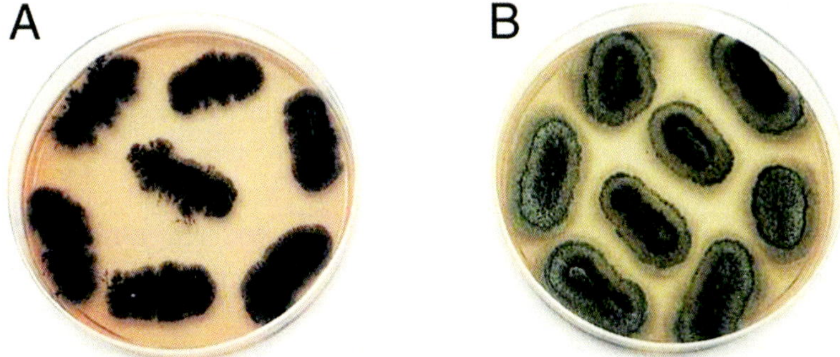

Fig. 22.5 Cultures of cotton roots (A) roots from nontreated seed showing *M. phaseolina,* and (B) roots from *Trichoderma virens* treated

Enzymes

More recent research into the possible mechanisms involved in biological control by *Trichoderma* spp. has lead to several alternative explanations for successful biocontrol. One idea that has been advanced is that enzymes such as chitinases and/or glucanases produced by the biocontrol agent are responsible for suppression of the plant pathogen. These enzymes function by breaking down the polysaccharides, chitin, and β-glucans that are responsible for the rigidity of fungal cell walls, thereby destroying cell wall integrity. Metcalf and Wilson (2001) described the colonization of onion roots, infected with *Sclerotium cepivorum,* by *T. koningii* (Tr5). Hyphae of the bio-control agent penetrated into infected epidermal and cortical tissue of the root to destroy the hyphae of the pathogen, with little or no damage to uninfected plant tissue. The authors ascribed this biocontrol phenomenon to production of endo and exo-chitinases by *T. koningii.* Baek *et al.* (1999) disrupted or over-expressed the gene coding for chitinase (cht42) in *T. virens* (Gv29-8). Transformants with reduced enzyme activity or over-expression of the enzyme gave significantly decreased or enhanced biocontrol activity, respectively, against *R. solani*–incited cotton seedling disease. However, the differences were not great and probably indicate that other factors were also at work. In a similar study, Woo *et al.* (1999) also disrupted chitinase (ech42) activity in *T. harzianum* (P1) and showed reduced bio-control activity against *Botrytis cinerea* on faba bean leaves. However, the transformant was as effective as the wild type against *P. ultimum* and exhibited enhanced activity against *R. solani* when compared with the control. The authors concluded that reactions between *T. harzianum* strains and various fungal hosts were based on different mechanisms. This, again, indicates that factors other than chitinase activity are important to the bio-control process. Transgenic lines were screened for fungal chitinase activity, and several high producers were identified. Homozygous lines were generated that retained high enzyme activity and expression of this gene in the roots, shoots, and leaves. Seedlings of these lines were screened against the seedling pathogens *Rhizoctonia solani* and *Thielaviopsis bassicola,* and detached leaves were screened against *Alternaria alternata*. Selected transgenic lines demonstrated significant levels of resistance against all three pathogens when compared with the parental control and a commercial cotton cultivar. Another interesting concept related to enzyme biosynthesis as a mechanism in the bio-control process is that advanced by Kapat *et al.*

(1998) and Elad and Kapat (1999) who suggested that bio-control of *B. cinerea* by *T. harzianum* (T39) might be due, in part, to the actions of *T. harzianum* produced proteases that inactivate the hydrolytic enzymes produced by *B. cinerea* on bean leaves. The protease enzymes break down hydrolytic enzymes into peptide chains and/or their constituent amino acids and thereby destroy their capacity to act on plant cells. The authors demonstrated that protease solutions produced by the bio-control fungus on faba bean leaves partially deactivated hydrolytic enzymes and reduced disease severity by 56 to 100% when the solutions were used to treat leaves infected with the pathogen. They also showed that the addition of protease inhibitors to drops containing mixed conidia of the two fungi, applied to faba bean leaves, nullified the bio-control effect of the *T. harzianum* conidia. Protease production by *T. harzianum* has also been associated with bio-control of the root-knot nematode *Meloidogyne javanica* on tomato plants. Sharon *et al.* (2001) showed that tomato plants treated with the biocontrol agent (T-203) and grown in nematode infested soil exhibited a drastic reduction in root galling when compared with the control. A transformed strain of *T. harzianum* (P2) containing multiple copies of a gene-encoding protease reduced root galling better than wild-type strains, and it was the only strain able to penetrate egg masses and the eggs inside. The authors suggested that improved proteolytic activity might be important for biological control of nematodes. Di Pietro *et al.* (1993) studied the synergistic effects of endochitinase and gliotoxin on conidial germination of *B. cinerea*. They found that treatment of *B. cinerea* conidia with the combination was much more effective than treatment with the enzyme or antibiotic alone. Lorito *et al.* (1996) further expanded this concept by combining a number of antifungal compounds with several kinds of hydrolytic enzymes and applying them to propagules of *B. cinerea* and *Fusarium oxysporum*. Synergism occurred in all cases, but the level depended on the antifungal activity of the enzyme. Synergism was lower when the enzyme was added after the antifungal compound, indicating that cell wall degradation was needed in order to establish the interaction.

Induction of defense responses in plants

Another mechanism proposed to explain biocontrol activity by *Trichoderma* spp. is that of induction of resistance in the host plant by treatment with the biocontrol agent. This concept is supported by the work of Yedidia *et al.* (1999) who demonstrated that inoculating roots of 7 day old cucumber seedlings in an aseptic hydroponic system with *T. harzianum* (T-203) spores to a final concentration of 105 per ml initiated plant defense responses in both the roots and leaves of treated plants. They also demonstrated that hyphae of the biocontrol fungus penetrated the epidermis and upper cortex of the cucumber root. The plant response was marked by an increase in per-oxidase activity (often associated with the production of fungitoxic compounds), an increase in chitinase activity, and the deposition of callose-enriched wall appositions on the inner surface of cell walls. Increased enzyme activities were observed in both roots and leaves. Interestingly, the plant defense became muted with time and began to resemble a symbiotic mycorrhizal association. Later, Yedidia *et al.* (2000) showed that inoculation of cucumber roots with *T. harzianum* (T-203) induced an array of pathogenesis related proteins, including a number of hydrolytic enzymes. Plants treated with a chemical inducer (2, 6-dichloroisonicotinic acid) of disease resistance displayed defense responses that were similar to those of plants inoculated with the biocontrol agent. In another study, Howell *et al.* (2000) demonstrated that seed treatment of cotton with biocontrol preparations of *T. virens* (G-6,G-11, G6-5) or application of *T. virens* culture filtrate to cotton seedling radicles induced synthesis of much higher concentrations of the terpenoids desoxyhemigossypol (dHG), hemigossypol (HG), and gossypol (G) in developing roots than those found in untreated controls. Gossypol was toxic only at high levels, but the pathway intermediates dHG and HG was strongly inhibitory to the cotton seedling pathogen, *R. solani*, at much lower concentrations. *Trichoderma* spp. was much more resistant to cotton terpenoids than were seedling disease pathogens. Biocontrol activity against *R. solani* was highly correlated with induction of terpenoid synthesis in cotton roots by *Trichoderma* species, even among strains of *T. virens* that were deficient for mycoparasitism and antibiotic production. In addition to terpenoid synthesis, treatment of cotton roots with *T. virens* also induced significantly higher levels of per-oxidase activity than that found in control roots. Per-oxidase activity and terpenoid levels in seedling hypocotyls were not significantly different from those found in the controls. In this case, plant defense responses appeared to be confined to the root system.

Metabolism of germination stimulants

Mechanisms involved in the biocontrol of pre emergence damping off of cotton seedlings incited by *P. ultimum* and /or *Rhizopus oryzae* Howell (2002) found that control by *T. virens* (G6, G6-5) or protoplast fusants of *T. virens*/*T. longibrachiatum* (Tvl-30, Tvl-35) was due to metabolism of germination stimulants released by the cotton seed. These compounds normally induced pathogen propagules to germinate. Disease control could be affected by wild-type strains or by mutant strains that were deficient for mycoparasitism, antibiotic production, and induction of terpenoid synthesis in cotton roots. The importance of metabolism of stimulatory compounds by the bio-control agent is further supported by the fact that cotton cultivars that do not produce pathogen propagule stimulants during germination are virtually immune to the disease. Again, artificial induction of pathogen propagule germination rendered these cotton cultivars susceptible to disease. Also, seed exudates cultured with dead *T. virens* bio-control preparation stimulated pathogen propagules to germinate, while exudates cultured with live bio-control preparation did not (Fig. 22.6).

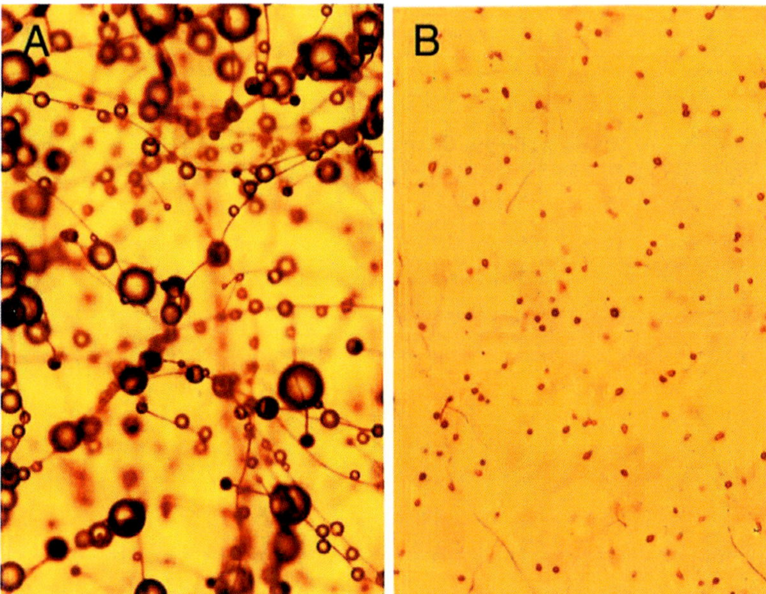

Fig. 22.6 Effect of seed exudates on sporangiospores: (A) exudate cultured with dead *T. virens* preparation stimulates spore germination and mycelial growth of Rhizopus oryzae, (B) exudate cultured with live *T. virens* preparation does not stimulate spore germination

Adjunct mechanisms

While perhaps not of primary importance as mechanisms in biological control, *Trichoderma* spp. exhibit other characteristics during interactions with host plants that may contribute to disease resistance or tolerance. These characteristics manifest themselves by increases in plant root and shoot growth, resistance to biotic and abiotic stresses, and changes in the nutritional status of the plant. These phenomena have been thoroughly reviewed by Harman (2000) who showed that seed treatment of corn planted in low nitrogen soil with *T. harzianum* (T-22) resulted in plants that were greener and larger in the early part of the growing season. At maturity, the treated plants had larger stem diameters and increased yields of grain and silage. In fields where nitrogen levels were sufficient, treatment with T-22 did not result in increased yields of grain or silage. However, the nitrogen required for optimum yields was lower for T-22 treated plants than for the untreated controls. Harman (2001) later reported a strong interaction between T-22 and the nitrogen-fixing bacterium *Bradyrhizobium japonicum*. Treatment of soybeans with the combination stimulated root growth more effectively than did either alone. Theoretically, the combination of a nitrogen-fixing bacterium and a fungus that enables the plant to utilize nitrogen more efficiently should decrease even further the nitrogen fertilizer requirements of the crop. More recently, Yedidia *et al.* (2001) showed that treatment of cucumber plants in soil with *T. harzianum* (T-203) resulted in large increases in root area and cumulative root length, and significant increases in dry weight, shoot length, and leaf area over that of the untreated control. In hydroponic culture, T- 203 inoculated cucumber roots contained significant increases in Cu, P, Fe, Zn, Mn, and Na. In the shoots of these plants, the concentrations of Zn, P, and Mn were increased. The authors considered that improvement of plant nutrition was directly related to the general beneficial growth effect on the root system of inoculation with *T. harzianum*.

Antifungal properties against pathogen- A case study

Application of *T. harzianum* as seed treatment significantly reduced the incidence of damping-off diseases some leguminous crops, *i.e.* lentil, pigeon pea, chickpea and, faba bean, when planted in a soil naturally infested with *Fusarium* spp. and *R. solani* (Abou-Zeid *et al.*, 2003). Haggag *et al.* (2006) Protease was produced by *Trichoderma harzianum* and *Talaromyces flavus* when it's were grown in the presence of casein. Proteases from the culture filtrate of *T. harzianum* and *T. flavus* were purified and tested for their antifungal activity against brown spot disease caused by *Botrytis fabae* on faba bean. The enzymes were purified in a 2-step procedure involving ammonium sulfate precipitation and Sephadex G-200 gel permeation chromatography. Both enzymes have an optimum temperature of 30 and optimal pH value of 6.5. T. flavus exhibited high levels of extracellular protease activity compared with *T. harzianum*. Germination and growth rate, extracellular polygalacturonase (P Gase) and carboxymethyl cellulase (CM Case) activities of *Botrytis fabae* were inhibited by the purified protease at a concentration of 40-120 U/ml. Growth and extracellular production of *B. fabae* were completely inhibited by the protease enzyme of *T. flavus* at a concentration of 80 U/ml, while protease from *T. harzianum* was effective at 120 U/ml. Proteases were effective in reducing brown spot disease severity and pathogen sporulation on faba bean leaves inoculated with *B. fabae*. Proteases at a low concentration of 80 U/ml inhibited *B. fabae* on leaf surface, as determined by Scanning Electron Microscope (SEM) examination. When enzymes were applied as a foliar spray on lentil plants grown under natural infested field conditions, the disease incidence was greatly reduced. Protease from *T. flavus* at 80 U/ml was more significant in reducing disease incidence on leaves and pod and increased pod yield /plant. These results indicated that protease can be a safe biodegradable biocontrol agent for control of brown *Fusarium* wilt of lentil. Reports on chitinases, glucanases, cellulase, ribosome-inactivating proteins, permatins and protease and the relative importance of any of these systems in the antagonistic process as antifungal agents are well documented (Dash *et* al., 2006, Dunne *et al.*, 2000, Elad, and Kapat, 1999, Flores *et al.*, 1997, Veenekar *et al.*, 2001). Proteases are one of the most important industrial enzymes accounting for nearly 60% of the total world-wide enzyme sales. However, few experiments have been conducted to evaluate this enzyme in controlling plant pathogens. Thus, study the antifungal effect of protease from *T. flavus* and *T. harzianum* against *B. fabae* in faba bean was investigated. Moreover, we have purified and characterized the proteases. Our results showed that *T. flavus* and *T. harzianum* produced high level of extracellular protease. Protease synthesis by the *T. flavus* occurred in the early stationary growth phases (72 h). Saturation with ammonium sulphate 60% afforded a precipitate with a purification fold of 3.8 and 2.5 for *T. flavus* and *T. harzianum* with 68.1 and 55.2% recovery of enzymatic activity, respectively. Protease was more active at pH range from 6 to 7 and at a temperature range from 2 to 30. *Bacillus spp.* and *Burkholderia* produced a protease with an optimal activity at 25 and pH of 6.0 (Metcalf and Wilson 2001, Howell *et al.*, 2000). The purified enzyme was inhibitory to germination and growth as well as extracellular PGase and CMCasae activity of *B. fabae*. Protease from *T. flavus* showed a more pronounced inhibitory effect than *T. harzianum*. The highest reduction of enzyme activity of *B. fabae* was achieved in the presence of protease of *T. flavus*, especially at 40 U/ ml. It can be concluded that protease from both bio-control agents can play a role in biological control of *B. fabae*. Under artificially infested conditions, growth, sporulation and disease incidence were greatly inhibited by protease treatment. SEM observation indicated that the 40 U/ ml protease degraded the mycelium of *B. fabae*. The results also showed that under natural conditions, disease incidence was greatly reduced by protease treatment. The highest reduction was recorded by protease from *T. flavus* at 80 U/ ml concentration .Whereas, complete inhibition by protease from *T. harzianum* needed the concentration of 120 U/ ml. Finally, our results strongly support that proteases plays an important role in the antagonism of *T. harzianum* and *T. flavus*. Different reports studies the direct use of antifungal protease enzyme to control plant pathogens by adding protease to the circulating water in the greenhouse or spraying plants in the field (Di Pietro *et al.*, 1993 and Harman, 2000). A considerable amount of recent research has been devoted to study of the lytic systems produced by *T. harzianum*, including chitinases (Chet, 1987), glucanases (Lorito *et al.*, 1998), and proteases (Bliss, 1951), and the relative importance of any of these systems in the antagonistic process. In addition to secreating cell wall degrading enzymes, *Talaromyces flavus* antagonizes *Sclerotium rolfsii*, *Verticillium dahliae* and others by parasitism and antibiosis (Elad and Kapat 1999; Howell and Stipanovic R. D. 1983, and Howell and Stipanovic 1995). In general, *T. flavus* produced high concentration of protease enzymes which appear to be more effective than protease from *T. harzianum*. This evidence indicates that differences between *T. flavus* and *T. harzianum* proteases may include sites of action influencing the biological efficacy. Clearly, the biochemical properties characterized *in vitro* and the antifungal activity of protease from *T. flavus* and *T. harzianum* indicated that this enzyme exhibits promising activity against the growth of *B. fabae*, rendering it a potentially useful agent for plant disease protection.

Conclusion

It can be concluded that the tested *Trichoderma* isolates reduced the growth of all the soil borne pathogens significantly and, therefore, can be incorporated for integrated disease management of soil borne plant pathogens. The degree of antagonism varied between and within species of *Trichoderma* against the soil borne plant pathogens. One of the things we have learned through research on bio-control mechanisms employed by *Trichoderma* sp. is that the screening methods developed early on to isolate potential bio-control agents were, at best, only marginally effective. Many strains were isolated because they were observed parasitizing plant pathogens or inhibiting their growth in culture. The enzymes and antibiotics produced by *Trichoderma* sp. that appear to be involved in bio-control are strongly influenced by the substrate on which the fungus is grown, and conditions in the laboratory probably occur only rarely in nature or not at all. Temperature also has a profound effect on the production and activities of enzymes and antibiotics associated with bio-control by *Trichoderma* spp.. What occurs at 78°F in a Petri-dish may not occur at all in the soil around a germinating seed at 60°F. The presence of other members of the soil micro flora also may influence bio-control activity by inhibiting the growth and development of the bio-control agent or by metabolizing its enzymatic and/or antibiotic products. This may not entirely negate the bio-control activities of *Trichoderma* sp., but it might well limit their efficacy in terms of the length of time that they are effective and the distance from the bio-control focal point that they exhibit influence. Given the above considerations, perhaps the best method for obtaining a potential bio-control agent might be one where the candidate *Trichoderma* sp. is isolated from areas of the plant and soil where it is expected to function in disease control, and where it is growing under conditions of temperature, moisture, and nutrient availability that approximate those found in nature. Isolates obtained in this manner can then be screened for bio-control efficacy. Since many of the mechanisms that have proven to be most important in biological control do not lend themselves to assay in a Petri dish, the screening procedure will probably entail treatment of the seed, soil, or plant with the bio-control agent. This can be followed by culture of the treated host plant in a pathogen-infested environment until disease symptoms manifest themselves. The level of symptom expression can be used as a measure of the disease control efficacy of the bio-control agent. The results of many years of research with *Trichoderma* sp. as bio-control agents has shown that not all the mechanisms and characteristics deemed necessary for optimum bio-control are found in the same organism. Very often those strains that have the capacity to produce enzymes and antibiotics that are associated with bio-control are not the ones that have good storage qualities or function well at temperature and moisture levels where pathogens flourish. Therefore, hybridization of different strains or species will be required in order to combine these beneficial characteristics. The production of hybrids from *Trichoderma* sp., many of which have no known sexual stage, will entail the use of transformation or protoplast fusion in order to obtain strains with optimal sets of characteristics. Once produced and screened, hybrids may yield strains with expanded host, temperature, and moisture parameters, and they may yield strains with better storage qualities than those exhibited by the parents. The mechanisms employed by bio-control agents to effect biological control of plant diseases are many and complex, and their use varies with the kind of bio-control agent, pathogen, and host plant involved in the interaction. Mechanisms are also influenced by the soil type, by the temperature, pH, and moisture of the plant and soil environment, and by other members of the micro flora. What we observe and define as bio-control may be the culmination of a number of different mechanisms working synergistically to achieve disease control. Our knowledge of the complexity of these systems is currently limited by our ability to perceive them, and a great deal of research will have to be undertaken in order to fathom exactly what is taking place during the bio-control process. As with so many other aspects of science, basic knowledge about the mechanisms involved in the bio-control process will be of immense value to that intent on developing new methods for utilizing bio-control agents.

References

Abou-Zeid NM, Arafa MK and Attia S. 2003. Biological control of pre and post emergence diseases on faba bean, lentil and chickpea in Egypt. *Egyptian J. Agric. Res.* 81(4): 1491–1503.

Aluko MO and Hering TF. 1970. The mechanism associated with the antagonistic relationship between *Corticium solani* and *Gliocladium virens*. *Trans. Br. Mycol. Soc.* 55: 173-179.

Anand S and Jayarama R. 2009. Biocontrol potential of *Trichoderma* sp against plant pathogens. International Journal of Agricultural science 2: 30-39.

Baek JM, Howell CR and Kenerley CM. 1999. The role of extracellular chitinase from *Trichoderma virens* Gv29-8 in the bio-control of *Rhizoctonia solani*. *Curr. Genet.* 35: 41-50.

Bhatty RS. 1988. Composition and quality of lentil (*Lens culinaris* Medik.): a review. *Canadian Institute of Food Sci. and Tech.* 21(2): 144-160.

Bliss DE. 1951. The destruction of *Armillaria mellea* in citrus soils. *Phytopathol.* 41: 665-683.

Bond DA, Jellis GJ, Rowland GG, Guen LD, Robertson SA and Khalil L. 1994. Present status and future strategy in breeding faba beans (*Vicia faba* L.) for resistance to biotic and a biotic stresses. *Euphytica* 73: 151–166.

Chet I. (eds.) 1987. *Trichoderma*–application, mode of action and potential as a bio-control agent of soil borne plant pathogenic fungi *In: I.* Chet (Ed. Innovative approaches to plant disease control. John Wiley and Sons, New York. pp. 137-160.

Chet J and Baker R. 1980. Indication of suppressiveness to *Rhizoctonia solani* in Soil. *Phytopathol.* 70: 994-998.

Chaudhary RG and Amarjit K. 2002. Wilt disease as a cause of shift from lentil cultivation in Sangod Tehsil of Kota, Rajasthan. *Indian J. Pulse Res.* 15: 193-194.

Coskuntuna A and Ozer N.2008. Biological control of onion basal rot disease using *Trichoderma harzianum* and induction of antifungal compounds in onion set following seed treatment. *Crop Protection* 27: 330–336.

Dash C, Vathipadiekal V, George SP and Rao M. 2002. Slow-tight binding inhibition of xylanase by an aspartic protease inhibitor kinetic parameters and conformational changes that determine the affinity and selectivity of the bio-functional nature of the inhibitor. *J. Biol. Chem.* 27(7): 17978-17986.

Dhingra OD and Sinclair JB. 1995. Basic Plant Pathology Methods. CRS Press Inc Boca Raton, Florida, 335 pp.

Di Pietro A, Lorito M, Hayes CK, Broadway RM and Harman G E. 1993. Endochitinase from *Gliocladium virens*: Isolation, characterization, and synergistic antifungal activity in combination with gliotoxin. *Phytopathol.* 83: 308-313.

Dunne C, Loccoz Y, Bruijn FO and Gara F. 2000. Overproduction of an inducible extracellular serine protease improves biological control of *Pythium ultimum* by *Stenotrophomonas maltophilia* strain W81. *Microbiology.* 146: 2069-2068.

Elad Y and Kapat A. 1999. The role of *Trichoderma harzianum* protease in the biocontrol of *Botrytis cinerea. Eur. J. Plant Pathol.* 105: 177-189.

Elad Y, Chet J and Katan J. 1980. *Trichoderma harzianum* a biocontrol effective against *Scleratium rolftsii* and *Rhizoctonia solani. Phytopathol.* 70: 119-121.

Elad Y and Chet I. 1983. Improved selective media for isolation of *Trichoderma* and *Fusarium* spp. *Phytoparasitica,* 11: 55-58.

Flores A, Chet I and Estrella A. 1997. Improved bio-control activity of *Trichoderma harzianum* by overexpression of the proteinase-encoding gene prb1. *Curr. Genet.* 31: 30-37.

Erskine W, Rihawe S and Capper BS, Park MS, Seo GS, Lee KH, Bae Yu. 1990. Variation in lentil straw quality. *Ann. Feed Sci. Tech.* 28: 61-69.

Etebarian HR. 2006. Evaluation of *Trichoderma* isolates for biological control charcoal coal stem rot in melon caused by *Macrophomia phaseolina. J. Agric. Sci. Technol.* 8: 243-250.

Haggag Wafaa M, Kansoh Amany L and Aly Amal M. 2006. Proteases from *Talaromyces flavus* and *Trichoderma harzianum*: purification, characterization and antifungal activity against brown spot disease on faba bean. *Plant Pathology Bulletin* 15: 231-239.

Hamdi A and Hassanein. 1996. Survey of fungal diseases in North Egypt. *LENS Newsletter* 23(1/2): 52-56.

Harman GE. 2000. Myths and dogmas of biocontrol: Changes in perceptions derived from research on *Trichoderma harzianum* T- 22. *Plant Dis.* 84: 377-393.

Harman GE. 2001. Microbial tools to improve crop performance and profitability and to control plant diseases. In: Int. Sympos. Biol. Control Plant Dis. New Century-Mode Action Application Technol. pp. 4-14.

Hassan AA. 1996. Eldar production of vegetable crops. *ELarabia Lil Nashr Wa Eltawzia.* Pp. 710.

Howell CR. 2002. Cotton seedling pre-emergence damping off incited by *Rhizopus oryzae* and *Pythium* spp. and its biological control with *Trichoderma* spp. *Phytopathol.* 92: 177-180.

Howell CR. 2003. Mechanisms employed by *Trichoderma* spp. in the biological control of plant diseases: The History and Evolution of Current Concepts. *Plant Disease* 87(1): 4-10.

Howell CR and Stipanovic RD. 1983. Gliovirin a new antibiotic from *Gliocladium virens*, and its role in the biological control of *Pythium ultimum*. *Can. J. Microbiol.* 29: 321-324.

Howell CR and Stipanovic RD. 1995. Mechanisms in the biocontrol of *Rhizoctonia solani* - induced cotton seedling disease by *Gliocladium virens*: Antibiosis. *Phytopathol.* 85: 469-472.

Howell CR, Hanson LE, Stipanovic RD and Puckhaber LS. 2000. Induction of terpenoid synthesis in cotton roots and control of *Rhizoctonia solani* by seed treatment with *Trichoderma virens*. *Phytopathol.* 90: 248-252.

Kapat A Z and Elad Y. 1998. Effect of two isolates of *Trichoderma harzianum* on the activity of hydrolytic enzymes produced by *Botrytis cinerea*. *Physiol. Mol. Plant Pathol.* 52: 127-137.

Khare MN. 1981. In: Diseases of Lentils, Eds.: Webb C and Hawtin G. Farnham Royal. UK: 1 C ARDA/CAB, pp. 163-172.

Kumar RN and Mukerji KG. 1996. Integrated disease management future perspectives, pp. 335-347. *In:* Mukerji KG,. Mathur B, Chamala BP and Chitralekha C. (Eds.), *Advances in Botany*. APH Publishing Corporation, New Delhi. *Microbiol.* 65: 1061-1070.

Lifshitz R, Windham MT and Baker R. 1986. Mechanism of biological control of Pre emergence damping-off of pea by seed treatment with *Trichoderma* spp. *Phytopathol.* 76: 720-725.

Lorito M, Woo SL, D'Ambrosio M, Harman GE, Hayes CK, Kubicek CP and Scala F. 1996. Synergistic interaction between cell wall degrading enzymes and membrane affecting compounds. *Mol. Plant-Microbe Interact.* 9: 206-213.

Lorito M, Woo SL, Fernandez IG, Collucci G, Harman GE, Pintor-Toros JA, Filippone E, Muccifora S, Lawrence C B, Zoina A, Tuzun S and Scala F. 1998. Genes from mycoparasitic fungi as a source for improving plant resistance to fungal pathogens. *Proc. Natl. Acad. Sci.* USA 95: 7860-7865.

Mehta RD, Patel KA, Roy KK and Mehta MH. 1995. Biological control of soil borne Plant Pathogens with *Trichoderma harzianum*. *Indian J. Mycol. Pl pathol.* 25: 126.

Metcalf DD and Wilson CR. 2001. The process of antagonism of *Sclerotium cepivorum* in white rot affected onion roots by *Trichoderma koningii*. *Plant Pathol.* 50: 249-257.

Nagamani A, manoharachary C, Agarwal DK and Chowdhry P N. 2002. Monographic contribution of *Trichoderma* Pers. ex Fr., Association Publishing Company, Post, New Delhi.

Nemat Alla MM, Shabana YM, Serag MM, Hassan NM and El-Hawary MM. 2008. Granular formulation of *Fusarium oxysporum* for biological control of faba bean and tomato *Orobanche. Pest Management Science* 64: 1237–1249.

Ranasingh N, Saurabh A and Nedunchezhiyan M. 2006. Use of *Trichoderma* in Disease Management, *Orissa Review*: pp 67-71.

Rehner S A and Samuels G J. 1994. Taxonomy and phylogeny of *Gliocladium* analyzed by large subunit ribosomal DNA sequences. *Mycol. Res.* 98: 625-634.

Rifai M. 1969. A revision of genus *Trichoderma*. *Mycological Papers* 116: 1-56.

Rojo FG, Reynoso MM, Ferez M, Chulze SN and Torres AM. 2007. Biological control by *Trichoderma* species of *Fusarium solani* causing peanut brown root rot under field conditions. *Crop Protection* 26: 549-555.

Sharon E, Bar-Eyal M, Chet I, Herra-Estrella A, Kleifeld O and Spiegel Y. 2001. Biological control of the root-knot nematode *Meloidogyne javanica* by *Trichoderma harzianum*. *Phytopathol.* 91: 687-693.

Stoilova S and Chavdarov P. 2006. Evaluation of lentil germplasm for disease resistance to fusarium wilt (*Fusarium oxysporum* f. sp. *lentis*) *Cent. Eur. Agr.* 7: 121-126.

Veenekar JV, Tanksale A, Ghatge M and Deshpande V. 2001. Novel bio-functional alkaline protease inhibitor: protease inhibitory activity as the biochemical basis of antifungal activity. *Biochem. Biophys. Res. Commun.* 28(5): 1018-1024.

Watts R, Dahiya J, Chaudhary K. (1988). Isolation and characterization of a new antifungal metabolite of *Trichoderma reseii*. *Plant and Soil* 107: 81-8.

Webster J and Lomas N. 1964. Does *Trichoderma viride* produce gliotoxin and viridin. *Trans. Br. Mycol. Soc.* 47: 535-540.

Weindling R. 1932. *Trichoderma lignorum* as a parasite of other soil fungi. *Phytopathol.* 22: 837-845.

Weindling, R. 1934. Studies on a lethal principle effective in the parasitic action of *Trichoderma lignorum* on *Rhizoctonia solani* and other soil fungi. *Phytopathol.* 24: 1153-1179.

Weindling R. 1941. Experimental consideration of the mold toxin of *Gliocladium* and *Trichoderma*. *Phytopathol.* 31: 991-1003.

Woo SL, Donzelli B, Scala F, Mach R, Harman GE, Kubicek, CP, Del Sorbo G and Lorito M. 1999. Disruption of the *ech42* (endochitinase-encoding) gene affects bio-control activity in *Trichoderma harzianum. P1. Mol. Plant-Microbe Interact.* 12: 419-429.

Yedidia I, Benhamou N and Chet I. 1999. Induction of defense responses in cucumber plants (*Cucumis sativus* L.) by the bio-control agent *Trichoderma harzianum. Appl. Environ. Microbiol.* 65: 1061-1070.

Yedidia I, Benhamou N, Kapulnik Y and Chet I. 2000. Induction and accumulation of PR proteins activity during early stages of root colonization by the mycoparasite *Trichoderma harzianum* strain T-203. *Plant Physiol. Biochem.* 38: 863-873.

Yedidia I, Srivastva AK, Kapulnik Y and Chet I. 2001. Effect of *Trichoderma harzianum* on micro element concentrations and increased growth of cucumber plants. *Plant Soil* 235: 235-242.

Zohary D and Hopf M. 2000. Domestication of Plants in the Old World: The origin and spread of cultivated plants in West Africa, Europe and the Nile Valley. Oxford University Press Inc., New York.

Citation

Singh HK, Shakywar RC, Singh Shyam and Singh AK. 2018. Trichoderma: A Potential Bio-Pesticide for Management of Lentil (*Lens Culinaris* Medikus) Diseases. In: Scientific Lentil Production. Singh AK, Bhakta N, Sangale UR, Manibhushan, Sundaram PK, Kumar S and Yasin JK. (eds.) Society for Upliftment of Rural Economy Varanasi, India. Pp. 385-409.

❑❑❑

Chapter 23

Lentil Rust-A Threat to Lentil Cultivation of Bihar & Practical Experiences of Farmer's Field

UR Sangle, AK Singh, Md Idris, Santosh Kumar and Mandhata Singh

SYNOPSIS

The tal area of Mokama was surveyed on 19.02.2015 for damage assessment caused by lentil rust. The disease was indentified on the basis of morphological and direct visual examination of fungal spores of recovered from infested plants. The cultural and morphological characteristics of pathogen indicate that the disease is caused by*Uromycesfabae* which is autoeocious in nature and most common in north India. At the time of infestation, the weather conditions were observed high humidity, cloudy, drizzling and temperature ranges from 25-28^{0}C. Such weather conditions favour the spread the diseases in epidemic form. The losses were assessed from 40-100 per cent. Some of fields were totally damaged. The infested plants were having green pods with shriveled seeds. In view of severity of the diseases, the farmers of the area were demanding for compensation from Government of Bihar. The visiting team submitted survey and damage assessment report to the government of Bihar and ICAR Research complex for Eastern Region, Patna for further necessary action.

Keywords: Lentil rust, farmer's interaction, weather, losses,*Lathyrus* and *Vicia*spp, Mokama (tal area).

Introduction

Lentil (*Lens culinaris*M.), is one of the ancient food legume crop. Lentil has originated from the South Western Asia region. The primitiveness of lentil is visible in Indian ancient literature (Singh *et al.*, 2014). After chickpea it is the second most important cool season grain legume (Singh *et al.*, 2011). Rust (*Uromycesfabae* Pers.) is a widespread foliar disease of lentil (Prasad and Verma, 1948). It is economically important disease capable of causing complete crop failure. The yield losses to the extent of 60-69 per cent have been reported in India (Singh *et al.*, 1986). The disease occurs during the flowering or early poding stage. High humidity and cloudy weather along with temperatures of 20-22^{0}C,are very much favorable for disease development.Rust is characterized by yellowish-white pycnidia and aecial cups on leaflets and on pods, singly or in small groups in a circular form. Later on, brown conidial pustules, oval to circular and up to 1mm in diameter develop on either surface of leaflets, branches, stem and pods. The telia, which are formed late in the season, are dark brown to black, elongated and present in mainly on branches and stems. In severe infections, 100 per cent losses the affected plant dries without forming any seeds in pods or with small shriveled seeds the same situation was observed in Sabajpura Experiment farm of ICAR RCER Patna in the project Improved Rice-Based Rainfed Agricultural Systems in Bihar State (IRRAS,2014) variety PL406 in rabi season 2014. The plant appear dark brown to blackish patches in the field. The disease first occurs during the flowering/early podding stage. Lentil rust occurs mostly in January-February, in the form of Pycnia and Aecia. Aeciopsores germinate at 17-22^{0}C and infect other plants forming either secondary aecia or uredia at 25^{0}C. At lower temperature, uredospores are probably an important means of survival in the absence of host. Uredo-mycelium is highly resistant to heat and sunlight and is probably important for continued development and survival of rust in hot, dry conditions. The disease generally starts from low lying patches in the field and radiates towards the border (Khare and Agarwal, 1978). Uredosori develop late in the season and are rapidly followed by telia. After harvest, aecia and uredia present on the plant die out, but teliosprores resist the heat and survives the summer as teliospores. It is also carried with seed as concomitant contamination. It may also perpetuate on weed hosts (*Lathyrus* and *Vicia* spp.) from where it may infect lentil by wind borne spores. High humidity and cloudy or drizzly weather with temperature of 20-22^{0}C favors' disease development (Khare, 1981). Good Agronomic Practices (GAP) - An efficient and eco-friendly tool for sustainable management of plant diseases under changing climate scenario (Singh *et al.*, 2012).

Fig. 23.1 Diseased lentil crop

Lentil rust observed in Experimental Research farm of ICAR RCER Patna

In severe infections, the affected plant dries without forming any seeds in pods or with small shriveled seeds the same situation was observed in Sabjpura Experiment farm of ICAR RCER Patna in the project Improved Rice-Based Rainfed Agricultural Systems in Bihar State (IRRAS) variety PL-406.In Rabi season 2013-14 lentil (*Lens culinaris* L.) variety PL-406 was evaluated for management of soil borne diseases in lentil through microbial like Trichoderma strainsS2. The strains Trichodermaharzianum Strain S2 was observed in management of root rot and wilt percentage of the lentil upto till flowering and pod formation of lentil crop very good/healthy condition. In this season Variety PL -406, totally damage by rust 100 per cent (*Uromycesfabae*) appears naturally in epidemic form in the month of February, 2014. This is not relation to microbial trial because *Trichoderma* could be manage by soil borne diseases (report submitted to the IRRAS project 2014). As per the farmers of Mokama block in the Patna distinct of Bihar experienced a serious disease problem in lentil crop and approached ICAR RCER Patna for diagnostic of problem. The Director ICAR RCER, Patna deputed three scientist namely Dr. AK Singh Principal Scientist (Agronomy) Mohammad Idris, Principal Scientist (Entomology) and Dr. U.R. Sangle, Senior Scientist (Plant Pathology). To assess the damage in lentil crop of

Mokamatal area. Both scientists proceeded for Mokama on 19.02.2015 for evaluation of insects pest and disease problems in lentil, lathyrus and chickpea. Mokama Tal is not a single tal but a group of seven tals covering an area of 1,062 square kilometer. Tal covers about 1,18000 ha. land and spreads in seven blocks- Barh, Mokama, Harnaut, Fatuha, Barahiya, Singhaul and Lakhisarai. As water receded, farmers of the area sowed mostly lentil, lathyrus and chickpea crops. It is a very old age practice.

Fig. 23.2 Healthy lentil and Lathyruscrop

Data regarding rust disease was recorded by using following 1-9 scoring scale at the time of survey

1 = No pustules visible (Highly Resistant),

3 = Few scattered pustules, usually seen after careful searching (Resistant),

5 = Pustules common on leaves and easily observed but causing no apparent damage (Moderately Resistant),

7 = Pustule very common and damaging, few pustules on petioles and stems (Susceptible),

9 = Pustule very extensive on all plant parts, some death of leaves and other plant parts (Highly Susceptible).

Different fields of the Tal area were surveyed and it is observed that lentil crop was severely (40-100%) damaged by the rust (*Uromyces*sp.) (Fig. 23.1and 23.2). The following farmers were interviewed for just to know the sown area, chemical applied and approximate percent loss (Table 23.1 and Table 23.2).

Table 23.1 Effect of different sowing dates and their percent loss in lentil in farmer's field

Area	Date of Sowing	Percent loss
2.5 ha	26/10/2014	40
1.0 ha	02/11/2014	30
2.75ha	03/11/2014	50
1.25 ha	03/11/2014	50
2.0 ha	06/11/2014	60

Cause of epidemic

Most of the farmers have been sowing traditional lentil variety in same field for long time without following any seed treatment techniques. The area get submerged under deep water in monsoon period and becomes unfit for cultivation of Kharif crop and cultivation of Rabi crops also suffers in productivity due to pests and diseases if water is not receded completely by their sown time i.e. 15th October. The late (after IInd week of November) sown crop experienced more damage in lentil than early (before 30th October) sown (based on farmers information) Farmers were applying more than one fungicides and insecticides in same spraying tank at a time and huge amount of spraying of different chemicals (Fig. 23.3).

Table 23.2 Showing name of interviewed farmers, village & blocks, cropping area, chemicals used and percent loss (approx.)

Name of Farmers	Village/ Block	Cropping area	Used Chemicals	Per cent loss (approx.)
Sh. Pran Shahi	Lakhan Chand, Mokama	7.5ha Lentil rest *Lathyrus* & Chickepea	Mancozeb, Fenval	50-60
Sh. Vinod Prasad Singh	Aonta, Mokama	2 ha Lentil	Mancozeb	50
Sh. Subodh Kumar	Aonta, Mokama	6.25 ha Lentil and 1.75 ha Lathyrus	Tilt (Propiconazole), Cantaf, Chlorpyriphos, Fenval	5-10
Sh Tripurari	Hathida, Mokama	2.0 ha lentil	Ist spray(3rd February, 2015 Cantaf2nd spray- Folicure (Tabuconazole)	50

Sh. Niraj Kumar	Aonta, Mokama	2.5 ha Lentil, 0.5 ha Bigha Lathyru, 0.25 ha Bigha Chickpea	Ist Campanion (Carbendazim 12% + Mancozeb 63%)IInd spray Propiconazole + Hume +Macoben (Mancozeb+ Carbendazim)+ Chlorpyriphos+ Cypermethrin	100
Sh. Tushan Singh	Aonta, Mokama	2.25 ha lentil	Campanion	50
Sh. Om Prakash	Marachi, Mokama	18.0 ha Lentil	1st spray (10 December, 2014: NPK+Mancozeb+Propenfos+ Tonic(Nitrobenzen)2nd Spray (10-11 January, 2015) NPK+ Profenocyper+Mancozeb+ Hexaconozole+(Frone)3rd Spray (8-9 February, 2015: Mancozeb + Tilt Acephate 4th Spray (19 February 2015): Folicure+Antracole+Profeno-cyper+ Streptocycline	50-100

Disease diagnosis

Identification based on morphology

Identification was carried out on the basis of morphology of fungal spores collected from infested plants. The cultural and morphological characteristics of the pathogen indicates that disease caused by *Uromycesfabae* (pers.) It is autoeocious in nature and most common rust of lentil in North India. It is known to occur on sweet pea (*Lathyrus* sp.) locally called as a Khesari and broad bean in the Indo-Gangetic plains.

Symptoms

The disease occurred in epidemic form in this February 2015 (Report submitted to the Government of Bihar). The earliest symptoms are the development of aecia in round or elongated clusters in February. All the spore stages develop on every green part of the plant including pods. The information of aecial stage is preceded by a slight yellowing which gradually turns brown.

Disease cycle and environmental relations

Infection is caused under relatively low temperature of 17-22^{0}C and uredia developed at 25^{0}C. The rust fungus perennates in the telial stage in the left over diseased plant trash or mixed with the seed as external contamination and infects the new crop in the next season.

Fig. 23.3 Indiscriminate use of pesticides

Probable reasons for rust incidence

The sudden outbreak of disease in this location was noticed first time during the winter season in 2nd week of February probably due to the drizzling rain in 3-4 days in the month of February, were as relative humidity was 90-95 percent and total sunshine hours were 4.1 accompanied temperature ranging from 20-28^{0}C, secondly due to increased area under lentil crop with all local cultivars from very old time thirdly growing single lentil crop in a year which helped the pathogenic inoculums in the soil plant debris.

Management of the disease

- Seed treatment Thiram @ 2g/kg seed
- Using resistant varieties like Pant 406, Pant 638, and L9-12, PL-639 varieties have been found suitable for controlling the wilt and rust diseases.
- Spraying Mancozeb (Dithane-M-45) @ 2g per litre of water + Add Sticker like Triton @ 1ml/lit. 2-3 spray will be sufficient and should be given at 15-20 days intervals.
- Affected plant debris should be destroyed by burning after harvest.

Genetical Method

Lentil is very susceptible to diseases and pests that cause significant yield loss. Lentil rust is widespread foliar diseases and also economically important because it is capable of causing complete crop failure. It is not only problem in Tal area but also occurred regularly in lentil growing areas of Bihar. The present study were carried out in Tal area of Mokama, district Patna. It is observed that the diseases, *U. fabae* caused 40-100 percent damage in Tal area. Some of the farmers (Table 23.2) of the area were also interviewed (Fig. 23.4). They revealed that this phenomenon occurred regularly while weather conditions like high humidity, cloudy, drizzling and temperature ranged from 25-28°C during March. To manage the disease, the farmers were spraying fungicides indiscriminately in addition with insecticides and growth promoter hormones as per the guidance of local pesticides shop keepers (KrishiSevaKendras). It is noticed that farmers were not well-versed with newer scientific technology to manage the rust diseases as well about the resistant varieties. They were using traditional varieties in same field year around which aggravated the problem manifolds. Team suggested to farmers that instead of using traditional varieties, select rust resistant lentil varieties and also follow recommended scientific plant protection measures at the time of sowing and on changing of weather conditions like cloudy, drizzling, high humidity and temperature ranging from 20-28°C which favors development of diseases.

Fig. 23.4 Interaction with farmers

References

IRRAS. 2014. Project report on Improved Rice-Based Rainfed Agricultural Systems in Bihar State (*IRRAS*)

Khare MN and Agarwal SC. 1978. Lentil rust survey in Madhya Pradesh, Paper presented in all India Pulses Workshop. Baroda. India. Pp. 1-3.

Khare MN. 1981. Diseases of lentil. (In) Lentils, pp163-72.Webb C and Hawtin G (EDs). Commonwealth Agricultural Bureau International, England, UK.

Prasad R and Verma UN. 1948. Studies on lentil rust, *Uromycesfabae* (Pers.) de Bary in India. *Indian Phytopathology* 1: 142-146.

Singh AK, SinghD, SinghAK, GadeRM and Sangle UR.2012. Good Agronomic Practices (GAP) - An efficient and eco-friendly tool for sustainable management of plant diseases under changing climate scenario. *J. Plant Disease Sci.* 7 (1): 1-8.

Singh AK, Singh KM, Bharati RC, Chandra N, Bhatt BP and PedapatiA. 2014.Potential of Residual Sulphur and Zinc Nutrition in Improving Powdery Mildew (Erysiphetrifolii) Disease Tolerance of Lentil (*Lens culunaris*L).*Communication in Soil Science and plant analysis* 45: 2807-2818.

Singh AK, Singh SS, Prakash V, Kumar S and Dwivedi SK. 2015. Pulses Production in India: Present Status, Bottleneck and Way Forward. *Journal of Agrisearch* 2(2): 75-83.

Singh K, Jhooty JS and Cheema HS. 1986. Assessment of losses in lentil yield due to rust caused by Uromucesfabae. *Lens.Newsletter* 13. 28.

Singh K. 1985. Effect of seed treatment on lentil rust (*Uromycesfabae*) development.*LENS Newsletter* 12: 26-27.

Singh, AK, Meena, MK Bharati RC. 2011. Sulphur and Zinc Nutrient Management in Rice Lentil Cropping System. International Conference on "Life Science Research for Rural and Agricultural Development. 27-29 December, 2011, CPRS Patna (Bihar). pp. 66-67.

Citation

SangleUR, Singh AK, IdrisM, Kumar S and Singh. 2018. Lentil Rust- A Threat to Lentil Cultivation of Bihar & Practical Experiences of Farmer's Field. In: Scientific Lentil Production. Singh AK, Bhakta N, Sangale UR, Manibhushan and Sundaram PK, edit Published by Society for Upliftment of Rural Economy Varanasi, India. Pp. 411-419.

□□□

Chapter 24

Survey of Wilt Complex Damage on Lentil (*Lens culinaris* Med.) in Bihar

UR Sangle, AK Singh, Santosh Kumar and Mandhata Singh

SYNOPSIS

A survey of 3 years data from 2011-2014 total 11 districts of different lentil growing blocks and villages covering 350 farmers' fields. Intensity of losses due to wilt complex problem were assessed and isolated the pathogen caused by lentil wilt and found mostly identified initially after 40-45 days sowing of the seed color rot/ root rot the seedlings. The main pathogens found associated with plant mortality at this stage were 15.68 per cent caused by *Sclerotium rolfsii* and at the time of flowering and pod formation 27.08% was damaged by wilt caused by *Fusarium oxysporium* f. sp. *lentis* were observed. In the year 2010-11 observed lentil was grown monocropping system in tal area due to continuous monocropping in tal area the wilt pathogen aggregated in manifolds. Intensity of losses due to wilt was found up to 40% was recorded in 2010-11, where the Tal lies in the Kiul-Harohar river basin and extends over an area of 1,062 square kilometres (410 sq m). However, in some places complete failure of crop was observed due to wilt.

Keywords: lentil; prevalence of wilt; root rot; *Fusarium oxysporum* f. sp. lentis.

Introduction

Agriculture in eastern region has been characterized as high potential low productivity enterprise as well as population concern, very high natural resources but low poverty of the people. As far as Bihar state is concerned, 90 per cent of its population lives in villages and depends on agriculture for its livelihood which causes continuous exploitation of soil and other natural resources (Singh *et al.*, 2008 and Singh *et al.*, 2012).Lentil is one of the oldest known proteins rich food pulse crop of India. Throughout in tal land area farmers grown lentil is the first most important rabi crop in Bihar. Owing to biotic stresses, the crop yields are below attainable levels (Deokaran *et al.*, 2014). Aggravated due to frequent droughts and floods soil beneficial bioagents totally destroy and harmful pathogen population increases that due to lentil diseases problem were found very much higher. Lentil suffered from many diseases caused by fungi, bacteria, viruses, nematodes and phaneroganic plant parasites. The major fungal diseases are wilt complex problem caused by the pathogens *Fusarium oxysporium* f. sp. *lentis* and *Sclerotium rolfsii*. (Sangle, 2012). Plant mortality caused by wilt, root rot and collar rot diseases are incited by a diverse group of pathogens such as *Fusarium oxysporum* f. sp. *lentis, Rhizoctonia bataticola* (*Macrophomina phaseolina*), *R. solani, Fusarium equisetii, F. semitectum, S. rolfsii*, and *Alternaria spp*., etc. (Agrawal 1998; Chaudhary and Kaur 2002). Lentil is grown as single crop in tal area of Bihar and due to continuous monocroping, the wilt pathogen aggregated in manifolds. Intensity of losses due to wilt was found upto 40 percent. However in some places complete failure of crop was observed (Sangle, 2012). The wilt is widespread in all lentil growing areas. In India 67 per cent losses from Delhi (Vasudeva and Srinivasan, 1952) and 75 per cent from Madhya Pradesh (Khare *et al*., 1979) were reported. The disease may cause complete crop failure under favorable conditions for disease development, and can be the major limiting factor for lentil cultivation in certain areas (Chaudhary and Kaur, 2002). In Tarai region of Uttarkhand state lentil wilt is a major problem. Lentil is generally grown rainfed and in order to harness better use of residual soil moisture, the crop is sown at the earliest opportunity. Many soil-borne pathogens invade the plant even at seedling stage due to congenial soil temperature and moisture. However, these soil-borne pathogens and the mortality caused by them in major lentil growing areas of India during 40-45 days of sowing seedlings, flowering to pod maturity are more crucial from a yield point of view.

Constraints in lentil production

Owing to biotic stresses, the crop yields are below attainable levels. Among the biotic factors, diseases are serious threat to lentil production in many parts of the world. Lentil wilt caused by *Fusarium oxysporium* f. sp. *lentis* is one of the main limiting factors to successful cultivation. In Bihar state in tal land 60% farmers cultivated Lentil crop under rainfed condition. Lentil was grown monocropping system in tal area due to continuous monocropping in tal aea the wilt pathogen aggregated in manifolds. Uncertain and erratic rainfall created formidable barrier for timely planting and crop growth. The crop required moderate railfall, while severe cold and frost are injurious. Lentil is grown in soils poor in fertility and moisture retention capacity. The crop is raised mostly by marginal and submarginal farmers and is grown unwillingly with no adequate fertilization and proper irrigation because of lack of support prices. Non availability of good high yielding varieties resistant varieties and insect pest and diseases problems are the major bottlenecks in realising higher yields (Singh *et al.,* 2014).

The Famous Tal Region of Mokama for cultivation of Lentil

The Tal is a Massive Land Segment in the Mokama under Patna district of Bihar. Many consider it as World's Largest Tal Region. It extends to the Backwater of Ganga during Monsoon and gives an impression of Ocean to Onlookers. During other Seasons, it provides a very fertile area for Agriculture. The Tal lies in the Kiul-Harohar river basin and extends over an area of 1,062 square kilometres (410 sq mi). It is a saucer shaped depression extending from Fatuha in the west to Lakhisarai in the east. Its width varies from 6 to 17 km. It runs close to and almost parallel to the right bank of the Ganges. The Harohar River which is the main outlet channel for the Tal area flows eastward and drains into the Kiul River. The entire Tal area undergoes submergence every year during the monsoon period from June through September. The special emphasis was made due to lentil single crop grown as monocropping system from many decades and on the basis of 3 years study in lentil crop itself intensity of losses due to wilt complex is very serious threat and this condition day by day are going to be worst.

Occurrence of disease and time

The major fungal diseases are wilt complex problem caused by the pathogens *Fusarium oxysporium* f. sp. *lentis* are observed and infected to the plants at the time of flowering and pod formation. The disease appeared in patches to patches in field. *Sclerotium rolfsii* is a most devastating soil borne pathogen was found in field after sowing of 40-45 days of the seeding the month of starting from mid December to last week of December in Bihar Eastern region of India. This disease is reported to cause heavy seedling mortality at early crop stages (Agrawal and Prasad 1997). The disease may be caused by either seed borne inoculums pretested in infected plant part left in soil. The disease can cause complete failure of the crop, especially in a warm spring and dry and hot summer. The optimum temperature for fungal growth is around 25^{0}C. Lentil wilt is favoured at soil pH 7.5 to 8.0, soil moisture of 25% water holding capacity and 20^0-30^{0}C soil temperature and no disease at 75% or more soil moisture.

History of the pathogen

Collar rots/root rot (*Sclerotium rolfsii*) in lentil

The collor rot is also called root rot (Pavgi and Upadhyay, 1967), occurs in almost every region where lentil is grown and is specially prevent in warm areas where high soil moisture along with high temperature (25^{0}C) at the seedling stage. The disease is economically important in some parts of North India; where the crop is grown after rice, causing death to the seedling in early stages of plant growth. The fungus survive in the form of sclerotia that remain viable in the primary source of inoculum and are capable of initiating infection with or without an additional food base (Punja, 1985). The fungus has an extensive host range mainly. The pathogen infects the plant at collar region causing yellowish-brown discolouration, and a rotting of tissue (Beniwal *et al.*, 1993). The seedling stage show damping, when infection occurred at advanced stage. Plant gradually turns pale, drop and dry. The white feathering grown of S. rolfsii generally associated with dirty white to brown coloured, mustard seed like sclerotia can be seen infected plant parts even rooted plant pathogen was isolated in culture media. Sometimes pathogen may move downwards causing root rot. The root system is poorly developed and side roots are destroyed hence plants are easily pulled out. The pathogen survives well in soil as sclerotia in presence of sufficient organic matter even under adverse weather conditions (Khare *et al.*, 1979). The disease is more severe when high soil moisture with high temperature 25-30^{0}C followed by good sunshine after rains (Khare *et al.*, 1981). Extensive plant to plant spread occurs in closely spread crops. The disease becomes severe if the stubble of the previous crop, such as paddy or sorghum is left in the field.

Wilt of lentil Fusarium wilt (Fusarium oxysporum f.sp.lentil)

In India wilt disease in lentil was first observed by (Padwick, 1941) in experimental field in Delhi and Karnal. (Vasudeva and Srinivasan, 1952) first authenticated and published record of vascular wilt of lentil and noted that about 67% disease incidences. In India (Khare *et al.*, 1979) reported that *F.oxysporum* is associated with the seeds of lentil and symptoms can be observed in field at three stages i.e., pre-emergence, post emergence and at maturity stages (Vasudeva and Srinivasan, 1952; Khare *et al.*, 1979). In pre-emergence seedling do not emerge above the soil surface as they die during or immediately after seed germination. At post emergence stage seedling infected 4-5 days after emergence exhibiting browning of radical, they remain stunted, droop, and finally collapse. In adult stage, plants are seen in isolated patches. The growth of infected plants are stunted, their leaves turn yellow and curl from lower to upper part of the plant, the crown droops and ultimately the plant dies. The roots of the affected plant turn yellowish-brown to dark-brown partially or wholly. The root nodules are poorly developed. (Khare *et at.*, 1976) reported that the *Fusarium* wilt characterized by stunting plant, yellowing and drooping of leaves, followed by drying and death of the whole plant. In few cases partial drying of the plant was also observed. The stem of the diseased plant mostly exhibited reddish discoloration. Apparently healthy roots reduced proliferation. The disease appear at flowering to late pod filling stage characterized by yellowing of leaves from lower parts and extending upwards, epinasty in some cases and finally yellowing and drying of whole plants. Dried leaflets of infested plants are not shed at maturity. The root system is poorly developed and exhibits no external rotting, drying or discoloration of roots. Secondary roots are proliferated in some cases above the affected level. The xylem vessels are discolored brown showing tyloses. The fungal mycelium can also be observed in stems of the infected plants (Khare, 1981). Seeds from plants affected in mid to late pod filling stage are often shriveled. Though *F. oxysporum* is highly host specific, great variability exist in it. Sharma and Agnihotri, 1972, identified its three isolates 'A' 'B' and 'C' on the basis of their virulence in lentil. (Bhalla *et al.*, 1984) identified a more virulent strain 82 A of *Fusarium oxysporum* causing complete necrosis of lentil seedling within 7 days.

Surveys were conducted during January-March 2010-11 and 2011-12 in different district and villages of Bihar. Samples were collected from five fields from each block of specific locations. Diseased plants were randomly collected and 20-25 plants were taken for isolation of pathogen. Percent mortality was calculated as following formula :

$$\text{Percent mortality} = \frac{\text{Number of infected plants in } 1m^2 \text{ area}}{\text{Number of total plants in } 1m^2 \text{ area}} x100$$

A roving survey of lentil fields was conducted in ten districts of different villages. On an average 5 fields of lentil in each villages were visited and the percent wilt incidence was recorded (Fig.24.1a and 1b). The data is presented in Table 24.1, it is revealed that heavy heavy disease incidence was noticed in Pandarak (38.68%) followed by Gaya (36.91%) and moderate in Maranchi (35.58%).The lowest disease incidence were Daniyawaha (16.47%), Nagarnausa(17.08%).In Bihar state in tal land 60% farmers cultivated Lentil crop and lentil wilt play a major role in reducing lentil production. When crop is totally fully mature and pod formation that time wilt complex pathogen active and destroy lentil crop wholly or sometimes patches to patches in main field heavy losses. Lentil was grown monocropping system in tal area due to continuous monocropping in tal area the wilt pathogen aggregated in manifolds. Intensity of losses due to wilt was found up to 40%was recorded recently by Sangle and Idris (2010). However, in some places complete failure of crop was observed (Fig. 24.1 a & 1b). Lentil is one of the oldest known protein rich food legumes. Now presently in tal land area farmers grown lentil is the first most important rabi crop in eastern India. Owing to biotic stresses, the crop yields are below attainable levels. Among the biotic factors, diseases are serious threat to lentil production in eastern India. Lentil wilt caused by *Fusarium oxysporium* f. sp. *lentis* is one of the main limiting factors to successful cultivation. On the basis of 3 years study different fields of the Mokama Tal area covering an area of 1,062 square kilometer. It contains seven different blocks, viz., Barh, Mokama, Harnaut, Fatuha, Barahiya, Singhaul and Lakhisarai where lentil crops grown as monocropping system. Even in Research farm trial intensity of losses due to wilt complex problem were assessed and isolated the pathogen caused by lentil wilt and found mostly identified initially after 40-45 days sowing root rot caused by *Sclerotium rolfsii* 15.68% damage the seedlings and at the time of flowering and pod formation 27.08% were damaged by wilt *Fusarium oxysporium* f. sp. *lentis* were observed.

Table 24.1 Percent mortality due to wilt complex pathogens in lentil in Bihar district's

District	Block/Villages	Percent mortality	
		Sclerotium rolfsii	*Fusarium oxysporium* f. sp. *lentis*
Patna	Fatuha block	12.12	18.93
	Bakhtiarpur	8.47	20.12
	Daniyawan	11.15	16.47
	Daniyawan Nagarnausa	15.47	17.08
	Tetipur	10.49	27.94
	Bihta Sadisopur	10.08	27.35
	Athmalgola Karauta	15.35	28.94
	Baktiyarpur Gayaspur Mahazi	18.12	26.02
	Bidhipur Narauli Baktiyarpur	12.85	26.68
	Dhanasupur	11.94	25.69
	Barh	18.12	36.25
	Athmalgola	15.68	36.22
	Auta Mokama	17.35	30.21
	Murarpur Mokama	14.12	27.85
	Hasanchak Barh	12.16	29.35
	Maner	15.68	28.05
	Pandarak	16.68	38.68
	Maranchi , Mokama	15.49	35.58
	Agawanpur Barh	18.21	30.51
	Punpun	17.50	26.68
	Naubatpur	15.20	32.24
	Bikram	11.24	24.74
	ICAR main field, Phulari Sharif	11.68	27.49
Nalanda	Chandi	15.68	27.50
	Harnaut	16.24	18.79
	Kundi Hilsa	17.91	28.91
	Noorsarai	20.12	32.2
Sitamarhi	Pupri, Bajpatti, Bokhra, Nanpur villages	11.51	29.68

Motihari	Barwa, Katha, Ramgarwa	14.23	26.68
Muzaffarpur	Gaighat, Muraul, Sakra, Sahebganj	13.58	32.24
Buxar	Dumraon, Chausha, Rajpur, Simri	16.64	35.02
Gaya	Tekari, Belganj, Khizarsarai	16.79	36.91
Kaimur	Bhabua, Bhagwanpur, Kudra, Nuawon	14.11	30.23
Vaishali	Hajipur, Raghopur, Vaishali. Bidupur, Lalganj, Jandaha. Patepur, Desri	11.68	32.64
Samastipur	Morwa, Pusa, Samastipur, Tajpur, Sarairanjan	15.91	30.24
Darbhanga	Hanuman Nagar, Hayaghat, Singhwara, Baheri, Darbhanga, Bahadurpur	14.00	22.40

*Average of five fields per village

The disease status was assessed by undertaking an intensive roving survey of different district of Bihar state Viz., Patna 25 villages, Nalanda 3 villages, Sitamarhi 32 villages, Motihari9 villages, Muzaffarpur10 villages, Buxar5 villages, Gaya 6 villages, Kaimur3 villages, Vaishali 5 villages, Samstipur 05 villages. Diseased samples from 5 fields per villages were collected during rabi crop season 2011-12 and the percentage mortality was find out. In Patna block highest mortality was observed in Pandarak (38.68%) followed by Barh (36.25%) and Atmalgola (36.22%). Lowest mortality was observed in Daniyawaha (16.47%) and Nagarnausa (17.08%). In Nalanda block highest mortality was assessed Nursarai (32.20%) followed by Kundi (28.91%) and Nalanda (27.50%). Other district of Bihar were surveyed and assessed highest mortality of lentil crop was Gaya (36.91%) followed by Buxar (35.02%) Muzaffarpur (32.24%) and lowest lentil mortality was calculated in the district of Motihari (26.68%) and Sitamarhi (29.68%) respectively. Initial stage of lentil crop mortality was isolated due to the pathogen *Sclerotium rolfsii* at seedling stage.

Isolations from wilted plants (initial and later stage) from different fields. The present survey was conducted during the flowering stage of the crop and this might be the reason for their relatively less recovery from the wilted samples of lentil. On the contrary, *F. oxysporium* f. sp. *lentis* causes maximum damage when the lentil is in flowering and podding stage. In throughout survey soil samples of rhizosphere of lentil plants were collected from different locations and isolated bioagent *Trichoderma* spp. Nos.6 isolates were identified out of tal area 4 nos. native *Trichoderma* spp was isolated. Further in the month of November-December 2011 survey were concluded that and found new report of initial stage of lentil crop mortality due to the pathogen *Sclerotium rolfsii* and different variation of the pathogens caused by wilt complex problem in lentil growing areas from 90 samples from different locality of Bihar.

References

Agrawal SC, Prasad KVV. 1997. Diseases of lentil. New Delhi: Oxford & IBH Publishing Co. Pvt. Ltd; p. 155.

Agrawal SC. 1998. IPM system in agriculture. New Delhi, India; Aditya Book Pvt. Ltd., Vol. 4,Pulses, Major diseases of lentil and their management, p. 427–450.

Beniwal SPS, Bayya B, Weigand. MK and Saxena MC. 1993. Field guide to Lentil Diseases and Insect Pests. International Centre for Agricultural Research in the Dry Areas, Aleppo, Syria, 106pp.

Bhalla MK, Nozzolillo C and Schneider EF.1984. Pathogenicity of soil fungi associated with a root rot of lentils. *Can.J.Pl.Path.* 6: 21-28.

Chaudhary RG and Amarjit K. 2002. Wilt disease as a cause of shift from lentis cultivation in Sangod Tehsil of Kota, Rajasthan. *Indian J. of Pulses Res.* 15: 193–194.

Deokaran, Singh SB and Ram Kewal. 2014. Effect of Zinc and Boron application on Yield of Lentil and Nutrient Balance in the Soil under Indo-Gangetic Plain Zones. *Journal of AgriSearch* 1(4): 206-209.

Khare MN, Agrawal SC and Jain AC. 1979. Lentil diseases and their control. Technical Bulletin; J.N.K.V.V., Jabalpur. Madhya Pradesh, India; p. 29.

Khare MN. 1981. Diseaes of lentil. (in) Lentils, pp. 163-72. Webb C and Hawtin G (Eds). Commonwealth Agricultural Bureau International, England, UK.

Pavgi MS and Upadhyay R. 1967. Root rots disease of lentil in Uttar Pradesh. *Sci. & Cult.* 33: 75-76.

Punja ZK. 1985. The biology, ecology and control of Sclerotium rolfsii. *Annual Review of Phytopathology* 23: 97-127.

Sangle UR, Md Idris and Singh SS. 2012. Production of lentil (Lens culinaris Medic.) decline due to wilt complex problem in rice wheat cropping system in Eastern Region of India. Paper presented in Pp213, 3rd global conference Plant Pathology for food security" organized by Indian society of Mycology and Plant Pathology and MPUAT Udaipur.

Sangle UR. 2011. Annual Report ICAR Research complex for Eastern Region Patna, Bihar India. pp. 21.

Sangle UR. 2012. Annual Report ICAR Research complex for Eastern Region Patna, Bihar India pp. 25.

Sharma KB and Agnihotri JP.1972. Pathogenicity and variability of lentil witl Fusarium. *Indian J. of mycology and Plant Pathology* 2: 170

Singh AK, Manibhushan, Chandra Naresh and Bharati RC. 2008. Suitable crop varieties for limited irrigated conditions in different agro climatic zones of India. *Int. J. Trop. Agri.* 26(3-4): 491-496.

Singh AK, Singh D, Singh AK, Gade RM and Sangle UR. 2012. Good Agronomic Practices (GAP) - An efficient and eco-friendly tool for sustainable management of plant diseases under changing climate scenario. *J. Plant Disease Sci.* 7(1): 1-8.

Singh AK, Singh KM, Bharati RC, Chandra N, Bhatt BP and Pedapati A. 2014.Potential of Residual Sulphur and Zinc Nutrition in Improving Powdery Mildew (Erysiphe trifolii) Disease Tolerance of Lentil (*Lens culunaris* L).*Communication in Soil Science and plant analysis* 45: 2807-2818.

Vasudeva, R. S. and Srinivasan, K.V. 1952. Studies on the wilt disease of lentil (*Lens esculenta* M). *Indian Phytopath.* 5: 23-32.

Citation

Sangle UR, Singh AK, Kumar Santosh and Singh Mandhata. 2018. Survey of wilt complex damage on lentil (*Lens culinaris* Med.) in Bihar. In: Scientific Lentil Production. Singh AK, Bhakta N, Sangale UR, Manibhushan, Sundaram PK, Kumar S and Yasin JK. (eds.) Society for Upliftment of Rural Economy Varanasi, India. Pp. 421-430.

❑❑❑

Chapter 25

Potential and Prospect of Plant Growth Promoting *Rhizobacteria* in Lentil

Shachi Shah, V Venkat Ramanan, AjayVeer Singh and Anil Kumar Singh

SYNOPSIS

Lentil is predominantly grown in Asia which accounts for 80 percent of global area and 75% of world production. Being major source of protein in the vegetarian diet system of the Indians, it has been an important component of agriculture since ancient times. Agriculture by default is heavily dependent on the use of chemical fertilizers and pesticides to achieve higher yields. This dependence is associated with problems such as environmental pollution, health hazards, interruption of natural ecological nutrient cycling and destruction of biological communities that otherwise support crop production. Hence, crop production and pest and disease management have to be achieved in shorter intervals of time with fewer detrimental inputs. Plant growth promoting rhizobacteria (PGPR) are a heterogeneous group of bacteria that can be found in the rhizosphere, at root surfaces and in association with roots can improve the extent or quality of plant growth directly and/or indirectly is the panacea. In last few decades a large array of bacteria including species of *Pseudomonas, Azospirillum, Azotobacter, Klebsiella, Enterobacter, Alcaligenes, Arthrobacter, Burkholderia, Bacillus, Rhizobium and Serratia* have reported to enhance plant growth. The direct promotion by PGPR entails either providing the plant with plant growth promoting substances that are synthesized by the bacterium or facilitating the uptake of certain plant nutrients from the environment. The indirect promotion of plant growth occurs when PGPR prevent deleterious effects of one or more phytopathogenic microorganisms. PGPRs are the potential tools for sustainable agriculture and trend for the future. For this reason, there is an urgent need for research to clear definition of what bacterial traits are useful and necessary for different environmental conditions and plants, so that optimal bacterial strains can either be selected and/or improved. Combinations of beneficial bacterial strains that interact synergistically are currently being devised and numerous recent studies show a promising trend in the field of inoculation technology. Applications of this technology have been investigated in lentils, maize, wheat, oat, barley, peas, canola, soy, potatoes, tomatoes, radicchio and cucumber. With the advent of PGPR technology and its use on crops, we can achieve higher productivity, better quality, improved nutrition, improved storage properties, increased pests and disease resistance and achieve higher prices for farmers in the global market place. By exploiting our knowledge of PGPR technology we have the opportunity to make the world better place to live.

Keywords: Plant growth promoting Rhizobacteria, Biological nitrogen fixation, phosphate solubilization, Siderophores, Induced systemic resistance, Inoculation Technology.

Introduction

The world's population is assumed to increase from 7 billion now to 8.3 billion in 2025. The world food requirement is expected to increase from 70 to 100 percent more by 2050. Since the world's population is continuously growing, food production should be increased. This should be done without damaging the environment further and with a decreased input of chemical hormones and fertilizers. This realization has resulted in an increasing interest in the use of microbes as sustainable and inexpensive alternatives for agrochemicals. Agriculture by default is heavily dependent on the use of chemical fertilizers and pesticides to achieve higher yields. This dependence is associated with problems such as environmental pollution, health hazards, interruption of natural ecological nutrient cycling and destruction of biological communities that otherwise support crop production. Hence, crop production and pest and disease management have to be achieved in shorter intervals of time with fewer detrimental inputs. Though the green revolution did increase food production, the productivity levels have remained low and increase was achieved at a cost of intensive use of water, fertilizer and other inputs which have caused problems of soil salinity, ground water pollution, nutrient imbalances, emergence of new pest and diseases and environmental degradation. To feed the ever increasing population globally more and more food has to be produced from less and less land, water and other natural resources. In this context, plant growth promoting rhizo-microorganisms are often novel and potential tools to provide substantial benefits to agriculture. PGPR and their interactions with plants are exploited commercially and hold great promise for sustainable agriculture. Applications of these associations have been investigated in maize, wheat, oat, barley, peas, canola, soy, potatoes, tomatoes, lentils, radicchio and cucumber. With the advent of PGPR technology and its use on crops, we can achieve higher productivity, better quality, improved nutrition, improved storage properties, increased pests and disease resistance and achieve higher prices for farm ers in the global market place. By exploiting our knowledge of PGPR technology we have the opportunity to make the world better place to live.

Lentil (*Lens culinaris*) Medik a name given by the German botanist medikus in 1778 (Cubero, 1981). The primary product of the cultivated lentil is the seed, which is a valuable human food product containing a high amount of protein (22.0-34.5%), carbohydrates (65%) and other minerals and vitamins (Yadav *et al.*, 2007). Lentil is predominantly grown in Asia which accounts for 80 percent of global area and 75% of world production. Being major source of protein in the vegetarian diet system of the Indians, it has been an important component of agriculture since ancient times. In India, lentil is the second most important winter season pulse crop after chick pea with an annual production of 571 thousand tons (National spot exchange, 2008-2009). Its seed is rich in proteins, minerals and vitamins for human nutrition and the straw is a valuable animal feed. However low average yields are largely due to cultivation in neglected marginal lands, inadequate fertilizer inputs as well as diseases. Plant growth promotion using PGPR (Plant Growth Promoting Rhizobacteria) is a desirable alternative to chemical fertilizers and chemical disease control.

1. Potential of Plant Growth Promoting Rhizobacteria

Plant Growth Promoting Rhizobacteria (PGPR) expression for the first time was propounded by Kloepper and Schroth 1978, and it was used especially for Fluorescent pseudomonads for a long time and just for kinds which indirectly and by controlling the plant pathogenic factor, improves the plant's growth. Microorganisms that colonize the rhizosphere can be classified according to their effects on plants and the way they interact with roots, some being pathogens whereas other trigger beneficial effects. Rhizobacteria inhabit plant roots and exert a positive effect ranging from direct influence mechanisms to an indirect effect. So, the bacteria inhabiting the rhizosphere and beneficial to plants are termed PGPR (Kloepper *et al.*, 1980). In the last few years, the number of PGPR that have been identified has seen a great increase, mainly because the role of the rhizosphere as an ecosystem has gained importance in the functioning of the biosphere. Various species of bacteria like *Pseudomonas, Azospirillum, Azotobacter, Klebsiella, Enterobacter, Alcaligenes, Arthrobacter, Burkholderia, Bacillus* and *Serratia* have been reported to enhance the plant growth (Kloepper *et al.*, 1989; Joseph *et al.*, 2007). PGPR promote growth through at least one mechanism; suppression of plant disease (termed Bioprotectants), improved nutrient acquisition (Biofertilizers), or phytohormone production (Biostimulants). Inoculant development has been most successful to deliver biological control agents of plant disease i.e. organisms capable of killing other organisms pathogenic or disease causing to crops. Various bacteria which are predominantly studied and increasingly marketed as the biological control agents includes the genera *Bacillus, Streptomyces, Pseudomonas, Burkholderia* and *Agrobacterium*.

Bashan and Holguin (1998) proposed the division of PGPR into two classes: biocontrol- PGPB (plant growth promoting bacteria) and PGPB. This classification may include beneficial bacteria that are not rhizosphere bacteria but it does not seem to have been widely accepted. According to Vessey (2003), numerous species of soil bacteria which flourish in the rhizosphere of plants, but which may grow in, on, or around plant tissues, and stimulate plant growth by a plethora of mechanisms are collectively known as PGPR. Gray and Smith (2005) have recently shown that the PGPR associations range in the degree of bacterial proximity to the root and intimacy of association. In general, these can be separated into extracellular (ePGPR), existing in the rhizosphere, on the rhizoplane, or in the spaces between cells of the root cortex, and intracellular (iPGPR), which exist inside root cells, generally in specialized nodular structures (Table 25.1).

Table 25.1 Classification of PGPR

PGPR forms	Definition	Mechanism of action
Biofertilizer	A substance that contains live microorganisms which, when applied on the seed, plant surface or soil, colonizes the rhizosphere and promote plant growth through increased supply of primary nutrients	Biological nitrogen fixation) and Utilization of insoluble Phosphorus
Phytostimulator	Microorganism, with the ability to produce phytohormones such as indole acetic acid, gibberellic acid, cytokinins and ethylene	Production of phytohormones
Biopesticide	Microorganisms that promote plant growth by controlling phytopathogenic agents	Production of antibiotics, siderophores, HCN, hydrolytic enzymes, Acquired and Induced systemic resistance

2. Mechanism of Action of Plant Growth Promoting Bacteria

PGPR may induce plant growth promotion by direct or indirect modes of action (Beauchamp, 1993; Kloepper, 1993; Kapulnik, 1996; Lazarovits and Nowak, 1997). Direct mechanisms include the production of stimulatory bacterial volatiles and phytohormones, lowering of the ethylene level in plant, improvement of the plant nutrient status (liberation of phosphates and micronutrients from insoluble sources; non-symbiotic nitrogen fixation) and stimulation of disease-resistance mechanisms (induced systemic resistance). Indirect effects originate for example when PGPR act like biocontrol agents reducing diseases, when they stimulate other beneficial symbioses, or when they protect the plant by degrading xenobiotics in inhibitory contaminated soils (Jacobsen, 1997). Based on their activities Somers *et al.* (2004) classified PGPR as biofertilizers (increasing the availability of nutrients to plant), phytostimulators (plant growth promoting, usually by the production of phytohormones), rhizoremediators (degrading organic pollutants) and biopesticides (controlling diseases, mainly by the production of antibiotics and antifungal metabolites). Bashan and Holguin (1998) proposed the division of PGPR into two classes: biocontrol-PGPB (plant-growth-promoting-bacteria) and PGPB. This classification may include beneficial bacteria that are not rhizosphere bacteria but it does not seem to have been widely accepted. When studying beneficial rhizobacteria, the original definition of PGPR is generally used: it refers to the subset of soil and rhizosphere bacteria colonizing roots in a competitive environment, *e.g.* in non-pasteurized or non-autoclaved field soils (Kloepper, 2003). Furthermore, in most studied cases, a single PGPR will often reveal multiple modes of action including biological control (Kloepper, 2003 and Vessey, 2003).

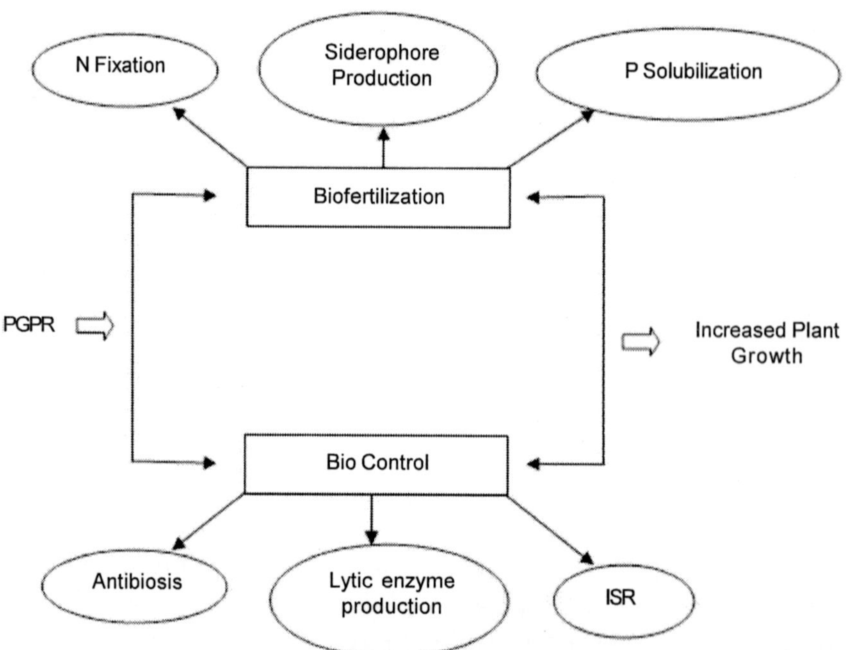

Fig. 25.1 Schematic Illustration of Important Mechanic For Plant Growth Promotion By PGPR

3. Direct Plant Growth Promotions

Direct plant-growth-promoting rhizobacteria enhance plant growth in the absence of pathogens. Depending on the mechanism used, several classes of bacteria that promote plant growth can be distinguished.

Table 25.2 Mechanisms employed by plant growth-promoting bacteria

Mechanisms	Effect on plant growth	Examples of bacterial species
Root-associated nitrogen fixation	Increase nitrogen content and biomass	*Azospirillum, Acetobacter, Azotobacter, Cyanobacteria, Herbaspirillium*
Production of plant hormones (auxin, giberellin, cytokinin)	Stimulate root branching, increase shoot and root biomass, and induce the reproductive cycle	*Azospirillum*
Phosphate solubilization	Increase biomass and P content	*Bacillus lichiniformis, Vibrio*
Inhabitation of plant ethylene synthesis	Increase root length	*Pesudomonas putida*
Production of signal molecules and enhance root permeability	Change in plant metabolism related to mineral absorption	*Azospirillum, Achromobacter*
Increase root permeability	Increase biomass and nutrient uptake	*Azospirillum*
Enhance general mineral uptake	Increase biomass and nutrient uptake	*Azospirillum*
Increase nitrite production	Increase formation of lateral roots	*Azospirillum*
Increase nitrate accumulation	Increase biomass and nitrate content	*Azospirillum*
Reduce heavy-metal toxicity	Protection against nickel toxicity	*Kluyvera*
Increase legume nodules or size	Increase biomass, N content, and reproductive yield	*Azospirillum*
Increase alder root nodules or size	Increase biomass and N content	*Frankia*
Increase frequency of infection by endomycorrhizal fungi	Increase biomass	*Pseudomonas*
Increase number of ectomycorrhizal root tips	Increase biomass	*Pseudomonas*
Increase temporary 'rain root' production in cacti	Improve survival of seeding during drought	*Azospirillum*
Increase resistance to adverse conditions (drought, salinity, compost toxicity)	Improve survival of seedlings and increase biomass	*Azospirillum*
Additive hypothesis	Increase biomass as a result of several small-magnitude mechanisms working in concert or at the same time	*Azospirillum*

3.1 Biofertilizers

The rhizobium-legume symbiosis is considered to be the major source of fixed nitrogen. It has been estimated that this symbiosis contributes more than 45 million metric tons of N per year to the terrestrial ecosystems (Vance, 2001). The current taxonomy of rhizobia includes 12 genera with more than 90 species (Weir, 2011) and it is still expanding. The best known rhizobia are those of the α–subclass of Proteobacteria (*Allorhizobium, Azorhizobium, Rhizobium, Mesorhizobium, Ensifer* (former Sinorhizobium) and Bradyrhizobium).

In addition, several beta-proteobacteria belonging to Burkholderia and Cupriavidus have been shown to nodulate plants (Moulin *et al.*, 2001). Rhizobia and other N-fixing bacteria share essential nod and nif genes encoding nodulation and nitrogen fixation functions, respectively (Zehr and Turnet, 2001). These genes are often carried on symbiotic plasmids which are highly transferable (Brom *et al.*, 2004).

a) **Symbiotic nitrogen fixation**: Symbiotic nitrogen fixation is a mutualistic relationship between a microbe and the plant. The microbe first enters the root and later on form nodules in which N_2 fixation occurs. Rhizobia have the ability to lay symbiotic interactions by the colonization and formation of root nodules with leguminous plants, where nitrogen is fixed to ammonia and make it available for the plant. So the amount of nitrogen fixed is high.

b) **Free-living nitrogen fixing:** Numerous nitrogen fixing microbes does not exhibit symbiotic interaction. These microbes are free living and survive on either plant residues or photosynthesise themselves. So, the amount of nitrogen fixed is comparatively small.

However, the process of symbiotic N2 fixation is limited only to legume crops and various trees and shrubs that form actinorrhizal roots with Frankia. On the other hand, non-symbiotic biological N2 fixation is basically carried out by free living diazotrophics, belonging to the genera like Azoarcus (Reinhold-Hurek *et al.*, 1993), *Azospirillum* (Bashan and de-Bashan 2010), *Burkholderia* (Estrada de los Santos *et al.*, 2001), Gluconacetobacter (Fuentes-Ramirez *et al.*, 2001) and Pseudomonas *et al.* (2006). Besides, combined inoculations of rhizobacterial species to improve the quality of soil are also seemed to be a potent area of research in present day agriculture. For instances, combined inoculations of Bradyrhizobium sp., with Pseudomonas striata have established enhanced nodule occupancy in soya bean resulting in more biological N_2 fixation (Dubey, 1996).

3.2 Phosphate solubilization

Phosphorus (P), the second important plant growth-limiting nutrient after nitrogen, is abundantly available in soils in both organic and inorganic forms (Khan *et al.*, 2009). Despite of large reservoir of P, the amount of available forms to plants is generally low. This low availability of phosphorous to plants is because the majority of soil P is found in insoluble forms, while the plants absorb it only in two soluble forms, the monobasic (H_2PO^{-4}) and the diabasic (HPO_2^{-4}) ions (Bhattacharyya and Jha, 2012). The insoluble P is present as an inorganic mineral such as apatite or as one of several organic forms including inositol phosphate (soil phytate), phosphomonesters, and phosphotriesters (Glick, 2012). To overcome the P deficiency in soils, there are frequent applications of phosphatic fertilizers in agricultural fields. Plants absorb fewer amounts of applied phosphatic fertilizers and the rest is rapidly converted into insoluble complexes in the soil (Mckenzie and Roberts, 1990). But regular application of phosphate fertilizers is not only costly but is also environmentally undesirable. This has led to search for an ecologically safe and economically reasonable option for improving crop production in low P soils. In this context, organisms coupled with phosphate solubilizing activity, often termed as phosphate solubilizing microorganisms (PSM), may provide the available forms of P to the plants and hence a viable substitute to chemical phosphatic fertilizers (Khan *et al.*, 2006). Of the various PSM(s) inhabiting the rhizosphere, phosphate-solubilizing bacteria (PSB) are considered as promising biofertilizers since they can supply plants with P from sources otherwise poorly available by various mechanisms (Zaidi *et al.*, 2009). Bacterial genera like *Azotobacter, Bacillus, Beijerinckia, Burkholderia, Enterobacter, Erwinia, Flavobacterium, Microbacterium, Pseudomonas, Rhizobium and Serratia* are reported as the most significant phosphate solubilizing bacteria (Bhattacharyya and Jha, 2012). Typically, the solubilization of inorganic phosphorus occurs as a consequence of the action of low molecular weight organic acids which are synthesized by various soil bacteria (Zaidi *et al.*, 2009). Conversely, the mineralization of organic phosphorus occurs through the synthesis of a variety of different phosphatases, catalyzing the hydrolysis of phosphoric esters (Glick, 2012). Importantly, phosphate solubilization and mineralization can coexist in the same bacterial strain (Tao *et al.*, 2008). Tomar *et al.* (1993) reported that inoculation with the phosphate-solubilizing bacterium B. firmus resulted in significant increase in seed yield in field trials on lentil (*Lens esculentus*) and black gram (Vigna mungo). Singh *el al*, (2010 & 11) reported phosphate solubilizing bacteria (PSB) from pea and mustard rhizosphere enhanced plant growth and seed yield enhancement of lentil (*Lens culinuris* Medik) under field conditions. They attributed the better performance of these bacterial isolates to its greater P solubilization activity and positive correlation with plant growth and seed yield of lentil.

3.3 Production of phytohormones

Plant growth and development is also regulated by phytohormone producing PGPR's. Phytohormones such as auxins and cytokine production by Plant growth hormones are organic compounds that influence the physiological processes in plants at extremely low concentrations. Production of phytohormones by inocula has been suggested as one of the most plausible mechanisms of action affecting plant growth. There are five classes of well-known phytohormones, namely auxins, IAA, cytokinins, ethylene and abscisic acid. Soil microbiotas, particularly the rhizosphere microflora, are potential sources of these phytohormones (Arshad and Frankenberger, 1998). PGPR's have been reported by many researchers, but evidence regarding production of gibberellins by the plant growth promotroy rhizobacteria is scanty. Yet, it has been reported to be produced by certain rhizospheric bacteria's like *Bacillus licheniformis* and *Bacillus pumilus*. Cytokinin encourages tissue expansion, cell division and cell enlargement in plant. *Pseudomonas fluorescens* isolated from the rhizosphere of soybean is reported to produce cytokinins. Auxin is an important phytohormone, which promotes root cell division, root initiation, and cell enlargement. *Indole-3-acetic acid* (IAA) produced by PGPR's are reported to increase root growth and length, modifying the plant (morphological functions) to uptake more nutrients from the soil. Gibberellins also can alter the plant morphology by the elongation of stem tissues (Das *et al.*, 2013).

4. Indirect Mechanisms

Plant diseases are responsible for annual crop losses on an average are at a total value of more than 200 billion. Resistant plants and chemicals are often used to control plant disease. Resistance does not exist against all diseases and the breeding of resistant plants takes many years. Moreover, acceptance of genetically engineered resistance is still a sensitive issue. The use of microbes to control diseases, which is a form of biological control, is an environment-friendly approach.

4.1 Suppression of plant diseases

The microbe is a natural enemy of the pathogen, and if it produces secondary metabolites, it does so only locally, on or near the plant surface, i.e., the site where it should act. In contrast, the majority of molecules of agrochemicals do not reach the plant at all. Moreover, the molecules of biological origin are biodegradable compared with many agrochemicals that are designed to resist degradation by microbes. The term biocontrol is used not only to control diseases in living plants but also to control diseases occurring during the storage of fruits (also called postharvest control). There are many examples of effective control of soilborne diseases by means of PGPR (Whipps 2001; Lucy *et al.,* 2004), and many bacterial strains have been shown to have potential for development as biocontrol agents on cereals. The biocontrol potential of Bacillus spp. as important agents to combat root and soilborne pathogens has been reported in many crops including chickpea (Landa *et al.,* 1997). Several Bacillus spp. isolated from the rhizosphere of chickpea had shown antagonistic activity against fusarium wilt caused by *Fusarium oxysporum.*

4.2 Modes of action of PGPR as biocontrol agents

For successful and sustainable biocontrol under field conditions, it is imperative that the mode of action of the Plant growth promoting strains being used is known. The mode of action involved will be a determining factor in the type of disease control strategy to be implemented.

4.2.1 Production of Antibiotics

Antibiotic production by biocontrol-PGPB is perhaps the most powerful mechanism against phytopathogens. The primary mechanism of biocontrol byPGPR involves the production of antibiotics such as phenazine-1-carboxyclic acid, 2, 4-diacetyl phloroglucinol, oomycin, pyoluteorin, pyrrolnitrin, kanosamine, zwittermycin-A, and pantocin. A cascade of endogenous signals such as sensor kinases, N-acyl homoserine lactones and sigma factors regulates the synthesis of antibiotics. The genes responsible for the synthesis of antibiotics are highly conserved. The antibiotics pertain to polyketides, heterocyclic nitrogenous compounds and lipopeptides have broad-spectrum action against several plant pathogens, affecting crop plants. In addition to direct ant pathogenic action, they also serve as determinants in triggering induced systemic resistance (ISR) in the plant system. Though antibiotics play a vital role in disease management, their role in biocontrol is questioned due to constraints of antibiotic production under natural environmental conditions (Siddiqui *et al.,* 2005).

4.2.2 Production of Siderophores

Iron, an element essential for microbial growth, is mostly unavailable because it is mainly present in soil in a hard to solubilize mineral form. To sequester iron from the environment, numerous soil microorganisms secrete low-molecular-weight, iron binding molecules, called siderophores, which have a high capacity for binding Fe 3+. The now-soluble, bound iron is transported back to the microbial cell and is available for growth. Siderophores produced by biocontrol-PGPB have a higher affinity for iron than the siderophores produced by fungal pathogens, allowing the former microbes to scavenge most of the available iron, and thereby prevent proliferation of fungal pathogens. Depletion of iron from the rhizosphere does not affect plant growth as plants can thrive on less iron than can microorganisms. Moreover, some plants can bind and release iron from bacterial iron siderophore complexes, and use the iron for growth. Thus, the plant benefits in two ways: from the suppression of pathogens and from enhanced iron nutrition, resulting in increased plant growth. Iron is an essential growth element for all living organisms. The scarcity of bioavailable iron in soil habitats and on plant surfaces foments a furious competition (Whipps, 2001). Under iron-limiting conditions PGPB produce low-molecular-weight compounds called siderophores to competitively acquire ferric ion (Whipps, 2001). Siderophores (Greek: "iron carrier") are small, high-affinity iron chelating compounds secreted by microorganisms such as bacteria, fungi and grasses (Neilands, 1952 and 1995, Miller and Marvin, 2008). Microbes release siderophores to scavenge iron from these mineral phases by formation of soluble Fe3+ complexes that can be taken up by active transport mechanisms. Many siderophores are non-ribosomal peptides, although several are biosynthesised independently. Siderophores are also important for some pathogenic bacteria for their acquisition of iron. Siderophores are amongst the strongest binders to Fe3+ known, with enterobactin being one of the strongest of these (Raymond *et al.*, 2003). Distribution of siderophore producing isolates according to amplified ribosomal DNA restriction analysis (ARDRA) groups, reveals that most of the isolates belong to Gram- negative bacteria corresponding to the *Pseudomonas* and *Enterobacter* genera, and *Bacillus* and *Rhodococcus* genera are the Gram-positive bacteria found to produce siderophores (Tian *et al.*, 2005).

4.2.3 Induction of Systemic Resistance

Induced systemic resistance (ISR) is the state of defensive capacity developed by the plant when stimulated by diverse agents including rhizobacteria. Once resistance is induced in plants, it will result in nonspecific protection against pathogenic fungi, bacteria, and viruses. The mode of action of disease suppression by nonpathogenic rhizosphere bacteria should be distinguished from pathogen induced systemic acquired resistance (SAR). Colonization of the plant root system by rhizobacteria can indirectly lead to reduced pathogen attack through induction of systemic resistance. PGPR elicit ISR in plants by increasing the physical and mechanical strength of the cell wall as well as changing the physiological and biochemical reactions of the host. This results in the synthesis of defense chemicals such as chitinase, peroxidase, and pathogenesis-related proteins.

5. Biocontrol of Lentil Using Plant Growth-Promoting Rhizobacteria

5.1 Biocontrol of the root-knot nematode *Meloidogyne javanica* on lentil using plant growth-promoting rhizobacteria

Several diseases affect lentil causing yield losses. Common fungal diseases of lentil are Fusarium wilt caused by *Fusarium oxysporum* f. sp. Lentis, rust caused by Uromyces fabae, and ascochyta blight caused by *Ascochyta lentis*. Siddiqui (2007) studied Biocontrol of the root-knot nematode Meloidogyne javanica on lentil using plant growth-promoting rhizobacteria (PGPR) namely *Pseudomonas putida, P. alcaligenes, Paenibacillus polymyxa* and *Bacillus pumilus* and root nodule bacterium Rhizobium sp. Pseudomonas putida caused greater inhibitory effect on the hatching and penetration of *M. javanica* followed by *P. alcaligenes, P. polymyxa* and *B. pumilus*. Inoculation of any PGPR species alone or together with Rhizobium increased plant growth both in M. javanica-inoculated and uninoculated plants. Inoculation of Rhizobum caused greater increase in plant growth than caused by any species of plant growth-promoting rhizobacteria in nematode-inoculated plants.

5.2 Biocontrol of the wilt disease caused by *Fusarium oxysporum* f. sp. lentis on lentil using plant growth-promoting rhizobacteria

Akhtar *et al.*, 2010 examined the effects of Bacillus pumilus, Pseudomonas alcaligenes, and Rhizobium sp. on wilt disease caused by *Fusarium oxysporum* f. sp. lentis and on the growth of lentil. Inoculation with *F. oxysporum* caused significant wilting, and reduced plant growth, the number of pods, and nodulation. Inoculation with *B. Pumilus* together with P. alcaligenes caused a greater increase in plant growth, number of pods, nodulation, and root colonization by rhizobacteria, and also reduced Fusarium wilting to a greater degree than did individual inoculation. Use of Rhizobium sp. resulted in a greater increase in plant growth, number of pods, and nodulation and reduced wilting more than *B. pumilus* did. Combined application of *B. pumilus* and *P. alcaligenes* with Rhizobium sp. resulted in the greatest increase in plant growth, number of pods, nodulation, and root colonization by rhizobacteria, and also reduced wilting in Fusarium-inoculated plants. Angaw *et al.*, 2011 studied the carrier-based (soil/FYM, 1:1) plant growth promoting rhizobacteria (PGPR). Isolates (*Bacillus subtilis, Klebsiella planticola* and *Proteus vulgaris*) were tested individually and in combination with *Bradyrhizobium japonicum* and *Rhizobium leguminosarum* biovar viciae under field conditions on lentil crops. Maximum nodulation in the lentil (Lens culinaris) was obtained with a combination of R. leguminosarum and *P. vulgaris* inoculation followed by a single inoculation with *Rhizobium* and *B. subtilis*. More detailed investigations of the relationships in various pathosystems, and interactions between microorganisms and host plants are needed in order to develop biocontrol methods for the related diseases.

6. Prospects of PGPR - Commercialization Inoculants Technology

A major problem for massive use of PGPR has been formulated for its commercial use. These include production in the scale of fermentation microorganisms with management of the quality, stability, and effectiveness of the product. B. subtilis has been assessed as of great potential for use in agriculture and has been used in the formulation of commercial products for agricultural use in several countries (Lazzareti and Bettiol, 1997). Several substances have been used in experimental formulations such as lactose, peptone, gum arabic and xanthan, cellulose, and others (Schisler *et al.*, 2004). This formulation may require a significant value to determine the effectiveness of the final product based on rhizobacteria such as the *B. subtilis*. Development of formulations with a potential PGP to ensure survival and activity in the field and compatibility with chemical treatment of seeds has been the focus of researches with application of PGPR in agriculture. The research among other things optimizes growth conditions before the formulation, development of vehicles, and appropriate technology for application (Date, 2001). In registration and marketing of products with PGPR, a large number of constraints are found (Mathre *et al.*, 1999). Carrier-based (soil/FYM, 1:1) plant growth promoting rhizobacteria (PGPR) isolates (*Bacillus subtilis, Klebsiella planticola and Proteus vulgaris*) were tested individually and in combination with Bradyrhizobium japonicum and Rhizobium leguminosarum biovar viciae under field conditions on soybean and lentil crops, respectively, under field conditions and was significant increase in growth was reported by Angaw Tsigie *et al.* (2011).

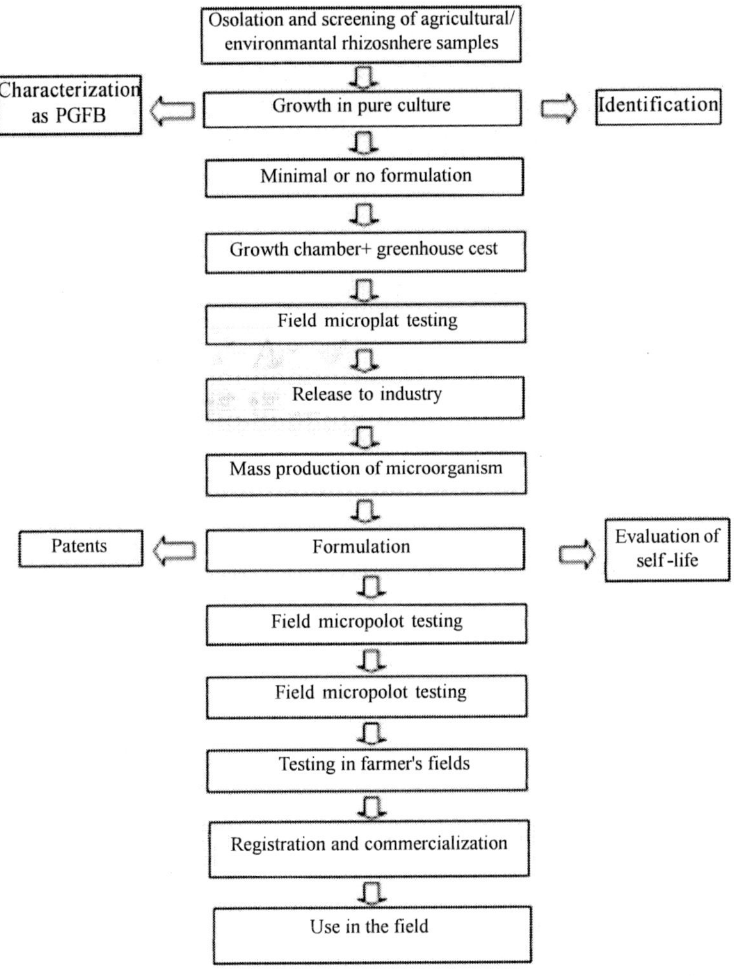

Fig. 25.2 Flow diagram of procedures for developing bacterial inoculants

Conclusion

The doubling time world's current growth is 54 years and we can expect the world's population to become 12 billion by 2054. This demographic growth has to be accompanied by an increase in food production. Thus, the humanity has to face a new challenge, by doing a good use of soils and developing new technologies mainly based in ecofriendly microorganisms that control pest and improve plant growth. Plant growth promoting rhizobacteria (PGPR) are a heterogeneous group of bacteria that can be found in the rhizosphere, at root surfaces and in association with roots, which can improve the extent or quality of plant growth directly and/or indirectly. In last few decades a large array of bacteria including species of Pseudomonas, *Azospirillum, Azotobacter, Klebsiella, Enterobacter, Alcaligenes, Arthrobacter, Burkholderia, Bacillus, Rhizobium and Serratia* have reported to enhance plant growth. The direct promotion by PGPR entails either providing the plant with plant growth promoting substances that are synthesized by the bacterium or facilitating the uptake of certain plant nutrients from the environment. The indirect promotion of plant growth occurs when PGPR prevent deleterious effects of one or more phytopathogenic microorganisms. The exact mechanisms by which PGPR promote plant growth are not fully understood, but are thought to include (i) the ability to produce or change the concentration of plant growth regulators like indoleacetic acid, gibberellic acid, cytokinins and ethylene (ii) asymbiotic N2 fixation (iii) antagonism against phytopathogenic microorganisms by production of siderophores, antibiotics (iv) solubilization of mineral phosphates and other nutrients. Some PGPR may promote plant growth indirectly by affecting symbiotic N2 fixation, nodulation or nodule occupancy. However, role of cyanide production is contradictory as it may be associated with deleterious as well as beneficial rhizobacteria. In addition to these traits, plant growth promoting bacterial strains must be rhizospheric competent, able to survive and colonize in the rhizospheric soil. Unfortunately, the interaction between associative PGPR and plants can be unstable. The good results obtained in vitro cannot always be dependably reproduced under field conditions. The variability in the performance of PGPR may be due to various environmental factors that may affect their growth and exert their effects on plant. Therefore, it is necessary to develop efficient strains in field conditions. One possible approach is to explore soil microbial diversity for PGPR having combination of PGP activities and well adapted to particular soil environment.

There is overwhelming evidence in the literature indicating that PGPR can be a true success story in sustainable agriculture. Presently, the absence of a universal magic PGPR bioinoculant formulation for each important field crop simply reflects the complexity of the interactions and of the molecular signal exchanges taking place in the soil-plant-organisms ecosystems. There are in the literature several examples of important synergism observed on plant growth when the inoculants used contain a mixture of organisms. To develop future beneficial inoculants for field grown crops, one approach should consider performing inoculation assays with a consortium containing a mixture of soil organisms instead of a single strain. A consortium could contain a mixture of PGPR stimulating plant growth at different growth stages, and showing one or more of the known PGPR mechanisms of action. It could also contain beneficial symbiotic organisms like AM fungi, rhizobia and mycorrhizae helper bacteria. Applications of these associations can be investigated in maize, wheat, oat, barley, peas, canola, soy, potatoes, tomatoes, lentils, radicchio and cucumber. With the advent of PGPR technology and its use on crops, we can achieve higher productivity, better quality, improved nutrition, improved storage properties, increased pests and disease resistance and achieve higher prices for farmers in the global market place. PGPRs are the potential tools for sustainable agriculture and trend for the future. For this reason, there is an urgent need for research to clear definition of what bacterial traits are useful and necessary for different environmental conditions and plants, so that optimal bacterial strains can either be selected and/or improved. Combinations of beneficial bacterial strains that interact synergistically are currently being devised and numerous recent studies show a promising trend in the field of inoculation technology. By exploiting our knowledge of PGPR technology we have the opportunity to make the world better place to live.

References

Ahemad M and Khan MS. 2010. Phosphate-solubilizing and plantgrowth-promoting Pseudomonas aeruginosa PS1 improves greengram performance in quizalafop-p-ethyl and clodinafop amended soil. Arch. Environ. *Contam. Toxicol.* 58: 361–372.

Das AJ, Manoj Kumar and Rajesh Kumar. 2013. Plant Growth Promotingm Rhizobacteria (PGPR): An Alternative of Chemical Fertilizer for Sustainable, Environment Friendly. *Research Journal of Agriculture and Forestry Sciences* 1(4): 21-23.

Angaw Tsigie and Kolluru VB.R. Tilak and Saxena AK. 2011. Field response of legumes to inoculation with plant growth-promoting rhizobacteria *Biol Fertil Soils* 47: 971–974.

Anonymous. 2011.Project coordinators Report. All India coordinated Research Project on MULLaRp (Rabi crops, 2010-11). Indian Instirute of Pulses Research Kanpur. Pp. 16.

Arshad, M., and Frankenberger, WT Jr. 1998. Plant growth regulating substances in the rhizosphere: Microbial Production and Functions. *Adv. Agron.* 62: 46-151.

Bashan Yand Holguin G .1998. Proposal for the division of plant growth-promoting rhizobacteria into two classiûcation: biocontrol-PGPB (plant growth-promoting bacteria) and PGPB. *Soil Biol Biochem* 30: 1225–1228.

Bhattacharyya PN, Jha DK. 2012. Plant growth-promoting rhizobacteria (PGPR): emergence in agriculture. *World J. Microbiol. Biotechnol.* 28: 1327–1350.

Brom S, Girard L, Tun-Garrido C, Garcia-de-los Santos A and Bustos P. 2004. Transfer of the symbiotic plasmid of Rhizobium etli CFN42 requires cointegration with p42a, which may be mediated by site-specific recombination. J. Bacteriol. 186: 7538–7548. Date RA (2001) Advances in inoculant technology: a brief review. *Austral J Exp Agric* 41: 321–325.

Cubero. JI. 1981. Origin, taxonomy and domestication In: Lentils (eds C.Webb & G.C. Hawtin). Commonwealth Agricultural Bureau, Slough, England.

Dubey SK. 1996. Combined effect of Bradyrhizobium japonicum and phosphate-solubilizing Pseudomonas striata on nodulation, yield attributes and yield of rainfed soybean (*Glycine max*) under different sources of phosphorus in Vertisols. *Ind J Microbiol.* 33: 61–65.

Estrada de los Santos P, Bustillos-Cristales MR, Caballero-Mellado J .2001. Burkholderia, a genus rich in plant-associated nitrogen fixers with wide environmental and geographic distribution. *Appl Environ Microb* 67: 2790–2798.

Fedoroff N, Battisti D, Beachy R, Cooper P, Fischhoff D, Hodges C, Knauf V, Lobell D, Mazur B, Molden D, Reynolds M, Ronald P, Rosegrant M, Sanchez P, Vonshak Aand Zhu J. 2010. Radically rethinking agriculture for the 21st century. *Science* 327(59): 833-834.

Glick BR. 1995. The enhancement of plant growth by free living bacteria. *Canadian Journal of Microbiology* 41(Suppl 2): 109–114.

Glick BR. 2012. Plant Growth-Promoting Bacteria: Mechanisms and Applications. Hindawi Publishing Corporation, Scientifica.

Godfray H, Beddington J, Crute I, Haddad L, Lawrence D, Muir J, Pretty J, Robinson S, Thomas S and Toulmin C. 2010. Food security: the challenge of feeding 9 billion people. *Science* 327(59): 812-818.

Gray EJ and Smith DL .2005. Intracellular and extracellular PGPR: commonalities and distinctions in the plant–bacterium signaling processes. *Soil Biol Biochem.* 37: 395–412.

Hille D. 2005. Encyclopedia of soils in the environment. Elsevier, Oxford, U.K. Vol. 1, pp. 103-115. 2200 p.

Joseph B, Patra RR and Lawrence R. 2007. Characterization of plant growth promoting Rhizobacteria associated with chickpea (*Cicer arietinum* L). *International Journal of Plant Production* 1 (Suppl 2): 141-152.

Kloepper JW, Leong J, Teintze M and Schroth MN. 1980. Enhanced plant growth by siderophores produced by plant growthpromoting rhizobacteria. *Nature* 286: 885–886.

Kloepper JW, Lifshitz R and Zablotowicz RM. 1989. Free-living bacterial inocula for enhancing crop productivity. *Trends in Biotechnology* 7 (Suppl 2): 39–43.

Landa BB, Hervas A, Bethiol W and Jimenez-Diaz RM .1997. Antagonistic activity of bacteria from the chickpea rhizosphere against *Fusarium oxysporum* f.sp.ciceris. *Phytoparasitica* 25: 305–318.

Lazzareti E and Bettiol W. 1997. Tratamento de sementes de arroz, trigo, feija˜o e soja com um produto formulado a base de ce´lulas e de metabo´ litos de *Bacillus subtilis. Sci Agricola* 54: 89–96.

Lucy M, Reed E and Glick BR. 2004. Applications of free living plant growth-promoting rhizobacteria. *Antonie Leeuwenhoek* 86: 1–25.

Mathre DE, Cook RJ and Callan NW. 1999. From discovery to use. Traversing the world of commercializing biocontrol agents for plant disease control. *Plant Dis.* 83: 972–983.

Miller Marvin J. 2008. Siderophores (microbial iron chelators) and siderophore-drug conjugates (new methods for microbially selective drug delivery). University of Notre Dame. Dame, 4/21/2008. http://www.nd.edu/~mmiller1/page2.htm

Mirza MS, Mehnaz S, Normand P. 2006. Molecular characterization and PCR detection of a nitrogen fixing Pseudomonas strain promoting rice growth. *Biol Fertil Soils* 43: 163–170.

Moulin L, Munive A, Dreyfus B, Boivin-Masson C. 2001. Nodulation of legumes by members of the beta-subclass of Proteobacteria. *Nature* 411: 948–950.

Neilands JB. 1952. A Crystalline Organo-iron pigment from a Rust Fungus (*Ustilago sphaerogena*). *Journal of the American Chemical Society* 74(Suppl 19): 4846–4847.

Neilands JB. 1995. Siderophores: Structure and Function of Microbial Iron Transport Compounds. *The Journal of Biological Chemistry* 270(Suppl 45): 26723–26726.

Raymond KN, Dertz EA and Kim SS. 2003. Enterobactin: An archetype for microbial iron transport. *Proceedings of the National Academy of Sciences* 100(Suppl 7): 3584–3588.

Okon Y and Labandera-Gonzalez CA. 1994. Agronomic applications of Azospirillum. In Improving Plant Productivity with *Rhizosphere Bacteria*. Commonwealth Scientific and Industrial Research Organization, Adelaide, Australia, Edited by Ryder MH, Stephens PM, Bowen GD 274–278.

Pretty J. 2008. Agricultural sustainability: concepts, principles and evidence. *Philosophical Transations of The Royal Society Biological Science* 363: 447-465.

Reinhold-Hurek B, Hurek T and Gillis M .1993. Azoarcus gen. nov., nitrogen-fixing Proteobacteria associated with roots of Kallar grass (*Leptochloa fusca* (L.) Kunth), and description of two species, Azoarcus indigens sp. nov. and Azoarcus communis sp. nov. *Int J Syst Bacteriol* 43: 574–584.

Saharan BS and Nehra V. 2011. Plant Growth Promoting Rhizobacteria: *A Critical Review Life Sciences and Medicine Research* 1: 18.

Schisler DA, Slininger PJ, Behle RW and Jackson MA. 2004. Formulation of *Bacillus* spp. For biological control of plant diseases. *Phytopathology* 94: 1267–1271.

Singh AV, Prasad B and Shah S. 2010. Screening plant growth promotory rhizobacteria for improving seed germination and seedling vigor of Lentil (*lens culinaris* Medik). *Env. Ecol.* 28: 2055-58.

Singh AV, Prasad B and Shah S. 2010.Influence of phosphate solubilizing bacteria for enhancement of plant growth and seed yield in lentil. *Journal of Crop and Weed* 7(1): 1-4.

Tian F, Ding Y, Zhu H, Yao L, Du B. 2009. Genetic diversity of siderophore-producing bacteria of tobacco rhizosphere. *Brazilian Journal of Microbiology* 40(Suppl 2): 276-284.

Tomar SS, Pathan MA, Gupta KP and Khandkar UR .1993. Effect of phosphate solubilizing bacteria at different levels of phosphate on black gram (*Phaseolus mungo*). *Ind J Agron.* 38: 131–133.

Vance CP. 2001. Symbiotic nitrogen fixation and phosphorus acquisition: plant nutrition in a world of declining renewable resources. *Plant Physiol.* 127: 390–397.

Vessey JK. 2003. Plant growth-promoting rhizobacteria as biofertilizers. *Plant Soil.* 255: 571–586.

WeirBS. 2011. The current taxonomy of rhizobia. New Zealand rhizobia website. http://www.rhizobia.co.nz/taxonomy/rhizobia.html.

Whipps JM. 2007. Complex multitrophic interactions in the plant environment can affect disease biocontrol. In: Proceedings of the XIV international plant protection congress. 2: 432–433.

Whipps JM. 2001. Microbial interactions and biocontrol in the rhizosphere. *Journal of Experimental Botany* 52(Suppl 1): 487-511.

Yadav SS, Stevenson PC, Rizvi AH, Manohar M, Gailing S and Mateljan G. 2007. Uses and consumption In: Lentil. An ancient crop for modern times (eds S.S. Yadav, DL McNeil and PC Stevenson). Spring, Dordrecht, The Netherlands.

Siddiqui ZA (ed.) 2005. PGPR: Biocontrol and Biofertilization, 67-109. Springer, Dordrecht. The Netherlands.

Zahran HH. 2001. Rhizobia from wild legumes: diversity, taxonomy, ecology, nitrogen fixation and biotechnology. *J. Biotechnol.* 91: 143–153.

Zehr JP and Turner PJ. 2001. Nitrogen Fixation: Nitrogenase genes and gene expression. In: Methods in Microbiology, Paul JH, ed. (New York: Academic Press), pp. 271–285.

Citation

Shah Shachi, Ramanan VV, Singh AV and Singh AK. 2018. *Trichoderma*: Potential and Prospect of Plant Growth Promoting *Rhizobacteria* in Lentil. In: Scientific Lentil Production. Singh AK, Bhakta N, Sangale UR, Manibhushan, Sundaram PK, Kumar S and Yasin JK. (eds.) Society for Upliftment of Rural Economy Varanasi, India. Pp. 431-451.

□□□

Chapter 26

Morpho-Physiological Responses of Lentil (*Lens culinaris* Medikus *subsp. culinaris*) to Abiotic Stresses

Sharad Kumar Dwivedi, Anil Kumar Singh and Nupur Malviya

SYNOPSIS

Lentil (*Lens culinaris* Medik), a pulse of global economic importance, has been long domesticated. Among the pulses, lentil is of special interest with 23.7% content of grain protein. In addition to protein, its seed is a rich source of minerals and vitamins as human food, while the straw serves as high-value animal feed. Not only that, its cultivation enriches soil nutrient status by adding nitrogen, carbon and organic matter, which promotes sustainable crop production system. However, lentil yield potential far below than the other cereal crops. The adaptability and productivity of cool season food legumes such as lentil (*Lens culinaris* Medik) is further limited by major abiotic stresses, including drought, heat, frost, chilling, water logging, salinity, and mineral toxicities. Among these salinity and drought stress is major constraint. Moreover, leaf temperature was increased, but leaves per plant and leaf chlorophyll content index (CCI) were decreased by increasing salinity. Proline content of leaves under severe salinity was significantly higher. Reduction in grain yield was related to less leaves per plant, low CCI and leaf water content and high leaf temperature under saline conditions. Foliar Relative Water Content (RWC) (%), K^+/Na^+ ratio, chlorophyll (chl) a, chl a/b ratio, stomatal conductance and net photosynthetic rate declined considerably in both crops under water stress. Leaf free proline level increased significantly and decreased marked in nodules of lentil. Plant growth traits and seed yield components reduced significantly. Water deficit affects many morphological features and physiological processes associated with plant growth and development. In water limitation conditions, plants close their stomata to avoid further water loss. Decrease in internal CO_2 concentration and inhibition of ATP synthesis lead to a decrease of net photosynthetic rate under imitated water condition. Responses of rhizobia to water deficiency concerns the morphological changes. In addition water deficit also results in restriction of nodule development and function.

Keywords: Abiotic stresses, drought, salinity, lentil, Morphology, physiology, yield attributes

Introduction

Lentil (*Lens culinaris* Medik) is an important pulse crop and has global economic importance. Among the pulses grown in India, lentil (*Lens culinaris* Medik) is of special interest with 23.7% content of grain protein. In addition to protein, its seed is a rich source of minerals and vitamins as human food, while the straw serves as high-value animal feed (Rasheed *et al.*, 2010). Not only that, its cultivation enriches soil nutrient status by adding nitrogen, carbon and organic matter, which promotes sustainable crop production system (Sarker *et al.*, 2004). However, lentil yield potential far below than the other cereal crops. In Asia, the yield of lentil remains low and average seed yield on a country basis is below 1.0 t/ha (SAIC, 2009). Further, the area under lentil cultivation in Asia has been decreasing at a faster rate because of increasing demand for staple grains like rice and wheat (Rahman and Ali, 2011). Lentil has been identified as a narrow adapted crop and the principal constraint of lentil production is its low yield potential because of undesirable plant type (Hanlan *et al.*, 2006). Lentil is one of the predominant winter pulse crops in India, which is extensively cultivated (1.4 m ha) in northern, east central and eastern parts of the country, being the largest lentil growing country in the world (FAO, 2010) The contribution of India to global lentil area and production is 39.52% and 42.42%, respectively. However, the productivity of lentil in India is very low (619 kg/ha) in comparison with the world average of 887 kg/ha because it is normally grown under rain fed conditions by resource poor farmers with dearth of potential highyielding cultivars (FAO, 2010). The adaptability and productivity of cool season food legumes such as lentil (*Lens culinaris* Medik) are limited by major abiotic stresses, including drought, heat, frost, chilling, water logging, salinity, and mineral toxicities. Agricultural productivity in arid and semi-arid regions of the world is very low due to accumulation of salts in soils (Ashraf and Sarwar, 2002; Munns, 2002). Total area under salinity is about 953 million ha covering about 8 percent of the land surface in the world (Singh, 2009). Several physiological pathways *i.e.* photosynthesis, respiration, nitrogen fixation and carbohydrate metabolism have been observed to be affected by high salinity (Chen *et al.*, 2008). Water is one of the most important environmental factors regulating plant growth and development. Apart from salinity the lentil crop plants is also sensitive to water stress, acknowledged a major constrain in crop production. Water deficit affects many morphological features and physiological processes associated with plant growth and development (Toker and Cagirgan, 1998).

The severity of these stresses is unpredictable. The identification of adaptive mechanisms to drought and salinity is of considerable importance, especially for legumes, as they play significant ecological and economic roles. Lentil (*L. culinaris* Medik) is considered valuable plant resources for diverse types of biological studies, including cytogenetic, mutation genetic, and genomic research (Balyan *et al.*, 2002; Talukdar, 2009). **Response of lentil crops under abiotic stress condition:** Lentil, a leguminous crop species is grown in arid and semi arid regions where yield loss due to water stress and salinity is very high. Plant response variably under this stress and try to cope up with the stress. Various morphological, physiological and biochemical traits were influenced under abiotic stress condition. Effect on molecular level was also observed under this abiotic stress condition. **Morpho-physiological responses of lentil crop under drought stress:** Plant responses to water limitation are very complex and include adaptive changes or deleterious effects (Chaves *et al.*, 2002). The effects of water limitation are observed in the form of phonological responses, morphological adaptations, physiological changes and biochemical adaptations. Plant reactions are affected by the amount of soil water directly or indirectly. All physiological processes like photosynthesis, transpiration, cell turgidity, and cell and tissue growth in plants are directly affected by water availability (Sarker *et al.*, 2005). Foliar RWC (%), K^+/Na^+ ratio, chl a, chl a/b ratio, stomatal conductance and net photosynthetic rate decreased significantly in lentil seedlings under water stress condition. The measurement of RWC under low soil moisture is of importance as high RWC appears to be a common trait in drought resistant species, as species that exhibit restricted changes in RWC per unit reduction in water potential are often considered to be relatively drought resistant (Saraswathi and Paliwal, 2011). Under water stress condition there is lowering of K^+ content and concomitant rise in Na^+ level. Similarly, reduction in chlorophyll "a" led to a decrease in the chl a/chl b ratio. The lowering of pigment composition along with impairment in stomatal conductance might be responsible for the reduction of the net photosynthetic rate in lentil cultivar. Stomatal conductance has long been considered an important selection criterion for drought resistance. Higher stomatal conductance in plants increases the CO_2 diffusion into the leaf thereby, favoring higher photosynthetic rates. Higher net CO_2 assimilation rates can in turn favor a higher biomass and higher crop yields (Saraswathi and Paliwal, 2011). Direct evaporation from soil is often a major loss of available water because it is not contributing to biomass production. In water limitation conditions, plants close their stomata to avoid further water loss. Decreasing internal CO_2 concentration and inhibition of ATP synthesis lead to a decrease of net photosynthetic rate under water limitation (Dulai *et al.*, 2006).

Fig. 26.1 Quality seed is foundation of good reap Lentil

The effect of water limitation on CO_2 assimilation rate, transpiration rate and water use efficiency has been investigated in many crops such as *Zea mays* (Ashraf *et al.*, 2007), *Brassica napus* L. (Kauser *et al.*, 2006) and mung bean genotypes (Ahmed *et al.*, 2002). Moreover, many studies have shown that water deficit leads to an increase in stomatal density (Zhang *et al.*, 2006), and a decrease in stomatal size (Spence *et al.*, 1986), indicating this may enhance the adaptation of plant to drought (Martinez *et al.*, 2007). Water deficiency is a major limiting factor of plant productivity and symbiotic nitrogen fixation in many arid regions. One of the immediate responses of rhizobia to water deficiency concerns the morphological changes (Shoushtari and Pepper, 1985; Busse and Bottomley, 1989). In addition water deficit also results in restriction of nodule development and function (Serraj *et al.*, 1999). Photosynthetic pigments are generally regarded as indicators of environmental stresses, whose accumulation altered when plants are in stressed conditions. Many researchers have reported the decreased chlorophyll contents under water stress (Suriyan *et al.*, 2009). Excessive accumulation of Na+ and Cl″ in the leaves has been considered highly harmful for normal metabolism of plants, and tolerant genotypes have the capacity of successful salt exclusion (Khurana *et al.*, 2008). The K^+/Na^+ ratio has been used as a discriminating factor between tolerant and sensitive genotypes, with a greater capacity of the former to block or reduce the uptake or exclude the excess amount of Na^+ and associated increase in K^+ content (Khurana *et al.*, 2008). Disruption in pigment composition may be due to severe impairment in pigment synthesis, structure and function, which is mainly orchestrated through stress induced membrane damage, as reported in different crop plants (Talukdar, 2011). However, this phenomenon is not universal. The nitrogen fixing ability of root nodules in leguminous plants is generally influenced by the availability of water to the host plant. If drought stress persists for long, this activity is reduced, and is often lost completely (Ramos *et al.*, 1999). Number of root nodules and DW of nodule decreased in crops with increasing intensity of water stress. No nodule was formed in lentil seedlings under 20 days stress period, Among the parameters responding to stress treatment, rapid accumulation of free proline content is one of the significant events in plants. The accumulation of cytosolutes, particularly proline, should avoid the detrimental effects of a low osmotic potential of the cells without interfering with protein synthesis (Anoop and Gupta, 2003). Proline content took reverse directions in the leaves and nodules of lentil; it increased significantly in the leaf but decreased markedly in nodules at 13 d and 20 d stress durations as compared with the control. However, the exact role of proline in abiotic stress tolerance is being debated. Presumably, increased amounts of free proline could be associated with more effective dehydration and drought avoidance mechanisms, as explained earlier (Ashraf and Iram, 2005). A positive correlation between proline over-accumulation and increasing salinity/drought tolerance has also been found in different crop plants, including transgenics that were engineered for overproduction of proline (Anoop and Gupta, 2003). In fact, proline synthesis has been associated with protein hydrolysis induced by water deûcit (Anoop and Gupta, 2003). A marked increase in proline content in the roots and nodules of *S. aculeate* was reported as an indicator of its high drought tolerance. Contrasting response of proline level to uniform exposure to water stress in lentil and, furthermore, in leaves and nodules indicated differential responses of plant parts to water-deficit conditions through osmotic adjustment mediated by proline accumulation. The cumulative effect of reduction in growth stage and physiological parameters under water stress was ultimately manifested in total dry mass accumulation and DTE of plants. Total DW (root + shoot) was reduced 1.5 fold in lentil under 13 d stresses and 1.9 fold under 20 d growth period. The retardation of growth usually comes as a cumulative effect of low water relation, stomatal nature, low photosynthesis and mineral nutrition, and other morpho-physiological features related to root and shoot growth as reviewed in different legumes, including groundnut (Reddy *et al.*, 2003) and mung bean (Kumar *et al.*, 2008). All these ultimately manifested in retardation of grain yield under water stress in the present crops with more severe effect on lentil.

The stomatal density of lentil was significantly affected by water limitation. An increase in stomatal density was observed under moderate drought, but a decrease occurred with drought severity. Several reports have shown that the stomatal density increase with water stress (Yang and Wang, 2001; Zhang *et al.*, 2006). Chlorophyll content of lentil leaves diminished with increasing of water limitation. Reduction in chlorophyll content under water stress could be related to increasing damage to chloroplasts by active oxygen species, pigment photo-oxidation and chlorophyll degradation (Terzi *et al.*, 2010) and (Makbule *et al.*, 2011). Water deficiency induced a significant decrease in nodulation. Water deficit also results in restriction of nodule development and function (Serraj *et al.*, 1999; Zahran, 1999). Reductions in leaf chlorophyll content index can strongly limit light absorption, photosynthesis and consequently grain yield. Chlorophyll reduction under severe water deficit may be partly resulted from low nutrient uptake. According to Rimski-Korsakov *et al.* (2009), water stress can affect nutrient uptake. Since nitrogen is an essential part of the chlorophyll molecule, any lack of it in the mineral nutrient supply of the plant can suppress chlorophyll formation. Water limitation produces a higher amount of stomata per square mm than plants which received normal amounts of water. These phenomena allowed the plant an increased precision and control over water loss and productivity, displaying the phenotypic plasticity of lentil.

Fig. 26.2 Lentil crop at grand growth phase

Morpho-physiological responses of Lentil crop response under salinity stress

High NaCl concentrations in the growth medium of plants generate primary and secondary effects that negatively affect plant growth and development. Primary effects are ionic toxicity and osmotic stress. Ionic toxicity occurs because high concentrations of Na+ and Cl-in the cytoplasm of cells disturb several biochemical and physiological processes, and osmotic stress is induced by the lowering of the water potential causing turgor reduction and cellular water loss. Secondary effects of NaCl stress include inhibitionof K^+ uptake, membrane dysfunction and generation of reactive oxygen species in the cells (Upadhyay and Panda, 2005). Salinity changes photosynthetic parameters, including osmotic and leaf water potential, transpiration rate, leaf temperature, and relative leaf water content (RWC). Salt also affects photosynthetic components such as enzymes, chlorophylls, and carotenoids. Changes in these parameters depend on the severity and duration of stress (Misraet *et al.*, 1997) and on plant species (Dubey, 1994). Chlorosis is a common response of salinity stress (Jamilet *et al.*, 2007) that can also limit the photosynthetic efficiency. Salinity can affect chlorophyll content through inhibition of chlorophyll synthesis or an acceleration of its degradation (Reddy and Vora, 1986).Reduction in chlorophyll concentrations is probably due to the inhibitory effect of the accumulated ions of various salts on the biosynthesis of the different chlorophyll fractions (Ali *et al.*, 2004). Reducing evaporation can help conserve soil moisture, save irrigation water, and reduce salt accumulation in surface layer of the soil (Yamanaka, 2004).

Performance of Yield and Yield attributes of Lentil under abiotic stress condition

Several reports have been made about the contribution of various yield components towards yield (Samad *et al.*, 2007; Mondal *et al.*, 2012). The yield components depend on some physiological traits. Positive correlations between pod number, seeds pod^{-1} and seed size with yield was noticed by Younis *et al.*, 2008. These attributes in turn depend on morpho-physiological characters (Dutta and Mondal, 1998). However, the information on the genotypic plasticity of morpho-physiological characters and its relationships with the major yield components in lentil are scarce. Important physiological attributes such as LAI, CGR, NAR and photo-assimilate production capacity and its efficient partitioning to economic yield etc. can address various constraints of a variety for increasing its productivity. A plant with optimum LAI and NAR may produce higher biological yield (Mondal *et al.*, 2012). The dry matter accumulation may be the highest if the LAI attains its maximum value within the shortest possible time (Mondal *et al.*, 2011). It is suggested that high partitioning efficiency (harvest index) would be advantageous for high yield. For lentil, all genotypes are indeterminate growth habit and branched (Erskine and Goodrich, 1991). Lentil produces disproportionately large vegetative growth and support lesser grains. The ratio of grains to vegetative matter in lentil is roughly 1:3 (Whitehead *et al.*, 2000). It has been suggested that by increasing photosynthetic efficiency, productivity could be increased (Shrestha *et al.*, 2006). Branch production is more important than plant height in achieving higher seed yield in lentil. Seed yield was positively and significantly correlated with branch number. Short plant stature with less number of branches is desirable to achieve higher seed yield in lentil. Significant genotypic influence on nitrate reductase (NR) activity and photosynthesis (Pn) in leaves and protein content in grains. Result further revealed that grain protein content had no relation with seed yield in lentil. This result is in agreement with Erskine *et al.* (1985) who reported that seed yield was negatively correlated with grain protein and positively correlated with seed size. Sarker *et al.* (2004) reported that seed size and protein content in seed was negatively correlated. Gill and Singh, (2012) observed a positive relation between seed yield and protein content. However, most of scientists reported that seed yield was negatively correlated with seed protein content in lentil. In addition to superior characters for yield components, a high yielding lentil genotype should possess a relatively higher growth rate at early growth stages and having capacity to better dry matter partitioning to economic yield.

Effort for improvement of lentil yield under abiotic stress condition

The lentil (*Lens culinaris* Medick.) is a lens-shaped grain legume well known as a nutritious food. It grows as an annual bushy leguminous plant typically 20-45 cm tall. Lentil seed is a rich source of protein, minerals (K, P, Feand Zn) and vitamins (Bhatty, 1988). A lot of research has been done to increase the present yield of grain legumes including lentil (*Lens esculenta* Moench). But so far, no breakthrough has occurred in the yield ceiling of these crops. Inspite of the best efforts for improving the lentil varieties, yield of this crop remain low. Several studies have been made to understand their performance which mainly includes the contribution of various yield components towards yield. The studies indicate the presence of However, the information on the genotypic plasticity of morpho-physiological characters and its relationships with the major yield components in lentil are scarce. On the other hand, component characters for yield are interdependent to each other while one character may express at the expense of other (Kakde *et al.*, 2005). The importance of correlation in any breeding programme is well documented for various crop species as if provides a basis for effective selection. Correlation index acts as a guide to the reliability of phenotypic and genotype values and determines success in crop improvement. International endeavours to improve lentil have been promising in some cases (Erskine, 1998), but it is not the case everywhere (Campbell CG, 1997). In recent times, number of desirable mutations, including growth habit, flower and seed coat color, stipule and leaf characteristics, etc., have been isolated through induced mutagenesis (Talukdar, 2009). Both the crops, however, experience different types of abiotic stresses, although lentil has remarkable capacity to withstand extreme environmental conditions (Campbell, 1997). In lentil, cadmium induced oxidative stress and its amelioration by calcium has been studied, (Talukdar, 2012). Gene action for some biometrical traits in *L. culinaris* medik was carried out under mid hill conditions of the northwestern Himalayas (Kumar *et al.*, 2011). Mulching is an efficient way to reduce evaporation, improve WUE (Hartkamp *et al.*, 2004) and maintain soil under stable temperature (Kar and Kumar, 2007). Plastic or straw mulch is an efficient practice, which can alter water distribution between soil evaporation and plant transpiration (Raeini-Sarjaz and Barthakur, 1997). Zhang *et al.* (2005) observed that mulching with straw reduced soil evaporation loss and increased water use efficiency of winter wheat in northern China. Number of agronomic and physiological strategies has been implicated to ameliorate the detrimental effects of drought on crop plants, but the possible role of straw mulch in inducing drought tolerance in plants by improving photosynthetic pigments, increasing nodulation and stomatal activities has been rarely studied. The purposes of this study were to determine the effect of water limitation treatment and wheat straw mulch on physiological properties of lentil and develop a relationship between physiological traits and yield of this legume. Among the pulses, lentil breeding is one of the important approaches to enhance yield and nutritional quality (Dixit *et al.*, 2011) and several molecular markers have been developed in this crop (Fikiru *et al.*, 2011).

Except for the response to sowing date, little is known about the morpho-physiological and biochemical characteristics to explain the causes of low yield of lentil. A lot of research has been done to increase the present yield of grain legumes including lentil (*Lens esculenta* Moench). But so far, no breakthrough has occurred in the yield ceiling of these crops. Inspite of the best efforts for improving the lentil varieties, yield of this crop remain low. Several studies have been made to understand their performance which mainly includes the contribution of various yield components towards yield. The studies indicate the presence of positive correlations between pod number, seeds/pod, seed size with yield (Yadav *et al.*, 2003; Younis *et al.*, 2008). These attributes in turn depend on morpho-physiological characters (Dutta and Mondal, 1998). However, the information on the genotypic plasticity of morpho-physiological characters and its relationships with the major yield components in lentil are scarce.

Conclusion

Lentil being legume is one of the rich source of quality protein (22-24%), but its yield potential far below than the other cereal crops. Lentil is very much responsive to abiotic stress and adopted to come up with the stress. Various morphological, physiological and biochemical traits were influenced under abiotic stress condition. Important physiological attributes such as LAI, CGR, NAR and photo-assimilate production capacity and its efficient partitioning to economic yield etc. can address various constraints of a variety for increasing its productivity. Chlorophyll content affect drastically under drought stress condition. The nitrogen-fixing ability of root nodules in lentil plants is influenced by the availability of water to the host plant. If drought stress persists for long, this activity is reduced, and is often lost completely A lot of research has been done to increase the present yield of grain legumes including lentil (*Lens esculenta*). But so far, no breakthrough has occurred in the yield ceiling of these crops. In spite of the best efforts for improving the lentil varieties, yield of this crop remain low. Mulching is an efficient way to reduce evaporation, improve WUE. Understanding lentil performance under abiotic stress in relation to yield and yield attributes is very important, which mainly pod number, seeds/pod and seed size and grain yield. These attributes in turn depend on morpho-physiological characters. However, the information on the genotypic plasticity of morpho-physiological characters and its relationships with the major yield components in lentil are still scarce.

References

Ahmed S, Nawata E, Hosokawa M, Domae Y, and Sakuratani T. 2002. Alterations in photosynthesis and some antioxidant enzymatic activities of mung bean subjected to water logging. *Plant Sci.* 163: 117-123.

Ali Y, Aslam Z, Ashraf MY and Tahir G R. 2004. Effect of salinity on chlorophyll concentration, leaf area, yield and yield components of rice genotypes grown under saline environment. *International Journal of Enginreeing Sicence and Technology* 3: 221-225.

Anoop N and Gupta AK. 2003. Transgenic indica rice cv IR-50 overexpressing *Vigna aconitifolia* delta (1)-pyrroline-5-carboxylate synthetase cDNA shows tolerance to high salt. *J Plant Biochem Biotechnol.* 12: 109-16.

Ashraf M and Iram A. 2005. Drought stress induced changes in some organic substances in nodules and other plant parts of two potential legumes differing in salt tolerance. *Flora.* 200: 535-46.

Ashraf M, Nawazish SH and Athar H. 2007. Chlorophyll fluorescence and photosynthetic capacity potential physiological determinants of drought tolerance in maize (*Zea mays* L.). *Pakistan J Bot.* 39: 1123-1131.

Ashraf MY and Sarwar G. 2002. Salt Tolerance Potential in members of brassicaceae. Physiological studies on water relations and mineral contents. In: Ahmad R and Malik KA (eds): Prospects for Saline Agriculture. Kluwer Academic Publishers Netherlands. pp. 237-245.

Balyan HS, Houben A and Ahne R. 2002. Karyotype analysis and physical mapping of 18S-5.8S-25S and 5S ribosomal RNA loci in species of genus *Lens* Miller (Fabaceae). *Caryologia* 55: 121-8.

Bhatty RS. 1988. Compositions and quality of lentil (*Lens culinaris* Medick.). Rev Canadian Institute. *Food Sci. Tech.* 21: 144-160.

Busse MD and Bottomley PJ. 1989. Growth and nodulation responses of *Rhizobium meliloti* to water stress induced by permeating and non-permeating solutes. *Appl. Environ. Microbiol.* 55: 2431-2436.

Campbell CG. 1997. Grass pea. *Lathyrus sativus* L.: Promoting the Conservation and Use of Underutilized and Neglected Crops. Rome: IPGRI Publication;. p. 18.

Chaves MM, Pereira JS, Maroco J, Rodrigues ML, Ricardo CPP, Osorio ML, Carvalho I, Faria T and Pineiro C. 2002. How plants cope with water stress in the field? Photosynthesis and growth. *Annual Bot.* 89: 907-916.

Chen HJ, Chen JY and Wang SJ. 2008. Molecular regulation of starch accumulation in rice seedling leaves in response to salt stress. *Acta Physiologiae Plantarum*. 30: 135-142.

Dixit GP, Katiyar PK and Singh BB. 2011. Characterization of lentil (*Lens culinaris* Medik.) varieties based on morphological traits. *J Food Legumes*. 24: 194-7.

Dubey RS. 1994. Protein synthesis by plants under stressful conditions, in Handbook of Plant and Crop Stress, M. Pessarakli, Editor., Marcel Dekker New York. pp. 277-299.

Dulai S, Molnar I, Pronay J, Csernak A, Tarnai R and Molnarlang M. 2006. Effects of drought on photosynthetic parameters and heat stability of PSII in wheat and in *Aegilops* species originating from dry habitats. *Acta Biologica Szegediensis* 50: 11-17.

Dutta RK and Mondal MMA. 1998. Evaluation of lentil genotypes in relation to growth characteristics, assimilate distribution and potentioal. *Lens Newsl* 25: 51-55.

Erskine W and Goodrich WJ. 1991. Variability in lentil growth habit. *Crop Sci.* 31: 1040-1044.

FAO. 2010. FAOSTAT statistical database of the United Nations Food and Agriculture Organization (FAO). Italy, Rome.

Fikiru E, Tesfaye and Bekele E. 2011. Morphological and molecular variation in ethiopian lentil (*Lens culinaris* Medikus) varieties. *Int J Genet Mol Biol*. 3: 60-7.

Gill RK and Singh M. 2012. Studies on inter relationship between protein content, grain yield and its component traits in lentil under different growth environments. *EM Int*. 18: 77-8.

Hanlan TG, Ball RA and Vandenberg A. 2006. Canopy growth and biomass partitioning to yield in short-season lentil. *Can J Plant Sci*. 86: 109-119

Hartkamp AD, White JW, Rossing WAH, Van Ittersum MK, Bakker EJ and Rabbinge R. 2004. Regional application of a cropping systems simulation model: crop residue retention in maize production systems of Jalisco, Mexico. *Agric Sys*. 82: 117-138.

Jamil M, Rehman S, Lee KJ, Kim JM, Kim HS and Rha ES. 2007. Salinity reduced growth PS2 photochemistry and chlorophyll content in radish. *Sci Agric.* 64: 111-118.

Kakde SS, Sharma RN, Khilkre AS and Lambade BM. 2005. Correlation and path analysis studies in lentil. *J. Soil Crops.* 15: 67-71.

Kauser R, Athar HUR and Ashraf M. 2006. Chlorophyll fluorescence: a potential indicator for assessment of water stress tolerance in canola (*Brassica napus* L.). *Pakistan J Bot*. 38(5): 1501-1509.

Khurana P, Vishnudasan D and Chhibbar AK. 2008. Genetic approaches towards overcoming water deficit in plants special emphasis on LEAs. *Physiol Mol Biol Plants*. 14: 277-98.

Kumar N, Singh S, Nandwal AS, Waldia RS and Sharma MK. 2008. Genotypic differences in water status, membrane integrity, ionic content, N2 fixing efficiency and dry matter of mungbean nodules under saline irrigation. *Physiol Mol Biol Plants*. 14: 363-8.

Kumar N, Sood BC, Sharma TR, Chahota RK and Sood S. 2011. Determination of gene action for some biometrical traits in *Lens culinaris* Medik. under mid hill conditions of northwestern Himalayas. *J Genet*. 90: 493-7.

Makbule S, Saruhan-Guler N, Durmus N and Guven S. 2011. Changes in anatomical and physiological parameters of soybean under drought stress. *Turkish J Bot*. 35: 369-377.

Misra AN, Sahu SM and Misra M.1997. Sodium chloride induced changes in leaf growth and pigment and protein contents in two rice cultivars. *Biol Plant*. 47: 257-262.

Mondal MMA, Fakir MSA, Islam MN and Samad MA. 2011. Physiology of seed yield in mungbean: growth and dry matter production. *Bangladesh J Bot.* 40: 133-138.

Mondal MMA, Puteh AB, Malek MA, Ismail MR, Rafii MY and Latif MA. 2012. Seed yield of mungbean (*Vigna radiata* (L.)Wilczek) in relation to growth and developmental aspects. *Sci World J.* 2012, Article ID 425168, 7 pages, doi:10.1100/2012/425168.

Munns R. 2002. Comparative physiology of salt and water stress. *Plant Cell Environ.* 25: 239-250.

Raeini-Sarjaz M and Barthakur NN 1997. Water use efficiency and total dry matter production of bush bean under plastic straw mulch. *Agric Forest Meteorol.* 87: 75-84.

Rahman MA and Ali MO. 2011. The causes of decrement in pulse production and its possible remedy. Proceedings of the International Conference on 'Pulses in South Asia' held at the Bangladesh Agricultural Research Institute, Gazipur-1701, Bangladesh during 17 -19 February 2011. pp. 91-94.

Ramos LM, Gordon AJ, Minchin FR, Sprent JI and Parsons R. 1999. Effect of water stress on nodule physiology and biochemistry of a drought tolerant cultivar of common bean (*Phaseolus vulgaris* L.). *Ann Bot.* 83: 57-63.

Rasheed M, Jilani G, Shah IA, Najeeb U and Iqbal T. 2010 Genotypic variants of lentil exhibit differential response to phosphorus fertilization for physiological and yield attributes. Acta Agr Scan Section-B: Soil Plant Sci 60: 485-493.

Reddy MP and Vora AB.1986. Changes in pigment composition, hill reaction activity and saccharides metabolism in bajra (Pennisetumtyphoides) leaves under NaCl salinity. *Photosynthetica.* 20: 50-55.

Reddy TY, Reddy VR and Anbumozhi V. 2003. Physiological responses of groundnut (*Arachis hypogea* L.) to drought stress and its amelioration: A critical review. *Plant Growth Regul.* 41: 75-88.

Rimski-Korsakov H, Rubio G and Lavado RS. 2009. Effect of water stress in maize crop production and nitrogen fertilizer fate. *J Plant Nutrition.* 32: 565-578.

SAIC. 2009. Annual report of SAARC Agricultural Information Centre (SAIC), Farmgate, Dhaka-1215.

Samad MA, Rahman, MK, Mondal MM A, Fakir MSA. 2007. Evaluation of some advanced lentil mutants in relation to growth, yield attributes and yield. *Bangladesh J Crop Sci.* 18: 117-122.

Saraswathi SG and Paliwal K. 2011. Drought induced changes in growth, leaf gas exchange and biomass production in *Albizia lebbeck* and *Cassia siamea* seedlings. *J Environ Biol*. 32: 173-8.

Sarker A, Erskine W and Saxena MC. 2004. Global perspective on lentil improvement. *In*: Masood A, Singh B, Kumar S, Dhar V (eds) Pulses in new perspective. Indian Institute of Pulses Research, Kanpur, India. pp. 543-550.

Serraj R, Sinclair TR and Purcell LC. 1999. Symbiotic N2 fixation response to drought. *J Exp Bot.* 50: 143-155.

Shoushtari NH and Pepper IL. 1985. Mesquite rhizobia isolated from the Sonoran desert: competitiveness and survival in soil. *Soil Biol Biochem*. 17: 803–806.

Shrestha R, Turner NC, Siddique KHM and Turner DW. 2006. Physiological and yield responses to water deficits among lentil genotypes from diverse origins. *Aust J Agric Res* 57: 903-915.

Singh G. 2009. Salinity related desertification and management strategies: Indian experience. *Land Degrade Develop*. 20: 367-385.

Spence RD, Wu H, Sharpe PJH and Clark KG. 1986. Water stress effects on guard cell anatomy and the mechanical advantage of the epidermal cells. *Plant Environment*. 9: 197-202.

Suriyan C and Chalermpol K. 2009. Proline accumulation, photosynthetic abilities and growth characters of sugarcane (*Saccharum officinarum* L.) plantlets in response to iso-osmotic salt and water deficit stress. *Agric Sci China*. 8: 51-58.

Talukdar D. 2009. Dwarf mutations in grass pea (*Lathyrus sativus* L.): Origin, morphology, inheritance and linkage studies. *J Genet*. 88: 165-75.

Talukdar D. 2011. Isolation and characterization of NaCl-tolerant mutations in two important legumes, *Clitoria ternatea* L. and *Lathyrus sativus* L.: Induced mutagenesis and selection by salt stress. *J Med Plant Res*. 5: 3619-28.

Talukdar D. 2012. A glutathione-overproducing mutant in grass pea (*Lathyrus sativus* L): Alterations in glutathione content, modiofications in antioxidant defense response to cadmium stress and genetic analysis using primary trisomics. *Int J Rec Sci Res*. 3: 234-43.

Terzi R, Saglam A, Kutlu N, Nar H and Kadloglu A. 2010. Impact of soil drought stress on photochemical efficiency of photosystem II and antioxidant enzyme activities of *Phaseolus vulgaris* cultivars. *Turkish J Bot.* 34: 1-10.

Toker C and Cagirgan M. 1998. Assessment of response to drought stress of chickpea (*Cicer arietinum* L.) lines under rain field conditions. *Turk J Agric For.* 22: 615–621.

Upadhyay RK and Panda SK. 2005. Salt tolerance of two aquatic macrophytesPistiastratiotes and Salviniamolesta. *Biol Plant.* 49: 157-159.

Whitehead SJ, Summerfield RJ, Muehlbauer FJ, Coyne CJ, Ellis RH and Wheeler TR. 2000. Crop improvement and the accumulation and partitioning of biomass and nitrogen in lentil. *Crop Sci.* 40: 110-120.

Yadav SS, Phogat DS, Solanki I S and Tomer TS. 2003. Character association and path analysis in lentil. *Indian J Pulse Res.* 16: 22-24.

Yamanaka T, Inoue M and Kaihotsu I. 2004. Effects of gravel mulch on water vapor transfer above and below the soil surface. *Agric Water Manag.* 67: 145-155.

Yang HM and Wang GX. 2001. Leaf stomatal densities and distribution in *Triticum aestivum* under drought and CO_2 enrichment. *Acta Phytoecologica Sinica.* 25: 312-316.

Younis N, Hanif M, Sadiq S, Abbas G, Asghar MJ and Haq MA. 2008. Estimation of genetic parameters and path analysis in lentil. *Pak J Agric Sci.* 45: 44-48.

Zahran HH. 1999. *Rhizobium*-legume symbiosis and nitrogen fixation under severe conditions and in an arid climate. *Microbiolo Molecular Biol Rev.* 63: 968-989.

Zhang X, Chen S, Liu M, Pei D and Sun H. 2005. Improved water use efficiency associated with cultivars and agronomic management in the north China plain. *Agron J.* 97: 783-790.

Zhang YP, Wang ZM, Wu YC and Zhang X. 2006. Stomatal characteristics of different green organs in wheat under different irrigation regimes. *Acta Agronomica Sinica.* 32: 70-75.

Zhang YP, Wang ZM, Wu YC and Zhang X. 2006. Stomatal characteristics of different green organs in wheat under different irrigation regimes. *Acta Agronomica Sinica.* 32: 70–75.

Citation

Dwivedi SK, Singh AK and Malviya N. 2018. Morpho-Physiological responses of lentil (*Lens culinaris* Medikus *subsp. culinaris*) to abiotic stresses. In: Scientific Lentil Production. Singh AK, Bhakta N, Sangale UR, Manibhushan, Sundaram PK, Kumar S and Yasin JK. (eds.) Society for Upliftment of Rural Economy Varanasi, India. Pp. 453-469.

□□□

Chapter 27

Perspective of Lentil Production in Abiotic Stresses Environment

Ved Prakash, Santosh Kumar and SK Dwivedi

SYNOPSIS

Abiotic stress is one of the most important environmental factors regulating plant growth and development. The sensitivity of crop plants to abiotic stress is acknowledged a major constrain in crop production. Legumes, including field pea, faba bean, lentil and chickpea, are very sensitive to Extreme low and high temperature, chilling and freezing temperatures, particularly at the flowering, early pod formation and seed filling stages. Lentil is a cool season crop with a relatively shallow root system (0.6 metres) and is moderately resistant to high temperature and drought.

Keywords: Abiotic stresses, cold and frost, drought and heat, lentil.

Introduction

Lentil is an important cool season food legume crop of rainfed agriculture and it is one of important pulse crops for diversifying cereal-based cropping systems worldwide. Lentil is high in fibre and protein, but low in fat and free of cholesterol (Podder *et al.*, 1999). In addition, it enriches soil fertility by fixing large quantities of atmospheric nitrogen (Bokhari and Ashraf, 1990). Lentils are well adapted to low-rainfall conditions (300–500 mm annual rainfall). Lentil is adapted to cool growing conditions, and the young plants are tolerant of spring frosts. This allows for early spring planting dates. Lentils have been grown extensively in the semi-arid parts of the world, where they have slightly lower yields, but good seed quality. High humidity and excessive rainfall during the season encourages vegetative growth, which prevents good yield and can reduce seed quality. The plants are relatively tolerant to heat and drought stress. The present productivity of lentil in India is very low as it is mostly grown as post-rainy season crop under receding soil moisture conditions during the winter season.

As a result, the crop invariably encounters terminal moisture stress, thus, leading to forced maturity and lower yield. In environments with typical Mediterranean type climates also, lentil crop often experiences terminal as well as intermittent droughts throughout the growing season (Silim *et al.*, 1993). Hence, drought has become an important major yield constraint in production of lentil (Fouad *et al.*, 2011). The intensity of drought stress varies from year to year, depending on the amount and distribution of rainfall and on spring and early summer temperatures (Soltani *et al.*, 2001). High leaf temperature is also a consequence of drought because plants lose the ability for transpirational cooling when water availability is limited (Lu *et al.*, 1997). By analyzing leaf temperature and CO_2 and H_2O exchange rates, it was concluded that the decline in photosynthesis at reduced leaf water potentials was due primarily to an increase in stomatal diffusion resistance and As a consequence of the reduction in transpiration rate, leaf temperature increases and the effects of drought and heat stresses frequently combine to scorch leaves (Mohammadian *et al.*, 2005).

Drought and Heat Stress

The optimal temperatures for the cool season pulses range between 10°C and 30°C. Temperatures that fall outside the optimum range cause stress. Daily maximum temperature above 25°C is considered as the threshold level for heat stress in cool season crops.

The critical temperature for heat tolerance seems to be higher in chickpea than, lentil and the reverse is true for cold tolerance. Lentil is particularly sensitive to heat at the full bloom stage (Fig. 27.1). A few days of exposure to high temperatures (30- 35°C), causes heavy yield losses through flower drop and pod abortion.

Marked reductions in seed yields after brief episodes of high temperatures during seed filling can diminish seed set, seed weight and accelerate senescence and reduce yield. Canopy temperature measurement with infrared thermometers has been an effective tool for irrigation scheduling in semi arid and arid conditions (Evett *et al.*, 2000). Canopy temperature can be an indicator of plant water status because a non stressed plant transpires, cooling its environment. Stomatal closure on a water stressed plant will suppress transpiration, raising its temperature. However, irrigation scheduling using this concept was impractical until hand held infrared thermometers became commercially available (Gardner *et al.*, 1992).

Fig. 27.1 Heat stress major abiotic stress being cool season crops

Drought is a major constraint in all pulse growing regions of India, particularly in the low rainfall areas. From a practical point of view, it is difficult to separate drought and heat stress and they generally occur simultaneously. The types of drought stress which affect pulse production in India are: lack of or low opening rainfall at planting, intermittent drought stress by breaks in winter rainfall; and the terminal drought stress, resulting from receding soil moisture and increasing high temperature in spring. Terminal drought is the most common form of drought stress experienced by cool season lentil in India. The severity of terminal drought stress depends upon the amount and distribution of rainfall, capacity of the soil to store moisture and the evaporative demand of the atmosphere. There are two major effects of drought on lentil productivity: failure to establish the desired plant stands; and reduction in growth and yield due to sub-optimal soil moisture availability. Lentil seeds absorb water equal to their weight in less than 36 hours and germinate soon after, but germination is reduced if dehydration occurs thereafter. This makes lentil crop sensitive to early season drought, particularly when planted shallow. Earlier studies suggest that shoot and root attributes at an early growth stage of plants are related to drought tolerance through dehydration avoidance, a mechanism that allows a crop plant to thrive in water-limited environments (Sarker *et al.,* 2005). Drought stress during vegetative stages of growth alone does not appear to cause a significant yield loss in most of pulses. Flowering is the most sensitive stage to drought. It is probable that high sensitivity to drought during the reproductive stage is due to the lack of new root growth in pulse crops at this stage of growth. Lentil is very sensitive to heat (Table 27.1) and considered relatively tolerant to drought. Potential losses in yield can range between 6 per cent and 54 per cent.

Cold and Frost

The major symptoms of cold season in cool lentil are reduced leaf production, area expansion, poor dry matter production, chlorosis and, finally necrosis of the older leaves (Fig. 27.2). A daily average temperature between 0⁰ C and 10⁰ C is considered a threshold level for cold or chilling stress in cool season lentil. Under Indian winter conditions reproductive organs, mainly flower buds and flowers, are most susceptible part of cool season pulses to cold or chilling injury. The reduced ovule fertilisation, associated with decline in pollen tube growth and ovule viability is the major cause for poor seed set in lentil at low temperatures.

Fortunately recent evaluation of germplasm both in India and overseas shows distinct differences in pod and seed set at low temperatures. These traits are currently being exploited in the breeding programs. As a result, frost and cold tolerance and cold resistance are major objectives of international lentil breeding programmes (Eujayl *et al.*, 1999). One of the major limitations to lentil production in this area and in West Asia is losses due to radiant frost, freezing and cold injury at the flowering stages. Up until the late 1990s, little research was directed towards the study of freezing tolerance in lentil at the reproductive stages. The effects of cold, frost and freezing on lentil plants include flower and pod abortion, damage to seed and injuries to vegetative tissues. Research has been carried out in the field under variable seasons to study the inheritance of radiation frost tolerance genes in lentil (Eujayl *et al.*, 1999). When performing studies on frost, it is important to observe the response of the plant at chilling and freezing temperatures separately (Stoddard *et al.*, 2006). In comparison to field studies, screening for frost tolerance under controlled environmental conditions provides enables the researcher to speed up the screening process, lessen the associated costs and screen numerous times a year to increase accuracy (Ali *et al.*, 1999). Levitt (1980) described cold acclimation as a process that occurs when plants are gradually exposed to colder temperatures, eventually resulting in the reduced temperature at which a plant is damaged by freezing range temperatures. Ali *et al.* (1999) showed that lentil plants that did not undergo prior cold acclimation were unable to survive freezing temperatures. Croser *et al.* (2003) described the changes in cellular structures and metabolic processes that occur during cold acclimation, including increases in the levels of sugars, soluble proteins, prolines and organic acids and changes in the composition of the lipid membrane.

Fig. 27.2 Watering to maintain ambient environment warm during cold weather in chilly winter

Waterlogging

Transient water logging is very common in winter on fine textured soils. Seed germination is very susceptible to water logging. Poor crop establishment is a common problem when water logging occurs at seedling emergence. Lentil is very sensitive to water logging (Table 27.1). The growth and development of the crop mostly depends on residual soil moisture after the end of monsoon (Miah and Rahman, 1998). However, erratic rain and poor surface drainage during early or later stages of crop growth may result in excess moisture or even transient flooding. As a result, the damage of the crop from flooding or excess soil moisture is not uncommon. Generally, lentil cannot withstand flooding, waterlogging, or soils with high salinity (Heuke, 1999; McVicar and Vandenberg, 2007).

Water logging depresses vegetative growth of plants but affects root growth more than shoot growth. Once seedlings are established, the sensitivity of plants to water logging stress is generally increases with advancing age (Fig.27.3). Thus the ability to survive and recover following water logging is dependent on the timing of the water logging event relative to the stage of growth. Ability to recover generally declines sharply as reproductive growth (flowering and pod filling stage) approaches. Seed yield loss resulting from water logging can be substantial. Limited data for lentil, field pea and chickpea suggest that reduction in yield can vary from negligible to almost 100 per cent, depending upon the species, stage of growth when water logged, duration, and the extent of root zone affected by water logging. Management practices to reduce the effects of water logging include the species choice paddock selection, time of sowing, seeding rate and drainage. Selection for genetic variation in water logging tolerance within a cool season crop has been limited.

Fig. 27.3 Water logging is fatal for lentil; very susceptible to water logging.

Table 27.1 Importance of major abiotic stresses affecting cool season pulse crops in India.

Stress	Lentil	Chickpea	Faba bean	Field pea
Cold	X	XXX	X	X
Frost	XX	XXX	XX	XXX
Water logging	XXX	XXX	X	XXX
Drought	XX	XX	XXX	XX
Heat	XXX	XX	XXX	XXX
Salinity	XX	XXX	XX	XX
Boron toxicity	XXX	XXX	XX	XX

(Very important XXX, important XX, not very important X)

Management of Abiotic Stresses

Abiotic stress can be reduced by choosing the most appropriate species and adjusting agronomy (sowing time, plant density, soil management) to ensure sensitive crop stages occur at the most favourable time in the season. For example choosing the optimum sowing time, species and cultivars with appropriate phenology can reduce the effect of frost and drought in lentil crop in dry land environments with terminal stress. One approach of dealing with stresses caused by extremes in the abiotic environment is to develop cultivars resistant to specific stresses. Breeding and selection for resistance to these stresses is often considered difficult because of the unpredictability of climatic conditions. The developmental stage of the plants and the duration of the stress have a strong influence on the effect of the abiotic stress, together with the ability of the plants to tolerate the stress themselves. Drought and heat escape through earliness in flowering and maturity is probably the characteristic most widely used by breeders for pulses and other crops to escape drought especially in low rainfall terminal drought environments. Currently lack of simple and accurate screening procedures to screen parental genotypes and breeding population for various abiotic stresses is the major bottleneck in the development of stress tolerant cool season pulse crops. Germplasm for resistance to various abiotic stresses are not widely available to breeders. Recent studies from overseas suggest that wild species of cool season pulses posses desired traits for a number of abiotic stresses. Future effort is required to identify desirable genes from this germplasm for transfer to adapted cultivars by conventional and/or biotechnological approaches to develop abiotic stress resistant cultivars.

Conclusion

Lentil is an important cool season food legume crop of rainfed agriculture and it is one of important pulse crops for diversifying cereal-based cropping systems worldwide. High humidity and excessive rainfall during the season encourages vegetative growth, which prevents good yield and can reduce seed quality. The plants are relatively tolerant to heat and drought stress. The present productivity of lentil in India is very low as it is mostly grown as post-rainy season crop under receding soil moisture conditions during the winter season. Daily maximum temperature above 25°C is considered as the threshold level for heat stress in cool season crops. The critical temperature for heat tolerance seems to be higher in chickpea than, lentil and the reverse is true for cold tolerance. Lentil is particularly sensitive to heat at the full bloom stage. Drought stress during vegetative stages of growth alone does not appear to cause a significant yield loss in most of pulses. Flowering is the most sensitive stage to drought. It is probable that high sensitivity to drought during the reproductive stage is due to the lack of new root growth in pulse crops at this stage of growth. Lentil is very sensitive to heat. Similarly under Indian winter conditions reproductive organs, mainly flower buds and flowers, are most susceptible part of cool season pulses to cold or chilling injury. The reduced ovule fertilization, associated with decline in pollen tube growth and ovule viability is the major cause for poor seed set in lentil at low temperatures. Lentil seed germination is very susceptible to water logging. Poor crop establishment is a common problem when water logging occurs at seedling emergence. Lentil is very sensitive to water logging.

References

Ali A, Johnson DL and Stushnoff C.1999. Screening lentil (*Lens culinaris*) for cold hardiness under controlled conditions. J. *Agric. Sci.* (Camb) 133: 313–319.

Bokhari MH and Ashraf M. 1990. Pulse Crops of Pakistan. The Biological Society of Pakistan, Lahore.

Croser JS, Clarke HJ, Siddique KHM and Khan TN.2003. Low temperature stress: implications for chickpea (*Cicer arietinum* L.) improvement. *Crit Rev Plant Sci*. 22: 185–219.

Eujayl I, Erskine W, Baum M and Pehu E. 1999. Inheritance and linkage analysis of frost injury in lentil. *Crop Sci.* 39: 639–642.

Fouad, M., Imtiaz, M., Kumar, S. and Malhotra R. 2011. Food security and climate change in dry areas. In *Proceedings of an International Conference* (ed. M Solh and MC Saxena), pp 244–254. International Centre for Agricultural Research in the Dry Areas, Aleppo, Syria.

Gardner BRD, Nielsen C and Shock CC. 1992. Infrared thermometry and the crop water stress index: I. History, theory, and baselines. *J. Prod. Agric.* 54: 462 466.

Heuke L. 1999. Emerging opportunity in agriculture. The Australian New Crop Newsl. Pobl., 11.

Levitt J. 1980. Response of plants to environmental stresses; chilling, freezing and high temperature stresses. Academic Press, New York.

Lu Z, Chen J, Percy RvG and Zeiger E. 1997. Photosynthetic rate, stomatal conductance and leaf area in two cotton species (*Gossypiumbarbadense* and *Gossypiumhirsutum*) and their relation with heat resistant and yield. *Aust. J Plant Physiol.* 24: 693-700.

McVicar R and Vandenberg A. 2007. Red lentil production. Saskatchewan Agriculture and Food. Website: hptt://www.agr.gov.sk.ac/.

Miah AA and Rahman MM. 1998. Agronomy of Lentil in Bangladesh. In : Lentil in South Asia - Proceedings of the seminar on lentil in South Asia, 11-15 March 1991, New Delhi, India pp. 128-138.

Mohammadian R, Moghaddamb M, Rahimianb H and Sadeghian SY. 2005. Effect of early season drought stress on growth characteristics of sugar beet genotypes. *Turkish J. Agric. Fores.* 29: 357-68.

Podder AK, Islam MZ, Rashid MH, Chowdhury S and Pasha MNU. 1999. Bio-fertilizer and our Agriculture. A bulletin in Bengali, BINA, Mymensingh, Bangladesh.

Sarker A, Erskine W and Singh M. 2005. Variation in shoot and root characteristics and their association with drought tolerance in lentil landraces. *Genet. Res. Crop Evol.* 52: 89–97.

Silim SN, Saxena MC and Erskine W. 1993. Adaptation of lentil to the Mediterranean environment. I. Factors affecting yield under drought conditions. *Exp. Agric.* 27: 155–175.

Soltani A, Khooie FR, Ghassemi-Golezani K and Moghaddam M. 2001. A simulation study of chickpea crop response to limited irrigation in a semi-arid environment. *Agri. Water Manag.* 49: 225-37.

Stoddard FL, Balko C, Erskine W, Khan HR, Link W and Sarker A. 2006. Screening techniques and sources of resistance to abiotic stresses in cool-season food legumes. *Euphytica* 147: 167-186.

Citation

Ved Prakash, Kumar S and Dwivedi SK. 2018. Perspective of Lentil Production in Abiotic stresses environment. In: Scientific Lentil Production. Singh AK, Bhakta N, Sangale UR, Manibhushan, Sundaram PK, Kumar S and Yasin JK. (eds.) Society for Upliftment of Rural Economy Varanasi, India. Pp. 471-479.

□□□

Chapter 28

Moisture-Dependent Engineering Properties of Lentil Seed

Raghuraj Singh and PK Sundaram

SYNOPSIS

Lentil is a legume crop and legumes are consumed as a good nutrient source of protein. The engineering properties like geometrical characteristics, density and mechanical characteristics of lentil seeds can be used to predict their processing ability and appropriate design of machines for different unit operations. The geometrical characteristics of lentil may be determined by random picking of seed/grain and their three linear dimensions are to be measured using a digital Vernier caliper with a good accuracy. The range of geometrical characteristics of red lentil seed/grains at moisture levels ranged from 9.1 to 21.1 per cent (wb). All properties except bulk and true densities increased with an increase in moisture content from 9.1 to 21.1% (wb) ($p < 0.05$).

Keywords: Angle of repose, Mechanical Properties, Porosity, Static coefficient of friction, Terminal velocity.

Introduction

Lentil is a legume crop and legumes are consumed as a good nutrient source of protein. It is also an important traditional dietary item in developing countries like India. They play an essential role in human nutrition as rich sources of starch, calories, certain minerals, and vitamins. Despite being nutritious, legumes, especially lentils, are underused in many developed countries. Its production in India is 9.40 lakh tonnes and Uttar Pradesh has highest share (43.62%) in its production (APEDA, 2011-12). The engineering properties like geometrical characteristics, density and mechanical characteristics of lentil seeds can be used to predict their processing ability and appropriate design of machines for different unit operations. These properties are useful in design of equipment for harvesting, processing, handling, plantation, transportation and storage etc. The behavior of this seed under technological treatments is influenced by its differentiated mechanical resistance (Nazari, 2009). A broad knowledge about both lentils' mechanical properties and factors affecting seed/grains hardness such as microstructure characteristics can be applied to the design of new equipment and processes.

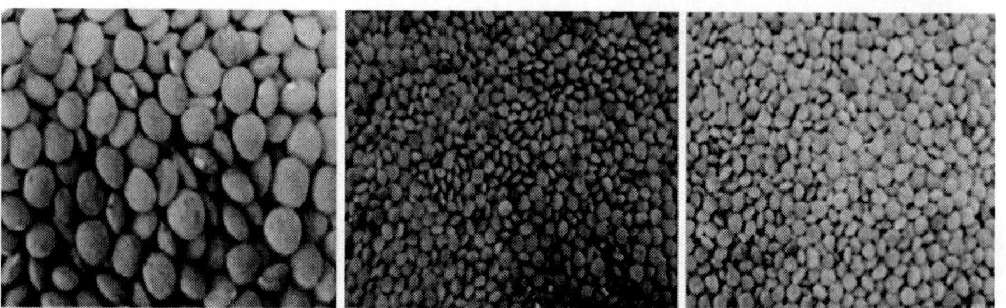

Fig. 28.1 Variability in shape size and seed coat colour lentil seeds

Geometrical, Gravimetrical Characteristics and Density

Geometrical Characteristics

The geometrical characteristics of lentil may be determined by random picking of seed/grain (100 or more number of seed/grains) and their three linear dimensions (length, L; width, W; and thickness, T) are to be measured using a digital Vernier caliper with a good accuracy. The seed/grains' arithmetic mean (D_a), geometric mean diameter (D_g) and sphericity may be calculated according to Mohsenin, 1986.

$$D_a = \frac{L + W + T}{3}$$

$$D_g = (L \times W \times T)^{1/3}$$

Degree of Sphericity (DS) in percent,

$$DS = \frac{(L \times W \times T)^{1/3}}{L} \times 100$$

The surface area of seeds was calculated from the following formula presented by Jain and Bal (1997):

Surface area, $S = \frac{\pi L \times \sqrt{WT}}{2L - \sqrt{WT}}$

Table 24.1 gives the range of geometrical characteristics of red lentil seed/grains at moisture levels ranged from 9.1 to 21.1 per cent (wb). All properties except bulk and true densities increased with an increase in moisture content from 9.1 to 21.1% (wb) ($p < 0.05$). The increase in the three axial dimensions is attributed to expansion or swelling as the result of moisture uptake in the intracellular spaces within the seeds (Gharibzahedi *et al.*, 2014). The diameter and thickness of lentil seed/grains increased linearly (Amin *et al.*, 2004). The sphericity of red lentil seed/grains increased from 81.85 to 82.16 percent (Gharibzahedi *et al.*, 2014).

Gravimetrical Characteristics

Thousand-seed weight

It is an important indicator of seed standard. The thousand-seed mass/weight may be determined by using an electronic balance to an accuracy of 0.001 g. To calculate 1000 grain mass/weight, mass/weight of 100 randomly selected grains from the bulk can be calculated and multiplied by a factor 10. The table 28.1 shows that thousand-seed mass increased from 51.23 to 51.82 g for red lentil seed/grains as the moisture content increased from 9.1 to 21.1%(wb) (Gharibzahedi *et al.*, 2014). It increased linearly (Amin *et al.*, 2004).

Bulk density (BD)

The bulk density of lentil seed/grains is determined by filling an empty known volume of scaled cylinder with the seed/grain and weighed. The weight of the seed/grains may be obtained by subtracting the weight of the scaled cylinder from the weight of the cylinder and seed/grains. The scaled cylinder may be shaken desired number of times to achieve uniformity in bulk density. The volume occupied is then recorded. The process is replicated as desired to reduce the error and the bulk density for each replication is calculated from the following relation and the average bulk density (BD) is determined.

Bulk density, BD $= \frac{\text{mass of a sample of seed/grains}}{\text{total volume of sample seed/grains}}$

True density (TD)

It is defined as the ratio between the mass of lentil seed/grain and the true volume of the seed/grains, using the liquid/toluene (C_7H_8) displacement method. Toluene may be used instead of water because it is absorbed by seed/grains to a lesser extent. The volume of toluene displaced is found by immersing a weighed quantity of seed/grains in the measured toluene (Mohsenin, 1986). Therefore, true density (TD) of lentil seed/grains is determined as follows:

$$\textbf{True density, TD} = \frac{\text{mass of a sample of seed/grains}}{\text{solid (true) volume of sample seed/grains}}$$

Porosity

It is an important parameter that affects the grain hardness, breakage susceptibility, drying rate and resistance to fungal development. The porosity of the bulk grain is computed from the values of the true density and the bulk density using the relationship given by Mohsenin, 1986.

$$\text{Porosity,}(\%) = \left[1 - \frac{\text{Bulkdensity}}{\text{Truedensity}}\right] \times 100 = \left[1 - \frac{BD}{TD}\right] \times 100$$

Table 24.1 shows variation in bulk density and true density for given condition of moisture levels. The porosity increased from 37.7 to 41.8% (wb) for red lentils with the increase in moisture content ($p < 0.05$). The increase in porosity might be attributed to the expansion and swelling of seeds, which may have created more voids between the seeds and increased the bulk volume (Gharibzahedi *et al.*, 2014; Carman,1996). The bulk density and true density decreased linearly but porosity increased linearly (Amin *et al*, 2004 and Carman, 1996).

Static coefficient of friction

The static coefficient of friction on galvanized iron sheet/rubber sheet surface is measured for lentil seed/grains by using inclined plane method. The material is kept on horizontally placed surface and slope is increased gradually. The angle at impending slip is recorded. The coefficient of static friction is computed by using following equation:

Coefficient of static friction = tan ¢

Where,

¢ = angle of static friction, degrees

Table 24.1 gives that static coefficient of friction for lentil seed/grains varied from 0.405 to 0.432 having moisture content range from 10.33 to 21.0% (wb) at galvanized iron sheet, respectively. It varied from material to material and depended on the roughness and wetness of the seeds (Amin *et al.*, 2004; Carman, 1996)

Angle of repose

The angle of repose of seed was determined by allowing the material to fall freely to form a heap cone. The diameter of the base circle and height of the heap are measured. The angle of repose is calculated by using the following expression:

$$\text{ø} = \tan^{-1}\left(\frac{2h}{D}\right)$$

Where, ¢ = Angle of repose, degrees

h = Height of cone, m

D = Diameter of the cone,

Angle of repose is surface and material dependent property. Table 28.1 shows that angle of repose for lentil seed/grains varied from 24.80° to 27.78° having moisture content range from 10.33 to 21.0% (wb), respectively (Amin *et al*, 2004).

Table 28.1 Engineering properties of lentil seed/grains

Engineering properties	Value	Condition(s)	Remark	Reference
Length (mm), L	4.07±0.01 to 4.13±0.03	M.C.=9.1 to 21.1% (wb)	Red lentil	Gharibzahedi *et al.*, 2014
	4.63±0.27	Not mentioned	Seyran-96	Gürsoy *et al., 2010*
Width (mm), W	3.93±0.06h to 3.99±0.02e	M.C.=9.1 to 21.1% (wb)	Red lentil	Gharibzahedi *et al.*, 2014
Thickness (mm), T	2.32±0.01 to 2.38±0.03	M.C.=9.1 to 21.1% (wb)	Red lentil	Gharibzahedi *et al.*, 2014
	2.50±0.15	Not mentioned	Seyran-96	Gürsoy *et al., 2010*
	2.18 to 2.48	M.C.=10.33 to 21.0% (wb)	-	Amin *et al*, 2004
Arithmetic meandiameter (mm), D_a	3.44±0.06 to 3.50±0.03	M.C.=9.1 to 21.1% (wb)	Red lentil	Gharibzahedi *et al.*, 2014
	3.84 to 4.06	M.C.=10.33 to 21.0% (wb)	-	Amin *et al*, 2004
Geometric meandiameter (mm), D_g	3.33±0.06 to 3.39±0.03	M.C.=9.1 to 21.1% (wb)	Red lentil	Gharibzahedi *et al.*, 2014
	3.40±0.17	Not mentioned	Seyran-96	Gürsoy *et al., 2010*
Sphericity (%)	81.85±0.77 to 82.16±0.68	M.C.=9.1 to 21.1% (wb)	Red lentil	Gharibzahedi *et al.*, 2014
	73.5±2.3	Not mentioned	Seyran-96	Gürsoy *et al., 2010*
Surface area (mm^2), S	34.89±1.74 to 36.15±1.67	M.C.=9.1 to 21.1% (wb)	Red lentil	Gharibzahedi *et al.*, 2014
Thousand (1000) seedsmass (g)	51.23±0.75 to 51.82±0.66	M.C.=9.1 to 21.1% (wb)	Red lentil	Gharibzahedi *et al.*, 2014
	36.45±5.0	M.C.=8.53%	Seyran-96	Gürsoy *et al., 2010*
	20.0 to 25.5	M.C.=10.33 to 21.0% (wb)	-	Amin *et al*, 2004
Bulk density (kgm"3)	822.2±4.4 to 718.5±4.9	M.C.=9.1 to 21.1% (wb)	Red lentil	Gharibzahedi *et al.*, 2014
	809	M.C.=8.53%	Seyran-96	Gürsoy *et al., 2010*
	832 to 768	M.C.=10.33 to 21.0% (wb)	-	Amin *et al*, 2004
	1190 to 935	M.C.=6.5 to 32.6% (db)	-	Carman, K. 1996

True density (kgm^{-3})	1320±5.4 to 1236.6±11.0	M.C.=9.1 to 21.1% (wb)	Red lentil	Gharibzahedi *et al.*, 2014
	1409	M.C.=8.53%	Seyran-96	Gürsoy *et al., 2010*
	1270 to 1212	M.C.=10.33 to 21.0% (wb)	-	Amin *et al*, 2004
Porosity (%)	37.7±0.2 to 41.8±0.1	M.C.=9.1 to 21.1% (wb)	Red lentil	Gharibzahedi *et al.*, 2014
	34.48 to 37.00	M.C.=10.33 to 21.0% (wb)	-	Amin *et al*, 2004
	27·4 to 32·0	M.C.=6.5 to 32.6% (db)	-	Carman, K. 1996
Static coefficient of friction	0.405 to 0.432	M.C.=10.33 to 21.0% (wb)	At Galvanized iron sheet	Amin *et al*, 2004
	0·374 to 0·532	M.C.=6.5 to 32.6% (db)	Rubber surface	Carman, K. 1996
Angle of repose, degree	24.80 to 27.78	M.C.=10.33 to 21.0% (wb)	-	Amin *et al*, 2004
Mechanical strength				
i) Rupture strength, N	138.62 (14.28) to 95.97 (2.32)	M.C.=9.5 to 21.1% (wb)	Red and green lentil	Gharibzahedi *et al.*, 2014
ii) Deformation, mm	0.377 (0.04) to 0.647 (0.04)	M.C.=9.5 to 21.1% (wb)	Red and green lentil	Gharibzahedi *et al.*, 2014
iii) Rupture energy, mJ	26.57 (4.17) to 31.47 (5.77)	M.C.=9.5 to 21.1% (wb)	Red and green lentil	Gharibzahedi *et al.*, 2014
Terminal velocity, m/s	7.78±0.23	Not mentioned	Seyran-96	Gürsoy *et al., 2010*
	10·95 to 12·06	M.C.=6.5 to 32.6% (db)	-	Carman, K. 1996

± Values represent standard deviation.
Mean (standard error).

Measurement of Mechanical Properties

The mechanical properties of lentil seed/grains are measured by a texture analyzer machine (Universal Testing Machine) equipped with a 50 N load cell or desired load cell and integrator. The measurement accuracy should be high (±0.001 N in force and 0.001 mm in deformation). The individual seed/grain is loaded between two parallel plates of the machine and compressed until rupture occurred as denoted by a bio-yield point in the force-deformation curve. The bio-yield point is detected by a break in the force-deformation curve. Once the bio-yield point is detected, the loading is stopped. Energy absorbed by the sample at rupture is determined by calculating the area under the force-deformation curve by means of a digital planimeter.

From table 28.1, the means of rupture force, deformation and rupture energy for lentil seed/grains may be compared for moisture content range from 9.5 to 21.1 percent (wb). The rupture force of both red and green lentil seeds decreased from 138.62 to 95.07 N with an increase in seed moisture content. This may be due to the fact that at higher moisture content, the seeds became softer and required lesser force. Deformation of lentil seed/grains at rupture increased from 0.377 to 0.647 mm as the moisture content of the seed increased as mentioned above. The energy absorbed at the rupture of seed/grains increased from 26.57 to 31.47 mJ with the increase in the moisture content range from 9.5 to 21.1% (wb) (Gharibzahedi *et al.*, 2014). Energy absorbed at seed rupture is a function of both force and deformation up to rupture point. At low moisture contents, the seed required high force to be ruptured and experienced low deformation, but at high moisture contents, the rupture force was low and the deformation was high. This suggests that seed rupture energy required during compressive loading increases as the moisture content of the seed increases.

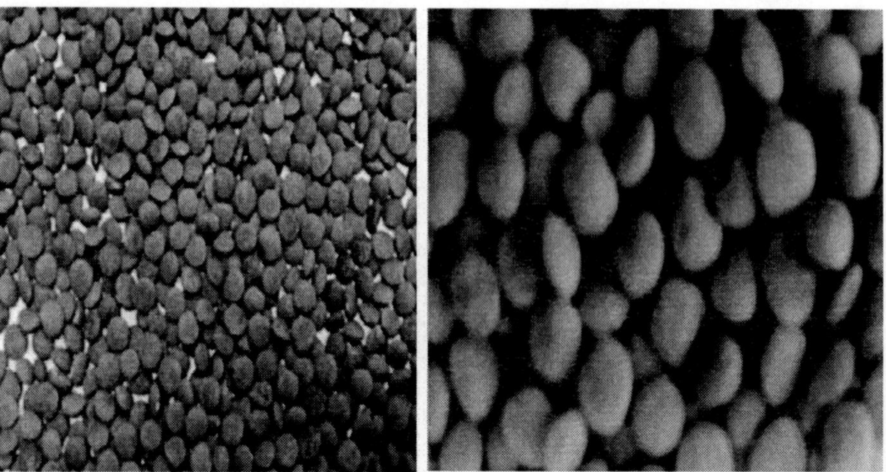

Fig. 28.2(a) Small seeded lentil **Fig. 28.2(b)** Bold seeded lentil

Terminal velocity

Terminal velocity is the constant velocity of a freely falling object in a medium at which the net gravitational acceleration force is equal to the resisting upward drag force. The terminal velocity of seed/grains of lentil is important property to select the air velocity for cleaning seed/grain from foreign material. The experimental setup is required to measure the terminal velocity. Terminal velocity increased from 10·95 to 12·06 m/s as moisture content increased from 6·5 to 32·6% (db) (Carman, 1996).

Behavior and Properties of Lentil Seed/Grains

Lentil is an important traditional dietary item in developing countries like India. The geometrical characteristics of lentil may be determined by random picking of seed/grain. The increase in the three axial dimensions is attributed to expansion or swelling as the result of moisture uptake in the intracellular spaces within the seeds. Thousand-seed mass increased from 51.23 to 51.82 g for red lentil seed/grains as the moisture content increased from 9.1 to 21.1% (wb). Porosity is an important parameter that affects the grain hardness, breakage susceptibility, drying rate and resistance to fungal development. The mechanical properties of lentil seed/grains are measured by a texture analyzer machine (Universal Testing Machine) equipped with a 50 N load cell or desired load cell and integrator. Terminal velocity increased from 10·95 to 12·06 m/s as moisture content increased from 6·5 to 32·6% (db).

Conclusion

The engineering properties like geometrical characteristics, density and mechanical characteristics of lentil seeds can be used to predict their processing ability and appropriate design of machines for different unit operations. The geometrical characteristics of lentil may be determined by random picking of seed/grain (100 or more number of seed/grains) and their three linear dimensions (length, L; width, W; and thickness, T) are to be measured using a digital Vernier caliper with a good accuracy. The mechanical properties of lentil seed/grains are measured by a texture analyzer machine (Universal Testing Machine) equipped with a 50 N load cell or desired load cell and integrator. The measurement accuracy should be high (±0.001 N in force and 0.001 mm in deformation). Lentil is an important traditional dietary item in developing countries like India. The geometrical characteristics of lentil may be determined by random picking of seed/ grain. The increase in the three axial dimensions is attributed to expansion or swelling as the result of moisture uptake in the intracellular spaces within the seeds.

References

Amin MN, Hossain MA and Roy KC. 2004. Effects of moisture content on some physical properties of lentil seeds. *Journal of Food Engineering* 65(1): 83-87.

Carman K. 1996. Some Physical Properties of Lentil Seeds. *Journal of Agricultural Engineering Research* 63(2): 87–92.

Gharibzahedi SMT, Djomeh ZE, Razavi SH and Jafari SM. 2014. Mechanical Behavior of Lentil Seeds in Relation to their Physicochemical and Microstructural Characteristics. *International Journal of Food Properties* 17: 545–558.

Gürsoy S and Güzel E. 2010. Determination of Physical Properties of Some Agricultural Grains. *Research Journal of Applied Sciences, Engineering and Technology* 2(5): 492-498.

Jain RK and Bal S. 1997. Properties of pearl millet. *J. Agric. Eng. Res.* 66: 85-91.

Mohsenin NN. 1986. Physical Properties of Plant and Animal Material. *Gordon and Breach Science Publications,* New York.

Nazari Galedar M, Mohtasebi SS, Tabatabaeefar A, Jafari A and Fadaei H. 2009. Mechanical behavior of pistachio nut and its kernel under compression loading. *Journal of Food Engineering* 95(3): 499–504.

Citation

Singh Raghuraj and Sundaram PK 2018. Moisture-Dependent Engineering Properties of Lentil Seeds. In: Scientific Lentil Production. Singh AK, Bhakta N, Sangale UR, Manibhushan, Sundaram PK, Kumar S and Yasin JK. (eds.) Society for Upliftment of Rural Economy Varanasi, India. Pp. 481-490.

□□□

Chapter 29

Lentil in India: An Overview

KM Singh and AK Singh

SYNOPSIS

Lentil *(Lens culinaris)* is an edible pulse. It is a bushy annual plant of the legume family, grown for its lens-shaped seeds. The lens (double-convex shaped) is so called because the shape of a lens is basically the same shape as lentils. Lens is the Latin name for lentil. With about 30% of their calories from protein, lentils have the third-highest level of protein, by weight, of any legume or nut, after soybeans and hemp. On average, about 70% of all world lentil production is consumed in the countries where they are produced. Lentil production for the major lentil producing nations has been trending upwards during the past 7 years, ranging from 2.1 million tones (Mt) in 2002-2003 to 2.9 Mt in 2009-2010. In India, Lentil production has always been important as it is the one of the most important *rabi* crops in the country. India has been producing lentil since 1st century AD and has always been an important producer of the crop. In fact, India was the largest producer of the Lentil crop in the world until recently Canada took over the lead leaving India at the second place.

Keywords: Indo-Gangetic Plain, Constraint, Export, Lentil, Production.

Introduction

Traditionally pulses have been considered important elements of cropping systems in the Indo-Gangetic Plains. They were popular because of their importance as a source of protein and ability to fix atmospheric nitrogen (N) and thus improve soil fertility (Joshi, 1998). With the introduction of irrigation and due to high profitability of alternative sources of soil nutrients in the form of inorganic fertilizers in the mid- 1960s, pulses were replaced or relegated to marginal lands. During the late 1960s and early 1970s, a large area under pulses in the Indo-Gangetic Plain (IGP) was substituted by high- yielding varieties of rice (*Oryza sativa*) and wheat (*Tritium aestivum*). The new technology of rice and wheat substantially changed the agricultural scenario and largely contributed to increase in agricultural production in the IGP.

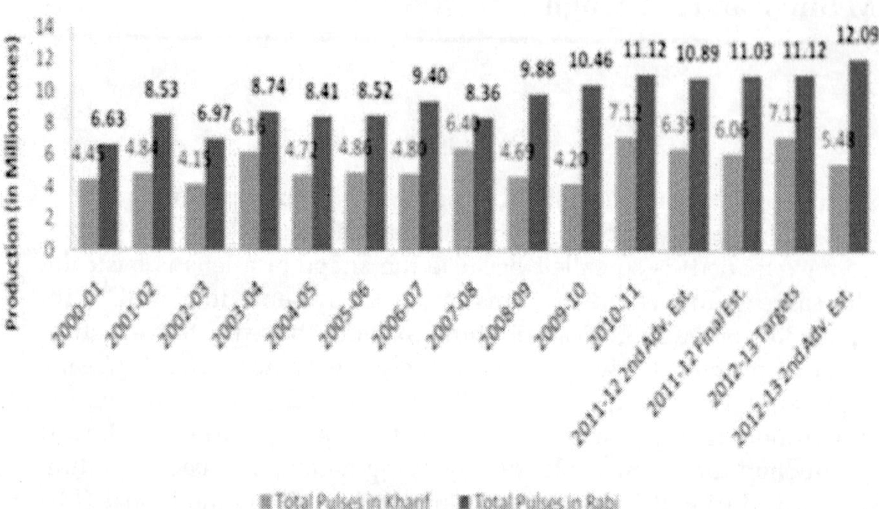

Fig. 29.1 Status of pulse production in India (2000-01 to 2012-13)
Source: http://agricoop.nic.in/imagedefault/trade/pulses%20profile.doc

With the passage of time, excessive use of chemical fertilizers and irrigation in rice and wheat to maintain their productivity has created an imbalance in soil fertility and threatened the sustainability of the most productive food grain belt in South Asia (Hobbs and Morris, 1996) Pulses are an effective source of reversing the process and can contribute significantly to achieving the twin objectives of increasing productivity and improving the sustainability of the rice and wheat-based cropping system in the IGP (Ahlawat *et al.*, 1998; Lauren *et al.*, 1998; Yadav *et al.*,1998). It has been reported that there is a scope of increasing area under lentil during the *rabi* season, as its cost per hectare is less with higher net returns than the competing crops like wheat, gram and mustard in water-deficit and resource-poor conditions. There are large returns for adoption of disease management (80 per cent increase in net return), and improved small-seeded varieties (about 40 per cent increase in net return) in lentil. The study has found that lentil-based cropping systems are profitable and also have high water productivity, hence are suitable for mostly un-exploited rice-fallows under water-deficit conditions (Reddy and Reddy, 2010).

Table 29.1 Area, Production and Yield of Lentil during 2010-11 and 2011-12 in major producing states in India

States	2011-12			2010-11		
	Area	**Production**	**Yield**	**Area**	**Production**	**Yield**
Uttar Pradesh	0.59	0.41	695	0 .59	0.48	814
Bihar	0.24	0.21	875	0.17	0.15	882
Madhya Pradesh	0.59	0.18	305	0.54	0.28	519
West Bengal	0.06	0.05	833	0.05	0.05	1000
Rajasthan	0.04	0.04	1000	0.03	0.02	667
Assam	0.02	0.01	500	0.02	0.01	500
Others	0.06	0.04	®	0.08	0.04	@
All India	**1.6**	**0.94**	**591**	**1.48**	**1.03**	**697**

Source: http://agricoop.nic.in/imagedefault/trade/pulses%20profile.doc

About Lentil

Lentil *(Lens culinaris)* is an edible pulse. It is a bushy annual plant of the legume family, grown for its lens-shaped seeds. It is about 40 cm (16 in) tall and the seeds grow in pods, usually with two seeds in each. Lentils have been part of the human diet since the aceramic (before pottery) Neolithic times, being one of the first crops domesticated in the Near East. Archeological evidence shows they were eaten 9,500 to 13,000 years ago. Lentil colors range from yellow to red-orange to green, brown and black. Lentils also vary in size, and are sold in many forms, with or without the skins, whole or split.

Fig. 29.2 Field view of lentil crop at flowering stage

Lentil in culture

The lens (double-convex shaped) is so called because the shape of a lens is basically the same shape as lentils. Lens is the Latin name for lentil. Lentils are mentioned many times in the Hebrew Bible, the first time recounting the incident in which Jacob purchases the birthright from *Esau* with stewed lentils (a "mess of pottage"). In Jewish mourning tradition, lentils are traditional as food for mourners, together with boiled eggs, because their round shape symbolizes the life cycle from birth to death. Lentils were a chief part of the diet of ancient Iranians, who consumed lentils daily in the form of a stew poured over rice. Lentils are also commonly used in Ethiopia in a stew-like dish called *kik*, or *kik wot*, one of the dishes people eat with Ethiopia's national food, injera flat bread. Yellow lentils are used to make a non-spicy stew, which is one of the first solid foods Ethiopian women feed their babies. In Pakistan, lentils are often consumed with *Roti*/bread or rice.

In India, lentils soaked in water and sprouted lentils are offered to gods in many temples. It is also a practice in South India to give and receive sprouted peas by women who perform *Varalakshmi Vratam*. It is considered to be one of the best foods because the internal chemical structures are not altered by cooking. In Italy and Hungary, eating lentils on New Year's Eve traditionally symbolizes the hope for a prosperous new year, most likely because of their round, coin-like form. In Shia narrations, lentils are said to be blessed by seventy Prophets, including Jesus and Mohammed. In "Cinderella", one of Grimm's Fairy Tales, Cinderella's stepmother assigns to Cinderella the task of fishing lentils out of ash. If she succeeds, she may go to the ball. The seeds of lentil require a cooking time of 10 to 40 minutes, depending on the variety—shorter for small varieties with the husk removed, such as the common red lentil – and have a distinctive, earthy flavor. Lentil recipes are used throughout South Asia, the Mediterranean regions and West Asia. They are frequently combined with rice, which has a similar cooking time. A lentil and rice dish is referred to in western Asia as *mujaddara or mejadra.* Rice and lentils are also cooked together in khichdi, a popular dish in the Indian subcontinent (India and Pakistan); a similar dish, kushari, made in Egypt, is considered one of two national dishes. Lentils are used to prepare an inexpensive and nutritious soup all over Europe and North and South America, sometimes combined with some form of chicken or pork. Dried lentils can also be sprouted by soaking in water for one day and keeping moist for several days, which changes their nutrition profile. Lentils with husk remain whole with moderate cooking; lentils without husk tend to disintegrate into a thick purée, which leads to quite different dishes.

Nutritional Status of lentil

With about 30% of their calories from protein, lentils have the third-highest level of protein, by weight, of any legume or nut, after soybeans and hemp. Proteins include the essential amino acids isoleucine and lysine, and lentils are an essential source of inexpensive protein in many parts of the world, especially in West Asia and the Indian subcontinent, which have large vegetarian populations. Lentils are deficient in two essential amino acids, methionine and cysteine. However, sprouted lentils contain sufficient levels of all essential amino acids, including methionine and cysteine. Lentils also contain dietary fiber, folate, vitamin B_1, and minerals. Red (or pink) lentils contain a lower concentration of fiber than green lentils (11% rather than 31%). Health magazine has selected lentils as one of the five healthiest foods. The low levels of Readily Digestible Starch (RDS) 5%, and high levels of Slowly Digested Starch (SDS) 30%, make lentils of great interest to people with diabetes. The remaining 65% of the starch is a resistant starch that is classified RS1, being a high quality resistant starch, which is 32% amylose. Lentils also have some anti-nutritional factors, such as trypsin inhibitors and relatively high phytate content. Trypsin is an enzyme involved in digestion, and phytates reduce the bio-availability of dietary minerals. The phytates can be reduced by soaking the lentils in warm water overnight. Lentils are a good source of iron, having over half of a person's daily iron allowance in a one cup serving.

Table 29.2 Nutritional value of lentil per 100 g dry weight

Energy	1,477 kJ (353 kcal)
Carbohydrates	60 g
Sugars	2 g
Dietary fibre	31 g
Fat	1 g
Protein	26 g
Vitamins	
Thiamine (B1)	(76%) 0.87 mg
Riboflavin (B2)	(18%) 0.211 mg
Niacin (B3)	(17%) 2.605 mg
Pantothenic acid (B5)	(42%) 2.120 mg
Vitamin B6	(42%) 0.54 mg
Folate (B9)	(120%) 479 ìg
Vitamin C	(5%) 4.4 mg
Trace metals	
Calcium	(6%) 56 mg
Iron	(58%) 7.54 mg
Magnesium	(34%) 122 mg
Phosphorus	(64%) 451 mg
Potassium	(20%) 955 mg
Sodium	(0%) 6 mg
Zinc	(50%) 4.78 mg
Other constituents	
Water	10.4 g

Units μg = micrograms, mg= milligrams, IU, International Units

Source: USDA Nutrient Database

Production

Lentil production for the major lentil producing nations has been trending upwards during the past 7 years, ranging from 2.1 million tones (Mt) in 2002-2003 to 2.9 Mt in 2009-2010. Among the main producers, production has been trending upwards in Canada, the US, and Australia, but has been highly variable and trending down in India, Bangladesh, Syria and Turkey. In the US, production increased sharply when lentils were first included under the loan program in 2002. Lentils are relatively tolerant to drought, and are grown throughout the world. The FAO reported that the world production of lentils for calendar year 2009 was 3.917 million metric tons, primarily coming from Canada, India, Turkey and Australia. About a quarter of the worldwide production of lentils is from India, most of which is consumed in the domestic market. Canada is the largest export producer of lentils in the world and Saskatchewan is the most important producing region in Canada. Statistics Canada estimates that Canadian lentil production for the 2009/10 year is a record 1.5 million metric tons.

Table 29.3 Top ten lentil producers during 2012

Top 10 Lentil Producing Countries (in metric tons)

Rank	Country	2010	2011	2012
1	Canada	1,947,100	1,531,900	1,493,620
2	India	1,031,600	943,800	950,000
3	Australia	140,000	379,659	463,000
4	Turkey	447,400	405,952	438,000
5	United States	392,675	214,640	240,490
6	Nepal	151,757	206,969	208,201
7	Ethiopia	80,952	128,009	151,500
8	China	125,000	150,000	145,000
9	Syria	77,328	112,470	130,229
10	Iran	100,174	71,808	85,000
	World	4,686,673	4,386,870	4,522,097

Source: UN Food & Agriculture Organization

Uses

On average, about 70% of all world lentil production is consumed in the countries where they are produced. Total world use has been trending upwards during the past 10 years. It is cultivated for its seed and mostly eaten as *dal*. Dal is seed that is decorticated and split. The primary product is the seed which has relatively higher contents of protein, carbohydrate and calories compared to other pulses and are the most desired crop because of its high average protein content and fast cooking characteristic in many lentil producing regions (Muehlbauer *et al.*, 1985). Lentils are generally used for food. They are canned or packaged, whole or split, for retail sale, or processed into flour. They are then used in soups, stews, salads, casseroles, snack food and vegetarian dishes. In southern Asia, split red lentils are used in curries. Lentil flour is added to cereal flour to make breads, cakes and baby foods. Lentils are often used as a meat extender or substitute because of the high protein content and quality. Lentils have a shorter cooking time than other pulses and do not need to be pre-soaked. Only a relatively small volume of low quality lentils are used for livestock feed when degrading factors such as chipping, wrinkling or staining make them undesirable for human food uses where visual attributes are important. However nutritional analysis indicates that they make an excellent feed. Husks, dried leaves, stems, fruit walls and bran (residues), can be fed to livestock. Lentil residues contain about 10.2% moisture, 1.8 % fat, 4.4% protein, 50% carbohydrate, 21.4% fiber, and 12.2% ash (Muehlbauer *et al.*, 1985). According to Muehlbauer *et al.* (1985), when production of forage crops fall below the level required in the market, lentil residue commands an equal or a better price than lentil seeds in some Middle Eastern countries. Green plants make valuable green manure. Seeds are a source of commercial starch for textile and printing industries (Kay, 1979).

Table 29.4 World: Major Lentil Exporters

Crop year	2004-2005	2005-2006	2006-2007	2007-2008	2008-2009	2009-2010f
			thousand tonnes			
Canada	372	576	853	811	973	1,395
United States	83	160	118	114	116	193
Australia	150	201	103	74	50	128
Turkey	171	118	147	130	25	44
India	137	0	0	0	0	0
Syria	71	0	0	0	0	0
Total	**984**	**1,055**	**1,221**	**1,129**	**1,164**	**1,760**

f: forecast

Source FAO, Statistics Canada, USDA and Global Trade Atlas

Yields and Economics

Seed yields range from 450-675 kg/ha in dry areas, may increase to 2000 kg/ha with irrigation, and yields over 3,000 kg/ha have been recorded. The straw-to-seed ratio in one cultivar was about 1.2:1 and in studies conducted on 28 cultivars in New Delhi, India, pulse yields ranged from 558 to 1,750 kg/ha, while dry matter yields ranged from 2,667 to 3,550 kg/ha (Duke, 1981). The major producer of lentils in the world is India, with about 1,160,000 hectares producing 850,000 MT in 1994; while World production was 2.875 million MT on about 3.36 million hectares during the same year (FAO, 1994). Other important producers are USA, Australia, Canada, Pakistan, Syria, Argentina, Chile, Turkey, Ethiopia and Spain. In these countries, yields have ranged from 637 to 1263 kg/ha. The highest yield, 5000 kg/ha was recorded in Germany (FAO) while the world production of lentil increased by about 65% over the past 25 years (FAO, 1996). In developing countries alone, lentil production and yield rose by 60% (Hulse, 1994). Major production increases have been recorded in Turkey and Canada. In the USA, lentil occupies about 60,000 ha, and from 1984 to 1993, 65, 000 tons of lentil have been produced (Muehlbauer, 1996). Among the agronomic problems, pod shedding, pod shattering, lodging, and the spreading growth habit are important constraints in production (Robertson *et al.*, 1996). Close to 80-85 % of lentil production in the USA is exported (Muehlbauer *et al.*, 1995 and Muehlbauer, 1996). Canada, Turkey and the USA are major exporters of lentil.

Production of Lentil in India

Lentil production in India has always been important as it is the one of the most important *rabi* crops in the country. India has been producing lentil since 1st century AD and has always been an important producer of the crop. In fact, India was the largest producer of the Lentil crop in the world until recently Canada took over the lead leaving India at the second place. Indian production of this crop hovers around 10 lakh metric tons per year that is cultivated on about 14 lakh hectares of land. Lentil crop is grown in India in the winter season in the following states:

- Uttar Pradesh
- Madhya Pradesh

- Bihar
- West Bengal
- Rajasthan
- Haryana
- Punjab
- Assam
- Maharashtra

Around 90% of the production comes from the top four states in the list pertaining to the eastern and the northern part of the country. The southern part of the country hardly contributes to India's total production. Uttar Pradesh accounts for the maximum production in the country contributing to around 45% of the country's production as well as for the maximum area under Lentil cultivation. The crop is both cultivated as a primary crop and a secondary crop in the country. Sagar, Jabalpur, Bundelkhand and Bhopal in Madhya Pradesh, *Tal* lands spread over south Bihar districts in Bihar, Kanpur in Uttar Pradesh and Kota in Rajasthan are the districts where Lentil is cultivated primarily.

Indian Lentil market

Indian subcontinent has a reputation of being an important player in the world's pulse scenario. Indian economy is largely affected from the fluctuations in world pulse demand and supply, as the country is one of the largest consumers of pulses due to a large population to feed to. In case of Lentil, India had been quite dominant country in the world production schedule as it was the largest Lentil producer till 2004 and had been leading since a very long time. Canada took over India's lead in 2005 with the help of a sharp rise in its production level. However, India still stands at the second place as far as the world production are concerned, producing around 10 lakh tons of Lentils every year. Uttar Pradesh in India is the state that contributes to around 45% share in the total production. The area which is utilized in the cultivation of Lentil in India, sums up to around 14 lakh hectares. India consumes almost 90-95% of its domestic produce to fulfil its domestic consumption demand as one of the most important constituent in the country's staple diet. The major trading centres of lentil in India are located at Jabalpur, Rewa, Bhopal, Sagar and Vidisha in Madhya Pradesh; Sultanpur and Bahraich in Uttar Pradesh; Bhatpara in West Bengal and Rajnandgarh in Chhattisgarh. The factor which influence the market, include, weather fluctuations, information flow regarding the supply of the pulse, price movements of the substitute pulses and production level in the main exporting countries Also, lentil is traded in the various Indian commodity exchanges like Multi Commodity Exchange of India, National Multi Commodity Exchange of India and National Commodity and Derivatives Exchange.

Constraints in lentil production

Terminal Drought

There has been a high degree of risk in pulses production in general and lentil production in particular. More than 87% of the area under pulses is presently rain fed and the mean rainfall of major pulse growing states of India is around 1,000 mm with coefficient of variation of the rainfall being 20-25 percent. Moisture stress is the oft-cited reason for crop failures. Terminal drought and heat stress results in forced maturity with low yields. Drought stress alone may reduce seed yields by 50% in the tropics (Reddy, 2009).

Excess soil moisture and humidity

Water-logging during seed germination and initial seeding growth phase is a constraint for pulses grown in rice-fallows. High humidity during vegetative growth stages (particularly during November to January) makes lentil susceptible to diseases such as BGM. This can happen in some years, where excessive rainfall occurs during the winter that can encourage excessive vegetative growth leading to lodging, and also encourage development of various leaf and root diseases (Subbarao *et al.*, 2001).

Temperature

Grain yield is mainly influenced by temperature. Cold is an abiotic stress, limiting the grain yield of pulse crops. All hot season pulses are sensitive to low temperatures, but generally these are not exposed to low temperatures. On the other hand, cool season pulses (chickpea) are often subjected to chilling temperatures especially in areas of north India. However there has not been much improvement in the development of chilling and frost tolerant varieties.

Soil acidity

Soil acidity was a serious constraint to legume introduction into rabi season as most pulses are highly sensitive to soil acidity as compared to cereals and other non-pulses crops (Choudhary and Pande 1986, Chong *et al.,* 1987).

Uncertainty in Rainfall

Uncertainty in rainfall is a serious abiotic constraint in lentil production. Poor drainage/water stagnation during the rainy season causes heavy losses on account of low plant stand and increased incidence of blight disease, particularly in the states of UP, Bihar, West Bengal, Chhattisgarh, MP and Jharkhand. Since most pulse crops are drought tolerant, most of the research efforts have been confined to develop genotypes and associated production technologies to suit dry land conditions. Consequently, germplasm suited to high rainfall and irrigated conditions are lacking (Reddy, 2009)

Fig. 29.3 Availability of quality seed of lentil is major bottleneck in realizing its production potential

Socio Economic Constraints

Lack of knowledge, non availability of improved seed, poor technical guidance were few socio economic institutional constraints (Subbarao *et al.,* 2001 and Pande *et al.,* 2003) 12 while seed storage, poor irrigation and poor marketing were socio economic infrastructural constraints in pulses production (Pande *et al.,* 2003). According to Reddy (2009) lack of knowledge, availability of seed, lack of technical knowledge regarding improved pulses cultivation as some of the important constraints in pulses production. Followed by lack of availability of suitable agro chemical and inputs are as an important institutional constraint. Among infrastructural constraints, in pulses production; seed storage and poor marketing of pulses are important constraints. Markets for pulses are thin and fragmented in comparison with rice and wheat. It is generally perceived that the government procurement for pulses is not as effective as it is for rice and wheat, and often, farmers do not realize the minimum price announced by the government. Also the price spread (i.e. the market margin) for pulses is much higher than that of rice and wheat (Joshi and Pande, 1996). Because of this, farmers do not benefit by the higher market prices of pulses, as most of the profits go to the traders, rather than to the farmers. Hence the farmers are less motivation for pulses cultivation and thus, the serious threat to sustainable agriculture system and the economy of the country. Problem of poor transportation is a very common problem faced by the growers. In addition, to the above, poor irrigation also affects the pulse crops in general and lentil in particular.

Point to be consider for lentil production

Lentil is best adapted to the Brown and Dark Brown soil zones. Excess moisture aggravates disease problems and greatly delays maturity. Lentil plants do not tolerate waterlogged soils and will die if flooded. Maturity becomes very important in areas where moisture is abundant, temperatures are cooler, and the growing season is shorter. Herbicide tolerance is another varietal characteristic to consider. Crop insurance coverage may also influence variety choice. Perennial weeds such as Canada thistle, dandelion, perennial sow thistle, and quack grass are very competitive in any crop and even more so in a noncompetitive crop like lentil. Lentil is susceptible to soil residues of many herbicides commonly used (in-crop and pre-seed).Crop rotation goes beyond disease considerations and weed control (including volunteer crop). Crop rotation can influence both the lentil crop and the crop(s) following lentil. Disruption of disease cycles is critical. Lentil fits well into a direct seeding crop production system. Use of high quality seed is the first step in establishing a rapidly emerging, vigorous stand and producing a high quality, profitable crop. Different fungicides control different species of fungal organisms, so it is important to know what organism is infecting your seed. Lentil inoculated with the proper *rhizobium* (bacterial) strain has the potential to fix up to 80 per cent of its nitrogen requirement through nitrogen fixation. In addition to fixing a substantial quantity of nitrogen during the growing season, lentil makes a positive contribution to the overall soil nitrogen level over multiple years. Following application of the inoculants, plant the inoculated seed into moist soil as soon as possible. Inoculants are available in different formulations including liquid, powdered, and granular. A soil test should be used to plan a fertility program for lentil.Generally, nitrogen fertilizer is not required if nitrogen fixation is optimized. In fields where lentil is grown for the first time, starter-nitrogen may be useful, particularly on low nitrogen soils.The recommended plant population for lentil is 130/m2 (12/ft.2). Lentil should be seeded early, as soon as the soil temperature at seeding depth is 5°C or greater, and the soil is not excessively wet. Seeding depth of 2.5 to 7.5 cm is advised. Lentil is a poor competitor against weeds. Resistance can build with each application; applications do not have to be consecutive year after year. Crop rotations that include cereal and oilseed crops can reduce the build-up of soil-borne pathogens specific to lentil. A disease decision support checklist for ascochyta and anthracnose has been developed by AAFC as a working tool for producers to help in establishing thresholds for fungicide application on lentil to control anthracnose and *ascochyta* in lentil. Early identification of infection sites is critical. The list of insects to monitor in lentil is not extensive. For red lentil, the basic quality parameters are dictated by milling requirements and are colour, size, condition of dehulled split seeds, or the condition of dehulled whole seeds having cotyledons attached (footballs). For green lentil the basic quality parameters are seed diameter, seed thickness and uniformity, color uniformity, and intact green seed coats without wrinkling or staining. Swathing will hasten dry down and prevent shattering. Chemical desiccation to dry down crop foliage and weeds will reduce the time from maturity to threshing readiness, reduce shatter losses, and result in improved quality if the seed is harvested before being exposed to wet weather. Glyphosate is not a desiccant. If applied, it should be for the purpose of weed control (most frequently perennial weeds but, also annual weeds in situations described above). Seed moisture higher than 18 per cent results in longer drying times and as lentil moisture approaches 20 per cent the crop is difficult to thresh without smashing the seed. Lentil seed at 14 per cent moisture or less and 15°C or less can be stored safely for up to 40 weeks.

Conclusion

Lentil is one of the most important and cheap source of vegetable protein. India is largest consumer of lentil. However, India was the largest producer of the Lentil crop in the world until recently Canada took over the lead leaving India at the second place.

References

Ahlawat IPS, Ali M, Yadav RL, Kumar Rao and Rego TJ. 1998. Biological nitrogen fixation and residual effects of summer and rainy season grain pulses in rice and wheat cropping systems of the Indo-Gangetic Plain. In JVDK Kumar Rao, C. Johansen (eds.) Residual Effects of Pulses in Rice and Wheat Cropping Systems of the Indo-Gangetic Plain. International Crop Research Institutes for Semi-Arid Tropics (ICRISTAT), Oxford and IBH Publishing Co. Pvt. Ltd, New Delhi, 1998. ISBN 81-204-1297-4, pp. 31-54.

Chong K, Wynne V, Elkan GH and Schneewies TG. 1987. Effects of soil acidity and aluminium content on Rhizobium inoculation, growth and nitrogen fixation of groundnut and other pulses. *Tropical Agriculture* 64: 97-104.

Choudhary, MA and Pandey RK. 1986. Extending limits to legume crop establishment in rainfed lowland rice. Pages 235-239 in Food pulses improvement for Asian farming systems: proceedings of an international Workshop held in Khon Kaen, Thailand, 1-5 September 1986 (Wallis, E.S., and Byth, D.E., eds.). ACIAR Proceedings no. 18. Canberra, Australia: Australian Centre for International Agricultural Research.

Duke JA. 1981. Handbook of pulses of world economic importance. Plenum Press, New York. p. 52-57.

Food and Agriculture Organization of the United Nations. 1996. Production Yearbook. Rome, Italy.

Hobbs P and Morris M. 1996. Meeting South Asias future food requirements from rice-wheat cropping systems: priority issues facing researches in the post green revolution era. Natural Resource Group Paper 96-101. Mexico: Centro International de Mejoramiento de Maiz y Trigo.

Hulse JH. 1990. Nature, composition and utilization of grain pulses. p.11-27. In: ICRISAT. Uses of tropical grain pulses: Proceedings of a consultants' meeting, 27-30 Mar, 1989. ICRISAT Center, India, Patancheru, AP. 502 324. India: ICRISAT.

Joshi PK. 1998. Performance of Grain Pulses in the Indo-Gangetic Plain. In J.V.D.K. Kumar Rao, C. Johansen (ed.) Residual Effects of Pulses in Rice and Wheat Cropping Systems of the Indo-Gangetic Plain. International Crop Research Institutes for Semi-Arid Tropics (ICRISTAT). Oxford and IBH Publishing Co. Pvt. Ltd, New Delhi, 1998. ISBN 81-204-1297-4. pp 3-12.

Joshi PK and Pande S. 1996. Constraints to and prospects for adoption of extra-short duration pigeon pea in northern India: Some socio-economic aspects. Pages 106-115 in Prospects for growing extra-short duration pigeon pea in rotation with winter crops: proceedings of the IARI/ICRISAT workshop and Monitoring tour, 16-18 October 1995, New Delhi, India. Patancheru 502 324, Andhra Pradesh, India: International Crops Research Institute for Semi-Arid Tropics.

Jumbunathan RHL, Blain KS, Dhindsa LA, Hussein K, Kogure L, Li-Juan and Youseef MM. 1994. Diversifying use of cool season food pulses through processing. pp. 98-112. In: F.J. Muehlbauer and W.J. Kaiser (eds.) Expanding the Production and Use of Cool Season Food Pulses. Kluwer Academic Publishers. Dordrecht, The Netherlands.

Kay D. 1979. Food pulses. Tropical Development and Research Institute (TPI). *TPI Crop and Product Digest* No. 3. p.48-71. UK.

Kumar S and Bourai VA. 2012. Economic Analysis of Pulses Production Their Benefits and Constraints- A Case Study of Sample Villages of Assan Valley of Uttarakhand, India. *IOSR Journal of Humanities and Social Science* 1(4): 41-53.

Lauren JG, Duxbury JM, Beri VS, Razzaque IIMA, Sattar MA, Pande SP, Bhattarai S, Mann RA and Ladha JK. 1998. Direct and residual effects from forage and green manure pulses in rice-based cropping systems. In: J.V.D.K. Kumar Rao, C. Johansen (ed.) Residual Effects of Pulses in Rice and Wheat Cropping Systems of the Indo-Gangetic Plain. International Crop Research Institutes for Semi-Arid Tropics (ICRISTAT). Oxford and IBH Publishing Co. Pvt. Ltd, New Delhi, 1998. ISBN 81-204-1297-4, pp. 55-82.

Metrological Department Dehradun, Govt. of India 2008-09.

Muehlbauer FJ, Short RW, Summerfield RJ, Morrison KJ and Swan DG. 1981. Description and culture of lentils. Cooperative Extension, College of Agriculture, Washington State University and USDA-ARS. EB 0957.

Muehlbauer FJ, Cubero JI and Summerfield R.J. 1985. Lentil (*Lens culinaris* Medic.). p. 266-311. In: R.J. Summerfield and E.H. Roberts (eds.), Grain Legume Crops. Collins, 8 Grafton Street, London, UK.

Muehlbauer FJ, Kaiser WJ, Clement SL and Summerfield RJ. 1995. Production and breeding of lentil. *Advances in Agronomy* 54: 283-332.

Muehlbauer FJ. 1996. Advances in the production of cool season pulses. *American Journal of Alternative Agriculture* 11: 71-76.

Reddy AA. 2009. Pulses Production Technology: Status and Way Forward, *Economic & Political Weekly*, 26 December, vol. xliv no 52, pp. 73-80.

Reddy AA and Reddy GP. 2010. Supply Side Constrains in Production of Pulses in India: A Case Study of Lentil. *Agricultural Economics Research Review 23*(January-June): 129-136.

Robertson LD, Singh KB, Erskine W, Ali M and Abd El Moneim. 1996. Useful genetic diversity in germplasm collections of food and forage pulses from West Asia and north Africa. *Germplasm Resources and Crop Evolution* 43: 447-460.

Subbarao GV, Kumar Rao, JVDK, Kumar J, Debb UK, Ahmed I and Harris D. 2001. Spatial Distribution and Quantification of Rice-Fallows in South Asia- Potential for Pulses. International Crops Research Institute for Semi-Arid Tropics (ICRISTAT), National Remote Sensing Agency (NRSA) and Department for International Development (DFID). pp 74-80. ISBN 92-9066-436-3.

Williams PC and Singh U. 1988. Quality screening and evaluation in pulse breeding. p. 445-457. In: R.J. Summerfield (ed.), World Crops: Cool Season Food Pulses. Kluwer Academic Publishers, Dordrecht The Netherlands.

Yadav FJ, Rl., Dwivedi BS, Gangwar KS and Prasad K. 1998. Over view and prospects for enhancing residual benefits of pulses in rice and wheat cropping systems in India. In: JVDK Kumar Rao C. Johansen (ed.) Residual Effects of Pulses in Rice and Wheat Cropping Systems of the Indo-Gangetic Plain. International Crop Research Institutes for Semi-Arid Tropics (ICRISTAT). Oxford and IBH Publishing Co. Pvt. Ltd, New Delhi, 1998. ISBN 81-204-1297-4, pp. 207-226.

Citation

Singh KM and Singh AK. 2018. Lentil in India: An Overview. In: Scientific Lentil Production. Singh AK, Bhakta N, Sangale UR, Manibhushan, Sundaram PK, Kumar S and Yasin JK. (eds.) Society for Upliftment of Rural Economy Varanasi, India. Pp. 491-507.

❑❑❑

Chapter 30

Performance of Lentil Cultivar Arun (PL-77-12) at Farmer's Field in Patna District: Practical Experience

AK Singh, SS Singh and UR Sangle

SYNOPSIS

Lentil is poora man meat and its a crop of marginilze farmers, often they heavly depend lentil crop as due to unfavourbale agroclimatic condition one lentil can be grown on residual moisture, if this crop is also fail, they are bound to be in deep trouble. This is personal experience of whole team when visited the lentil growing areas where onset of reproductive phase of this cropunusally delayed. This report reveals some of possible reasoning for the delayed flowering of lentil. Later on the crop was harvested with moderate yield due to less time avaialabe for seed reproductive phase.

Keywords: Climate change, Delayed flowering, Lentil, Nutrient management.

Introduction

Lentil (*Lens culinaris* M.), is cool season annual legume crop. Lentil (*Lens culinaris* M.), is one of the ancient food legume crop (Deokaran *et al.*, 2014). After chickpea it is the second most important cool season grain legume (Singh *et al.*, 2011). At global level, India ranks first in area and second in production, the other countries growing lentil are Ethiopia, North Africa and to lesser extent in southern Europe, USA, Australia and Canada. Being a grain legume, it has the ability to fix atmospheric nitrogen and do carbon sequestration to improve soil health (Singh and Bhatt, 2015).

Term of reference

The department of agriculture, government of Bihar has received complains that during *Rabi* 2012-13 lentil producing farmers in Goai village of Naubatpur Block, Patna districts, have got the seed of other cultivars /varieties in name of two lentil varieties namely Arun (PL-77-12) and Mallika(K-75) provided by Bihar Rajy Beej Nigam (BRBN) Bihar. Farmers claim/complains that the performance of Arun (PL-77-12) did not match with the standard one. The government of Bihar has requested ICAR-RCER, Patna and ARI, BAU, Patna to depute lentil scientist for on spot inspection and report to the government. The state Government was instructed to facilitate the visit of scientist. The Director ICAR-RCER, Patna has deputed the team of scientists on the recommended of Dr SS Singh, Head, Crop Research, Division, constituting Agronomist Dr A. K. Singh, and Plant Pathologist Dr U. R. Sangle, vide Note dated 25.02.2013. The team was assigned to report about the quality & germination of supplied lentil seed and truthfulness of claimed lentil cultivar/variety.

Fig. 30.1 Field inspection benig conducted by the team

Fig. 30.2 Sample collection by the team

Date of visit

The ICAR-RCER team visited the site along with District Agriculture officer (Patna) and Sub-divisional Agriculture officer (Danapur Subdivision), Govt. of Bihar who had requested to ICAR RCER, Patna to join the team of ARI, BAU, Bihar. The team along with Sub-divisional Agriculture Officer (Danapur Sub-division), Govt. of Bihar proceeded for Naubatpur Block, Patna district on 25.02.2013 and visited four sites for verification of variety used and its performance. The observations on different growth parameters were recorded at the sites of farmer's field of Naubatpur block of Patna districts.

Following members were also accompanied the ICAR-RCER team:

1. Sh. Ganesh Ram, District Agril. Officer, Patna
2. Sh. Niranjaj Kumar Bhardwaj, Block Agril. Officer, Naubatpur
3. Sh. DN Paswan, Sub Divisional Agriculture Officer (Danapur)
4. Dr. Ajay Kumar, Chief Scientist (Agronomy)
5. Dr. Arvind Kumar, Chief Scientist (Plant Pathology)
6. Dr. Dilip Kumar, Chief Scientist (Soil Science)

Team visited the lentil growing sites and discussed with the farmers for seeking their viewpoint on these aspects. Following farmers of Goai village of Naubatpur Block, Patna, actively participated in the discussion.

1. Shri Bramha Dev Vema
2. Shri Jitendra kumar
3. Shri Ramesh Sharma
4. Shri Ram Mohan Sharma
5. Shri Surendra Sharma
6. Shri Mahesh Kumar
7. Shri Santosh Kumar
8. Shri Satya Narayan Singh
9. Shri Rajendra Singh
10. Shri Ram Kishor Sharma

Fig. 30.3 Plant sample collect from farmers field (a) and actual plant sample from research field(b)

Ideal plant type of lentil variety Arun (PL-77-12)

The phenotypic characteristic of the acclaimed variety *i.e.* Arun (PL-77-12) are entirely different from those of other contemporary cultivars still in cultivation (Singh *et al.*, 2008). The cultivar had unique pale green leaf colour and semi spreading plant type, medium stature (Fig. 30.5a & 30.1b and Fig. 30.6) along with typical and peculiar feature i.e. pronounced leaflet tendril as compared to other genotypes. The tendril characterises the needle like structure at the terminal point of leaflet (Fig. 30.3a & Fig. 30.3b). Seeds belong to medium size category (Fig. 30.4a & Fig. 30.4b). The gestation period of this cultivar is about 120-130 days. Yield potential of the said cultivar is 1.8 to 2.0 t/ha. Ideal sowing time for this cultivar is second fortnight of October to first fortnight of November for whole of the Bihar state.

Actual plant type at farmer's field of lentil variety Arun (PL-77-12)

Team observed and found that the phenotype of plant (as stated above) was up to the mark and was having proximity with the said cultivar within the accepted range (Fig. 30.3a & Fig. 30.3b). Other lentil cultivars were grown in the vicinity of this plot having distinct characters with onset of reproductive stages, having comparable dark green leaf colour (Fig. 30.6). There is lack of sure supply of quality seed in general and pulse crop in particular (Singh *et al.*, 2015).

Fig. 30.4(a) Medium bold seeded Arun (PL-72-12)

Fig. 30.4(b) Small seeded Mallika (K-75)

Evaluation of true plant type and farmer's field plant type of Arun (PL-77-12)

Plant samples were collected from affected farmers fields and the same were compared with plant type of variety Arun (PL-77-12) grown on 22.11.2012 at Research Farm of ICAR-RCER, Patna (**Fig.** 30.3 & Fig. 30.6). the comparison was made with the crop sown at the research farm. The growth stage of crops of farmer's field was similar to the crops sown at the experimental farm of ICAR RCER, Patna.

Fig. 30.5(a) Leaflet tendril in the Plant sample collect from farmers field

Fig. 30.5(b) Leaflet tendril in actual plant sample from research field

Possible reasons for delayed / no onset of reproductive phase

1. Since the variety Arun (PL-77-12) is a medium duration cultivar having 120-130 days duration, delay in planting results into poor flowering and podding. Under extreme delayed sowing conditions sometimes, lentil may be unable to bear flower completely. At ICAR RCER research farm, under delayed planting condition (planting was done on 18th and 22nd November, 2012), all the lentil cultivars had already entered in the reproductive phase and podding and pod filling is still going on, whereas it is worth to mention that cultivars Arun (PL-77-12) did not bear flowers as on 26.02.2013. This clearly indicates its nature and response to the prevailing agro-climatic conditions (Fig 30.3a & Fig. 30.3b & Fig. 30.4).
2. Farmers revealed that this is traditionally rice-lentil growing belt. They simply broadcast lentil seeds and harvest. No use of fertiliser may lead to acute deficiency of secondary nutrient i.e. sulphur and micronutrients especially boron and molybdenum apart from zinc (Singh *et al.*, 2012 and Singh and Bhatt, 2015).
3. However, keeping in view the extended favourable weather conditions, limited anthesis and pod formation may be realized in due course of time.
4. Most importantly, there is urgent need to compare prevailing weather parameters during the current crop season with those of previous one to ascertain the role of changing climate, if any.

Outpt of the Visit

1. Poor germination was neither reported by farmers not observed by the visiting team across the farmer's field during the field survey conducted on 25.02.2013. Hence, quality of supplied lentil seeds of said cultivar may not be confirmedly said to be substandard.
2. Lentil cultivar Arun (PL-77-12) possesses distinctive features like leaflet tendril and pale yellow leaf colour, which are different, visible, unique and easy to identify from the other lentil genotypes. The cultivar under field condition resembles the features of lentil cultivar Arun (PL-77-12).
3. Delayed planting along with short supply of mineral nutrition, gradual build-up of biotic pests and unfavourable agro-climatic

condition, individually and collectively responsible for delayed or no onset of reproductive phase as indicated by poor or no flower initiation so for particularly in case of lentil cultivar Arun (PL-77-12).

Recommendations

1. The delayed planting of cultivar Arun (PL-77-12) of lentil should be avoided to overcome the problems of delayed/ no flowering. To see the effect of delayed planting, an experiment is being conducted at ICAR RCER, research farm. The crop stand condition of cultivar in question is similar to that on the farmer's field till today. The clear-cut effect of delayed planting will emerge after the results of the on-going experiment are obtained.
2. There is urgent need to undertake the balanced nutrient management with reference to sulphur, boron; molybdenum and zinc.
3. There is urgent need to adopt crop rotation and cropping system to avoid build-up of plant pathogen and host specific parasitic weed.
4. Farmers are advised to adopt the full package of agronomic practices especially with regard to time of planting as per the recommendation.

Fig. 30.6 Difference in the leaf colour and other morphologial characters

Effects of delayed onset of reproductive phase on lentil production

Ducring the visit very poor flowering was noticed and as the foarmers feer for no flowering may be happens and consequently no seed setting may be takesplace, was a simple fear of falure, as the same situation was also happens at ICAR Resaecr complex for eastern region Patan. Peropdiacal observation was recorded and ultimately crop came in to reproductive phase with delay of 7-12 days at both the places at our research farm as well as affected farmer's field. Due to delay in onset of reproductive phase resulted negative impact on lentil production and 15-25 % lower lentil seed production was recoded due to less time availave for seed setting and force maturity.

Conclusion

Poor germination of lentil may be attributed by delayed planting along with short supply of mineral nutrition, gradual build-up of biotic pests and unfavourable agro-climatic condition, individually and collectively responsible for delayed or no onset of reproductive phase as indicated by poor or no flower initiation so for particularly in case of lentil cultivar Arun (PL-77-12). Since cultivar in question has been recommended for normal sowing being long duration crop, hence delayed planting should be ignore. It is worth to mentiond that at both the place at our research farm as well as affected farmers field, delayed onset of reproductive phase caused forced maturity and lower lentil production was recoded as compared to normal years

References

Deokaran, Singh SB and Ram Kewal. 2014. Effect of Zinc and Boron application on Yield of Lentil and Nutrient Balance in the Soil under Indo-Gangetic Plain Zones. *Journal of AgriSearch* 1(4): 206-209.

Singh AK and Bhatt BP 2015. Late sown lentil performance in response to foliar application of zinc. *Bangladesh Journal of Botany* 44(1): 125-128.

Singh AK, Bhatt BP, Sundaram PK, Chndra N, Bharati RC and Patel SK. 2012. Faba bean (*Vicia faba* L.) phenology and performance in response to its seed size class and planting depth. *Int. J. of Agril. & Stat. Sci.* 8(1): 97-109.

Singh AK, Manibhushan, Chandra Naresh and Bharati RC. 2008. Suitable crop varieties for limited irrigated conditions in different agro climatic zones of India. *Int. J. Trop. Agri.* 26(3-4): 491-496.

Singh AK, Singh SS, Prakash V, Kumar S and Dwivedi SK. 2015. Pulses Production in India: Present Status, Bottleneck and Way Forward. *Journal of Agrisearch* 2(2): 75-83.

Singh AK, Meena, MK Bharati RC. 2011. Sulphur and Zinc Nutrient Management In Rice Lentil Cropping System. International Conference on "Life Science Research for Rural and Agricultural Development. 27-29 December, 2011, CPRS Patna (Bihar). pp. 66-67.

Citation

Singh AK, Singh SS and Sangle UR. 2018. Performance of lentil cultivar Arun (PL-77-12) at farmer's field in Patna District: Practical Experience. In: Scientific Lentil Production. Singh AK, Bhakta N, Sangale UR, Manibhushan, Sundaram PK, Kumar S and Yasin JK. (eds.) Society for Upliftment of Rural Economy Varanasi, India. Pp. 509-517.

❑❑❑

Chapter 31

Database Management System- A Tool for the Development and management of Lentil

Manibhushan and N Chandra

SYNOPSIS

Database management system (DBMS) is software for the storage, management and retrieval of data without redundancy of data. Database has been developed using normalization technique to assure data consistency but no data redundancy. A DBMS allows for concurrent access by multiple users, each with their own specific role. Some users only need to view the data, some contribute to adding new data, while others design and manage the database - all at the same time. A DBMS provides automated methods to create, store and retrieve data. A database management system on lentil pulse crop has been developed using Oracle and Visual Basic software. This information system is able to store data into database, manage data and retrieve information as and when required. Database has been designed to follow third normal form. Database information system is working properly and data is being stored and managed very efficiently. This information system may be useful for scientists as well as farmers.

Keywords: Database, Lentil, Management system.

Introduction

Pulses are not only important sources of proteins but also offer vitamins and minerals, popularly known as "Poor man's meat" and "rich men vegetable". Pulses are the staple source of dietary protein to the majority of Indian population and contribute significantly to the nutritional security of the country. India is the largest producer and consumer of pulses in the world. For an active normal body pulses requirement is about 40 g per day or 14.6 kg per person per year. Now, India is in the world leader in production, consumption and import as well in case of pulses India imports 2-3 Mt of pulses on regular basis and 9.2 Mt of vegetable oils during 2010-11. By the 2050, India as a whole will be able to sustain their production. Indian will produced plenty with respect to both the non-food commodity *i.e.* oilseeds and pulses, to make us self-sufficient in the entire food commodity during 2050 (Singh *et al.*, 2013). Lentil (*Lens culinaris* Medic.) is a winter season crop which thrives well in cool conditions. In India, two types of lentils are commonly recognized in trade, comparatively bold seeded is known as masur or malka masur and small seeded is known as musri.

In India, lentil is cultivated in four agro-climatic regions viz., North hill zone, North West plain zone, North east plain zone and central zone. It occupies 1.51 million ha area with production of 0.96 million tones. In India it was grown in 1.48 m ha with production of 1.30 m ton during 2010-11 (Singh and Bhatt, 2015).

Fig. 31.1 Lentil seeds & plant at pre-flowering stage showing

Database Management System

Database Management System (DBMS) is software that is used to store data into a database, manipulate and retrieve data from a database. A database is a collection of data that are stored in tables in a well structured and defined manner in which these data can be utilized by DBMS for the manipulation and retrieval of data according to the requirement of users (Manibhushan *et al.*, 2014).

Purpose of database management system: A database management system (DBMS) is a software tool to organize the data properly in a suitable database. The database may be large or small. A DBMS contains of three elements that are physical database,

database engine and database scheme. DBMS makes it possible for the users to insert data into database, update data, delete data and edit data from a database properly. DBMS makes a database almost free from data redundancy and makes database consistent and secure. DBMS also provides concurrent access of data from a database by multiple users and provides integrity also. The development and management of a database in an organization is mainly done by database administrator (DBA) and other computer specialists. DBMS has the following properties:

Entity and attributes: DBMS has been developed using real world data or entity. As for example, a student database may use student as an entity and student roll number, name, address, course of study, class as their attributes.

Tables based on relation: Different tables are created for different entities and a relationship is established among various entities. This model is known as Entity- Relationship (E-R) model of a database system and this model is very popular and established. The logical association among entities called relationship. Relationships are mapped with entities in various ways that is known as mapping cardinalities are of mainly four types that are one to one, one to many, many to one and many to many. There are a variety of data modeling techniques that can be used but the entity relationship model approach has the widest acceptance (Chen, 1976). The concept of relational approach was investigated by Codd (1970). The relational approach is based on the theory of relational algebra (Ullman, 1982). An example of E-R model has been shown in the Fig. 31.2.

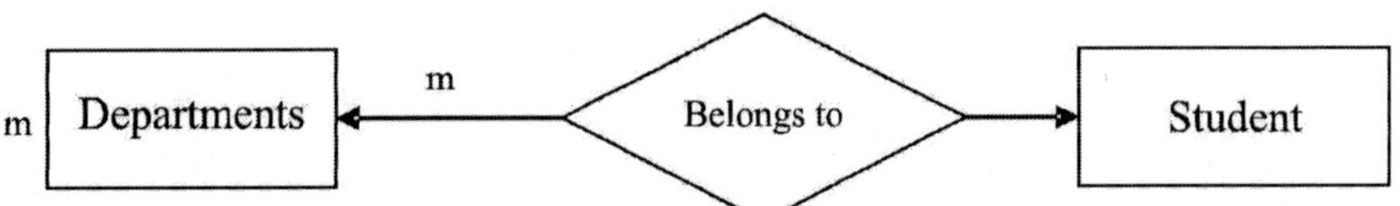

Fig. 31.2 E-R Model representing a student's relationship with the departments (many to many)

Less redundancy: Rule of normalization is followed by DBMS in creating databases. Due to normalization, data redundancy has been removed.

Consistency: Database remains consistent due to normalization and data consistency is very necessary to design a database.

ACID Property: ACID stands for Atomicity, Consistency, Isolation and Durability. ACID property has been followed by DBMS to design a good and healthy database which has no data redundancy. So that data entry, updation, deletion and modification become very easy and convenient.

Relational model of a database

The most popular data model used by DBMS is the relational model. In this model of database, data are stored in tables known as relations and relations has been normalized. Relationship sets are defined as subsets of the cross product of two or more entities (Tsichritzis and Lochovaky, 1982). So it contains data and the relationship among data. Data is arranged in a relation represented in a two dimensional tables and data is inserted into the tables in the form of tuples. A tuple is created by one or more than one attributes which are used as basic building blocks of a database. There can be any number of tuples in a table but the tuple contains fixed and same attributes with different values. A tuple is represented by a row and the attribute is represented by a column of the table. Student roll number, name, addresses are the attributes of the table 'student'.

Data Definition Language and Data Manipulation Language

Data Definition Language (DDL): A set of definitions expressed by a special language known as data definition language (DDL). DDL is expressed in SQL (Structured Query Language). An example of DDL has been expressed as following:

Create table student (student_roll char(10), integer).

In this definition, a table student has been created using SQL.

Data Manipulation Language (DML): DML is a language that enables users to access or manipulate data from database. Data manipulation mainly involved data insertion, retrieval, deletion, modification and updation of data in a database. A query is a statement requesting the retrieval of data. The part of DML that is responsible for data retrieval is known as query language. An example of query in SQL is following:

Select student.student_name

From student

Where student.student.roll=18.

The above statement is for the selection of a student name from the student table whose roll number is 18.

Normalization: Normalization is a method to remove data redundancies and bring database to a consistent state in which data entry, updation, deletion and management become easy and convenient from a database. Normalization helps a database to work properly. Normalization is of different types that are first normal form (1 NF), second normal form (2 NF), third normal form (3 NF), Boyce-Codd normal form (BCNF), fourth normal form (4 NF) and fifth normal form (5 NF). So normalization rule is followed in relational database creation for the proper functioning of a database.

Development of Information System of lentil crop: Lentil is a major international pulse crop. Lentils are grown in cool temperature zones of the world and in the winter season in India. Lentil occupies 1.51 million hectare area with a production of 0.95 million tones in India (FAOSTAT, 2009). It is mainly cultivated in Uttar Pradesh, Bihar, Madhya Pradesh, Chhatisgarh and West Bengal. These states together account for 80-90% total area and production of lentil (Rao *et al*, 2011). Average yield of lentil in Bihar is found to be 12-15 quintal/ hectare. Lentil crop can be grown in a wide range of soils but it is more sensitive to water logging and pH<6.5 than the other pulse crop that are chickpea, fababean and peas (Tang and Thompson, 1966). So district wise information system of a state will be very useful for increasing the production of lentil crop. So a district wise information system of Bihar state has been developed to manage the area and production of lentil crop. Some screen shots of information system on lentil crop are shown with dummy data in Fig 30.3 and Fig 30.4. The information system for lentil crop has been developed using Oracle and Visual Basic. The database has been developed in Oracle that uses SQL and the graphic user interfaces (GUI) has been developed in Visual Basic softwares. This information system (database management system) is working properly and the data entry, updation, deletion and modification are being done properly. Data has been inserted into database and updation, deletion and modification of data are done from the database. This database management system is very useful to maintain a database of lentil crop. This information system has been tested with dummy data and it is working properly. Database management system or expert system related to agriculture has been useful to improve crop management and crop production (Rafea, 1999). This technology may also be appropriate for keeping indigeneous knowledge (Rafea, 1995).

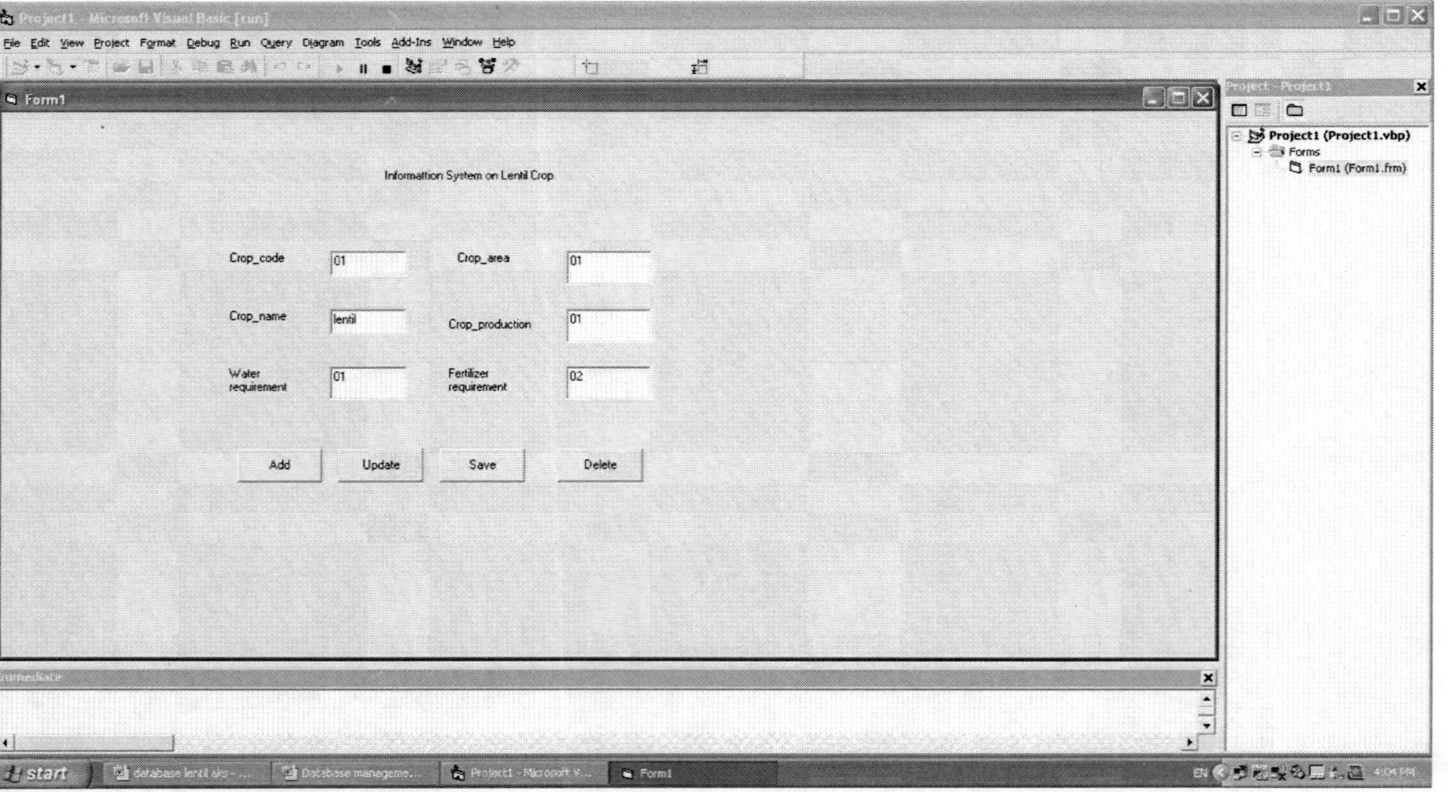

Fig. 31.3 Graphic User Interface for the management of data related to area, production, etc. for the lentil crop

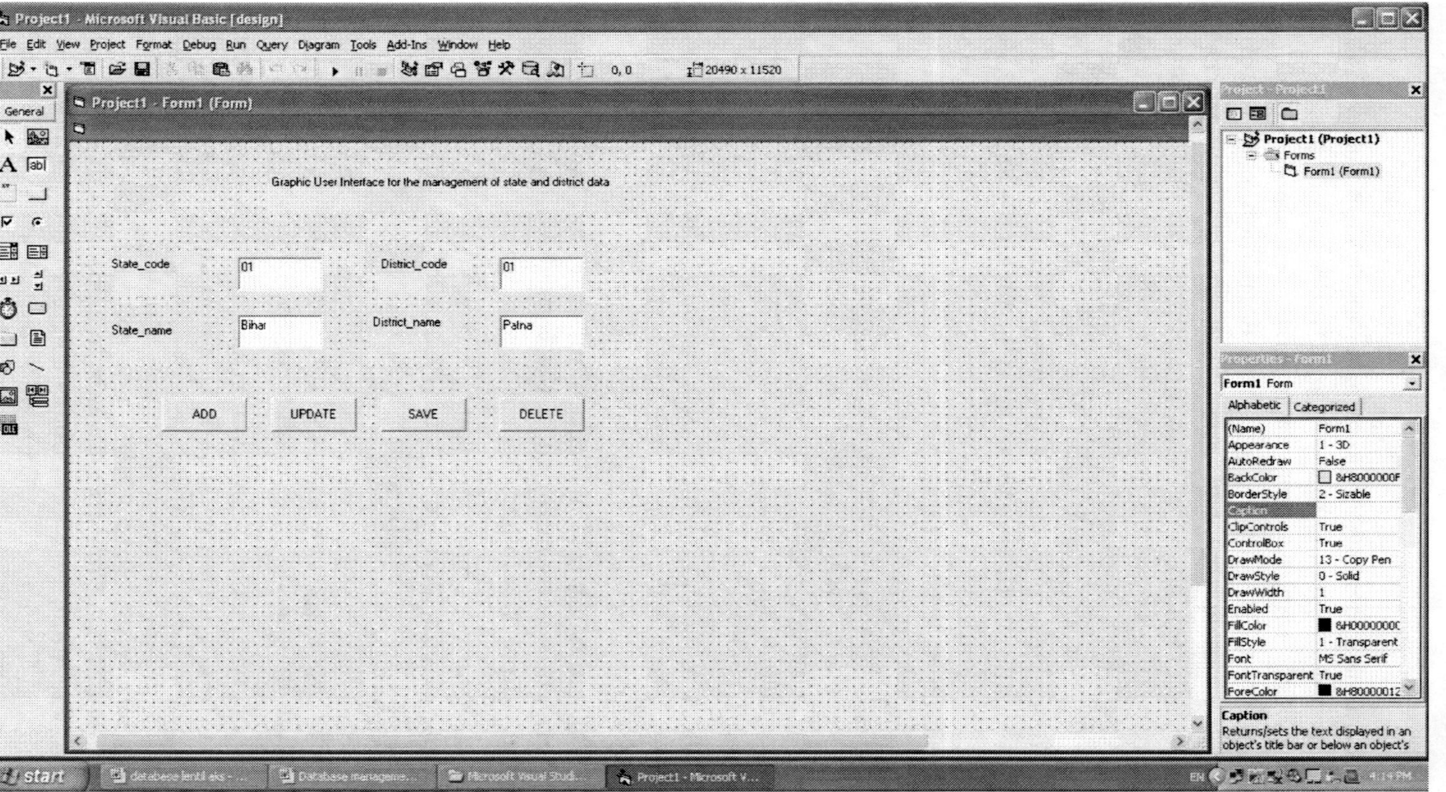

Fig. 31.4 Graphic User Interface for the management of data related to state and district names

Conclusion

Database management system or information system is very useful tool to keep and manage data properly in to a database. This type of information plays an important role for the analysis of data and getting information as and when required and this information may be useful in planning and development of agriculture and other enterprises. Relational model is widely acceptable and popular model to create a database properly. The database is also maintained properly. Information system on lentil crop will be very useful for the storage of data related to area and production of this crop. This information system is also very useful for the management of data and retrieval of required information which may be used in planning and increase the production of lentil crop. So this information system may be useful for researchers as well as farmers.

References

Chen P. 1976. The Entity- Relationship model- towards a unified view of data. *Association for Computing Machinery Transactions on Database Systems* 1(1): 9-36.

Codd EF.1970. A relational model of data for large shared data banks. *Communications of the Associations for Computing Machinery* 13(6): 377-387.

FAOSTAT. 2009. Agricultural Data: agriculture and food trade. In Food and Agricultural Organization of the United Nations, Statistics Division (http://faostat.fao.org).

Manibhushan, Upadhya A, Singh AK, Batta RK, Singh Anil Kumar, Rahman A, Kumar S. 2014. Decision support system for design, layout and cost estimation of pressurized irrigation system. *Int. J. Agril. & Stat. Sci.* 10(1): 231-236.

Rafea A. 1995. Expert systems as a tool for information technology in agriculture International informatics access 1995. Singapore.

Rafea A. 1999. Natural resource conservation and crop management expert systems. In G Kersten *et al.* (ed) Decision support systems for sustainable development. IDRC and Kluwer Acad. Publ., Boston.

Rao Sarveshwara D, Pooja Dhuppar, Subhash C. Biyan and Basanti Chintapalli. 2011. Emerging Trends in Lentil Crop Production System, Profitability and Global Concern Proceedings of the International Conference on Economics and Business: Analysis and Application EBA DEI, Agra.)

Singh AK and Bhatt BP 2015. Late sown lentil performance in response to foliar application of zinc. *Bangladesh Journal of Botany* 44(1): 125-128

Singh AK, Manibhushan, Bhatt BP, Singh KM and Upadhyaya A. 2013. An Analysis of Oilseeds and Pulses Scenario in Eastern India during 2050-51. *Journal of Agril. Sci.* 5(1): 241-249.

Tang C and Thompson BD. 1996. Effect of solution pH and bicarbonate on the growth and nodulation of grain legume species. Plant *and Soil* 186: 321-330.

Tsichritzis DC and Lsochovsky LC. 1982. Data models, Prentice Hall, Eagle wood cliffs, New Jersy.

Ullman JD. 1982. Principles of database systems. Computer Science Press, Rockville-Maryland.

Citation

Manibhushan and Chandra N. 2018. Database management system- A tool for the development and management of lentil. In: Scientific Lentil Production. Singh AK, Bhakta N, Sangale UR, Manibhushan, Sundaram PK, Kumar S and Yasin JK. (eds.) Society for Upliftment of Rural Economy Varanasi, India. Pp. 519-527.

❑❑❑

Chapter 32

Constraints in Banka District to Achieve the Optimum Yield in Lentil Crop

Sunita Kushwah, Parasnath and Pankaj Kumar Yadav

SYNOPSIS

Lentil is an important cop in district Banka in few areas. But there are so many constraints faced in production of lentil by the farmers. Erratic rainfall, undulated topography, heavy soils and sudden change in temperature, infestation of disease and insect pest are major problems reported in the district. Schedule tribe population has a wide range it is 15 percent of the total population. Their socioeconomic behavior is also not suitable to adopt the lentil cultivation. Out of 11 blocks 4 blocks covered by forest area and rest 6 blocks with undulated topography. Major constraints reported to rise up in temperature and chilling temperature not prevailing long since last few years. By the best management practices the problem can be sought out and lentil production will increases in Banka district. Rainfall and temperature are important climatic inputs for agricultural production, especially in the context of climate change.

Keywords: Lentil, Bihar, Constraints, problems, pulses production.

Introduction

Banka is one of the thirty-eight districts of Bihar state, India. The district headquarter of Banka is situated in Banka town. The district has been established on February 21, 1991. Earlier it was the richest and largest Sub-Division of the district Bhagalpur. In Bihar major lentil growing districts are Patna, Shekhpura, Nalanda, Lakhisarai, Jahanabad and Arwal. Major contribution of Taal areas of ganges. Banka district contribution is not at par. Area under lentil cultivation in Banka district is 4030 ha, production 26195 q and productivity is 6.5 q/ha (District Census, 2011). It is very low. The district has some forested area under block Banka, Belhar, Katoriya, Chandan and

Baunsi. The wood of Banka range lie on the Hill slopes, those in the other two ranges lie in undulating land. Among the fruits trees those of Mango and Jack fruits are common. Pulses are known as grain legume are a group of 12 crops that includes chick pea, pigeon pea, moong, urd, lentil, khesari, beans and peas. They are high in protein, fibre and various vitamins and provide amino acids. They are most popular in developing countries but are increasingly becoming recognized as an excellent part of the healthy diet through out the world (Reddy, 2015). The United Nation declared 2016 as an international year of pulses, given its importance in food and nutrition security and sources of income and employment to small and marginal farmers in developing countries. Historically India is the largest producers, consumer and importer of the pulses. Pulses production in India is about 19 million tones, every year India is importing 3-4 million tones of pulses to meet out its annual demand for 22-23 million tones. Lentil is one of the early domesticated plant species, as old as those of einkorn, emmer, barley and pea (Harlan, 1992 and Singh *et al.*, 2011). Lentil (*Lens culinaris* L.) is an important rabi pulse crop of India which is being grown on an area of about 1.5 m ha with annual production of 1.0 m tones. It is mainly cultivated in Uttar Pradesh, Madhya Pradesh, Chhattisgarh, Jharkhand, Bihar and West Bengal which together contribute more than 80% area and production of this crop. It is generally grown as rain fed crop during rabi season after rice, maize, pearl millet or *kharif fallow*. It is an ideal pulse crop for rice fallow situation of eastern part of the country. In the last three decades, the area under lentil has increased by 85%, production by 151% and productivity by 34%. The increased in area has been mainly in Uttar Pradesh, Madhya Pradesh, Rajasthan and Bihar. In Bihar area increased in other district excluding district Banka. Lentil is highly nutritious pulse and its grains are mostly used as *dal* (whole or dehulled) and in various other preparations. It contains 25% protein, 0.7% fat, 2.1% mineral, 0.7% fibre and 59% carbohydrate. It is also rich in phosphorus and carotene.

Major factors effecting the production of lentils are as under

Delayed sowing

Lentils grow best in cool weather. Lentil seeds will germinate in 10 days at 68°F. Lentils require 80 to 110 days to come to harvest (Carter and Materne, 1997). It appears that the sowing date of November 21 had higher seed production potential than other sowing dates because of allowing the plants to complete their vegetative growth and to optimally use the resources as well as increasing their photosynthesis potential and dry matter accumulation. As well, the population of 72 plants/ m^2 had higher seed production potential than other plant densities because of increasing the number of pods/ m^2. Therefore, sowing date of November 21 with the plant density of 72 plants/m^2 is recommended for the cultivation of lentil (Seyyed *et al.*, 2014). Maximum yield in lentil obtained if crop sown from 30 October to 20th November. But in Banka district late harvesting paddy till 15 December decreases the area of lentil due delayed sowing.

Non availability of suitable variety and undulated topography

In Banka district late sowing of lentil crop has done but presently, late sown high yielding varieties are not available so farmers are planted traditionally low yielding varieties which provide less profit. (Datta *et al.*, 2013). Several high yielding varieties resistant to rust have been released for northern hills and plains. Traditionally northern belt has been the area of small seeded lentil. Concerted efforts were made to develop large seeded varieties and consequently several of large seeded varieties suitable for northern plains were developed and released for cultivation. Total geographical area of the district is 305621 ha, net sown area is152371 ha, forest area 43310 ha and land not available for cultivation is 41258 ha. It is found that 50 percent land is available for cultivation. Only 10 percent land has proper leveling rest is undulated. So water stagnation is a big issue and proper moisture availability is a serious problem faced by the crop during germination.

Soil type

Lentil required well drained to medium textured soils for its cultivation (Singh *et al.*, 2011 and Laik *et al.*, 2014) but majority of soil type in Banka district are heavy clay soils and few of the parts having sandy loam. Lentil crop can not tolerate water logged condition. Two or three ploughing are required for field preparation. First ploughing should be done MB plough and later 1-2 ploughing should be done with desi plough in normal cultivation of lentil. But in soil characteristics of district Banka is different in this situation different management practices adopted.

In Banka district most of the soils are acidic in nature i.e. covered by lateritic and latosols. In this soils increases solubility and toxicity of

Al, Mn and Fe, deficiency of Ca and Mg reduce availability of P and Mo and also reduces microbial activity with decreasing pH, are characteristic's feature of these soils. Optimum pH for cultivation of lentil is 5.5 to 7.0 slightly acidic to neutral. Deficiency of Ca and Mg directly affects plant growth as both these nutrients are essential for cell wall formation, enzyme activities, establishment of cell membranes and carbohydrate metabolism. Poor soil condition like structure, aggregate stability etc arising due to Ca deficiency affects root growth as well as poor water retention in plants and can significantly reduce plant growth. Azotobactor population was found to be nil in these soils. Ozdemir (2002) Ca and Mo deficiency affects nodulation and nitrogen fixation. In this soil application of liming materials increases the availability of Ca, P as well as nitrogen fixation, mineralization and activity of microorganism. Some patches of Banka soil comes under soil order vertisols are montmorillorite rich clays with characteristics shrinking/ swelling properties having high clay content (>30%) and cracks (at least 1 cm wide reaching a depth of 50 cm or more). In this soil land preparation is difficult, low fertility status, poor drainage and soil erosion are major factors that stands in the way of achieving higher yield of lentil in these soils of Banka district. The soils become extremely hard on drying and require significantly high amount of power for tillage which most often is not possible for majority of the resource poor farmers on wetting, the soil again become very plastic and sticky. Nitrogen is often deficient in this soil while potassium is variable. Lentil is a well adapted plant that grows in a wide range of soil types. However, the heavy textured soils causes yield reduction, whereas sandy-loam soils are the most suitable for lentil growth. The view supported by Sehirali (1988).

Management strategies for soil

1. Land shaping
2. Dry season tillage
3. Improved cropping system
4. Fertilizer management

Rain fed agriculture

Out of 152371 ha, only 115698 ha area is irrigated by the different sources in Banka district. Last 15 years back there was surplus recharged ground water was available but now a days due low water table only 20 percent area is irrigated, rest is rain fed. Therefore, timely irrigation is not available to the crop. Lentil crop required 1-2 light irrigation so timely irrigation could not available to the crop. It causes low production (District Census, 2011).

Change in Climate

Last few years it is observed that temperature required during flowering and seed setting was not as per the need. There is a lot of literature studying the effects of temperature and rainfall on crop yield. Erskine and El Ashkar quantified the effect of rainfall on lentil seed yield and found that rainfall accounted for 79.8% of the variance of seed yield (Erskine *et. al.*, 1993). 12 major Californian crops and found rainfall was able to explain more than 60% of the observed variability in yields for most crops (Lobell *et al.*, 2007). It is found that not only the seasonal rainfall totals and their season-to-season variability were important, but also the "within season" variability had a major effect on crop productivity, which implies that monthly data is needed in crop production analysis (Cooper *et al.*, 2008). High temperature (30-35 C^0) reported during flowering time so that it causes heavy flower drop and poor grain setting in the lentil. Forced flowering and maturity obtained low yield in lentil in Banka last few years (Rong and Brady, 2012).

Erratic rainfall

Lentils require a minimum of 350mm rainfall a maximum of 550mm; in the higher rainfall areas good drainage is essential; water logging will have a great effect on yields and disease spread. Last one decade's rainfall pattern is recorded very erratic Carter and Materne (1997). Annual rainfall of this district is 1170 mm but it gradually decreases last few years. It is recorded that sometimes during off season there was heavy rainfall recorded in the district in the month of Feb and March. This is the high time for maturity of crop. Therefore, farmers are not ready to plant lentil on commercial basis.

Weed Infestation

Weed control is the most important management procedure. In many countries, herbicides are used for weed control in lentils fields (Sehirali, 1988). Weed control is an important agricultural practice that affect seed yield. Yenish *et al.* (2009) reported that weeds may reduce lentil yield by 20-80%. In lentil crop weed infestation causes low yield due competition for food and water in between crop and weed. In Banka district in lentil crop Cyprus rotundas locally known as motha and is also affected by parasitic flowering plants like *Cuscuta* sp. and *Orobanche* sp. (Ydogan *et al.*, 2008). Presently, suitable package of practices not available for control these weeds.

Infestation disease

Lentil crop is prone to diseases like rust, wilt and root rot. These biotic stresses have been reported to reduce the productivity of lentil crop by 20-25%. The other important constraints are lower productivity in rice fallows, terminal stress, and low biomass production. Several diseases affect lentil causing yield losses. Common fungal diseases of lentil are Fusarium wilt caused by *Fusarium oxysporum* f. sp. *Lentis*, rust caused by *Uromyces fabae*, and ascochyta blight caused by *Ascochyta lentis*. Bacterial disease caused by *Mycobacterium insidiosum* also affect lentil. Several viral diseases affect lentil, including pea enation mosaic virus, bean yellow mosaic virus and pea seed borne mosaic virus (PSbMV). Among them, PSbMV is potentially dangerous for lentil (Muehlbauer *et al.*, 1995).

Socioeconomic factors

Out of total population 1600000, ST male population is 45901 and female 44321. Most of the rain fed area cultivated by the tribal population. They are not cultivating pulses and cash crop in general. Their livelihood basically depends upon the forest produce and millets. So their farming practices not supported to grow the lentil on commercial basis (Srivastava *and Dayawanti,* 2010).

References

Carter J and Materne M. 1997. Lentil Growers Guide: A Guide to the Production of Lentils (Agriculture Victoria -Horsham), (ISBN 0 7306 6681 6).

Cooper PJM, Dimes J, Rao KPC, Shapiro B, Shiferaw B and Twomlow S.2008 Coping better with current climatic variability in the rain-fed farming systems of sub-Saharan Africa: an essential first step in adapting to future climate change? *Agriculture, Ecosystems and Environment*. 126(1-2): 24–35.

Datta SK, Sarkar MAR and Uddin FMJ. 2013. Effect of variety and level of phosphorus on the yield and yield components of lentil. *Int. J. Agril. Res. Innov. & Tech.* 3(1): 78-82.

District Census 2011. Census 2011.co.in. 2011. Retrieved 2011-09-30.

Erskine W and El Ashkar F. 1993. Rainfall and temperature effects on lentil (*Lens culinaris*) seed yield in Mediterranean environments. *Journal of Agricultural Science.* 121(3): 347–354.

| access-date= requires | url= (help) Harlan JR. 1992. *Crops and man* Second Ed. American Society of Agronomy, Agricultural Science Madison, Wisconsin, USA.

Laik R, Sharma S, Idris M, Singh AK, Singh SS, Bhatt BP, Saharawat Y. 2014. Integration of conservation agriculture with best management practices for improving system performance of the rice–wheat rotation in the Eastern Indo-Gangetic Plains of India. *Agriculture, Ecosystems & Environment* 195: 68-82.

Lobell DB, Cahill K N, Field CB.2007. Historical effects of temperature and precipitation on California crop yields. *Climatic Change.* 81(2): 187–203.

Muehlbauer FJ, Kaiser WJ, Clement SL and Summerfield RJ. 1995. Production and breeding of lentil. *Advances in Agronomy,* 54: 283-332.

Ozdemir S. 2002. *Grain legume crops* (p. 142). Hasad Publishing, Istanbul, Turkey.

Reddy AA. 2015. Pulses production trends and strategy to become self sufficient. *Indian farming.* 65(6): 2-10.

Rong-Gang Cong and Mark Brady. 2012. The Interdependence between Rainfall and Temperature: Copula Analyses *Scientific World Journal.* 2012; 2012: 405675. Published online 2012 November 13.

Sehirali S. 1988. *Grain legume crops.* Ankara University, Faculty of Agricultural Engineering, Ankara, Turkey 1089 (314), p. 435.

Seyyed GM, Mohamad J, Seghatoleslami1and Mohamad RD. 2014. Effect of Sowing Date and Plant Density on Yield and Yield Components of Lentil (*Lens culinaris* cv. Sistan). *Annual Research & Review in Biology.* 4(1): 296-305.

Singh AK, Meena MK, Bharati RC. 2011. Sulphur and Zinc Nutrient Management In Rice Lentil Cropping System in International Conference on "Life Science Research for Rural and Agricultural

Development organised by MSET Ranchi Jharkhand at Patna India. pp 66-67.

Srivastava and Dayawanti. 2010. "States and Union Territories: Bihar: Government". India 2010: A Reference Annual (54th ed.). New Delhi, India: Additional Director General, Publications Division, Ministry of Information and Broadcasting (India), Government of India. pp. 1118–1119. ISBN 978-81-230-1617-7.

Ydogan A, Karagul V and Gurbuz A. 2008. Effects of different sowing dates on yield and yield components of green and red lentils. *Journal of Field Crops Central Research Institute*, 17(1-2): 25-33.

Yenish JP, Brand J Pala M and Haddad A. 2009. The lentil: botany, production and uses. In W. Erskine, F. J. Muehlbauer, A. Sarker, & B. Sharma (Eds.), *Weed Management* (Chapter 20, pp. 326).

Citation

Kushwah S, Parasnath and Yadav PK. 2018. Constraints in Banka district to achieve the optimum yield in lentil crop. In: Scientific Lentil Production. Singh AK, Bhakta N, Sangale UR, Manibhushan, Sundaram PK, Kumar S and Yasin JK. (eds.) Society for Upliftment of Rural Economy Varanasi, India. Pp. 529-536.

□□□

Chapter 33

Protection of Farmers' Variety Through Protection of Plant Varieties and Farmers' Right Act, 2001: A Case Study on Lentil Growers of Bihar State

Sanjeev Kumar and Anand Kumar

SYNOPSIS

India is among the first countries in the world to have passed legislation granting Farmers' Rights in the form of Protection of Plant Varieties and Farmers' Rights Act,2001. India's experience important due to its International contribution to negotiations on Farmers' Rights, its position as a centre of biodiversity, and the complexities of agriculture in India within which the country is attempting to implement rights. India's law is unique not only because of its far reaching rights for farmers, but also in that it simultaneously aims to protect both breeders and farmers. This attempt to evolve a multiple rights system could, however, pose several obstacles to utilization and exchange of plant genetic resources among the farmers. India has framed a unique legislation, but still faces the task of implementation. This should serve as a signal internationally that establishing legislation is insufficient to effectively promote Farmers' Rights.

Keywords: Farmers' Rights, Lentil, Bihar, PPVFR.

Introduction

India is one of the world's leading producers of Rice, Wheat, Maize as well as legume crops, failing to implement Farmers' Rights in India would be a heavy loss for all the farmers who need to Farmers' Rights to protect their livelihoods, secure their access to resources, protect their rights to seeds, and, above all, lift them out of poverty.Lentil (*Lens culinaris* sp. culinaris) is an important food legume with various uses as food and feed because of its protein-rich grains and straw. Globally, it is cultivated as a rainfed crop on 3.85 million hectares (m ha) area with 3.59 million tonnes (mt) production (Erskine *et al.*, 2011). The major geographical regions of lentil production are South Asia and China (44.3%), North America (41%), Central and West Asia and North Africa i.e. CWANA (6.7%), Sub-Saharan Africa (3.5%) and Australia (2.5%). Lentil (*Lens culinaris* L.) is an important rabi pulse crop of Indo Gangetic Plain which is being grown on an area of about 1.51 million hectare area with a production of about 1.0 million tonnes (FAOSTAT, 2011). It is mainly cultivated in Uttar Pradesh, Madhya Pradesh, Chhattisgarh, Bihar and West Bengal. These states together account for 80% and 90% of total area and production of lentil respectively (Rao *et al.*, 2011).

Bihar is one of the major states in India for pulse production covering an area of about 6 lakhs hectare in which lentil and moongbean contributes 57% of total share. Cultivated Lentil (*Lens culinaris*Medik.), a self-pollinating diploid (2n=14), is an annual legume crop has been grown as an important food source for over 8,000 years. In this section, we will discuss about the Protection of Plant Varieties and Farmers' Rights Act, 2001 and the need of implementation of the act particularly in pulse crops viz; chickpea, pigeonpea and lentil with some case studies of lentil landraces growers of Bihar state.

Landraces and their importance

Crop biological diversity has been and continues to be the basis of our food supply and good nutrition, providing humans with various and nutritious foods, and other products and services. In particular, the phenotypic and genetic variation present within a crop species – that is, crop infra-specific diversity (hereafter referred to as crop diversity)–allows farmers and scientists to adapt a crop to heterogeneous and changing environments. This adaptive capacity is illustrated by the diffusion of a great number of crops from their centre's of origin to completely new and different environments. The conservation of infra-specific diversity of crops has been a worldwide concern for many decades due to the worry that a great amount of this diversity would disappear with agricultural and economic development, that is, genetic erosion. This concern has led to the collection and conservation of seeds and planting material in gene banks, that is, ex- situ conservation. While genetic erosion certainly has occurred worldwide (Food and Agriculture Organization of the United Nations [FAO],2010), a large amount of crop diversity is still retainedin developing countries by smallholder farmers. This is particularly true for crops in their centers of domestication and diversity where farmers continue to grow land-races.

A **landrace** is a domesticated, regional ecotype; a locally adapted, traditional variety of a domesticated species of a plant that has developed over time, through adaptation to its natural and cultural environment of agriculture and pastoralism, and due to isolation from other populations of the species.

Characteristics of landrace

- It is morphologically distinctive and identifiable (i.e., has particular and recognizable characteristics or properties), yet remains "dynamic".
- It is genetically adapted to, and has a reputation for being able to withstand, the conditions of the local environment, including climate, disease and pests, even cultural practices.
- It is not the product of formal (governmental, organizational, or private) breeding programs, and may lack systematic selection, development and improvement by breeders.
- It has a historical origin in a specific geographic area, usually it will have its own local name(s), and will often be classified according to intended purpose.
- Where yield (e.g. of a grain or fruit crop) can be measured, a landrace will show high stability of yield, even under adverse conditions, but a moderate yield *level,* even under carefully managed conditions.
- At the level of genetic testing, its heredity will show a degree of integrity, but still some genetic heterogeneity (i.e. genetic diversity).

The word *landrace* literally means 'country-breed' (German: *Landrasse*) and close cognates of it are found in various Germanic languages. The term was first defined (in German) by Kurt von Rümker in 1908, and more clearly described (in Dutch) in 1909 by U.J. Mansholt, who wrote that landraces have better "stability of their characteristics" and "resistance capacity to tolerate adverse influences" but lower production capacity than cultivars, and are apt to change genetically when moved to another environment. H. Kiessling added in 1912 that a landrace is a mixture of phenotypic forms despite relative outward uniformity, and a great adaptability to its natural and human environment. In 2005, a "working definition" of plant landraces was proposed: "a dynamic population(s) of a cultivated plant that has historical origin, distinct identity and lacks formal crop improvement, as well as often being genetically diverse, locally adapted and associated with traditional farming systems." Another definition, dating to 1975, of the term *landrace* as used in botany (and by extension in agriculture, horticulture, anthropology, etc.) was provided by J. R. Harlan: "Landrace populations are often highly variable in appearance, but they are each identifiable morphologically and have a certain genetic integrity. Farmers usually give them local names. A landrace has particular properties or characteristics. Some are considered early maturing and some late. Each has a reputation for adaptation to particular soil types according to the traditional peasant soil classifications, e.g. heavy or light, warm or cold, dry or wet, strong or weak. All components of the [plant] population are adapted to local climatic conditions, cultural practices, and disease and pests. "But most important, they are genetically diverse. They are balanced populations - variable, in equilibrium with both environment and pathogens and genetically dynamic".Similarly, *landrace* may be defined as a "local cultivar that has been improved by traditional agricultural methods". Landraces lack formal, scientific breeding for genetic improvement (e.g. uniformity and stability) is characteristic of landraces; such sources clearly distinguish landraces from cultivars).

The first systematic crop improvement effort in lentil was made in 1924, when a collection of mixed market samples from West Bengal, Bihar, Orissa, the then united Punjab (Punjab, Haryana, Himachal Pradesh and Western Punjab in Pakistan), Madhya Pradesh, Maharashtra and Tamil Nadu were evaluated and 66 pure lines were selected at the erstwhile the Imperial Agriculture Research Institute, Pusa, Bihar. Thcse pure lines were designated 'Type 1' to 'Type 66' (Shaw and Bose 1929). Many of these selections like 'Type 8' and 'Type 36' were released by various state departments of agriculture in their respective states, where they are still cultivated as in states like Uttar Pradesh and Bihar. Richharia (1957) selected 'BR 20'. 'BR 24', 'BR 25' and 'BR 37' from the local germplasm, out of which 'BR 25' is still cultivated in some pockets of Uttar Pradesh. Since last decades, lentil genotypes namely ARUN, L-4076, MalaviyaVishwanath (HUL-57), MALIKA (K-75), NARENDRA MASOOR-1 (NDL-92-1), NOORI (IPL-81), PANT L-406, PANT L-639, PANT LENTIL-4, PRIYA (DPL-15), SHERI, DPL-62 and few local genotypes are most popular cultivars of Bihar state. In Bihar, Taal area of Patna district, Nalanda district, Lakhisarai district are the predominantly growing area for lentil production where farmers grow the improved varieties as well as their landraces.

Biodiversity and conservation of Landraces

Due to their adaptation to the local environment, some farmers using scientifically "improved" domesticates also continue to raise landraces, because the latter often exhibit benefits, ranging from lower cost, and cultural (e.g. culinary) preference, to superior hardiness in a less-than-ideal climate, and better disease resistance. There may be more variety-specific pluses; a plant landrace may have, e.g., lower fertilizer requirements, or something about a plant or animal product's texture, color or ease of use might be a major factor.Landraces are often free from many intellectual property and other regulatory encumbrances. However, in some jurisdictions, a focus on their production may result in missing out on some benefits afforded to producers of genetically selected and homogenous organisms, including breeders' rights legislation, easier availability of loans and other business services, even the right to share seed or stock with others, depending on how favorable the laws in the area are to high-yield agribusiness interests. The irony of the matter is that the poorest farmers are the stewards of genetic diversity." Protecting farmer interests and protecting biodiversity is at the heart of the International Treaty on Plant Genetic Resources for Food and Agriculture (the "Plant Treaty" for short), under the Food and Agriculture Organization of the United Nations (FAO), though its concerns are not exclusively limited to landraces.

Farmers' Variety

A variety which has been traditionally cultivated and evolved by the farmers in their fields; or is a wild relative or land race or a variety about which the farmers possess the common knowledge are covered under this category. The criteria for distinctiveness, uniformity and stability for registration of a farmers' variety and variety about which there is a common knowledge has been notified by the Central Government (The Gazette of India, 2009c). Any person who applies for registration under clause (c) of Section 14 of the Act shall submit half of the quantity of seed material specified for a new variety in the respective crop species, divided into five equal numbers of packets for the purpose of field test and also for storing in the National Gene Bank and the seed supply procedures shall be such as may be specified in the Journal. Field test is conducted for confirming distinctiveness, uniformity and stability at the test Centres. The farmers' variety along with reference varieties and other similar variety are evaluated in the paired row test. The length of the row and plant population is kept such as specified in the Journal. A replicated trail is conducted for one season at two locations with limited purpose of confirming the distinctiveness, following the descriptors such as specified in the Journal. The uniformity levels for Farmers' variety for the respective species shall not exceed double the number of off-types such as specified in the Journal. If the variety meets the uniformity criteria, it is deemed to have met the stability criteria.

Why Protection of Farmers' variety is required?

Genetic erosion is the main threat to landraces and has been referred to in the literature as:

- The loss of a crop, variety or allele diversity
- The reduction in richness (in the total number of crops, varieties or alleles)
- The reduction in evenness (i.e. of genetic diversity)

Why are Farmers' variety threatened?

There are numerous factors that negatively impact plant species and their populations which will result in taxonomic (species, subspecies, and varietal) and genetic diversity erosion, and eventually extinction.

The main factors that contribute to the genetic erosion of Farmers' variety include:

- Changes in agricultural practices and land use;
- Use of pesticides and herbicides;

- Replacement of traditional varieties with modern, uniform cultivars which lead to a genetic bottleneck; once Farmers' variety have been replaced by modern cultivars, unless the Farmers' variety is conserved ex situ, the unique combination of genetic diversity is unavailable to breeders; as a consequence, the total number of different varieties grown is reduced and/or cultivars grown by farmers become increasingly similar to each other;
- Type of variety and seed certification system associated with the enforcement of plant breeders' rights, which limits the sale of crop seed unless the variety is included in the national or regional varietal list; Farmers' variety growers do not usually register their varieties since this process is relatively expensive and generally returns limited value to individual farmers; therefore, as it is illegal to grow non-registered varieties in many countries, farmers are inadvertently encouraged to switch to registered varieties and their Farmers' variety material is lost;
- Simplification of silvi-agriculture productive processes due to high manpower costs;
- Subsidy schemes that promote the use of uniform varieties;
- Incentives given by, for instance, government agricultural advisory services, such as the free distribution of modern cultivars;
- Constant decrease of rural populations due to migration and emigration;
- Research programmes that ignore Farmers' variety and their associated knowledge and uses;
- Ageing of farmers and the unsuccessful passage of Farmers' variety and associated knowledge from one generation to the next;
- Lack of education of the unique value of Farmers' variety as a local, national and global resource;
- Changes in consumption habits;
- Food standards that limit entry of Farmers' variety and products into markets.

Why protection of Farmers' variety has become an important

issue?

The breeding activities and exploitation of new varieties are the decisive factors for improving rural income and their overall economic development. Since the process of plant breeding is long and expensive, it is important to provide aneffective system of plant variety protection with an aim to encourage the development of new varieties of plants for the benefit of society.

Defining Farmers' Right

In 1989, the Food and Agricultural Organisation (FAO) of the United Nations defined farmers' rights "as rights arising from the past, present and future contributions of farmers in conserving, improving, and making available plant genetic resources, particularly those in the centres of origin/diversity" (Resolution 5/89). Two reasons are generally given in the literature for the protection of farmers' rights as defined by FAO. First, it has been held that protection of farmers' rights creates incentives for the preservation of crop genetic resources. This is called the "conservation" argument. The conservation rationale behind "farmers' rights" lies in the fact that it can provide a framework within which farmers/farming communities can appropriate some of the global values of agro-biodiversity, thus creating a structural incentive for the retention of plant genetic resources. Secondly, it has been held that protection of the rights of farmers in developing countries helps them face the onslaught of the intellectual property rights regimes ofthe developed countries. This is referred to as the "equity" rationale. There is a fundamental asymmetry between rewards accruing to plant genetic resources that form the basis of development of new varieties and rewards accruing to new varieties that are the products of research. Farmers' rights provisions represent an attempt to redress this asymmetry in the reward structure. An important feature of the concept of farmers' rights in the contemporary context is that it recognises two distinct roles of farmers: farmers as cultivators and farmers as conservers. The former role refers to those farmers engaged in cultivation but not conservation, while the latter practice in situ conservation. The part played by farmers as innovators, however, has yet to gain adequate international attention. The Indian legislation may be the first to explicitly recognise the farmers' role as an innovator. Section 2(k) of the Protection of Plant Varieties and Farmers' Rights Act, 2001, defines a farmer to be a cultivator (cultivating the land himself or through direct supervision), or one who conserves and preserves "any wild species or traditional varieties," or a breeder who adds value to such wild species and traditional varieties "through selection and identification of their useful properties." The Act also recognises farmers' varieties as those which have been traditionally cultivated and evolved by farmers in their fields, or those that are wild relatives, or land races of a variety about which farmers possess common knowledge. (Section 2(l)). In the Indian law, the abstract concept of farmers' rights takes the form of nine specific rights . It is noteworthy that rights guaranteed under the Act are rights that accrue to farmers as cultivators and as plant breeders. Sections 39 to 46 of the Protection of Plant Varieties and Farmers' Rights Act, 2001, provide for the following rights: Rights on seeds. These reflect customary practices which acknowledge the traditional rights of farmers in agriculture. They include the right to save seeds and use them for sowing, exchanging, sharing, and selling to other farmers.

Right to reward and recognition. This right recognises the role of farmers in preserving and developing agro-diversity. The National Gene Fund presents awards to selected farmers or communities in this regard.

Right to benefit-sharing. Farming or tribal communities that contributed to the development of a new crop variety are entitled to an equitable sharing of the benefits earned from it.

Right to compensation for losses. The Act requires that registered varieties be sold with a declaration regarding their agronomic performance and the conditions needed to ensure successful cultivation. This right guarantees compensation to farmers who are victims of exaggerated claims regarding the performance of newly registered varieties. The right thus serves as a means to prevent unfair marketing practices by breeders and seed sellers.

Right against undisclosed use of traditional varieties. This provision protects farmers when a commercial breeder makes undisclosed use of a traditional variety. Farming communities are seldom able to detect the use of "their" varieties or traditional knowledge in the breeding of a new variety. In such situations, any third party who has a reasonable knowledge of the use of traditional methods in the breeding of the new variety is eligible to file a claim for compensation on behalf of the concerned local or tribal community.

Right to access to seed. If farmers are to benefit from scientific crop improvement, it is self-evident that they need access to seeds. While allowing a plant breeder the exclusive right to the commercial production and marketing of seeds of the varieties registered in their names, the Act also directs that breeders meet farmers' demand for seeds at reasonable prices.

Right to free services. The Act exempts farmers from paying any fees at any stage of the registration of a variety.

Right to protection against accusations of infringement. A farmer cannot be prosecuted on charges of infringement of another's title to a variety if he or she affirms ignorance of the legal provision deemed to have been violated by him or her.

What are Farmers' rights?

1. Farmer who has bred or developed a new variety shall be entitled for registration and other protection under PPV&FR Act, 2001 in the same manner as a breeder of a variety.
2. Farmer who is engaged in the conservation of genetic resources of land races and wild relatives of economic plants and their improvement through selection and preservation shall be entitled in the prescribed manner for recognition andreward from the Gene Fund provided that material soselected and preserved has been used as donors of genes in varieties registered under this act.
3. Farmer shall be entitle to save, use, sow, re-sow, exchange and share or sell his farm produce including seed of a variety protected under this act in the same manner as he was entitled before the coming into force of this act provided that the farmer shall not be entitled to sell branded seed of a variety protected under this act.

Duration of protection of a registered plant variety

The duration of protection of registered varieties is different for different crops which are as below:

1. For trees and vines - 18 years.

2. For other crops - 15 years.
3. For extant varieties - 15 years from the date of notification of that variety by the Central Government under section 5 of the Seeds Act, 1966.

Under PPV and FR Act, as such those plant varieties present in wilderness cannot be registered, under PPV&FR Authority. However, any traditionally cultivated plant variety which hasundergone the process of domestication / improvement through human interventions can be registered and protected subjected to fulfillment of the eligible criteria.

Prerequisites for filing an application form for registration of plant variety: For registration of a plant variety the following prerequisites has to be completed:

1. Denomination assigned to such variety.
2. Accompanied by an affidavit that variety does not contain any gene or gene sequences involving terminator technology.
3. Complete passport data of parental lines with its geographical location in India and all such information relating to the contribution if any, of any farmer (s), village, community, institution or organization etcin breeding, evolving or developing the variety.
4. Characteristics of variety with description for Novelty, Distinctiveness, Uniformity and Stability.
5. A declaration that the genetic material used for breeding of such variety has been lawfully acquired.
6. A breeder or other person making application for registration shall disclose the use of genetic material conserved by any tribal or rural families for improvement of such variety.

The application for registration of a variety is to be made in the form prescribed by the PPV & FR Authority.

1. Form I - for registration of new variety, extant variety and farmer's variety and
2. Form II - for essentially derived varieties (EDVs) and transgenic

Registration of Plant Varieties

For notified Crop species

Application (fiding)
With all details as required in the application

↓ Section 14 and 23

Initial Browsing / Preliminary examination

↓ Section 20

If application in proper shape
Acknowledgement receipt issued

↓

File sorting done
Unique Number allotted to each application

↓ Allication issued for examination

↓ Simultaneous data entry in digital application
Denomination search for novelty and Distinctivity

↓

Examination Report submitted to Registrar
Report discussed and verified

↓

Communication made for submission of lacking information

↓ (Reply with in 30 days)

Application accepted for further processing

↓ Section 20

Applicants asked to submit DUS fee and seeds

↓

Passport data advertised in the PVI of India for opposition
(With in three months) Section 21

↓ No Opposition

↓ Notice sent to the applicant
regarding clarification of opposition made

↓ (With in two months)

Applicant fails to submit clarification within
specified time frame / Registrar not satisfied with reply
Application is deemed to be rejected

↓

Applicant satisfies the Registrar
Application accepted and
carried forward for Registration

→ DUS/Grow-out/On site test carried out

↓

DUS report in accordance to
data provided by applicant

↓

Registration done
Section 24 (1) / 23 (4)

↓

Certificate granted
Section 24(2)/ 23(8)

→ Protection

→ 15 yr (6 +)

→ 18 yr (9 +)

→ Certificate of
Entitlement
Section 28 (4)

varieties.

3. Technical Questionnaire attached with Form I andForm II - for detailed information of the concerned variety. These filled application forms must be accompanied by the fee prescribed by the Authority.

Application for registration of a variety can be made by

1. Any person claiming to be the breeder of the variety;
2. Any successor of the breeder of the variety;
3. Any person being the assignee or the breeder of thevariety in respect of the right to make such application;
4. Any farmer or group of farmers or community of farmers claiming to be breeder of the variety;
5. Any person authorized to make application on behalf of farmers and
6. Any University or publicly funded agricultural institution claiming to be breeder of the variety.

Registration of plant varieties can be made in the office of Registrar, PPV&FRA, New Delhi. The address of the Office is: Registrar, Protection of Plant Varieties and Farmers' Rights Authority, Govt. of India, Ministry of Agriculture, NASC Complex, DPS Marg, Opposite Todapur, New Delhi– 110 012. Also any information regarding the protection, application, fee structure, etc. can be obtained from this office. Completed forms in triplicate, with fee/charges should be submitted to the Registrar with enclosures, TQ, affidavits in his New Delhi Office or can be sent by Registered Post. In the first instance do not send DUS test fee or seed.

Cost of registering a plant variety: The fee structure as defined by the PPV & FR Authority is as below:

Rupees

A. Form Charges

1. Application Form Charges 200/-

B. DUS test fee*

1. Rice, Wheat, Maize, Sorghum, Pearl millet, 20,000/-

Pigeon pea, Mungbean, Urdbean, Chickpea,

Fieldpea, Lentil, Rajmash

C. Annual Fee

The Authority shall notify the amount separately inthe official Gazette (in the PVJ of India).

II. Other charges:

A. Registration Charges

1. For Extant Variety of Common Knowledge where no DUS test 2,000/- is needed.

2. Registration Certificate charges 5,000/-

B. DUS testsite visit charges/person/visit 500/-

(To be retained at DUS test site after issue of receipt)

Exemptions provided under the PPV &FR Act, 2001

1. Farmers' Exemption:Farmer shall be entitled to produce, save, use, sow, re-sow, exchange, share or sell his farm produce including seed of a variety protected under this Act.
2. Researcher's Exemption: (i) the use of registered variety for conducting experiment. (ii) the use of variety as an initial source of variety for the purpose of creating other varieties.

Acts of infringement of the rights provided to the registered breeder under the Act

1. If a person who is not a breeder of a variety registered under this act or a registered agent or a registered licensee of that variety, sells, exports, imports or produces such variety without the permission of itsbreeder or within the scope of a registered license or registered agency without their permission of the registered license or registered agent.
2. If a person uses, sells, exports, imports or produces any other variety giving such variety, the denomination identical with or deceptively similar to the denomination of a variety already registered under this act in such a way that it causes confusion in the mind of general people in identifying the registered variety.

Lentil: An important legume crop of semi-arid region

Lentil (*Lens culinaris*Medikus) is an important grain legume particularly, in the Indian sub-continent and South East Asia. Approximately, half of the world's area (48%) and 37% of production of lentil is contributed by South Asia. Among South Asian lentil growing countries, India alone accounts for 78.56% of acreage and 77.98% of production but lags well behind world's productivity of 966 kg/ha (Rahman *et al.*, 2005). Poor yield level of lentil is mainly because it is grown on residual soil moisture under rainfed conditions by resource poor farmers with traditional cultivars. Traditionally grown Indian lentils are although, fairly adapted to the local environmental conditions but susceptible to many biotic and abiotic stresses (Kumar *et al.*, 2009). In India, lentil occupies 1.56 m ha area, contributes 1.05 million tones production with 678 kg/ha productivity (DAC, 2011-12) and about 65% of lentil area is under the varieties of *microsperma* (small seeded) type, mainly grown in Punjab, Haryana, Uttar Pradesh and Bihar whereas about 35% of area is under *macrosperma* (large seeded) type varieties grown mostly in Madhya Pradesh and adjoining districts of Uttar Pradesh (Tickoo *et al.*, 2005). Traditionally cultivated *microsperma*type Indian lentils are more polymorphous, more pigmented, characterized by early flowering and maturity, low biological yield, short stature and small seeds (Erksine *et al.*, 1998) with *pilosae*traits *i.e.* pubescence on the vegetative organs and short/rudimentary tendril at the tip of leaf (Vandenberg and Slinkard, 1989). On the contrary, *macrosperma*lentil genotypes are adapted to long day conditions and possesses many desirable traits like upright growth habit, profuse foliage and large seed size but generally, yields lower than the small seeded ones (Majumdar, 2011; Tickoo *et al.*, 2005). As the seed size increases, the number of seeds per pod and pods per plant decreases considerably which adversely affect the seed yield (Singh *et al.*, 2009). Furthermore, small seeded lentils are better adapted to dry environments whereas large seeded varieties have greater cold tolerance (Erskine, 1996). Lentils of *microsperma*and *macrosperma*types have diverse geographical evolution as well as genetic makeup and judicious use of both of these in hybridization programmes can be helpful in developing high yielding, biotic and abiotic stress resistant genotypes with wider adaptability as well as advantageous in the broadening of the existing genetic base of lentil (Lal *et al.*, 2000; Chahota *et al.*, 2007; Kumar *et al.*, 2009).

The lentils are an important source of essential amino acids, fatty acids and trace mineral. In the last 50 years the cultivated area of lentil has been reduced due, in part, to low productions because it has received less attention in breeding programs than other legumes and crops. In many regions of Bihar the farmers cultivate lentil only for their own consumption. They sow local varieties, populations that consist of a mixture of genotypes, result of a long and lasting unconscious selection by these farmers which has theoretically resulted in a better adaptation to the land and climatic conditions of the sowing areas. These autochthonous populations have great importance since they have a lot of genes for resistance to pest, diseases and tolerance to adverse environmental conditions (Cubero *et al.*, 2006). Landraces are collected and preserved by gene banks and they represent a valuable contribution to establish the core collection. These collections have a representative variation of the cultivated species and in order to be able to use them in future breeding programs, they should be characterized and evaluated. Morphological characters have been frequently used in order to know the diversity in germplasm collections (Erskine & Choudhary, 1986; Saker *et al.*, 1999; Toklu *et al.*, 2009).

In case of Pulses, productivity in the Bihar state is significantly more preponderant than overall India productivity. In the year 2006-07, Bihar was having productivity of 735 kg/ha while India total productivity was about 616 kg/ha. Chickpea, Pigeonpea and Lentil are the major pulses grown in this state. In the last three decades, the area under lentil has increased by 85%, production by 151% and productivity by 34%. The increased in area has been mainly in Uttar Pradesh, Madhya Pradesh, Rajasthan and Bihar. In Bihar, the yield of lentil has remarkably increased since last decade (DOA, Govt. of Bihar, 2013-14). The local variety is much suitable for the farmers because of its adaptability to longer date of sowing.

A Case Study: Farmers' Right for Lentil growers of Bihar State

This study intends to analyze 'why we need a law that concern with rights of farmers' and whether such a protection helps in conservation of protection of plant varieties in Bihar. Farmer is the first link in the food security especially in countries like India where 70% of population engaged in agrarian based activities. The vast majority of farmers still act as stewards and innovators of genetic diversity [Ramanna, 2006]. Recognizing and rewarding them is significant considering their vast practical knowledge in conserving and improving plant genetic resources to adapt to climate change and environmental changes. Therefore farmers' knowledge, skill and innovation have to be protected in the parlance of Intellectual Property Rights. The study will also look into the history of evolution of the parlance of Intellectual Property Rights. The study will also look into the history of evolution of concept of farmers' rights in India till its materialization. It is highly significant to look into reasons behind adoption of a *sui generis* legislation in India for protection of plant varieties. Indian legislation has become a history by synchronizing both farmers' rights and breeders' rights in our legislation. This paved way for protection of traditional plant varieties and wild species disappearing due to inadequate protective measures. However this study is also an endeavor to point out inadequacies in the legislation in providing full proof protection to the farmers' agricultural knowledge in India.

Farmers' rights are very crucial in state like Bihar to ensure present and future food security where the farmers are majorly responsible for development of vast genetic diversity resources through keen observation and intelligence. The farming communities across the world have been following, since time immemorial, the practice of sharing of knowledge and resources. Sharing of seeds among farmers, for example, constitutes perhaps the most important part in these traditional agricultural practices. But unfortunately the developers of this wealth remained invisible in trade and commerce and their resources have been utilized by Private Agrarian Sector. They were not either consulted or benefited or not even informed while detraction of theircenturies'worth toil and results. With the advent of Intellectual Property Regime, ownership rights vested with the Multi-national Companies (MNCs). They deliberately used the resources of poor and uneducated farmers of developing nations for their monetary gain. The poor farmer remained poorer in this bargain and they lost their means of livelihood as they have to buy seeds at fixed prices maintained by the private companies. This sudden shift from 'commons' approach to 'anticommons' approach has affected socio-economic political status of the farmers. This point is analyzed in 'Conceptual Analysis of 'Farmers' Rights'. Immediately after realization of Gregor Mendel's contribution to plant hybridization, the more technology and infrastructure convenient western nations developed commercial breeding and established monopoly on plant varieties. The debate between developed nations and developing nations reached at a turning point when Food and Agriculture Organization came up with a Commission on Plant Genetic Resources set up under Food and Agriculture Organization to study about the materialization of the concept 'farmers' rights. TRIPs (Trade Related Aspects of Intellectual Property) Agreement to which India is a signatory mandated minimum protection for 'technologically developed plant varieties' [Ramanna, 2006].

In Bihar, Patna is located on the south bank of river Ganga (25.350N 85.120E). It lies in Indo-Gangetic plain & has natural fertile soil with water bodies i.e. Ganga, sone & Punpun river. The soil of the district is loamy is considered as plain except that of Tal & Diara land. Tal land is spreaded in Mokama area of this district in an area of about 1050 square kilometers. Tal land soils are gray to dark gray in colour, medium heavy to heavy in texture, poor in drainage & neutral to slightly alkaline in reaction. They crack during summer and swell on the onset of monsoon. Practically no irrigation exists in these areas. Only *rabi*crops are grown under rainfed situation. Among *rabi*crops, lentil & gram are the predominant ones followed by wheat, rape-mustard, pea,lythrus etc. Apart part Tal area, the lentil is cultivated in rainfed area of Lakhisarai and Nalanda district.Lakhisari district is located just adjacent to Tal area of Mokama, spreaded over 1228 square kilometer. This area is rainfed and lentil grows in this area as rainfedrabi crop.Nalanda, is famous all over the world for the the ancient International Monastic University established in 5th century BC, which taught Vedas, Logic, Grammar, Medicine, Meta-Physics, Prose Composition and Rhetoric. It is spread over the area of 2,367 sq. kms. Nalanda falls under Agro Climatic Zone III B where agriculture is the main source of occupation. The major crop grown in the district is paddy, wheat, Maize, Pulses (Lentil and Pigeonpea), oilseeds and vegetables (especially potato, onion and cauliflower). In these districts, farmers predominantly grow lentil landraces and till date about 30 landraces named after their cultivation area are being reported. Apart from these three districts, lentil also cultivated in Araria, Arwal, Bhagalpur, Banka, Katihar, Kishanganj area as minor rabi crop.

Numerous lentil landraces, many of them not cultivated at present, have been sampled from different region of Bihar, and preserved at the seed bank of the National Bureau of Plant Genetic Resource, New

Delhi. Recently several lentil cultivars were developed through single plant selection from several local landraces collected from different lentil growing regions of Bihar. Until 30 years ago, lentil landraces were still local agro-ecotypes, usually named after their cultivation area. Socio-economic changes, however, led to a dramatic reduction of lentil landraces cultivation and to the disappearance of local populations. To stem this loss of genetic variation, conservation and reconnaissance of existing biodiversity are fundamental. One primary objective of conservation genetics is to assess, maintain and enhance the genetic variation available in the landraces, often using molecular markers. Notwithstanding the recent trends of reduction of genetic diversity in Bihar, we expected that Bihar lentil landraces still represent important genetic resources. Farmers, particularly in the southeastern part of the country still grow on a small scale landraces selected for their adaptation to local environmental conditions. These landraces could still play a significant role in lentil breeding. However, no detailed information is available on the genetic diversity Bihar lentil landraces.

The present case study was studied and farmers of different districts were acquainted with Crop biological diversity, land races and Farmers' variety through KrishiVigyanKendras. The farmers were empowered with knowledge to protect their variety through Plant Protection Variety and Farmers' Act, 2001.

Sixty fivefarmers were filled application to office of PPV and Farmers' Right Authority, New Delhi for registration of his traditionally cultivated pulse variety includes Lentil, Gram and Pigeonpea. Out of sixty five, ten lentil growers farmers, filled application to office of PPV and Farmers' Right Authority, New Delhi for registration of his traditionally cultivated lentil variety. So that theyshall be entitle to save, use, sow, re-sow, exchange and share or sell his farm produce including seed of a variety protected under this act in the same manner as he was entitled before the coming into force of this act provided that the farmer shall not be entitled to sell branded seed of a variety protected under this act.

S.No.	Name of the farmers/ Address	Crop	Scientific Name	Denomination
1.	Sri Om Prakash Singh Vill.- Marachi, Block- Mokama, Dist.-Patna	Lentil	*Lens culinaris* L.	Mokama local Masoor
2.	Sri Umesh Manjhi Vill. Dhanakdov, Block- Goswari, Dist.- Patna	Lentil	*Lens culinaris* L.	Titaki Masoor
3.	Sri Niranjan Singh, Vill.– Sankarwa Tola, Block- Mokama, Dist.- Patna	Lentil	*Lens culinaris* L.	Desala Masoor
4.	Sri Biresh Kumar, Vill.+ P.O.-Junnaidi, Dist.- Nalanda	Lentil	*Lens culinaris* L.	Maghi Masoor
5.	Sri Ramparbodh Singh Vill. –Langhora, Panchyat- Poari, Dist.- Nalanda	Lentil	*Lens culinaris* L.	Titaki Masoor
6.	Sri Harnandan Singh Vill.+ P.O.-Barabasti, Dist.- Nalanda	Lentil	*Lens culinaris* L.	Maghi Masoor
7.	Sri Binod Kumar Yadav Vill.-Bhairwa, Post- Gonawa, Dist.- Nalanda	Lentil	*Lens culinaris* L.	Titaki Masoor
8.	Sri Subhash Chandra Sharma, Vill.+P.O.- Kharthua, Dist.- Nalanda	Lentil	*Lens culinaris* L.	Mithiki Masoor
9.	Sri Ram Saran Roy, Dist.- Shekhpura	Lentil	*Lens culinaris* L.	Desala Masoori
10.	Sri Tej Narayan Singh, Dist.-Shekhpura	Lentil	*Lens culinaris* L.	Titaka Masoori

Conclusion

The present legislation to protect plant varieties introduced an 'anticommons' approach in the case of genetic bio-diversities in India. Though the Act provides ownership rights to both farmers and breeders, the MNCs have stronger say while poor farmers would struggle to assert their legal claim. Considering the socially, economically and educationally weak situations of the local and tribal communities, without explicit provisions to protect customary agricultural practices and their varieties, the Act would fail to implement one of its objectives to protect the interest of traditional farming communities and farmers. In reality, the inclusion 'Farmers' Rights' would remain as a decorative title in the Act. In the case of benefit sharing, the process should be clearly laid down and case-by-case negotiation must be held as the value of each plant variety varies according to its commercial viability. There is healthy collaboration for research between public sector and farmers. So if the law could provide incentives for sharing of knowledge that would promote research and innovation. At the same time the farmers must have acess to information about the technology used in hybridization. Express Research Exemption has to be provided to farmers also. Along with this to strengthen the small and marginal farmers, they should be provided with access to infrastructure and interest reductions in credit and other developmental accessories. Farmers must be assisted with advisors during negotiation process of access and benefit sharing. For farmers and village communities, awareness generation and information empowerment is a must through vernacular press, radio, television and the Internet. To ascertain the ownership rights of farmers the village/grass root level democratic institutions should cooperate and if possible notify the varieties belong to each farmer in the Panchayat.

References

Anitha R. 2006. Farmers' Rights in India: A Case Study, Farmers' Rights Project, *FNI report 6/2006,* The Fridtjof Nansen Institute 2006.

Chahota RK, Kishore, Dhiman N; Sharma TR; Sharma SK. 2007. Predicting transgressive segregants in early generation using single seed descent method-derived *micro-macrosperma*gene pool of lentil (*Lens culinaris* Medikus). *Euphytica,* 156: 305-310.

Cubero JI. 1984. Taxonomy, distribution and evolution of lentil and its wild relatives.(in) *Genetic Resources and their ExploitatiorzChickpea, Fababean, urzdLenlils,* pp 187-204. Witco~nl~e J R and Erskine W (Eds). MartinusNijhhoffl W. JemkPut~lithers, the Hague-Boton-Lancaster.

DAC (2010-11) *http://agricoop.nic.in/agristatistics.htm*

Erskine W and M Choudhary. 1986. Variation between and within lentil landraces fromYemen Arab Republic. *Euphytica* 35: 695-700.

Erskine W. 1996. Seed-size effects on lentil (*Lens culinaris*) yield potential and adaptation to temperature and rainfall in West Asia. *The Journal of Agricultural Science,* 126: 335-341.

Erskine W, Chandra S, Chaudhry M, Malik IA, Sarker A, Sharma B, Tufail M and Tyagi MC. 1998. A bottleneck in lentil: widening its genetic base in South Asia. *Euphytica*, 101: 207–211.

Erskine W, Chalidra S, Chaudhury M, Sarkar A, Sharnla, Tufail M and Tyagi MC. 2001. Broadening the genetic base L:flentil in South Asia. *(in) Broadening the Genetic Base of' Crop Production*, pp 341-9. Cooper H D, Spillane C and Hodgkin T (Eds). Commonwealth Agricultural Bureau International, Wallingford, Oxon, UK.

Government of India. 2001. Report of the Joint Committee on the Rotection of Plant Varieties and Farmers' Rights Bill, 1999, 2000 Government of India, The Protection of Plant Varieties and Farmers' Rights Act, 2001, Act No. 52 of 2001 New Delhi: Akalank Publications.

Harlan, JR, de Wet JMJ and Stemler ABL .1976.Plant domestication and indigenous African agriculture. In: Harlan JR, de Wet JMJ and Stemler ABL (eds) Origin of African plant domestication. Mouton, The Hague, Netherlands.

Kumar N, Chahota RK and Sood BC. 2009. Component analysis for seed yield and yield traits in *microsperma* ×*macrosperma*derivatives of Lentil (*Lens culinaris*Medik.)*Agric. Sci. Digest*, 29: 163-168.

Lal C, Sharma SK and Chahota RK. 2000. Combining ability and heterosis for growth parameters in *microsperma*×*macrosperma*lentil. *Crop Research*, 19: 115-121.

Majumdar DK. 2011. Lentil. In: Pulse Crop Production: Principles and Technologies. pp. 398-436.

Pratibha B, Sanjeev S and Dhillon BS. 2004. The Protecton of Plant varieties and Farmers Right Act of India.*Current Science*, 86(3): 396.

Protection of Plant Varieties and Farmers' Rights Authority of India, Government of India, *Plant Variety Journal of India*, various issues, available at http://plantauthority.gov.in/ publications.htm, viewed on August 2, 2012.

Rahman MM, Mishra SK, Malik RM, Haqqani AM, Neupane RK, Rahman ML and Yadav NK. 2005. Trends in South Asian production, consumption and trade of food legumes. In: Kharakwal M.C. (Ed.), Food Legumes for Nutritional security and Sustainable Agriculture, ISGPB, New Delhi, India 1: 84-100.

Richharia RH. 1957. Plant Breeding and Genetics in India, vol 1.Scientific Book Co., Patna, Bihar, India.

Singh S, Singh I, Gill RK, Kumar S and Sarker A. 2009. Genetic studies for yield and component characters in large seeded exotic lines of lentil. *Journal of Food Legumes,* 22: 229-232.

Tickoo JL, Shari B, Mishra SK and Dikshit HK. 2005. Lentil (*Lens culinaris*) in India: Present status and future perspectives. *Indian Journal of Agricultural sciences,* 75: 539-67.

Toklu F, Karaköy T, Hakli E, Bicer T, Brandolini A, Kilian B, Özkan H 2009. Genetic variation among lentil (*Lens culinaris* Medik) landraces from Sout east Turkey. *Plant Breed.* 128: 178-186.

Citation

Kumar S and Kumar A. 2018. Protection of Farmers' Variety Through Protection of Plant Varieties and Farmers' Right Act, 2001: A Case Study on Lentil Growers of Bihar State. In: Scientific Lentil Production. Singh AK, Bhakta N, Sangale UR, Manibhushan, Sundaram PK, Kumar S and Yasin JK. (eds.) Society for Upliftment of Rural Economy Varanasi, India. Pp. 537-559.

❑❑❑

Chapter 34

Estimation of Evapotranspiration and Water Requirement for Irrigation Scheduling of Lentil Crop (*Lens culinaris*) in *Kurnool*, Andhra Pradesh

J Veeranna and AK Mishra

SYNOPSIS

Water resources estimation is an integral part of planning, development and management of water resources of any country. The estimation of water resources is based on several hydrological and meteorological parameters. Due to overexploitation of available water resources, it has become very important to define appropriate strategies for planning and management of water in semiarid conditions for successful growing of Lentil crop (Lens culinaris) as it is highly profitable. To achieve effective water allocation and planning, the information about Lentil crop water requirements, irrigation withdrawals, soil types and climate conditions were gathered from the study area i.e. Kurnool district of Andhra Pradesh. The main objectives of the study area wereto estimate the Lentil crop water requirement (i.e., evapotranspiration) in semi-arid agro climatic conditions of Kurnool, A. P. and to estimate the climatic water deficit, net irrigation requirement (NIR) and gross irrigation requirement (GIR) under different rain fed and irrigated conditions with six different growing dates with interval of 10 days. Based on the water deficit analysis, recommended supplemental irrigation scheduling were developed using CROPWAT 8.0 model to reduce water stress. The results showed that the best sowing dates were last week of September to 1st week of October, which gives best utilization of rainfall as effective rain fall. The net and gross irrigation requirement(NIR and GIR) varied from a minimum of 67.7 and the maximum of 106.9 mm and 96.7 to 169.2 mm; respectively. Thus by adopting a proper sowing date and irrigation scheduling criteria, it is possible to save 39.2 mm of water as NIR for early sown crop. Hence, the Lentil crop can be successfully grown in the semi-arid condition of Kurnool district of A.P. as well as similar other areas in the Telangana and Andhra Pradesh which often suffer from the want of water. Farmer can fetch high return by selling Lentils than other crops.

Keywords: CROPWAT; Crop Water Requirement; Evapotranspiration; Effective Rainfall and Irrigation Schedule. (*Based on the M. Tech. (Agricultural Engineering) Dissertation of the First Author).

Introduction

Climatic water demands, that govern the crop water requirement, are highly location specific hence, required to be establish with high accuracy and precision. Depending up on the climatic water demand alternatively refereed as reference evapotranspiration (ET_0) all other components of root zone water balance may be determined accurately. Present study was attempted to estimate thsese components of root zone water balance using standard methodologies with FAO CROPWAT model 8.0 for Lentil crop (Lens culinaris). Lentil (Lens culinaris) is a pulse crop predominantly grown in rain fed conditions with low yield in our country. Lentil name was derived from 'Lens'/ having the 'lens' shaped seeds, is one of the important and most nutritious rabi pulses. Lentil is a typically dry land crops, can be grown under irrigation with careful water management. Excess water in lentil crops can cause problems including delayed maturity, increased disease, and lower yields. It has the potential to cover the risk of rain fed farming. It is also used as a cover crop to check the soil erosion in problem areas (Othman *et al.*, 2016; Tiwari *et al.*, 2016). Lentil seeds also provide a source of starch for textiles and printing. Lentil residues form important livestock feed. Lentil floor is used for thickening of soups. It is mixed with wheat flour in bread and cake production.

Lentil Production: Indian scenarios

India is currently producing 14.76 million tons of pulses from an area of 23.63 million hectare, which is one of the largest pulses producing countries in the world. However, about 23 million tonsof pulses are imported annually to meet the domestic consumption requirement (G.O.I, 2015; Donna, 2016). Thus, there is need to increase production and productivity of pulses in the country by more intensive interventions. The important lentil-growing countries of the world are India, Canada, Turkey, Bangladesh, Iran, China, Nepal and Syria. The five most important pulse crops depending upon their contribution in national production viz., Chickpea (39%), Pigeonpea (21%),

Mungbean (11%), Urdbean (10%) and Lentil (7%). The area production and productivity of major pulses in India is reported in Table 34.1. The production of Lentil in the decreasing order for following stares are Madhya Pradesh (40.53%), Uttar Pradesh (37.60%), Bihar (10.80%), Andhra Pradesh (7.21%) and West Bengal (4.00%). (Donna, 2016; Tiwari *et al.*, 2016).

Table 34.1 The Area, Production and Productivity of major pulse crops in India during 2016-17

Particulars	Area	Per cent	Production	Per cent	Productivity
Chickpea	73.7	38.71	58.9	48.28	799.19
Pegionpea	36.3	19.07	27.6	22.62	760.33
Mungbean	34.4	18.07	14	11.48	406.98
Uradbean	31	16.28	14	11.48	451.61
Lentil	15	7.88	9.5	7.79	633.33
Total	190	100	124	101.64	651.2

Source: Commodity Profile for Pulses-March, 2015, Govt. of India

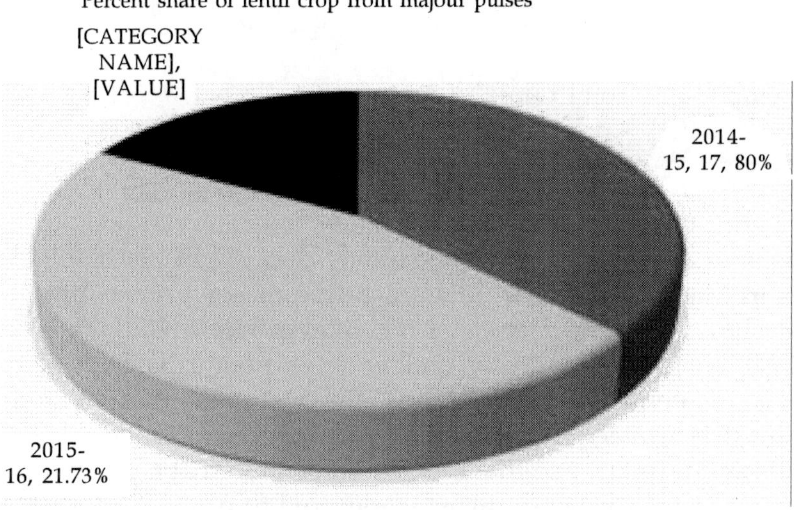

Fig. 34.1 The total share of Lentil crop in India from major pulses.

Study area: *Kurnool district*, Andhra Pradesh, India

Climatic Condition and soil characteristics

The growth of lentil crop is adversely affected at temperatures above 27°C. Hence, it is grown as a winter season crop in semi-arid tropics. The present study area has average minimum and maximum temperature, relative humidity, wind speed as 22.1 °C, 34 °C, 50 %, 7.5 hours; respectively (Table 34.2). Lentils require a minimum of 350 mm and maximum of 550 mm rainfall.In the higher rainfall areas, good drainage is essential. Waterlogging will have a great effect on yields and disease spread. Drought and severe or prolonged hot weather can cause loss in yields through pod cracking. The study area receivesan average low rainfall around 400-450 mm.The total effective rainfall were also presented in the Table 34.2. Effective rainfall was computed according to the "USDA Soil Conservation Service Method" formula. (Smith, 1992; FAO, 1998; FAO, 2009) using FAO CROPWAT 8.0 model by using the equation 34.1 and 34.2.

$$\text{Rainfall excess (Pe)} = \frac{(P*(125-0.2*3*P))}{(125)} \quad \text{if rainfall} \leq \frac{250}{3}\ mm \qquad (34.1)$$

$$\text{Rainfall excess} = \frac{125}{3} + 0.1 * P \qquad \text{if rainfall} \geq \frac{250}{3}\ mm \qquad (34.2)$$

where, P = Total rainfall (mm)

Soil data

Lentil crop is raised on light loams and alluvial soils in upper India, and on well-drained, moderately deep, light black soils, loam soils in Madhya Pradesh, Maharashtra and some parts of Andhra Pradesh (A.P). (Some districts of A.P. like Anthapur, Kadapa, Kurnool and Chitoor). It is also grown on low lands, poor soils. In Punjab, it is often cultivated on inundated lands. The crop can withstand moderate alkalinity however, it can't tolerate waterlogging. Soil should be made friable and weed free so that seed could be placed at a uniform depth. In case of light soils, less tillage is required to prepare an ideal seed-bed. In heavy soils, one deep ploughing followed by 2-3 cross harrowing should be given. After harrowing, the field should be levelledby giving a gentle slope for easy irrigation. There should be proper moisture in the soil at the time of sowing for proper germination of seeds.

Table 34.2 Long term climatic parameters and the estimated reference evapotran spiration ET_o using FAO-PenmainMotieth equationof study area (*Kurnool*, A.P).

Month	Min Temp °C	Max Temp °C	Humidity %	Wind km/day	Sun hours	Rad MJ/m²/day	ETo mm/day	Rain (mm)	Effective rain (mm)
January	17	31.3	47	104	8.8	18.7	3.99	0	0
February	19.3	34.3	37	112	9.3	21.2	4.91	0 0	
March	22.5	37.5	30	121	9.7	23.5	5.93	0	0
April	26	39.3	34	138	9.2	23.6	6.66	0	0
May	27.2	40	37	225	8.3	22.2	7.93	25	24
June	25	35.6	54	354	5.8	18.3	7.19	50	46
July	23.8	32.5	64	363	4.4	16.2	5.74	75	66
August	23.5	32.1	63	302	4.9	16.9	5.47	100	84
September	23.3	31.9	65	207	5.5	17.3	4.83	8	7.9
October	22.4	32.4	61	95	8.7	20.7	4.58	107	88.7
November	19.2	31	56	78	7.7	17.6	3.73	67	59.8
December	16.6	30.3	51	69	8.4	17.7	3.4	0	0
Average	22.1	34	50	181	7.5	19.5	5.36	432	376.4

Table 34.3 The soil date used with CROPWAT 8.0 model for Lentil crop (*Lens culinaris*)

Soil Data	Red Sandy Loam
Total available soil moisture (mm/meter)	140
Maximum Rain infiltration rate (mm/day)	30
Maximum rooting depth (cm)	60
Initial soil moisture depletion (as % TAM)	0
Initial available soil moisture (mm/meter)	140

Time of sowing

The sowing of seed mainly depends on the Antecedent Moisture Condition (AMC) of soil. The performance of the crop greatly depends on rainfall, irrigation condition, type of soil, storage capacity of soil and infiltration condition. In the present steady, the different sowing time was used based on soil moisture and rainfall condition to proper irrigation scheduling. The different sowing dates are presented in the Table 34.3. Early sowing generally increases potential yields, but also increases the likelihood of crop lodging and delay in planting causes reduction in yield. Hence, one has to make it sure that there is good soil moisture before sowing and that seed has been inoculated. The reduction in yield could be minimized up to a certain extent by relatively closer spacing and use of higher seed rate. In central zone, where moisture is a limitation, early planting in mid-October is recommended to ensure proper germination. In north-west plains zone, end of October and in north-east zone. Sowing depth should be 4-6 cm. Under late-sown conditions, the row spacing should be reduced to 20 cm. Lentil seed should be sown at a depth of 3-4 cm in heavy soil. Target plant density should be 120-150 plants/m^2. However, the higher plant density for lower rainfall areas and short season environments is recommended. An ideal germination percentage is 80 %, if less, sowing rates may need to be increased to compensate. The seeding rate is expressed in the equation (34.3) as given by Peter, 2005.

$$\text{Seed rate}\left(\frac{\text{kg}}{\text{ha}}\right)=\frac{\text{Target plant density}\left(\frac{\text{plant}}{\text{m}^2\ \text{X}\ 100}\right)\text{X seed weight (grams) X 10}}{\text{Germination per centage X Establishment per centage}} \qquad (34.3)$$

Cropping Systems

Generally, Lentil is grown mixed with barley, toria, rape seed and mustard crops. It is also raised as an intercrop in autumn sugarcane. Ali *et al.*, 2012; Laik *et al.*, 2014 reported that two lines of lentil may be sown 30 cm apart in the center of 2 sugarcane rows. The distance between sugarcane and lentil rows should be 30 cm. Intercropping of linseed and lentil (2:1), lentil and mustard (4-6:1) in Bundelkhand region of Uttar Pradesh is also promising. Lentil is relatively more shade loving than gram. Hence, it performs better in mixed and intercropping systems. Generally lentil is grown after the harvest of *kharif* crops or as the sole crop of the year. The present study was having Groundnut-Lentil cropping system. The irrigation scheduling of Lentil crop was presented as estimated using the FAO CROPWAT 8.0 in this study area. The other lentil-based rotations are given below:

- Groundnut - Lentil; Sorghum - Lentil
- Pearlmillet - Lentil; Maize - Lentil
- Cotton - Lentil; Kharif fallow - lentil (rainfed areas)
- Rice - lentil; Mustard - Maize (fodder).
- Rice-lentil is the most common rotations.

Influence of crop type on crop water needs

The influence of the crop type on the crop water need is important in two ways. The crop type has an influence on the daily water needs of a fully grown crop i.e. the peak daily water needs of a fully developed Lentil crop will need more water per day than a fully developed mustard crop. Second one is that the crop type has an influence on the duration of the total growing season of the crop. There are short duration crops, e.g. peas, with a duration of the total growing season of 90-100 days and longer duration crops, e.g. melons, with a duration of the total growing season of 120-160 days. There are, many perennial crops that require more water than Lentil.

Calculation of crop water requirements

The calculation of Lentil crop water requirements was done using FAO Penman-Monteith equation. The basic formula for the calculation reads as follows (34.4):

i) **ET_{crop} = K x ETo** 34.4

Where, ET_{crop} is the water requirement of a given crop in mm per unit of time (mm/day, mm/month or mm/season).,K_c = the "crop factor", ET_o = the "reference crop evapotranspiration" in mm per unit of time (mm/day, mm/month or mm/season).

ii) **ETo reference crop evapotranspirations**

The reference crop evapotranspiration ET_o (sometimes called potential evapotranspiration, PET) is defined as the rate of evapotranspiration from a large area covered by green grass which grows actively, completely shades the ground and which is not short of water. The rate of water which evapotranspirates depends on the climate. The highest value of ET_o is found in areas which are hot, dry, windy and sunny whereas the lowest values are observed in areas where it is cool, humidand cloudy with little or no wind.ET_o is calculated by FAO-Penman-Monteith equation.

FAO Penman-Monteith equation

The FAO Penman-Monteith method is used to estimate ETo based on the original Penman Monteith equation and the equations of the aerodynamic and surface resistance can be given as in the equation (34.5). (Smith, 1992; FAO, 1998; FAO, 2009; Shah *et al.*, 2015).

$$ET_o = \frac{0.408\Delta(R_n - G) + \gamma \frac{900}{T+273} U_2 (e_s - e_a)}{\Delta + \gamma(1 + 0.34U_2)} \qquad (34.4)$$

Where, ET_o = reference evapotranspiration (mm/day); R_n= net radiation at the crop surface ($MJ.m^{-2}.d^{-1}$); G = soil heat flux density ($MJ.m^{-2}.d^{-1}$); T = mean daily air temperature at 2 m high (°C); u_2 = wind speed at 2 m high ($m.s^{-1}$); e_s = saturation vapor pressure (kPa); e_a = actual vapor pressure (kPa); e_s-e_a = saturation vapor pressure deficit (kPa); " = slope vapor pressure curve ($kPa.°C^{-1}$); = psychrometric constant ($kPa.°C^{-1}$). The equation uses standard climatological records of solar radiation (i.e. sunshine), air temperature, humidity and wind speed.

The reference evapotranspiration of Lentilcropwascalculated by FAO Penman-Monteith method, using decision support software CROPWAT 8.0 developed by FAO, based on FAO Irrigation and

Drainage Paper 56 (FAO 1998). The FAO CROPWAT program (FAO 2009) incorporates procedures for reference crop evapotranspiration and crop water requirements and allow the simulation of crop water use under various climate, crop, and soil conditions (www.fao.org). The equation uses standard climatological records of solar radiation (sunshine), air temperature, humidity and wind speed. To ensure the integrity of computations, the weather measurements should be made at 2 m (or converted to that height) above an extensive surface of green grass, shading the ground and not short of water.

Selection of Crop Factor K_c for Lentil crop

In order to obtain the crop water requirement (ET_{crop}) the reference crop evapotranspiration (ETo) must be multiplied by the crop factor (K_c). The crop factor (crop coefficient) varies according to the growth stage of the crop. There are four growth stages to distinguish: (i) Theinitial stage: when the crop uses little water (ii) Thecrop development stage, when the water consumption increases (iii) Themidseason stage, when water consumption reaches a peak (iv) Thelate season stage, when the maturing crop once again requires less water. The different crop Kc values were given in the Table (34.4). Smith, 1992; FAO, 1998; FAO, 2009; Shah *et al.*, 2015.

Table 34.4 The crop data file used with CROPWAT 8.0 model for Lentil crop (*Lens culinaris*)

Stage (Days)	Initial Stage	Develo-pment Stage	Mid-Season Stage	Late Season	Total
Length of crop growth period	20	30	20	30	100
Kc Value	0.3	0.65	0.95	0.7	0.45
Rootzone Depth (m)	0.3		0.6		
Critical deplition (fraction)	0.45	0.45	0.45	0.5	
Yield response (fraction)	0.2	0.8	0.6	0.2	0.7
Crop height	0.20	0.35	0.4	0.45	

Irrigation Scheduling and water requirement of Lentil crop

The efficiency of water in agricultural production is generally low. Only 40 to 60 % of the water is effectively used by the crop, the rest of the water is lost in the system or in the farm either through evaporation runoff, or by percolation into the groundwater. Irrigation scheduling, if properly managed can offer a good solution to improve water efficiency in the farm. Irrigation scheduling involves deciding when and how much water to apply to a field. Good scheduling will apply water at the right time and in the right quantity in order to optimize production and minimize adverse environmental impacts. Bad scheduling will mean that either not enough water is applied or it is not applied at the right time, resulting in underwatering, or too much is applied or it is applied too soon resulting in over watering (Shah *et al.*, 2015; Othman *et al.*, 2016). Under or overwatering or deficit conditions can lead to reduced yields, lower quality and inefficient use of nutrients. In the present study the irrigation scheduling makes sure that water is consistently available to the plant and that it is applied according to crop requirements. The irrigation scheduling was prepared using CROPWAT 8.0 model, irrigation at 100% critical depletion, irrigation at fixed interval per stage and method of irrigation application, refill soil moisture content to 100% to field capacity. For the present study the standard procedure was followed for irrigation scheduling of Lentil crop under most acceptable criterion.

Calculation of irrigation water requirements

Net irrigation water requirement (NIWR) or (NIR) is the quantity of water necessary for crop growth. It is expressed in mm per year or in m^3/ha per year (1 mm = 10 m^3/ha). It depends on the cropping pattern and the climate.Crop water requirements (CWR in mm) for a given crop, i, are given in the equation (34.6):

$$CWRi = \sum_{t=0}^{T} (Kci.ETo\text{-}P) \qquad (34.6)$$

Where, K_{ci} is the crop coefficient of the given crop during the growth stage t and where T is the final growth stage.

Each crop has its own water requirements. Net irrigation water requirements (NIWR) in a specific scheme for a given year are thus the sum of individual crop water requirements (CWR_i) calculated for each irrigated crop i. Multiple cropping (several cropping periods per year) is thus automatically taken into account by separately computing crop water requirements for each cropping period. By dividing by the area of the scheme (S. in ha), a value for irrigation water requirements is obtained and can be expressed in mm or in m^3/ha (1 mm = 10 m^3/ ha). Therefore, The NIWR or NIR in mm is given in the equation (34.7)

$$NIWR = \frac{\sum_{i=1}^{n} CWR_i.S_i}{S} \quad (34.7)$$

Where, S_i is the area cultivated with the crop i in ha.

The cropping intensity (C_i) of the scheme can be defined as equation (34.8):

$$C_i = \frac{\sum_{i=1}^{n} S_i}{S} \quad (34.8)$$

The FAO's CROPWAT software (version 8.0) was used to compute NIWR for the study area and the model was run for six different scenarios of sowing dates; six sowing dates were simulated (2 each) for early sowing, normal sowing and late sowing date conditions. Gross irrigation water requirement (GIWR) is the amount of water to be extracted (by diversion, pumping) and applied to the irrigation scheme. It includes NIWR plus water losses: The GIWR in mm was given in the equation (34.9).

$$GIWR = \frac{NIWR}{E} \quad (34.9)$$

Where, E is the global efficiency of the irrigation system.

Advantages of Irrigation Scheduling (IS) in Lentil crop

- IS enables farmers to schedule watering to minimize crop water stress and maximize yields
- IS reduces farmer's costs of water and labour through less irrigation, thereby making maximum use of soil moisture storage.
- IS lower fertilize costs by holding surface runoff and deep percolation (leaching) to a minimum.
- IS increases net returns by increasing crop yields and crop quality.
- IS minimizes water-logging problems by reducing the drainage requirements.

Table 34.5 The calculated crop evapotranspiration ET_c, effective rainfall and irrigation requirement predicted by CROPWAT 8.0 model for Lentil crop

Month	Decade	Stage	Kc (mm/ day)	ETc (mm/ dec)	ETc (mm/ dec)	Eff rain (mm/dec)	Irr.Req.
Oct	2	Init	0.3	1.38	1.4	3.5	1.4
Oct	3	Init	0.3	1.29	14.2	29.8	0
Nov	1	Deve	0.31	1.23	12.3	24.1	0
Nov	2	Deve	0.46	1.7	17	21.4	0
Nov	3	Deve	0.67	2.41	24.1	14.3	9.8
Dec	1	Mid	0.87	2.99	29.9	0.1	29.7
Dec	2	Mid	0.93	3.05	30.5	0	30.5
Dec	3	Late	0.92	3.24	35.6	0	35.6
Jan	1	Late	0.79	2.99	29.9	0	29.9
Jan	2	Late	0.62	2.49	24.9	0	24.9
Jan	3	Late	0.49	2.09	14.6	0	14.6
				Total	234.4	93.2	176.5

Table 34.6 Irrigation scheduling, sowing, net and gross irrigation requirements for lentil crop

Planting Date	Earliest sowing date (01-Sep)	Early sowing date (10-Sep)	Normal sowing date-1 (20-Sep)	Normal sowing date-2 (01-Oct)	Late sowing date (10-Oct)	Latest sowing date (20-Oct)
Gross Irrigation (mm)	96.7	159.3	111.6	170.1	169.2	229.8
Net Irrigation (mm)	67.7	115.3	78.2	119.4	118.4	160.9
Actual water use by crop (mm)	261.5	251.5	241.9	234.7	232.2	232.3
Potential water use by crop (mm)	261.5	251.5	241.9	234.7	232.2	232.3
Rainfall (mm)	182.1	178.2	178.2	170.4	144.9	102.7
Effective rainfall (mm)	169.9	135.6	125.6	102.7	78.7	53.9
Total Rain loss (mm)	12.2	42.6	52.6	71.3	66.3	48.8
Moisture deficit at harvest (mm)	23.9	4.7	38.2	12.9	35.1	17.5
Actual Irrigation requirement (mm)	91.6	116.3	116.3	132	153.6	178.4
Efficiency rain (%)	93.3	76.1	70.5	59	54.3	52.5

The study area hasreceived annual rainfall around 400-450 mm. The maximum rainfall is received during the months from August to November as shown in Table (34.2). The results showed that more amount of water results into effective rainfall, this can be used for crop growth. The excess water can be stored as infiltration in the show soil depth that can be used for crop production when there is absence of rain or irrigation supply or deficit irrigation in the field. Figure (34.2) shows that the six different sowing dates with interval of 10 days sowing periods namely; earliest sowing dates starts from 1st September, early sowing date, normal sowing date-1, normal sowing date-2, late sowing and latest sowing up to 20 October as shown in Table (34.6). The results speaks that effective rainfall can be maximum during the crop growing periods due to field preparation as results of less runoff.

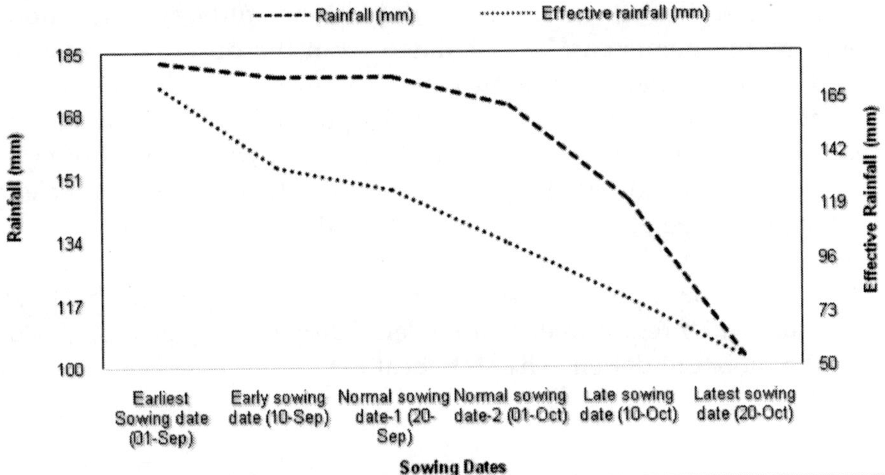

Fig. 34.2 The relation between rainfall and effective rainfall of study area.

The sowing date and period play vital role in the crop production and yield as well as the productivity of crop.It mainly depends on the Antecedent soil Moisture Condition and soil moisture storage in the soil root zone. Figure (34.3) shows that the six different sowing dates with interval of 10 days sowing periods namely earliest sowing dates starts from 1st September, early sowing date, normal sowing date-1, normal sowing date-2, late sowing and latest sowing upto 20 October. The results speaks that the maximum utilization of rainfall is on the normal sowing dates that is 1st October to 10 October as shown in Table (34.6). The earliest or latest sowing date's results into less efficiency of rain due to more runoff and rainfall periods.

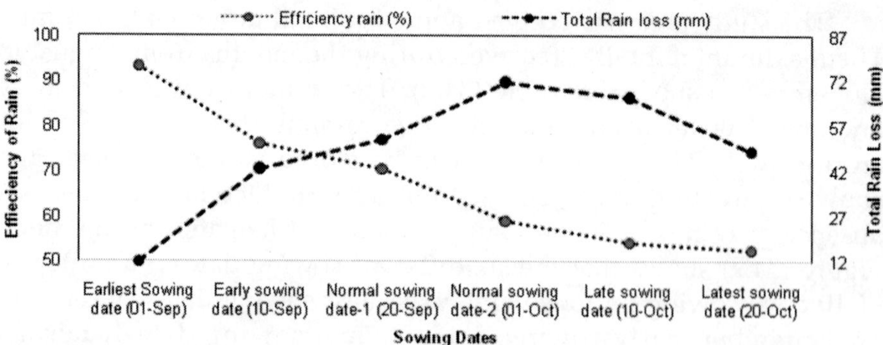

Fig. 34.3 The relation between efficiency of rain and total rain loss for the study area.

Establishment of Irrigation Water Requirements of Lentil crop

Based on the effective rainfall data in the study area, irrigation water requirements can be calculated from the difference between effective rainfall and the total water requirement. The results are presenting in Figure (34.4). The net irrigation and gross irrigation requirement are in increasing orderthrough growing period of crop. It is mainly depends on the growth stages of Lentil, crop coefficient, variety, soil type etc. Figure shows the NIR and GIR during the six different sowing date's starts from September 1st to October 20 with interval of 10 days period as shown in Table (34.6). Figure (34.5) shows the actual water requirement for the lentil crop during the crop period and the moisture deficit. The statistical data ispresented in the Table (34.5).

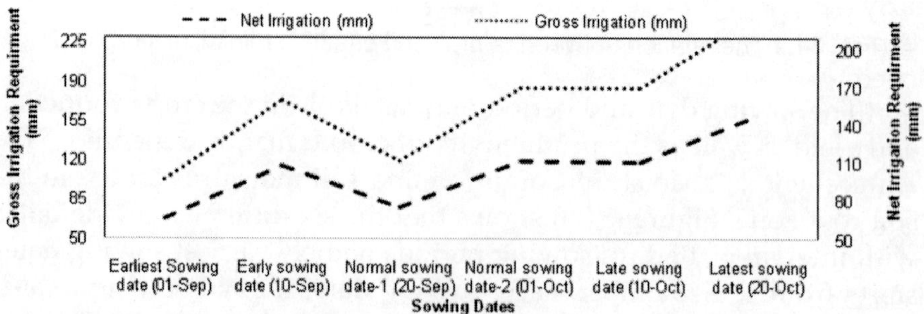

Fig. 34.4 Relation between gross and net irrigation requirement of Lentil crop (GIR and NIR).

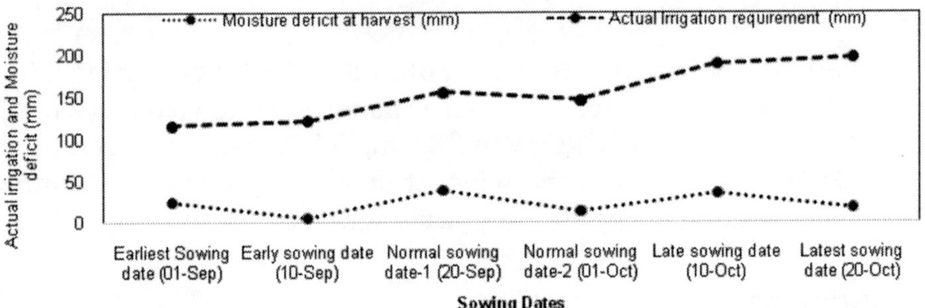

Fig. 34.5 The relation between actual irrigation requirement and moisture deficit at harvesting time of lentil crop.

Conclusion

An attempt was made to estimate the crop water requirement, net irrigation requirement (NIR) and the gross irrigation requirement (GIR) of Lentil crop (Lens culinaris) grown in semi-arid climatic region of Kurnool district of Andhra Pradesh. This district receives very low scanty of rainfall varies from 400-450 mm. The FAO CROPWAT 8.0 model was used with actual field crop information clubbed with the average long term climatic parameters for calculating the irrigation requirement and evapotranspiration of Lentil crop. The results showed that 176.5 mm/deci.e. 5.36 mm/day under the six different crop sowing dates staring from 1st September to 20 October with lag of 10 days period. The FAO CROPWAT 8.0 has simulated the NIR and GIR of Lentil crop under these sowing dates. The results revealed that the less NIR, GIR requirement was observed under normal date-1 and early sowing dates i.e. 78.2, 115.3 mmand 111.6, 159.3, 170.0 mm respectively.Based on the above information the proper irrigation scheduling is planned to obtain the maximum production, yield under the limited water availability. The study concluded that the Lentil crop,one of the highly nutritional crop, needs relatively less water requirement and it can tolerate the water scarcity so that the normal dates for sowing are mid-September to mid-October. Further, it can be safely propounded that Lentil crop can be grown in A.P. and Telanagna states very successfully with low water availability. Hence, this crop is highly recommended to be adopted in semi-arid conditions like Kurnool district of Andhra Pradesh.

Acknowledgement

Authors are indebted to many authorities for providing facilities to conduct the research work as well as many learned scientists whose inputs are built in FAO CROPWAT 8.0 model over the long period of time. Help received from the scientist of Water Technology Centre, IARI, New Delhi (WTC) are duly acknowledged.

References

Ali RI, Awan TH, Ahmad MM, Saleem U and Akhtar M. 2012. Diversification of rice based cropping system to improve soil fertility, sustainable productivity and economics. *Journal of Animal and Plant Sciences* 22(1): 108-12.

Donna Fleury P. 2016. International year of pulses. Saskatchewan pulse growers. 1-4.

FAO. 2009. CROPWAT 8.0 for Windows User Guide. Rome, Italy.

FAO. 1998. Crop Evapotranspiration: Guidelines for Computing Crop Water Requirements. FAO Irrigation and Drainage Paper 56. Rome, Italy.

Governament of India, 2015 (G.o.I.) Commodity Profile for Pulses-March, 2015, Ministry of Agriculture. Govt. of India. (https://india.gov.in/topics/agriculture/irrigation).

Laik R, Sharma S, Idris M, Singh AK, Singh SS, Bhatt, BP, Saharawat Y, Humphreys, E and Ladha JK. 2014. Integration of conservation agriculture with best management practices for improving system performance of the rice wheat rotation in the Eastern Indo Gangetic Plains of India. *Agriculture, Ecosystems & Environment* 195(6): 68-82.

Othman AH, and Dahim MH. 2016. Scenario based agriculture land use for achieving sustainable irrigation water requirements in Saudi Arabia: A case study of Najran region. *Journal of Current Research in Science*, 4(1): 17.

Peter Matthews. 2005 Germination testing and seed rate calculation, NSW Department of Primary Industries, Temora, and Di Holding, NSW Department of Primary Industries, Wagga, Australia, 1-20.

Shah PV, Mistry RN, Amin JB, Parmar AM, and Shaikh MR. 2015. Irrigation scheduling using CROPWAT 8.0. *International Journal of Advance Research in Engineering, Science and Technology* (IJAREST). 2(4): 1-10.

Smith M. 1992. CROPWAT: A Computer Program for Irrigation Planning and Management, by FAO Irrigation and Drainage. Paper No. 46, Rome.

Tiwari AK and Shivhare AK. 2016. Pulses in India: Retrospect and Prospects. Government of India, Ministry of Agriculture and Farmers Welfare (Department Of Agriculture, Cooperation and Farmers Welfare), Directorate of Pulses Development Vindhyachal Bhavan Bhopal (M.P.) 1(2): 1-317.

Citation

Veeranna J and Mishra AK. 2018. Estimation of Evapotranspiration and Water Requirement for Irrigation Scheduling of Lentil Crop (Lensculinaris) in *Kurnool*, Andhra Pradesh. In: Scientific Lentil Production. Singh AK, Bhakta N, Sangale UR, Manibhushan, Sundaram PK, Kumar S and Yasin JK. (eds.) Society for Upliftment of Rural Economy Varanasi, India. Pp. 561-577.

❑❑❑

Chapter 35

Improved Equipment and Technologies for Lentil Cultivation in India

Tapan K Khura, Satish D Lande, HL Kushwaha, Prem K Sundaram, Roaf A Parray and Ram Chandra

SYNOPSIS

Mechanising lentil cultivation is one of the important aspects to enhance the production, productivity and area under this crop. The traditional methods of cultivation mainly depend on the human and animal energy; poor water management and post-harvest operations and storage methods which increase the cost of production and storage lose. On an average, human and bullock (or machine) labor constituted 84%, 80%, 81% and 78% share of the total operational expenses for chickpea, lentil, black gram and pigeon pea production, respectively.With the intervention of the improved machinery, equipment and technologies the not only production will be increased but also more area can be brought under the crop to meet the pulse demand of the country. In case of pulse the post-harvest loses is about 25-30 percent. The improved post-harvest machinery for different unit operations would increase the shelf life of the produce.

Keywords: Mechanisation, Lentil, Seed drill, post-harvest, milling.

Introduction

Most of the Indian are vegetarian, milk and pulses are the main source of proteins in their diet. Moreover, 15-30 percent of the daily protein needs are supplied from edible legumes or pulses. Lentils are rich in proteins and offer the third highest level of proteins from plant-based foods, after soybeans and hemp and has extremely rich soluble fiber which forms a gel-like substance in the digestive tract helping in removing bile from the body. Lentil occupied fourth place in production of pulses in the world. In India, annual production of pulses ranges from 12 to 17 million tonnes. India is second largest producer of with 0.95 million tonnes lentils followed by Canada (FAO, 2012). Lentil is an important rabi pulse crop after gram grown in India and four major lentil producing states are Uttar Pradesh, Bihar, Madhya Pradesh and West Bengal and contributed more than 80 percent lentil production and UP is leading with 43 percent alone. Its share in the acreage and production of total rabi pulses is about 12 per cent, whereas in the overall pulses production, its share is about 6 per cent. Kumar (1998) projected the demand and supply of pulse crops for the year 2020, predicted that the domestic supply would be 9 per cent short of domestic demand under most optimistic scenario, and about 26 per cent under the pessimistic scenario. Hence, even by considering only the projected domestic demand, ignoring the export potential, there is an urgent need for increasing the supply of lentil.

Mechanisation of cultivation of lentil is one of the important inputs to enhance the production, productivity and area under the crops. The traditional methods of cultivation mainly depend on the human and animal energy; poor water management and post harvest operations and storage methods whichincrease the cost of production and storage lose. On an average, human and bullock (or machine) labor constituted 84%, 80%, 81% and 78% share of the total operational expenses for chickpea, lentil, black gram and pigeonpea production, respectively. In developed nations the lentil cultivation has been mechanized, however, India on the path of mechanization of the crop with slows pace. With the intervention of the improved machinery, equipment and technologies the not only production will be increased but also more area can be brought under the crop to meet the pulse demand of the country. In case of pulse the post harvest loses is about 25-30 percent. The improved post harvest machinery for different unit operations would increase the shelf life of the produce and also add value on the product. Some of the improved machinery and technologies for lentil is given below for reduce human drudgery; cost of production, post harvest loses and enhances water use efficiency, production and productivity.

Improved equipment and machines for Lentil cultivation

Seed bed preparation

The seedbed name indicates that the seed should be placed in the proper soil and moisture conditions to start its germination and then established on it. One of the important and energy intensive operations in seed bed preparation is tillage. The tillage equipment manipulates the soil structure, destroy weeds, and retention soil moisture by mulching and percolation of rain water in it. In India animal and tractor drawn tillage equipment are popular. Depth of the ploughing proportional to energy requirement for the operation therefore, it is necessary that the root penetration and root density of the crop at different depth to be analysed before starting the tillage operation along with soil structure and texture, rain fall of the region etc. The root penetration of the lentil is within 150200 mm from the surface.

- Improved Animal Drawn Plough: The traditional animal drawn plough varies with soil and size of the animals of the regions. The traditional wooden plough largely disturb the soil with little inversion therefore, weed intensify in the initial stages of the crops. However, with the use of the improved animal drawn Iron plough the soil inversion is highest and pulverisation also improves.

  Tractor Drawn Tillage Machinery

  Disc Plough: Considering the fact that lentils root development is small, not so deep ploughing should be done. The ploughing operation should be carried out during summer particularly in clay soil so the clods can be disaggregated by the meteorological agents. Lentil isgrown in rabi season, in rain fed area and in dryland agriculture the in-situ moisture retention is very important. The soil is open up by primary tillage (MB plough or Disc plough) conserve and retain the rainwater at the root zone. In India the Disc plough (Fig.35.1a) is preferred over the MB plough as the soil is hard during the summer ploughing, inversion of soil is not so important and draft requirement by MB plough is more. The ploughing is followed by two or three harrowing to create a fine tilth and level planting bed.

  Offset Disc Plough: At the beginning of winter harrowing should be done. Disc harrow,secondary tillage implement are used for cutting the soil to as shallow depth (100-120mm) for smoothening, creating proper soil tilth and pulverizing the soil as well as cut the weed. It also breaks the soil clods and levels the seed bed. Animal drawn disc harrow and tractor drawn disc arrow (Fig.35.1b) may be used for the preparation of seed bed.

a. Disc Plough b. Offset disc harrow

Fig. 35.1 Disc plough and harrow

- Tyne cultivator and patella

The tyne cultivator (Fig.35.2) is commonly used secondary tillage tool in Nother part of our country. It is used both as secondary tillage and primary tillage tool. The cultivator stirs the soil, breaks the clods. The tynes fitted on the frame combs the soil at shallow depth for better aeration in the field. The weeds are destroyed satisfactorily. Patela can be used behind the cultivator and sometime used as a separate operation. The wooden plank(patella) used for smoothening the soil, crushing the clods, packing and levelling the cultivated soil. The curved spikes attached behind the wooden board help in collection and removal of weeds.

Fig. 35.2 Tyne cultivator

Sowing Methods

Lentil is considered one of the most sensitive leguminous crops to wet a well as to over-dry conditions.Timely planting is key factor for getting potential yield of improved varieties of lentil but delay in sowing cause reduction in yield. Middle of October is the most suitable time for sowing lentil. Sowing should be done in rows 30 centimetre apart. The optimum seed rate for normal sown crop is 30-40 kg per hectare. Lentil seeds should be sown at a shallower depth (30-40 mm). This could be done either by using a following equipment. El-Berry and Ahmed (1991) and El-Sahrigi and Abo Habaga (1993) reported that an increase of about 10% in yieldis possible due to machine sowing, due to accuracy of seed placement,depth and seeding rate.

A. *Broadcasting:* It is one of the oldest and simple method of sowing seed. The Distribution of seed depends on the skill of the person. However, as the prices of seed rising and more hybrid seed are on demand the broadcasting has been slowly changing towards mechanized sowing. Broadcasting is the process of random scattering of seeds on the surface of seed beds. In some regions. Lentil seed is broadcasted in before harvesting the standing crop so that the residualmoisture from the field of previous crop can be used for germination. However, in some cases after preparing the seed bed seed broad casted and after broadcasting seeds are covered by plankers or some other devices. This process is unscientific but labour requirement for swing this method is low and often preferred by farmers. The scattered seeds are prone to birds. Lentil seeds need moisture for germination so soil around the seed compacted to get the soil moisture.

B. *Seed drill/direct drilling:* Drilling consists of dropping the seeds in furrow lines in a continuous flow with a uniform rate and covering them in soil.

❐ Animal Drawn Seed-cum-Fertilizer Drill

CIAE 3-row Seed-cum-Fertilizer Drillis a simple, light weight and compact machine to sow lentil and drill fertilizer in black soil under rainfed condition. The fluted roller metering mechanism meter the seed as per the pre set seed rate and the shoe type furrow opener with non clogging boot place the seed at desired depth. The effective field capacity, field efficiency, draft and labour requirement were 0.10-0.28 ha/h, 65-75%, 450N and 515 man-h/ha, respectively.

- Tractor Drawn Seed drill

  The seed cum fertilizer drill consist of seed and fertilizer box with metering units, furrow opener and power transmitting wheel. The machine can be operated by 35 hp tractor. The Seed box capacity (cm^3): 63310-10141 Fertilizer box capacity (cm^3): 6624110308 Weight (kg): 224-328. The fluted roller type metering device can also be suitable for metering lentil seeds.

- Zero-Till Drill for Lentil Cultivation

  The nine row tractor mounted Pantnagar zero-till seed cum fertidrill of can be used for seeding lentil in the field without seed bed preparation (Fig.35.3). Singh and Singh (2006) observed that about 67.8 per cent more yields can be achieved by seeding lentil with zero till drill as compared to conventional method. Vegetative growth was more in tillage field as compared to zero till, but the number of pods/plant were more in zero till drill as compared to reduced tillage and conventional tillage.

Fig. 35.3 No till lentils; planting in rice residue

Weeding and inter row cultivation

Lentils compete poorly with weeds for light, water, and nutrients. During early stages of vegetative growth and in cool weather, lentil growth rates are slow and weeds can quickly overgrow the crop. If not adequately controlled, weed infestations can reduce yields by as much as 75 percent. Although the period of crop growth during which competition is most deleterious varies in different locations, competition from weeds is usually serious and requires some form of control in order to produce good seed yields. The weed can be controlled mechanical, chemical, crop rotation and also with integrated approach. Mechanical methods of weed control are simple and easy.

The weeding tools and equipment are categorized based on their power source such as manual, animal and power operated. Manual method is most effective but is slow and costly.Hand hoeing gave higher weed control i.e. 95.77 and 98.12 % of broad and narrow leaved weeds, respectively and produced lowest weed dry matter(Chaudhary *et al.*, 2011).

- Manual weeding Tools

  Small hand held traditional weeding tools is "Khurpi" used by the farmers. These tools are operated in squatting posture and have very low work output. Spades or chopping hoes work on the principle of impact and have straight, curved or pronged blades. Weeds are removed by digging, cutting and up rooting. These are operated in the bending posture. The operation is normally slow and tiring. Long handle tools have a soil working tool mounted at the end of a 1.5 to 2 metre long wooden/bamboo handle. These tools are operated in push or push-pull or pull mode and in standing posture (Fig. 35.4). The common shapes of blades on these weeders are straight,convex, V-shape, sweep, serrated, etc. These are designed to work under friable soil moisture conditions and give high work output at the early stages of crop growth when weeds are small.

Fig. 35.4 pushorpush-pull type manual weeders

- Animal Drawn weeder

  Improved animal drawn inter-culture and weeding implements could be used efficiently in oilseedcrops for saving time. However, widerrow spacing (above 30cm) should be provided for movement of animals and implement. The row to row spacing should be accurate and straight for successful weeding. Animal drawn tools reduce thecost ofoperation and time.

- Single row hoes

  Animal drawn single row hoes are most widely used by the farmers of different states. Straight orslightly curved blades are commonly used in single row hoes. The size of blade can be changed as per crop row spacing.

- Multi-row implements

  Multi-row units are widely used in Gujarat and other states for wide coverage and timely weeding operation. The threetine cultivator or "Triphali","Akola"hoe,"Bardoli" hoe and animal drawn sweeps of different designs are some of the new designs of animal drawn weeders.

- Power weeder

  a) Self-Propelled Power Weeder

  The self-propelled weeder is operated by 3-hp petrol-start-kerosene-run engine (Fig.35.5). Two steel lugged drive wheels tread (spacing between two wheels) can bead justed from 400 to 650 mm to suit the row-to-row spacing of different crops. The V-shaped weeding sweeps of 150 and 200 mm size are fitted on tynes. The spacing between sweeps is easily adjusted bysliding the tynes on tool mounting bar to suit the crop row spacing. For operation, the tynes are adjusted to the raised position so they do not dig into the ground and the weeder is transported to the field. The tynes are adjusted according to the row spacing of the crop and depth of operation.

Fig. 35.5 Self Propelled Power Weeder

b) Light Weight Power Tiller

This machine consists of a 5hp diesel engine mounted on the powertiller (Fig. 35.6). The rotary blades and ground wheels are connected to engine with the help of belt-pulleys and chain-sprockets. The clutch provided on the tiller for turning the machine to right or left. The machine can be used both for weeding and field preparation.

Fig. 35.6 Power tiller with rotary tiller

Plant protection equipment

Protection of crops from insects, pests, diseases, and weeds has been the main concern of farmers after having grown it by using costly seed, fertilizer and water. With the advance of the pesticide application technology and the availability of information on it, the farmers interest in improvement of spraying quality has also increased considerably, both among big and small growers. Knapsack sprayers are very widely used because of its simplicity in construction and operation besides flexibility in interchanging of the tips of the standard conetype nozzles, giving a wide range of spray outputs, pattern and quality at low cost.

Irrigation

The crop is mostly grown in rainfed areas. Pod formation followed by flower initiation is critical stages for irrigation. Irrigation is important especially under late sown conditions due to poor root development. The crop requires 200 mm of water depending on soil and climate. Lentil requires 1-2 irrigations depending upon the rains during the growing season. First irrigation is applied 6 weeks after sowing and the second at flowering or pod formation stage.

Harvesting

Gathering of useful parts of the plant after maturity is known as harvesting. In cases of pulses harvesting is carried out 10 to 15 days after the grain has reached physiological maturity. The physical characteristics of maturity of letile are when the upper leaves dry and pods pale yellow. The lentil harvested when the moisture content in the ranges of 15-20 per cent.

A. Manual harvesting

In traditional agricultural systems lentil crop is harvested by manually with sickles or by machines when the pods are ripe but not yet open. It takes about 80-100 man-h/ha to cut the crops by sickle.

B. Combine harvesting

Mechanized harvesting of legumes is usually done with combine harvesters which is designed for wheat and paddy. During combining, the following points must be taken care for better product quality. Straight-cutting is preferred against swath cutting to reduce shattering losses. Specialized pick-up reels, lifter and flex headers are required for effective harvesting. The speed of cylinder speed is kept between a slow speed to prevent seed damage and a faster speed to prevent cylinder plugging; is usually between 250 and 500 rpm. The concave clearance usually set to be wider since the crop threshes easily. Tailings should be kept to a minimum, and grain and return elevator chains should be adjusted correctly to avoid splitting or cracking seed. The fan speed should be set where an acceptable clean sample is produced and excessive fan speed will blow lentil seed out the back of a combine. The capacity of the combine harvester generally varies from 0.9 to 1.1 h/ha. In India mechanized harvesting of lentil is minimal.

Pre-drying

During this the harvested products is dried in order to ready for the threshing operation. At harvesting stage the plants may contain green plant materials and all grain may not have reached a uniform degree of maturity and may have moisture content too high. Hence pre-dying is necessary. The pre-drying can be done two methods: allowing the matured crop to stand on the field before harvesting but it has got many disadvantages. Other method, the newly harvested crop piles in the field or on threshing floor for prolonged exposure to hot air or sun drying for reduce moisture.

Threshing

In threshing operation the grain is separated from the plants. This operation can be carried out by manual or animal or machines. This operation should be done carefully to reduce the breakage of grain or protective husks, thus reducing the quality of product and losses due to action of insects and moulds.

A. Manual and animal/tractor treading

   Manual threshing is done by striking heaves of the crop spread over threshing floor with a flail or s stick. By hand threshing, a worker can obtain 15-40 kg of product per hour. Threshing is also done by trampling by bullocks where draught animals are available. Large quantity of crops spread over the threshing floor with about 20-40 cm thick layer of sheaves and treading the animal over it. Tractor can also be used 2-3 times over the spreaded crops to obtain grain from crop.After threshing, seed is cleaned and dried in sun to bring moisture content to 12% for safe storage.

B. Mechanical thresher

   It is a batch type thresher suitable for threshing lentil the form of single ear or bunch of ears. Single Ear Head Thresher/multicrop thresher developed by CIAE, Bhopal consists of feed inlet, beater type threshing cylinder, aspirator column, light seed gathering chamber and aspirator blower. It is 2-3 times faster over manual threshing. The unit is also helpful for cleaning of grain sample during threshing experiment. The seed recovery, threshing efficiency and cleaning efficiency were 99-100%, 99-100% and 95-100%, respectively. The capacity of the machine was 300-450 kg/h.

Post-harvest operations

After threshing the crop, the collected grains remains generally at higher moisture content then desired for safe storage of grains (12-14%). Drying is the important post harvest operation during which the grain/product is brought to safe moisture level. The aim of this unit operation is to lower the moisture content of the grain for safe storage and ready for further processing.

The grain can be dried in two methods: natural drying and artificial drying.

Natural Drying

In this method of drying grains exposed to natural convective air. The threshed grains is spread in thin layers on the floor, where it exposed to air. Duration for getting safe moisture level is depending on the thickness of layer, temperature and humidity of the air circulated and ambient condition. For uniform drying the grain must be stirred in regular interval. The low tunnel solar dryer can be used for drying. Advantages of this method are such as low initial cost, reduce the cost of drying and flexible in capacity.

Artificial Drying

In this method the heated air (dryers or unheated air (dehumidifiers) is circulated through a grains. This method is popular in semi-humid and humid conditions where natural drying cannot be used. Artificially heated air is forced to circulate through a mass of grain to absorb release moisture from grain mass. The unheated dehumidified air is blown through the grains until the moisture content reduce to the desired level. The artificial dryers are of two type batch type and continuous type.

a) Batch type dryer

   It is suitable for drying small and medium quantity of grains. The grains to be dried spread on the space of the dryer and once safe moisture level obtained it is taken out and next batch of grain is put in the dryer. These are floor dryer, bag dryer and box dryers. By drying the pulse grains in a thin layer on a concrete floored threshing yard or on a black tarpaulin or black polyethylene sheet under the hot sun, the grain temperature gets raised to 52°C. This will effectively kill all the life-stages of bruchids and remove the existing field- infestation

C) Continuous dryers

   Generally the continuous dryers are required complex infrastructure, special equipment and planning. Where large quantities of produce are to be handled continuous dryer is advisable. It can be classified as vertical dryer, belt dryer and rotary dryer. Continuous flowing vertical dryers are suitable for quick drying and for large volume of grain mass. Continuous flowing belt dryers are suitable for small chaffy seeds and quick drying. Rotary dryers can be used for chaffy grained with moderate moisture content and of large capacity.

Storage

Lentil can remain in edible condition for several years, if properly stored. Lentile suffer both qualitative and quantitative losses during storage. The milling losses in insect damaged grains are higher. The factor affect the deteriorations of grain during storage is the interactions of temperature, humidity and moisture. One of the traditional method where freshly threshed pulse grains should be dried in the sun for 3-5 days. After which it should be cooled and stored in suitable metal, plastic bins, large earthen pots, brick, and cement storage structure with tight lids. While storing the grains in the containers, farmers can apply a thick layer of about 3 cm of locally available sieved sand (without any soil particles) on top of the grains and tightly close the lid.

Milling

The removal of the outer husk and splitting the grain into two equal halves is known as milling of pulses. Lentilis converted into dal by decutilating and splitting. The Carborundum emery rollers are used for dehusking and burr grinders for splitting. Basic processes in dal milling are cleaning, dehusking, splitting, separation and storage (Fig. 35.7). Rewetting and drying is done to loosen portions of husk sticking after repeated rolling.

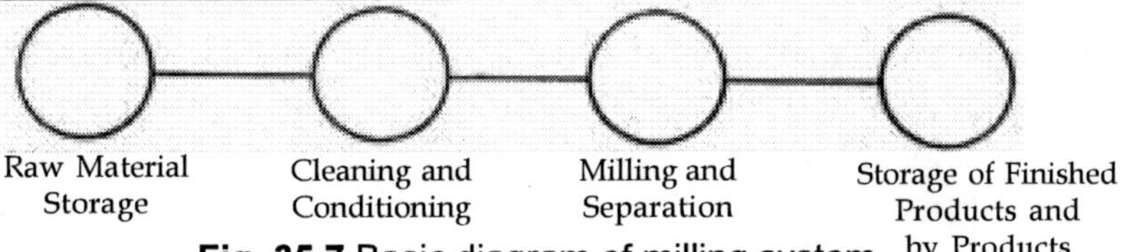

Fig. 35.7 Basic diagram of milling system

The traditional milling of pulses is classified as dry milling and wet milling. Both the processes consist of two basic steps: (i) Preconditioning (ii) milling by dehusking and splitting of the grains. It takes 3 to 7 days for complete processing of a batch of 20 to 30 tonnes of pulses into dals and milling losses are also quite high in the traditional method of milling of pulses.

Precon-ditioning

A. **Cleaning and grading**

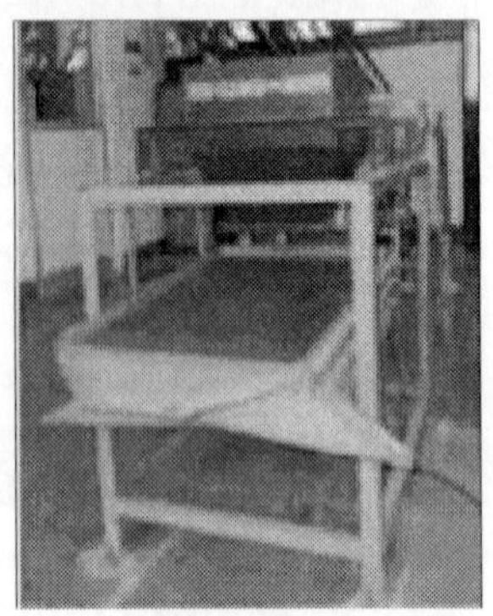

Fig.35.8

During harvesting and threshing operation lentil come along with dust, chaff, grits etc. It has to be cleaned before using for next unit operations. The reel type (Fig.35.8) or rotating sieve type cleaner-cum grader are used for separating grain from impurities and grading grain according to size.

B. **Pitting**

Fig.35.9 Pulse pitting machine

Before soaking in oil or water the clean lentile are passed through an emery roller machine that made cracks and scratches on the husk. Which facilitate penetration of oil/water for the loosening of husk. Tapperness in clearance is provided between the emery roller and cage (housing) from inlet to outlet (Fig.35.9). As the material is passed through the narrowing gap cracking and scratching of husk takes place by friction between grains and emery. Some of the lentils are dehusked and split during this operations which are then separated by sieving.

C. **Pre-treatments**

Sometime lentils grains soaked in water for 30 to 60min or water sprayed over the heap of lentil grains and kept it covered with jute bags or any other material for tempering. Then the tempered lentil removed from the stock and spread on the drying floor to allow to dried upto 10-12% moisture content.

D. **Conditioning of pulses**

Conditioning of pulses is done by alternate wetting and drying. After sun drying for a certain period, 3-5 per cent moisture is added to the pulse and tempered for about eight flours and again dried in the sun (Fig.35.10). Addition of moisture to the pulses can be accomplished by allowing water to drop from an overhead

tank on the pulses being passed through a screw con-veyor. The whole process of alternate wetting and drying is continued for two to four days until all pulses are sufficiently conditioned. Pulses are finally dried to about 10 to 12 per cent moisture content.

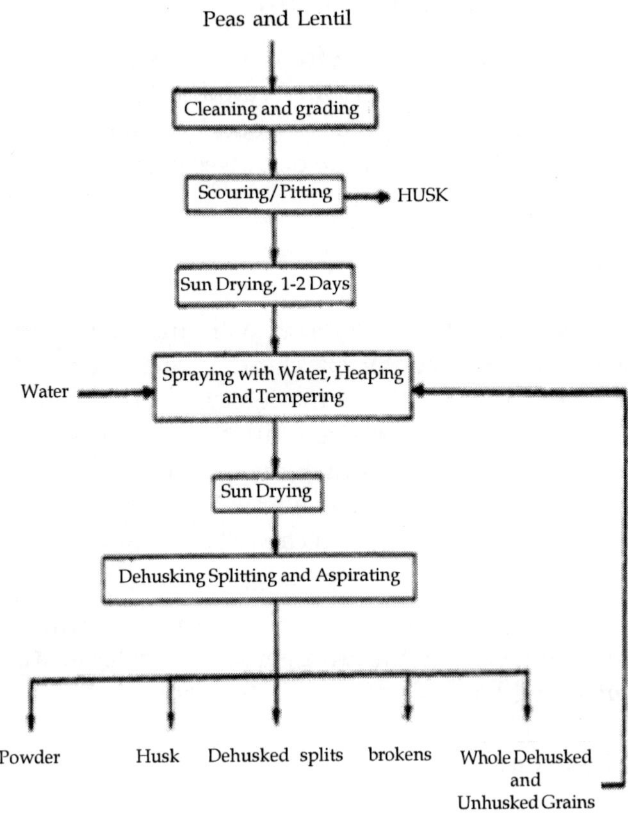

Fig. 35.10 Flow diagram for traditional lentil milling

D. Dry Milling of Pulses

Emery rollers, known as Gota machine are used for the dehusking of conditioned pulses About 50 per cent pulses are dehusked in a single operation (in one pass). Dehusked pulses are split into two parts also, the husk is aspirated off and dehusked, split pulses are separated by sieving. The tail pulses and unsplit dehusked pulses are again conditioned and milled as above The whole process is repeated two to three times until the remaining- pulses are dehusked and split.

E. Polishing

Polish is given to the dehusked and split pulses by treating them with a small quantity of oil and/or water.

Machinery for Lentile Processing

Pulses are processed in India for converting the raw material into variety of products. Machinery is available for production of dal dehusked splits, roasted and pulse flour and polished pulses. Selection of machine depends upon the type of pulses crop to be processed and the product desired.

Hand operated machines

A. *Domestic dal mill:* It consists of two circular stones mounted one on the other in horizontal plane and is one of the oldest existing unit. Upper stone is rotated manually and the material fed from the central opening to get size reduction. This manually operated unit is commonly used for making splits of variety food legumes. The traditional dal mill has been modified by KVIC consisting of grading and splitting unitand capacity of 4 quintal/day capacity.

B. *Pulse dehusking machines:* It consists of emery coated inverted metal cone fixed to a vertical shaft rotating inside a chemical wiremesh screen with the arrangement for clearance adjustments. This machine has a capacity of 50 to 60kg splits/hour at 60 to 70rpm and gives 75-80% dal yield with increased recovery of splits by about 3 to 5 per cent.

Power operated machines

Lentil falls in category of easy-to-mill type of pulses. Additional moisture is added after cleaning and grading process, followed by tempering and sun drying. To split the dehusked lentil grains , roller mills, under runner disk sheller, attrition mill (Chakki), elevator and hard surface and impact sheller are used.

A. *Roller machine:* It is similar to dehusking units, with emery roller made of abrasivecarborundum of diameter roller cylindrical mill of a 200-300 mm diameter to give 13.5 to 14 m/s speed at 850-900 rpm and dal (splits) recovery 74.75% with capacity 100 kg per hour can be operated by 1.5kw electric motor. It is suitable for small scale production (8 to 10q per day) of dal.

B. *Disk runner Machines:* It consists of two horizontal disks with 1.2cm emery coating. One of the disks is fixed and other disk rotates. This rubbing action causes splitting of the dehusked whole. The capacity of the machine depends on the speed and size of the disk.

Some of the efficient and economically viable mini dal mill developed by IARI New Delhi (Fig. 35.11), PDKV Akola, GBPUA&T Pantnagar, CFTRI Mysore and IIPR Kanpur.

Fig. 35.11 PUSA Mini Dal Mill

References

Anonymus. 2000. Saskatchewan Pulse Growers Pulse production manual second edition.

Bailey KL, Mortensen K and Lafond GP. 1992. Effects of tillage systems and crop rotations on root and foliar diseases of wheat, flax and peas in Saskatchewan. Canadian Journal of Plant Science72: 583-591.

BIS standards Agricultural produce milling Machinery - milling of pulses – flow Diagram

Bond D,Lawes D, Hawtin G, Saxena M and Stephenes J. 1985. Faba bean (Vicia faba L.).In: Summerfield R.J.and Roberts E.H.(eds.), Grain Legume crops. William Collins sons Co. Ltd. 8 Grafton street, London, WIX 3LA, UK, pp. 199-265.

Chakraverty A. 2000. Third Edition. Post Harvest Technology of Cereals, Pulses and Oilseeds.

El-Berry AM and Ahmed MH.1991. Comparison of two mechanical planting systems for wheat production under desert condition . Misr J. Agric. Eng. 7(1): 40-46.

El-Sahrigi AF and Abohbaga M M.1993. Effect of seed bed of some soil Fauna 5 th int. con. On mechanization and energy in Agriculture 1114 Oct 1993 Kusadas, Tuekive.

Erskine W, WIliams PC, Nakkoul H.1991. Splitting and dehulling lentil (lens-culinaris) -effects of seed size and different pretreatments. Journal of the Science of Food and Agriculture57: 77-84.

Gowda CLL, Srinivasan S, Gaur P M and Saxena K B.2013.Enhancing the productivity and production of pulses in India. In Climate Change and Sustainable Food Security 2013 (eds Shetty, P. K., Ayyappan, S. and Swaminathan, M. S.), National Institute of Advanced Studies, Bangalore and Indian Council of Agricultural Research, New Delhi, ISBN: 978-81-87663-76-8.

Kadambot S, Stephen L, Bob F and Chris V. 1997. Sowing depth for chickpea, faba bean,lentil and field pea, Farm note Agriculture Western Australia, South Perth and Merredin, and the Centre for Legumes in Mediterranean Agriculture No 45/97.

Kepnar R A, Bainer R and Barger EL. Principles of Farm Machinery.(3e) CBS Publishers and Distributors. Pp. 321.

Kumar S, Gopal R, Singh, Piggin C, Haddad A, Ahmed S and Kumar R.2011. No-till lentil: An option for profitable harvest in dry areas.Grain Legumes 57: 39-42.

Kumar P.1998. Food Demand and Supply Projections for India, Agricultural Economics Policy Paper 98–01, Indian Agricultural Research Institute, New Delhi.

Materne and Reddy AA.2007. Commercial cultivation and profitability In: The Lentil an Ancient Crop for Modern Times, Eds: SS Yadav, David McNeil and Philip.

Reddy AA.2004. Consumption pattern, trade and production potential of pulses.Economic and Political Weekly 39(44): 4854-4860.

Reddy AA.2009. Pulses production technology: Status and way forward.Economic and Political Weekly 44(52): 73-82.

Singh R D and Singh PM.2006. Performance of Zero-Till Drill for Lentil Cultivation at Farmer's Fields, Research Note. Journal of Agricultural Engineering 43(1): 70-73.

Tayade NH, Thakare SK and Nikhade JS.2010. Use of constant pressure valve for manually operated Knapsack Sprayer.International Journal of Agricultural Engineering 3(2): 328-330.

Wolf IA. 1983. Handbook of Processing and Utilization in Agriculture. (2 Vol. set). CRC Press, Oxford & IBH Publishing & Co. Pvt. Ltd., New Delhi.

http://www.indiaagronet.com/indiaagronet/post_harvest/ pulses. htm#Machinery for pulses processing

https://www.nabard.org/english/dal3.aspx

http://oar.icrisat.org/7101/1/Enhancing The Productivity-2013.pdf

http://pnwsteep.wsu.edu/tillagehandbook/chapter2/index.htm

Agritech.tnau.ac.in/postharvest/pht_pulses. html

http://www.pea-lentil.com/core/files/pealentil/uploads/files/Chapter4.pdf

Agricultural produce milling Machinery - milling of pulses – flow Diagram IS 13793 : 1993 September 1993.

Citation

Khura TK, Lande SD, Kushwaha HL, Sundaram PK, Parray RA and Chandra R. 2018. Improved Equipment and Technologies for lentil cultivation in India. In: Scientific Lentil Production. Singh AK, Bhakta N, Sangale UR, Manibhushan, Sundaram PK, Kumar S and Yasin JK. (eds.) Society for Upliftment of Rural Economy Varanasi, India. Pp. 579-597.

□□□

Appendices

Appendix 1: All India Area, Production & Yield of total Pulses

TOTAL PULSES

Year	A	P	Y
1950-51	19.09	8.41	441
1951-52	18.78	8.42	448
1952-53	19.84	9.19	463
1953-54	21.73	10.62	489
1954-55	21.91	10.95	500
1955-56	23.22	11.04	476
1956-57	23.32	11.55	495
1957-58	22.54	9.56	424
1958-59	24.31	13.15	541
1959-60	24.83	11.80	475
1960-61	23.56	12.70	539
1961-62	24.24	11.76	485
1962-63	24.27	11.53	475
1963-64	24.18	10.07	416
1964-65	23.88	12.42	520
1965-66	22.72	9.94	438
1966-67	22.12	8.35	377
1967-68	22.65	12.10	534
1968-69	21.26	10.42	490
1969-70	22.02	11.69	531
1970-71	22.24	11.82	524
1971-72	22.15	11.09	501
1972-73	20.92	9.91	474
1973-74	23.43	10.01	427
1974-75	22.03	10.02	455
1975-76	24.45	13.04	533
1976-77	22.98	11.36	494
1977-78	23.50	11.97	510

1978-79	23.66	12.18	515
1979-80	22.26	8.57	385
1980-81	22.46	10.63	473
1981-82	23.84	11.51	483
1982-83	22.83	11.86	519
1983-84	23.54	12.59	548
1984-85	22.74	11.96	526
1985-86	24.42	13.36	547
1986-87	23.16	11.71	506
1987-88	21.27	10.96	515
1988-89	23.15	13.85	598
1989-90	23.41	12.86	549
1990-91	24.66	14.26	578
1991-92	22.54	12.02	533
1992-93	22.36	12.82	573
1993-94	22.25	13.30	598
1994-95	23.03	14.04	610
1995-96	22.28	12.31	552
1996-97	22.45	14.24	635
1997-98	22.87	12.98	567
1998-99	23.50	14.91	634
1999-2K	21.12	13.42	635
2000-01	20.35	11.08	544
2001-02	22.01	13.37	607
2002-03	20.50	11.13	543
2003-04	23.46	14.91	635
2004-05	22.76	13.13	577
2005-06	22.39	13.38	597
2006-07	23.19	14.20	612
2007-08	23.63	14.76	625
2008-09	22.09	14.57	659
2009-10	23.28	14.66	630
2010-11	26.28	18.24	694
2011-12	24.78	17.21	694
2012-13	23.26	18.34	789
2013-14	25.21	19.78	785
2014-15		17.38*	

Source: http://www.iipr.res.in

UNIT

Area = Mha; Prod = Mt; Yield = kg/ha

Appendix 2: Import of Pulses by India

Year	Quantity (000′ t)	Value (Rs. In crore)	Unit Value (Rs. Per kg.)
1980-81	172.96	29.76	1.72
1981-82	128.07	44.34	3.46
1982-83	102.36	36.68	3.58
1983-84	227.90	82.85	3.64
1984-85	235.39	100.70	4.28
1985-86	431.44	189.06	4.38
1986-87	624.79	233.66	3.74
1987-88	612.40	272.02	4.44
1988-89	755.56	1190.01	–
1989-90	469.90	228.35	4.86
1990-91	1273.43	481.17	3.78
1991-92	312.61	254.77	8.15
1992-93	382.62	334.38	8.74
1993-94	627.96	566.85	9.03
1994-95	554.08	593.37	10.71
1995-96	485.65	685.55	14.12
1996-97	654.79	890.34	13.60
1997-98	1008.15	1194.67	11.85
1998-99	563.53	708.81	12.58
1999-2K	252.82	358.25	14.17
2000-01	350.57	500.06	14.26
2001-02	2232.29	3163.72	14.17
2002-03	1995.33	2741.05	13.74
2003-04	1723.33	2284.87	13.26
2004-05	1339.45	1777.58	13.27
2005-06	1695.65	2476.25	14.60
2006-07	2270.97	3891.91	17.14
2007-08	2835.05	5374.94	18.96
2008-09	2474.11	6216.40	24.66
2009-10	3509.58	9813.37	
2010-11	2698.66	7149.62	
2011-12	3364.80	8931.24	
2012-13	3839.30	12733.70	
2013-14*	3049.29	10550.97	

Source: http://www.iipr.res.in

Appendix 3: Export of Pulses from India

Year	Quantity (000′ t)	Value (Rs. In crore)	Unit Value (Rs. Per kg.)
1980-81	1.09	0.35	3.21
1981-82	0.95	0.34	3.58
1982-83	1.71	0.87	5.09
1983-84	6.37	0.66	1.04
1984-85	3.79	2.56	6.75
1985-86	0.57	0.46	8.07
1986-87	5.37	4.77	8.88
1987-88	9.18	6.41	6.98
1988-89	10.10	11.5	11.39
1989-90	13.14	15.98	12.16
1990-91	15.11	17.93	11.87
1991-92	25.78	39.04	15.14
1992-93	34.31	53.44	15.58
1993-94	43.60	73.59	16.88
1994-95	50.71	90.41	17.83
1995-96	61.36	131.81	21.48
1996-97	55.15	131.58	23.86
1997-98	170.76	360.80	21.13
1998-99	103.90	223.03	21.47
1999-2K	192.17	419.65	21.84
2000-01	244.26	537.58	22.01
2001-02	161.64	369.43	22.84
2002-03	148.08	345.02	23.21
2003-04	153.88	328.60	21.36
2004-05	277.18	602.57	21.94
2005-06	447.44	1115.27	24.92
2006-07	250.70	773.34	30.85
2007-08	164.20	526.41	32.06
2008-09	136.27	510.22	39.64
2009-10	99.92	407.35	40.77
2010-11	208.02	865.74	
2011-12	174.20	1065.84	
2012-13	201.70	1280.00	
2013-14*	343.49	1737.00	

Source: http://www.iipr.res.in

Appendix 4: Target and Achievement of Pulses in five Year Plan

During Xth Five Year Plan **(in million tonnes)**

Year	Target	Achievement
2002-03	16.00	11.13
2003-04	15.00	14.91
2004-05	15.30	13.13
2005-06	15.15	13.39
2006-07	15.15	14.20
Xth Plan	**76.60**	**66.76**

Source: http://www.iipr.res.in

During XIth Five Year Plan **(in million tonnes)**

Year	Target	Achievement
2002-03	15.50	14.76
2003-04	15.50	14.57
2004-05	16.50	14.66
2005-06	16.50	18.24
2006-07	17.50	17.21
XIthPlan	**81.00**	**79.44**

Source: http://www.iipr.res.in

During 2012-13 to 2014-15 **(in million tonnes)**

Year	Target	Achievement
2012-13	18.24	18.34
2013-14	19.00	19.27*
2014-15	19.50	

Source: http://www.iipr.res.in

Appendix 5: Availability and Demand of Pulses in India

(in million tonnes)

Year	Production	Imports	Total Availability	Demand for pulses **
2006-07	14.20	2.26	16.46	NA
2007-08	14.76	2.83	17.59	16.77
2008-09	14.57	2.48	17.05	17.51
2009-10	14.66	3.51	16.51	18.29
2010-11	18.24	2.69	20.93	19.08
2011-12	17.09	3.36	20.45	19.91
2012-13	18.34	3.84*	22.18	
2013-14	19.27			

Source: http://www.iipr.res.in

Appendix 6: Demand and Supply of Pulses in India

(in million tonnes)

Year	Demand	Supply
2007-08	16.77	14.76
2008-09	17.51	14.57
2009-10	18.29	14.66
2010-11	19.08	18.24
2011-12	19.91	17.09
2012-13		18.34
2013-14		

Source: http://www.iipr.res.in

Appendix 7: All Indian Area, Production & Yield of Lentil

LENTIL

Year	A	P	Y
1970-71	0.75	0.37	497
1971-72	0.63	0.30	481
1972-73	0.83	0.37	447
1973-74	0.93	0.41	440
1974-75	0.95	0.46	480
1975-76	0.94	0.46	492
1976-77	0.85	0.39	461
1977-78	0.88	0.38	436
1978-79	1.01	0.45	443
1979-80	0.85	0.32	377
1980-81	0.93	0.47	498
1981-82	0.95	0.50	525
1982-83	1.00	0.49	491
1983-84	0.94	0.53	566
1984-85	0.98	0.55	557
1985-86	1.09	0.66	607
1986-87	1.09	0.66	607
1987-88	1.04	0.66	635
1988-89	1.08	0.73	679
1989-90	1.11	0.71	635
1990-91	1.19	0.85	717
1991-92	1.19	0.80	674
1992-93	1.20	0.79	657
1993-94	1.18	0.75	632
1994-95	1.16	0.78	678
1995-96	1.26	0.71	569
1996-97	1.37	0.96	702
1997-98	1.29	0.80	632
1998-99	1.39	0.94	675
1999-00	1.46	1.08	738
2000-01	1.48	0.92	619
2001-02	1.47	0.97	664
2002-03	1.38	0.87	634
2003-04	1.39	1.04	743
2004-05	1.47	0.99	675
2005-06	1.51	0.95	628
2006-07	1.47	0.91	621
2007-08	1.31	0.81	622
2008-09	1.38	0.95	693
2009-10	1.48	1.03	697
2010-11	1.60	0.94	591
2011-12	1.56	1.06	679
2012-13	1.42	1.13	797

Source: http://www.iipr.res.in

Appendix 8: Area, Production & Yield of Lentil in Assam

ASSAM

Year	A	P	Y
1970-71	15.7	6.3	401
1971-72	6.7	2.9	433
1972-73	14.3	6.5	455
1973-74	8.6	3.8	442
1974-75	12.1	5.5	455
1975-76	10.1	4.1	406
1976-77	11.5	5.3	461
1977-78	11.8	5.3	449
1978-79	13.2	5.9	447
1979-80	13.6	6.1	449
1980-81	14.8	6.2	419
1981-82			
1982-83			
1983-84			
1984-85			
1985-86			
1986-87			
1987-88			
1988-89			
1989-90			
1990-91			
1991-92	17.3	8.1	468
1992-93	16.1	7.7	478
1993-94	15.0	7.4	493
1994-95	16.1	8.4	522
1995-96			
1996-97	19.0	10.10	528
1997-98	19.3	10.2	528
1998-99	25.3	13.3	526
1999-2K	21.4	11.1	519
2000-01	20.3	10.7	527
2001-02	21.3	11.2	526
2002-03	20.0	11.0	550
2003-04	22.0	12.0	545
2004-05	20.5	11.5	561
2005-06	18.6	9.9	532
2006-07	20.0	11.0	550
2007-08	22.0	12.0	545
2008-09	21.7	11.0	507
2009-10	21.4	10.8	505
2010-11	23.9	11.8	494
2011-12	23.0	11.5	500
2012-13	28.9	17.0	586

Source: http://www.iipr.res.in

Appendix 9: Area, Production & Yield of Lentil in Bihar

BIHAR

Year	A	P	Y
1970-71	140.7	90.5	643
1972-73	150.7	60.9	404
1973-74	141.6	65.7	464
1974-75	145.9	98.3	674
1975-76	152.1	88.6	582
1976-77	156.6	88.4	564
1977-78	162.9	82.1	504
1978-79	166.0	88.1	531
1979-80	154.6	82.3	532
1980-81	200.4	128.4	641
1981-82	152.9	99.8	653
1982-83	152.9	112.5	736
1983-84	165.3	114.8	695
1984-85	173.1	125.1	723
1985-86	166.5	128.6	772
1986-87	170.4	117.6	690
1987-88	168.1	131.2	780
1988-89	170.2	133.5	784
1989-90	181.7	137.2	755
1990-91	187.9	167.7	892
1991-92	182.0	150.7	828
1992-93	169.0	144.5	855
1993-94	180.5	128.8	714
1994-95	166.9	159.4	955
1995-96	167.2	97.1	581
1996-97	173.0	163.7	945
1997-98	173.9	128.5	739
1998-99	183.5	184.3	1004
1999-2K	175.4	173.8	991
2000-01	172.2	169.0	981
2001-02	172.6	137.8	798
2002-03	179.6	156.6	872
2003-04	171.0	159.8	935
2004-05	178.7	127.0	711
2005-06	162.5	114.5	705
2006-07	166.5	117.3	750
2007-08	162.6	129.0	793
2008-09	163.6	128.6	785
2009-10	171.1	150.5	880
2010-11	238.6	214.7	900
2011-12	168.5	171.6	1018
2012-13	159.7	183.2	1147

Source: http://www.iipr.res.in

Appendix 10: Area, Production & Yield of Lentil in Chhattisgarh

CHHATTISGARH

Year	A	P	Y
2000-01	12.6	3.4	270
2001-02	15.9	5.2	327
2002-03	15.0	4.6	307
2003-04	18.0	6.2	344
2004-05	17.2	4.4	256
2005-06	18.5	6.0	324
2006-07	16.5	5.4	327
2007-08	17.3	5.8	335
2008-09	15.6	5.1	327
2009-10	16.0	4.9	306
2010-11	13.9	4.2	302
2011-12	14.9	5.0	336
2012-13	12.9	4.2	326

Source: http://www.iipr.res.in

Appendix 11: Area, Production & Yield of Lentil in Haryana

HARYANA

Year	A	P	Y
1970-71	26.1	5.1	195
1971-72	29.2	11.8	404
1972-73	38.3	14.9	389
1973-74	32.7	13.9	425
1974-75	27.6	11.3	410
1975-76	25.6	11.9	465
1976-77	22.6	10.3	456
1977-78	27.9	13.0	466
1978-79	21.4	14.0	654
1979-80	21.4	14.0	654
1980-81	28.6	13.6	476
1981-82	24.3	11.5	473
1982-83	19.4	9.7	500
1983-84	19.1	9.4	492
1984-85	16.6	9.6	578
1985-86	26.3	14.2	540
1986-87	23.9	14.1	590
1987-88	14.7	7.0	476
1988-89	17.6	10.3	585
1989-90	16.1	9.6	596
1990-91	18.3	11.0	601
1991-92	14.4	8.9	618
1992-93	12.5	7.5	600
1993-94	11.3	7.6	632
1994-95	9.3	6.6	710
1995-96	11.1	7.7	694
1996-97	11.0	9.0	811
1997-98	13.8	11.2	862
1998-99	7.3	5.4	740
1999-2K	7.9	5.1	646
2000-01	7.4	4.9	662
2001-02	8.9	7.8	876
2002-03	5.4	5.1	944
2003-04	6.3	5.1	810
2004-05	8.6	8.2	953
2005-06	5.4	4.3	796
2006-07	4.2	4.2	1000
2007-08	4.1	3.2	780
2008-09	4.3	3.2	744
2009-10	5.0	4.0	800
2010-11	3.3	2.6	788
2011-12	3.3	2.6	788

Source: http://www.iipr.res.in

Appendix 12: Area, Production & Yield of Lentil in Himachal Pradesh

HIMACHAL PRADESH

Year	A	P	Y
2005-06	0.7	0.1	143
2006-07	0.7	0.4	571
2007-08	0.6	0.4	667
2008-09	0.7	0.2	286
2009-10	0.7	0.3	413
2010-11	0.6	0.4	563
2011-12	0.6	0.5	825

Source: http://www.iipr.res.in

Appendix 13: Area, Production & Yield of Lentil in Jammu & Kasmir

JAMMU & KASMIR

Year	A	P	Y
1976-77	1.2	0.7	583
1977-78	1.2	0.8	667
1978-79	1.2	0.6	500
1979-80	0.8	0.5	625
1980-81	1.0	0.5	500
1981-82	1.0	0.4	400
1982-83	0.7	0.3	429
1983-84	0.5	0.2	400
1984-85	0.4	0.1	250
1985-86	0.5	0.5	1000
1986-87	0.6	0.2	333
1987-88	0.7	0.3	429
1988-89	0.7	0.3	429
1989-90	0.5	0.2	400
1990-91	0.4	0.2	500
1991-92	0.3	0.1	333
1992-93	0.3	0.1	333
1993-94	0.3	0.1	333
1994-95	0.2	0.1	500
1995-96	0.2	0.1	500
1996-97	0.2		
1997-98	0.4		
1998-99	0.4		
1999-2K	0.8	0.1	
2000-01	0.1	0.1	1000
2001-02	0.1	0.0	0
2002-03	0.2	0.1	500
2003-04	0.2	0.1	500
2004-05	0.2	0.1	500
2005-06	0.2	0.1	500
2006-07	0.2	0.1	500
2007-08	0.1	0.0	0
2008-09	0.2	0.7	583
2009-10	0.9	0.6	617
2010-11	0.6	0.3	563
2011-12	0.5	0.3	613
2012-13			
2013-14			

Source: http://www.iipr.res.in

Appendix 14: Area, Production & Yield of Lentil in Jharkhand

JHARKHAND

Year	A	P	Y
2000-01	6.4	4.1	641
2001-02	6.3	3.8	603
2002-03	21.8	14.0	642
2003-04	22.2	16.3	734
2004-05	16.7	11.2	671
2005-06	16.7	11.2	671
2006-07	19.9	15.2	764
2007-08	22.7	16.9	744
2008-09	19.5	15.8	810
2009-10	22.9	12.8	559
2010-11	20.9	17.2	822
2011-12	48.7	40.9	840
2012-13	40.8	45.7	1121
2013-14			

Source: http://www.iipr.res.in

Appendix 15: Area, Production & Yield of Lentil in Madhya Pradesh

MADHYA PRADESH

Year	A	P	Y
1970-71	271.5	108.9	401
1971-72	265.1	103.2	389
1972-73	294.4	112.7	382
1973-74	345.3	121.2	351
1974-75	349.5	149.5	428
1975-76	325.4	143.5	441
1976-77	300.4	114.9	382
1977-78	310.2	100.4	324
1978-79	323.8	120.6	372
1979-80	282.5	72.4	256
1980-81	273.1	107.5	394
1981-82	301.7	112.5	373
1982-83	301.2	118.4	393
1983-84	288.2	126.3	438
1984-85	291.3	114.2	392
1985-86	309.3	121.4	392
1986-87	302.8	130.2	430
1987-88	308.0	149.0	484
1988-89	317.7	151.3	476
1989-90	315.1	145.3	461
1990-91	334.6	166.4	497
1991-92	347.5	160.5	462
1992-93	375.0	176.6	471
1993-94	398.3	190.8	479
1994-95	429.2	205.7	479
1995-96	499.7	229.0	458
1996-97	512.1	245.9	480
1997-98	479.4	206.1	430
1998-99	513.0	245.0	478
1999-2K	534.8	279.3	522
2000-01	489.2	206.5	422
2001-02	500.2	240.4	481
2002-03	466.9	181.0	388
2003-04	478.8	240.4	502
2004-05	497.7	247.7	498
2005-06	564.1	287.5	510
2006-07	522.1	259.3	497
2007-08	467.8	205.9	440
2008-09	531.4	268.4	505
2009-10	541.3	284.4	525
2010-11	590.5	177.9	301
2011-12	620.5	230.0	371
2012-13	570.5	333.4	584

Source: http://www.iipr.res.in

Appendix 16: Area, Production & Yield of Lentil in Maharashtra

MAHARASHTRA

Year	A	P	Y
1970-71	12.0	3.2	267
1971-72	12.8	3.2	250
1972-73	10.0	2.2	220
1973-74	12.1	4.0	331
1974-75	13.4	5.1	381
1975-76	15.4	5.8	377
1976-77	15.6	5.7	365
1977-78	16.1	5.6	348
1978-79	13.9	4.8	345
1979-80	9.1	3.9	429
1980-81	9.3	2.4	258
1981-82	15.7	4.7	299
1982-83	13.7	3.6	263
1983-84	21.0	4.7	224
1984-85	15.2	4.1	270
1985-86	16.3	4.6	282
1986-87	8.5	2.3	271
1987-88	12.4	3.7	298
1988-89	9.5	4.8	505
1989-90	9.1	4.0	440
1990-91	7.9	3.2	405
1991-92	3.9	1.4	359
1992-93	9.8	2.5	255
1993-94	12.0	5.1	425
1994-95	14.1	5.5	390
1995-96	10.4	4.9	471
1996-97	10.1	4.9	485
1997-98	7.0	3.0	429
1998-99	7.4	3.1	419
1999-2K	8.5	4.0	471
2000-01	6.0	2.0	333
2001-02	8.0	4.0	500
2002-03	6.0	2.0	333
2003-04	8.0	3.0	375
2004-05	5.5	1.1	200
2005-06	8.7	3.6	414
2006-07	9.0	3.7	411
2007-08	10.0	5.0	500
2008-09	8.0	3.0	375
2009-10	8.8	3.2	364
2010-11	4.0	2.0	500
2011-12	3.1	1.4	452
2012-13	3.5	1.4	400

Source: http://www.iipr.res.in

Appendix 17: Area, Production & Yield of Lentil in Punjab

PUNJAB

Year	A	P	Y
1970-71	13.3	5.3	398
1971-72	13.3	5.4	406
1972-73	17.4	6.6	379
1973-74	18.7	7.6	406
1974-75	20.0	9.0	450
1975-76	14.6	6.9	473
1976-77	13.1	6.5	496
1977-78	0.0	0.0	0.0
1978-79	20.3	10.0	493
1979-80	0.0	0.0	0.0
1980-81	20.1	8.9	443
1981-82	19.8	7.5	379
1982-83	14.4	6.5	451
1983-84	7.1	3.9	549
1984-85	6.8	3.9	574
1985-86	11.6	7.5	647
1986-87	13.1	9.6	733
1987-88	9.8	6.1	622
1988-89	11.0	8.0	727
1989-90	9.4	6.1	649
1990-91	9.3	6.4	688
1991-92	7.3	4.6	630
1992-93	6.0	3.8	633
1993-94	5.7	3.8	666
1994-95	5.4	3.9	722
1995-96	5.9	3.8	644
1996-97	5.6	3.3	579
1997-98	5.2	3.4	654
1998-99	4.3	2.8	651
1999-2K	3.9	2.6	667
2000-01	4.7	3.1	660
2001-02	4.3	3.0	698
2002-03	4.0	2.7	675
2003-04	3.6	2.3	639
2004-05	3.3	1.6	485
2005-06	2.0	1.0	500
2006-07	2.1	0.8	381
2007-08	1.2	0.6	500
2008-09	1.1	0.7	636
2009-10	1.1	0.7	636
2010-11	1.1	0.7	636
2011-12	1.0	1.0	1000
2012-13	0.7	0.5	714

Source: http://www.iipr.res.in

Appendix 18: Area, Production & Yield of Lentil in Rajasthan

RAJASTHAN

Year	A	P	Y
1970-71	9.1	7.3	802
1971-72	9.9	7.8	788
1972-73	10.2	8.7	853
1973-74	13.3	9.4	707
1974-75	12.7	9.1	717
1975-76	14.1	6.5	461
1976-77	12.5	6.8	544
1977-78	16.3	9.6	589
1978-79	22.8	11.8	518
1979-80	15.7	9.4	599
1980-81	15.6	9.9	635
1981-82	17.3	10.6	613
1982-83	18.3	12.6	689
1983-84	20.7	12.4	599
1984-85	19.5	17.4	892
1985-86	22.1	19.8	896
1986-87	16.6	17.1	1030
1987-88	10.4	7.7	740
1988-89	11.6	9.7	836
1989-90	10.0	6.5	650
1990-91	14.4	12.4	861
1991-92	12.0	9.4	783
1992-93	17.1	14.7	860
1993-94	13.3	10.7	805
1994-95	12.1	10.6	876
1995-96	17.8	14.8	831
1996-97	31.8	32.3	1016
1997-98	33.2	40.0	1205
1998-99	49.6	52.6	1060
1999-2K	41.1	50.6	1231
2000-01	21.7	25.8	1189
2001-02	17.0	15.4	906
2002-03	5.9	5.7	966
2003-04	26.1	28.8	1103
2004-05	28.8	28.6	993
2005-06	18.5	16.8	908
2006-07	14.0	12.9	921
2007-08	14.8	10.5	709
2008-09	20.6	17.9	869
2009-10	27.0	24.2	896
2010-11	44.1	38.4	872
2011-12	31.9	35.9	1125
2012-13	27.6	30.4	1100

Source: http://www.iipr.res.in

Appendix 19: Area, Production & Yield of Lentil in Tripura

TRIPURA

Year	A	P	Y
1976-77	0.5	0.2	400
1977-78	0.5	0.2	400
1978-79	0.6	0.2	333
1979-80	0.8	0.3	375
1980-81	1.0	0.3	300
1981-82	0.8	0.2	250
1982-83	0.9	0.3	333
1983-84	0.7	0.2	286
1984-85	0.6	0.2	333
1985-86	0.6	0.2	333
1986-87	0.6	0.2	333
1987-88	0.6	0.3	500
1988-89	0.7	0.3	429
1989-90	0.8	0.4	500
1990-91	0.8	0.4	500
1991-92	0.8	0.4	500
1992-93	0.9	0.5	556
1993-94	0.6	0.3	500
1994-95	0.5	0.2	400
1995-96	0.4	0.2	500
1996-97	0.5	0.2	400
1997-98	0.5	0.1	200
1998-99	0.3	0.2	667
1999-2K	0.3	0.2	667
2000-01	0.2	0.2	1000
2001-02	0.4	0.2	500
2002-03	0.4	0.2	500
2003-04	0.3	0.2	667
2004-05	0.4	0.2	500
2005-06	0.3	0.2	667
2006-07	0.4	.0.2	500
2007-08	0.	0.2	500
2008-09	40.3	0.2	667
2009-10	0.4	0.2	500
2010-11	0.4	0.2	500
2011-12	0.4	0.3	750
2012-13			
2013-14			

Source: http://www.iipr.res.in

Appendix 20: Area, Production & Yield of Lentil in Uttar Pradesh

UTTAR PRADESH

Year	A	P	Y
1970-71	166.4	100.5	604
1971-72	175.5	112.6	642
1972-73	186.6	117.5	623
1973-74	199.0	93.1	468
1974-75	210.8	93.8	445
1975-76	230.1	121.7	520
1976-77	221.5	190.0	492
1977-78	240.2	114.6	477
1978-79	300.4	144.8	482
1979-80	237.9	85.0	357
1980-81	276.4	157.2	569
1981-82	338.8	217.0	640
1982-83	382.6	203.1	531
1983-84	350.2	219.4	626
1984-85	384.0	231.9	604
1985-86	444.5	315.6	710
1986-87	466.0	323.4	694
1987-88	435.0	291.1	669
1988-89	461.9	352.4	764
1989-90	490.3	347.2	708
1990-91	540.4	430.5	796
1991-92	541.6	415.2	767
1992-93	543.2	394.6	726
1993-94	494.0	366.6	742
1994-95	460.0	358.0	778
1995-96	487.4	317.0	650
1996-97	554.0	449.2	811
1997-98	505.7	365.4	723
1998-99	545.5	395.4	725
1999-2K	585.5	484.6	828
2000-01	651.7	408.1	626
2001-02	625.0	500.0	800
2002-03	583.9	449.9	771
2003-04	556.7	504.8	907
2004-05	613.0	501.4	818
2005-06	607.9	434.9	715
2006-07	616.0	434.0	705
2007-08	507.0	368.0	726
2008-09	521.0	460.0	883
2009-10	592.4	475.8	803
2010-11	586.0	411.0	701
2011-12	573.0	505.0	881
2012-13	495.0	441.0	891

Source: http://www.iipr.res.in

Appendix 21: Area, Production & Yield of Lentil in Uttarakhand

UTTARAKHAND

Year	A	P	Y
2000-01	15.7	13.0	828
2001-02	14.9	7.8	523
2002-03	14.7	7.3	497
2003-04	14.2	7.9	556
2004-05	14.2	7.9	550
2005-06	20.5	9.2	449
2006-07	14.0	6.0	429
2007-08	16.3	9.3	571
2008-09	17.0	6.0	353
2009-10	19.0	12.0	632
2010-11	12.2	9.0	740
2011-12	12.0	10.0	883

Source: http://www.iipr.res.in

Appendix 22: Area, Production & Yield of Lentil in West Bengal

WEST BENGAL

Year	A	P	Y
1970-71	94.9	46.0	485
1971-72	108.2	52.2	483
1972-73	104.7	40.7	389
1973-74	149.3	85.6	573
1974-75	157.0	74.0	471
1975-76	145.4	69.3	477
1976-77	97.8	45.7	467
1977-78	93.6	52.4	560
1978-79	121.6	45.0	370
1979-80	112.4	46.4	413
1980-81	94.6	30.3	320
1981-82	72.9	32.3	443
1982-83	76.4	32.6	427
1983-84	70.3	42.4	603
1984-85	75.0	40.6	541
1985-86	93.0	50.1	539
1986-87	84.5	44.6	528
1987-88	80.6	63.9	793
1988-89	81.7	63.3	775
1989-90	78.8	49.7	631
1990-91	73.8	53.2	721
1991-92	62.0	42.4	684
1992-93	54.0	38.3	709
1993-94	52.4	26.7	510
1994-95	42.1	25.4	603
1995-96	44.1	30.4	689
1996-97	51.3	43.2	842
1997-98	51.2	35.0	684
1998-99	51.9	35.6	686
1999-2K	82.0	67.5	823
2000-01	76.0	68.5	901
2001-02	71.4	37.5	525
2002-03	68.7	43.2	629
2003-04	69.4	53.2	767
2004-05	62.7	37.9	604
2005-06	61.5	47.0	764
2006-07	64.2	42.7	665
2007-08	58.7	44.8	763
2008-09	49.7	32.4	652
2009-10	51.7	47.1	911
2010-11	57.4	53.4	929
2011-12	59.3	41.2	695
2012-13	64.0	61.5	960

Source: http://www.iipr.res.in

Appendix 23: Area, Production and Yield of Lentil in Africa

AFRICA

Year	A	P	Y
1971	180.08	121.47	675
1972	198.41	131.17	661
1973	201.17	135.84	675
1974	204.11	134.90	661
1975	205.18	129.13	629
1976	159.13	122.89	772
1977	133.85	81.65	610
1978	131.50	80.43	612
1979	119.10	72.62	610
1980	104.58	57.58	551
1981	107.19	73.40	685
1982	126.30	85.75	679
1983	151.41	83.97	555
1984	144.93	66.85	461
1985	148.81	81.73	549
1986	146.56	111.78	763
1987	156.64	84.30	538
1988	150.30	100.06	666
1989	111.47	68.15	611
1990	111.79	80.65	721
1991	108.22	132.12	1221
1992	108.74	65.92	606
1993	115.25	65.62	569
1994	114.14	83.78	734
1995	127.25	63.86	502
1996	151.28	91.52	605
1997	117.17	73.08	624
1998	137.80	74.92	544
1999	121.59	55.91	460
2000	126.40	68.77	544
2001	147.28	85.36	580
2002	130.58	87.39	669
2003	115.61	76.15	659
2004	116.18	76.16	656
2005	135.61	76.89	567
2006	146.27	99.59	681
2007	141.01	97.76	693
2008	154.61	111.83	723
2009	154.49	156.65	1014
2010	133.02	121.92	917
2011	190.22	191.72	1008
2012	191.99	187.71	979
2013	180.43	188.08	1042

Source: http://www.iipr.res.in

Appendix 24: Area, Production and Yield of Lentil in Americas

AMERICAS

Year	A	P	Y
1971	105.48	79.38	753
1972	96.64	72.55	751
1973	105.60	70.18	665
1974	100.62	82.08	816
1975	105.45	95.19	903
1976	101.35	89.98	888
1977	139.86	98.39	703
1978	185.48	146.15	788
1979	195.75	145.83	745
1980	245.76	185.78	756
1981	235.72	196.67	834
1982	242.17	219.82	908
1983	154.75	138.72	896
1984	171.50	116.30	678
1985	198.58	150.01	755
1986	286.32	320.45	1119
1987	391.24	461.91	1181
1988	243.25	158.74	653
1989	209.18	201.36	963
1990	256.33	316.60	1235
1991	368.55	477.69	1296
1992	396.23	473.82	1196
1993	459.04	494.12	1076
1994	518.11	604.45	1167
1995	439.93	571.05	1298
1996	389.99	489.18	1254
1997	426.57	512.78	1202
1998	461.94	589.68	1277
1999	598.45	866.20	1447
2000	806.67	1080.99	1340
2001	766.27	712.12	929
2002	468.63	465.24	993
2003	618.77	610.69	987
2004	871.33	1122.75	1289
2005	986.60	1420.56	1440
2006	741.32	856.95	1156
2007	719.52	919.12	1277
2008	829.89	1170.54	1410
2009	1150.19	1792.86	1559
2010	1612.17	2354.39	1460
2011	1186.54	1766.36	1489
2012	1207.66	1795.16	1486
2013	1116.77	1116.77	1900

Source: http://www.iipr.res.in

Appendix 25: Area, Production and Yield of Lentil in Asia

ASIA

Year	A	P	Y
1971	1289.48	728.07	565
1972	1384.90	831.39	600
1973	1410.60	681.72	483
1974	1573.59	886.50	563
1975	1619.84	921.82	569
1976	1769.99	1111.70	628
1977	1813.74	1069.83	590
1978	1753.53	968.92	553
1979	1844.37	989.08	536
1980	1673.44	909.26	543
1981	1815.79	1124.11	619
1982	2136.93	1416.82	662
1983	2250.98	1502.71	668
1984	2123.32	1465.35	690
1985	2142.50	1567.38	732
1986	2427.73	1941.28	800
1987	2597.67	2000.86	770
1988	2754.60	2280.16	828
1989	2749.74	1709.39	622
1990	2706.93	2076.81	767
1991	2685.63	1972.61	735
1992	2733.70	1989.38	728
1993	2780.24	2168.36	780
1994	2744.76	2068.95	753
1995	2701.55	2173.32	645
1996	2835.13	2082.24	669
1997	2786.19	2093.24	457
1998	2734.68	2048.61	563
1999	2767.24	1834.15	775
2000	2783.49	2013.43	771
2001	2872.18	2153.06	643
2002	2833.10	2220.14	815
2003	2674.12	2137.86	678
2004	2673.76	2284.75	700
2005	2765.27	2290.30	828
2006	2741.34	2299.79	839
2007	2674.37	2115.72	791
2008	2174.57	1443.99	664
2009	2240.48	1798.99	803
2010	2376.52	2031.78	855
2011	2462.76	2000.48	812
2012	2672.76	2173.78	813
2013	2807.73	2250.34	801

Source: http://www.iipr.res.in

Appendix 26: Area, Production and Yield of Lentil in Europe

EUROPE

Year	A	P	Y
1971	142.80	120.51	844
1972	147.82	104.76	709
1973	158.56	156.82	989
1974	146.71	125.93	858
1975	145.06	96.41	665
1976	129.95	57.81	445
1977	108.01	91.91	851
1978	110.87	89.31	806
1979	109.96	76.81	699
1980	96.63	88.66	918
1981	102.61	53.00	517
1982	105.67	65.89	624
1983	114.52	77.63	678
1984	125.00	96.08	769
1985	129.73	97.45	751
1986	147.87	90.07	609
1987	181.16	136.95	756
1988	159.00	113.88	716
1989	124.67	92.36	741
1990	136.77	82.01	600
1991	98.84	68.51	693
1992	54.87	30.26	552
1993	46.21	33.14	717
1994	44.92	31.18	694
1995	51.05	22.63	443
1996	58.76	45.23	770
1997	49.33	31.44	637
1998	42.66	30.00	703
1999	37.74	25.92	687
2000	42.85	42.26	986
2001	42.48	34.56	814
2002	50.02	43.51	870
2003	58.21	48.38	831
2004	58.38	49.54	849
2005	59.60	35.17	590
2006	45.60	44.66	982
2007	37.32	33.36	894
2008	36.49	32.22	883
2009	49.04	38.89	793
2010	69.88	70.89	1014
2011	89.85	85.54	952
2012	101.97	76.68	752
2013	75.75	87.80	1159

Source: http://www.iipr.res.in

Appendix 27: Area, Production and Yield of Lentil in Oceania

OCEANIA

Year	A	P	Y
1985	0.50	1.00	2000
1986	1.00	2.00	2000
1987	1.50	3.00	2000
1988	4.00	5.00	1250
1989	5.50	9.00	1636
1990	6.40	9.07	1418
1991	5.57	7.88	1415
1992	7.06	13.00	1843
1993	7.85	8.02	1022
1994	7.80	6.50	833
1995	9.40	20.00	2128
1996	18.62	39.60	2127
1997	57.65	37.70	654
1998	66.50	40.00	602
1999	75.50	104.00	1377
2000	117.50	164.00	1396
2001	158.50	267.00	11685
2002	165.50	68.00	411
2003	84.87	107.11	1262
2004	129.42	55.60	430
2005	128.76	212.80	1653
2006	153.80	37.50	244
2007	130.42	131.85	1011
2008	117.79	66.10	561
2009	104.63	144.65	1381
2010	143.40	143.80	1003
2011	220.52	384.23	1742
2012	208.17	465.87	2238
2013	146.10	327.30	2240
2014			

Source: http://www.iipr.res.in

Appendix 28: Area, Production & Yield of Total Pulses in Andhra Pradesh

ANDHRA PRADESH

Year	A	P	Y
1970-71	1450.9	449.5	310
1971-72	1342.8	379.6	283
1972-73	1344.3	299.4	223
1973-74	1405.6	395.3	281
1974-75	1434.5	417.7	291
1975-76	1447.8	421.1	291
1976-77	1300.1	352.1	271
1977-78	1302.3	332.5	255
1978-79	1421.3	405.6	285
1979-80	1366.7	328.0	240
1980-81	1445.7	414.5	287
1981-82	1429.5	466.6	326
1982-83	1458.0	585.0	401
1983-84	1498.3	566.1	378
1984-85	1345.6	501.8	373
1985-86	1350.2	629.0	466
1986-87	1422.7	616.8	434
1987-88	1481.0	644.0	435
1988-89	1488.6	692.4	465
1989-90	1571.7	682.0	434
1990-91	1631.9	695.5	426
1991-92	1645.2	791.8	481
1992-93	1588.4	739.0	465
1993-94	1559.3	677.0	434
1994-95	1600.1	672.5	420
1995-96	1610.6	771.2	479
1996-97	1615.5	838.0	519
1997-98	1565.0	515.7	330
1998-99	1587.0	827.0	521
1999-2K	1644.9	800.2	486
2000-01	1902.2	1054.2	554
2001-02	1920.0	1137.7	593
2002-03	2100.1	1061.6	506
2003-04	2185.0	1239.0	567
2004-05	1803.8	1019.0	565
2005-06	1781.7	1376.0	772
2006-07	1984.0	1347.0	679
2007-08	2113.0	1697.0	803
2008-09	1771.0	1448.0	818
2009-10	1932.0	1429.0	740
2010-11	2130.0	1439.0	676
2011-12	1931.0	1347.0	646
2012-13	1949.0	1623.0	833
2013-14	1672.0	1551.0	928
2014-15*	1367.0	1136.0	831

Source: http://www.iipr.res.in

Appendix 29: Area, Production & Yield of Total Pulses in Andman & Nicobar Islands

ANDMAN & NICOBAR ISLANDS

Year	A	P	Y
1976-77	1.6	0.6	375
1977-78	1.6	0.6	375
1978-79	0.8	0.3	375
1979-80	0.6	0.3	500
1980-81	0.8	0.4	500
1981-82	0.7	0.3	429
1982-83	0.5	0.3	600
1983-84	0.5	0.3	600
1984-85	0.6	0.3	500
1985-86	0.7	0.4	571
1986-87	0.6	0.3	500
1987-88	0.6	0.3	500
1988-89	0.3	1.8	600
1989-90	2.5	0.9	360
1990-91	1.7	0.8	471
1991-92	1.8	0.9	500
1992-93	1.6	0.6	375
1993-94	1.8	0.8	444
1994-95	2.1	1.0	476
1995-96	2.0	1.0	500
1996-97	2.0	1.0	500
1997-98	2.4	1.7	708
1998-99	0.7	0.5	714
1999-2K	0.9	0.3	333
2000-01	0.7	0.4	571
2001-02	0.4	0.2	500
2002-03	1.3	0.6	462
2003-04	0.7	0.4	571
2004-05	0.7	0.4	571
2005-06	0.8	0.4	500
2006-07	0.7	0.4	571
2007-08	2.2	1.3	609
2008-09	2.1	1.2	571
2009-10	2.9	1.8	600
2010-11	2.6	1.2	442
2011-12	1.8	1.0	541
2012-13	1.3	0.7	538

Source: http://www.iipr.res.in

Appendix 30: Area, Production & Yield of Total Pulses in Arunachal Pradesh

ARUNACHAL PRADESH

Year	A	P	Y
1990-91	5.5	4.2	764
1991-92	5.5	4.5	718
1992-93	5.8	5.3	914
1993-94	5.8	5.3	914
1994-95	6.8	5.6	824
1995-96	6.1	5.7	934
1996-97	6.7	6.7	1000
1997-98	6.5	6.6	1046
1998-99	7.2	6.8	944
1999-2K	4.9	4.9	1000
2000-01	6.1	6.8	1115
2001-02	6.8	7.1	1044
2002-03	7.3	7.7	1055
2003-04	6.8	7.4	1088
2004-05	6.5	6.8	1046
2005-06	7.7	8.3	1078
2006-07	7.7	8.3	1078
2007-08	7.7	8.3	1078
2008-09	8.5	9.0	1059
2009-10	8.9	9.7	1090
2010-11	8.6	9.1	1055
2011-12	9.5	10.5	1105
2012-13	9.5	10.6	1116

Source: http://www.iipr.res.in

Appendix 31: Area, Production & Yield of Total Pulses in Assam

ASSAM

Year	A	P	Y
1970-71	84.8	32.2	380
1971-72	86.3	30.9	358
1972-73	98.1	48.0	489
1973-74	95.0	44.7	471
1974-75	97.0	39.6	408
1975-76	90.6	35.6	393
1976-77	93.6	37.2	398
1977-78	100.2	38.1	380
1978-79	103.5	42.4	410
1979-80	102.6	41.0	400
1980-81	113.2	47.1	416
1981-82	117.2	52.2	445
1982-83	126.4	52.8	418
1983-84	130.6	50.6	387
1984-85	138.1	59.5	431
1985-86	140.5	66.2	471
1986-87	144.4	60.5	419
1987-88	132.0	60.9	461
1988-89	121.8	51.1	420
1989-90	113.1	51.0	451
1990-91	113.4	48.5	428
1991-92	116.8	53.3	456
1992-93	109.2	51.1	468
1993-94	109.2	57.0	522
1994-95	108.7	59.4	546
1995-96	106.9	57.1	534
1996-97	119.6	68.4	572
1997-98	117.9	64.5	547
1998-99	126.9	69.2	545
1999-2K	116.0	64.0	552
2000-01	111.5	62.3	559
2001-02	117.5	65.7	558
2002-03	111.0	60.0	541
2003-04	115.0	64.0	557
2004-05	107.6	61.4	571
2005-06	100.5	54.0	537
2006-07	106.0	59.0	557
2007-08	113.0	63.0	558
2008-09	113.7	64.5	567
2009-10	115.3	64.6	560
2010-11	126.4	70.1	555
2011-12	119.7	68.6	574
2012-13	141.2	84.4	598
2013-14	150.1	104.3	695
2014-15*	98.0	59.0	602

Source: http://www.iipr.res.in

Appendix 32: Area, Production & Yield of Total Pulses in Bihar

BIHAR

Year	A	P	Y
1970-71	1644.8	987.4	600
1971-72	1544.3	889.3	576
1972-73	1474.6	656.0	445
1973-74	1556.4	684.8	440
1974-75	1554.0	866.9	558
1975-76	1531.3	822.4	537
1976-77	1423.8	702.3	493
1977-78	1471.3	672.5	457
1978-79	1423.1	703.0	494
1979-80	1244.1	634.6	510
1980-81	1367.8	833.0	609
1981-82	1261.8	770.7	611
1982-83	1189.0	702.0	591
1983-84	1217.8	834.0	685
1984-85	1196.7	798.3	667
1985-86	1233.1	887.4	720
1986-87	1256.3	875.8	697
1987-88	1224.0	821.0	671
1988-89	1136.8	780.4	686
1989-90	1206.4	835.9	693
1990-91	1175.7	915.8	779
1991-92	1083.0	850.4	785
1992-93	979.6	693.3	708
1993-94	995.3	735.5	739
1994-95	959.6	789.6	823
1995-96	921.8	560.7	608
1996-97	931.1	745.4	8001
1997-98	910.7	665.7	731
1998-99	893.4	792.3	887
1999-2K	865.3	746.9	863
2000-01	717.2	620.7	865
2001-02	694.2	547.0	788
2002-03	697.8	560.9	804
2003-04	691.3	562.6	814
2004-05	657.9	466.9	710
2005-06	596.9	446.8	749
2006-07	607.0	438.2	722
2007-08	607.6	497.1	818
2008-09	585.7	469.1	801
2009-10	564.9	472.4	836
2010-11	605.0	556.0	918
2011-12	538.3	509.3	946
2012-13	515.8	542.6	1052
2013-14	500.0	522.0	1044
2014-15*	521.9	538.0	1031

Source: http://www.iipr.res.in

Appendix 33: Area, Production & Yield of Total Pulses in Chhattisgarh

CHHATTISGARH

Year	A	P	Y
2000-01	668.5	267.5	400
2001-02	866.7	448.4	517
2002-03	781.8	365.7	468
2003-04	955.1	580.7	608
2004-05	932.2	367.8	395
2005-06	949.7	453.2	479
2006-07	908.7	493.6	546
2007-08	915.6	536.8	586
2008-09	859.8	498.6	580
2009-10	808.9	488.7	604
2010-11	856.0	536.0	626
2011-12	804.8	496.5	617
2012-13	926.5	648.7	700
2013-14	839.3	482.1	574
2014-15*	823.8	531.3	645

Source: http://www.iipr.res.in

Appendix 34: Area, Production & Yield of Total Pulses in Delhi

DELHI

Year	A	P	Y
1976-77	6.8	4.8	706
1977-78	5.3	3.6	679
1978-79	6.7	4.0	597
1979-80	2.0	0.8	400
1980-81	3.3	1.9	576
1981-82	2.2	1.4	636
1982-83	1.5	1.0	667
1983-84	3.8	1.8	474
1984-85	2.9	1.4	483
1985-86	4.3	3.0	698
1986-87	2.8	1.6	571
1987-88	2.0	0.9	450
1988-89	2.5	0.9	360
1989-90	2.1	1.0	476
1990-91	2.4	1.1	458
1991-92	1.8	0.9	500
1992-93	1.7	0.9	529
1993-94	1.5	1.0	667
1994-95	1.5	1.0	667
1995-96	0.9	0.3	333
1996-97	0.4		
1997-98	0.8	0.2	250
1998-99	0.4	0.1	250
1999-2K	0.4	0.1	250
2000-01	0.7	0.4	571
2001-02	0.6	0.4	667
2002-03	0.4	0.3	750
2003-04	0.1	0.8	800
2004-05	0.3	0.3	1000
2005-06	0.5	0.5	1000
2006-07	0.4	0.7	1750
2007-08	0.4	0.7	1750
2008-09	0.3	0.7	2333
2009-10	0.4	0.8	2008
2010-11	0.5	0.8	1611
2011-12	0.4	0.7	1939
2012-13	0.3	0.7	2333
2013-14			

Source: http://www.iipr.res.in

Appendix 35: Area, Production & Yield of Total pulses in Goa

GOA

Year	A	P	Y
1991-92	6.0	5.1	850
1992-93	6.4	4.6	719
1993-94	7.3	5.1	699
1994-95	8.5	4.8	565
1995-96	10.1	5.4	535
1996-97	9.9		
1997-98	10.3	7.8	757
1998-99	9.9	9.5	960
1999-2K	10.2	9.0	882
2000-01	10.8	9.6	879
2001-02	9.4	8.4	894
2002-03	9.4	7.5	798
2003-04	9.8	9.2	939
2004-05	10.4	9.4	904
2005-06	11.0	11.5	1045
2006-07	12.0	16.3	1358
2007-08	11.4	11.3	991
2008-09	9.9	10.2	1030
2009-10	7.9	8.5	1082
2010-11	7.6	8.0	1057
2011-12	9.9	8.3	836
2012-13	10.0	9.0	900

Source: http://www.iipr.res.in

Appendix 36: Area, Production & Yield of Total pulses in Dadar & Nagar Haveli

DADAR & NAGAR HAVELI

Year	A	P	Y
1976-77	7.3	2.6	356
1977-78	7.7	2.8	364
1978-79	7.3	2.7	384
1979-80	6.8	2.7	397
1980-81	6.4	3.0	422
1981-82	6.5	2.6	462
1982-83	6.1	2.6	426
1983-84	5.7	2.6	456
1984-85	6.0	2.7	450
1985-86	7.1	3.3	465
1986-87	6.6	4.3	652
1987-88	6.2	4.3	694
1988-89	7.9	5.8	734
1989-90	6.1	4.7	770
1990-91	6.8	5.4	794
1991-92	7.0	5.8	829
1992-93	6.2	4.9	790
1993-94	4.7	3.6	829
1994-95	8.0	6.8	790
1995-96	5.6	4.1	766
1996-97	3.5	2.7	857
1997-98	5.6	5.6	732
1998-99	3.7	3.2	771
1999-2K	5.9	4.5	1000
2000-01	5.7	5.0	865
2001-02	8.0	6.6	763
2002-03	6.5	6.0	877
2003-04	6.0	5.1	825
2004-05	6.5	5.4	923
2005-06	6.5	5.6	850
2006-07	6.5	5.6	831
2007-08	6.5	5.6	862
2008-09	6.4	5.5	859
2009-10	5.9	4.9	829
2010-11	6.8	6.1	898
2011-12	4.3	4.0	930
2012-13	5.5	5.0	930

Source: http://www.iipr.res.in

Appendix 37: Area, Production & Yield of Total Pulses in Daman & Diu

DAMAN & DIU

Year	A	P	Y
1990-91	1.5	0.6	400
1991-92	1.4	0.7	500
1992-93	1.5	0.7	467
1993-94	1.3	1.0	769
1994-95	1.3	1.0	769
1995-96	1.3	1.0	769
1996-97	1.3	1.0	769
1997-98	1.3	1.0	769
1998-99	1.3	1.0	769
1999-2K	1.3	1.3	1000
2000-01	1.3	1.3	1000
2001-02	1.3	1.1	846
2002-03	1.3	1.0	769
2003-04	1.3	1.1	846
2004-05	1.3	1.2	923
2005-06	1.3	1.1	846
2006-07	1.3	1.1	846
2007-08	1.3	1.1	846
2008-09	1.3	1.1	846
2009-10	1.3	1.1	846
2010-11	1.3	1.1	846

Source: http://www.iipr.res.in

Appendix 38: Area, Production & Yield of Total Pulses in Gujarat

GUJARAT

Year	A	P	Y
1970-71	423.2	165.2	390
1971-72	429.4	161.3	376
1972-73	379.2	112.0	295
1973-74	420.4	165.2	393
1974-75	279.7	91.5	327
1975-76	447.7	179.6	401
1976-77	448.1	186.8	417
1977-78	451.5	149.2	330
1978-79	545.7	230.7	423
1979-80	541.5	193.6	358
1980-81	553.6	266.5	481
1981-82	662.0	349.5	528
1982-83	726.0	423.0	652
1983-84	802.1	558.7	697
1984-85	828.5	555.9	671
1985-86	756.2	338.3	447
1986-87	707.9	235.3	332
1987-88	482.7	142.7	296
1988-89	771.8	489.6	634
1989-90	890.5	567.4	637
1990-91	931.6	626.6	673
1991-92	838.8	393.7	469
1992-93	974.0	648.1	665
1993-94	888.7	538.1	605
1994-95	901.4	518.6	575
1995-96	841.0	456.8	543
1996-97	920.3	663.8	721
1997-98	895.1	613.3	685
1998-99	862.0	633.5	735
1999-2K	700.4	405.6	579
2000-01	635.1	190.7	300
2001-02	730.4	379.8	520
2002-03	697.7	327.2	469
2003-04	832.6	622.4	748
2004-05	710.2	479.3	675
2005-06	777.0	547.0	704
2006-07	1000.0	593.0	593
2007-08	881.0	743.0	843
2008-09	784.0	609.0	777
2009-10	733.0	517.0	701
2010-11	852.0	720.0	845
2011-12	951.0	783.0	823
2012-13	660.0	572.2	867
2013-14	813.0	729.0	897
2014-15*	654.0	599.0	916

Source: http://www.iipr.res.in

Appendix 39: Area, Production & Yield of Total pulses in Haryana

HARYANA

Year	A	P	Y
1970-71	1140.7	813.2	713
1971-72	1208.4	684.3	566
1972-73	1065.9	589.8	553
1973-74	1080.4	482.7	447
1974-75	780.9	374.3	479
1975-76	1193.8	952.0	797
1976-77	1102.8	866.9	786
1977-78	1217.0	1005.4	826
1978-79	1133.4	1084.3	957
1979-80	600.5	345.8	756
1980-81	803.8	504.7	628
1981-82	557.0	344.2	311
1982-83	558.1	314.7	564
1983-84	712.4	359.5	505
1984-85	678.4	356.7	526
1985-86	832.7	677.9	814
1986-87	687.2	479.4	698
1987-88	264.8	111.9	423
1988-89	725.6	672.2	926
1989-90	600.3	428.6	714
1990-91	736.5	540.2	733
1991-92	384.0	270.2	704
1992-93	468.1	330.2	705
1993-94	476.9	469.2	984
1994-95	463.8	493.6	1064
1995-96	432.5	412.4	974
1996-97	417.4	345.0	827
1997-98	422.0	374.7	888
1998-99	427.1	353.2	827
1999-2K	133.4	76.2	571
2000-01	159.9	99.4	622
2001-02	186.5	148.1	794
2002-03	126.6	82.9	655
2003-04	199.0	143.1	719
2004-05	184.2	146.0	793
2005-06	189.8	118.1	621
2006-07	170.0	140.0	824
2007-08	169.0	101.8	602
2008-09	182.9	177.8	972
2009-10	132.0	100.0	758
2010-11	199.0	184.0	925
2011-12	180.0	127.0	706
2012-13	162.9	1130.4	800
2013-14	152.6	125.1	819
2014-15*	179.5	144.0	802

Source: http://www.iipr.res.in

Appendix 40: Area, Production & Yield of Total Pulses in Himachal Pradesh

HIMACHAL PRADESH

Year	A	P	Y
1970-71	70.8	31.6	446
1971-72	68.5	29.1	425
1972-73	68.9	28.2	409
1973-74	76.1	40.9	537
1974-75	68.2	33.1	485
1975-76	71.3	31.3	442
1976-77	65.7	35.8	545
1977-78	76.3	39.4	516
1978-79	86.5	44.9	519
1979-80	63.2	24.4	386
1980-81	58.6	20.0	341
1981-82	54.9	12.0	219
1982-83	40.0	9.0	235
1983-84	43.6	11.4	262
1984-85	41.6	7.4	178
1985-86	43.8	12.8	294
1986-87	46.1	9.2	200
1987-88	44.9	4.6	102
1988-89	41.4	7.3	176
1989-90	40.8	10.7	262
1990-91	43.3	13.5	312
1991-92	39.3	13.1	333
1992-93	37.2	12.2	328
1993-94	37.4	8.6	230
1994-95	38.2	12.9	338
1995-96	37.2	10.7	287
1996-97	35.1	11.4	325
1997-98	34.9	12.6	361
1998-99	35.2	12.9	366
1999-2K	31.5	14.8	470
2000-01	31.3	19.7	629
2001-02	30.8	11.1	360
2002-03	33.6	16.7	497
2003-04	29.3	9.0	307
2004-05	32.3	18.1	560
2005-06	27.5	19.6	713
2006-07	31.0	28.9	932
2007-08	33.9	36.0	1062
2008-09	31.0	23.5	758
2009-10	30.4	20.7	681
2010-11	30.0	14.6	487
2011-12	32.3	35.6	1102
2012-13	32.6	46.1	1413
2013-14	28.9	51.0	1764
2014-15*	32.0	40.1	1256

Source: http://www.iipr.res.in

Appendix 41: Area, Production & Yield of Total pulses in Jammu & Kashmir

JAMMU & KASHMIR

Year	A	P	Y
1970-71	46.7	30.4	651
1971-72	49.9	29.1	583
1972-73	46.8	28.8	615
1973-74	52.0	31.6	608
1974-75	50.5	31.8	630
1975-76	50.1	23.8	475
1976-77	46.4	21.6	466
1977-78	53.8	26.5	493
1978-79	54.5	36.6	672
1979-80	47.8	29.5	617
1980-81	53.6	35.0	653
1981-82	48.8	31.7	650
1982-83	49.0	30.0	623
1983-84	47.5	31.2	657
1984-85	47.5	30.8	648
1985-86	44.8	28.5	636
1986-87	44.9	30.3	675
1987-88	44.2	18.0	407
1988-89	39.1	24.8	634
1989-90	38.8	23.9	616
1990-91	40.5	24.5	605
1991-92	38.2	20.6	539
1992-93	34.2	18.1	529
1993-94	34.2	19.2	561
1994-95	44.0	22.9	520
1995-96	32.9	15.9	483
1996-97	31.3	17.0	543
1997-98	32.6	18.4	564
1998-99	32.0	18.2	569
1999-2K	28.8	14.1	490
2000-01	27.4	12.9	471
2001-02	26.9	12.7	472
2002-03	28.8	14.0	486
2003-04	27.6	13.3	482
2004-05	30.7	15.1	492
2005-06	26.8	13.5	504
2006-07	28.1	14.2	505
2007-08	30.3	15.4	508
2008-09	30.6	14.2	464
2009-10	29.7	13.6	456
2010-11	46.0	21.8	474
2011-12	27.6	14.0	507
2012-13	26.7	14.2	530
2013-14	25.9	13.1	508
2014-15*	26.6	14.3	538

Source: http://www.iipr.res.in

Appendix 42: Area, Production & Yield of Total Pulses in Jharkhand

JHARKHAND

Year	A	P	Y
2000-01	166.3	96.0	825
2001-02	59.1	43.3	733
2002-03	165.7	133.5	806
2003-04	266.0	135.0	508
2004-05	273.0	160.0	580
2005-06	291.1	172.3	567
2006-07	376.4	258.1	686
2007-08	410.0	301.8	736
2008-09	387.6	280.7	724
2009-10	315.7	223.7	709
2010-11	407.0	267.0	656
2011-12	538.8	492.3	914
2012-13	587.0	609.3	1038
2013-14	566.8	578.6	1021
2014-15*	574.8	557.6	970

Source: http://www.iipr.res.in

Appendix 43: Area, Production & Yield of Total Pulses in Karnataka

KARNATAKA

Year	A	P	Y
1970-71	1131.1	403.6	357
1971-72	1259.2	466.1	343
1972-73	1056.3	238.9	226
1973-74	1333.8	570.9	428
1974-75	1347.5	547.5	406
1975-76	1438.5	625.7	435
1976-77	1058.5	335.8	317
1977-78	1322.7	605.5	458
1978-79	1512.4	626.6	414
1979-80	1241.7	637.8	414
1980-81	1485.7	462.5	311
1981-82	1567.5	622.9	397
1982-83	1585.0	519.0	348
1983-84	1573.5	619.0	393
1984-85	1645.5	580.4	353
1985-86	1583.6	488.9	309
1986-87	1657.2	589.5	356
1987-88	1799.0	694.0	386
1988-89	1625.8	462.6	285
1989-90	1646.3	545.0	331
1990-91	1621.5	551.7	340
1991-92	1753.9	665.9	380
1992-93	1659.4	562.6	339
1993-94	1523.0	630.3	414
1994-95	1648.3	620.7	377
1995-96	1519.4	688.1	453
1996-97	1776.4	722.2	407
1997-98	1681.7	496.5	295
1998-99	1820.0	746.9	410
1999-2K	1920.5	848.7	442
2000-01	2046.7	956.2	467
2001-02	1893.1	751.5	397
2002-03	2060.6	693.6	337
2003-04	1874.3	569.2	304
2004-05	2106.0	792.0	376
2005-06	1981.0	964.0	487
2006-07	2369.0	893.0	377
2007-08	2383.0	1265.0	531
2008-09	2087.0	972.0	466
2009-10	2479.0	1118.0	452
2010-11	2699.0	1497.0	555
2011-12	2332.0	1061.0	455
2012-13	2269.0	1259.3	555
2013-14	2498.0	1600.5	641
2014-15*	2286.0	1414.0	619

Source: http://www.iipr.res.in

Appendix 44: Area, Production & Yield of Total Pulses in Kerala

KERALA

Year	A	P	Y
1970-71	39.5	14.0	354
1971-72	37.8	13.1	347
1972-73	37.6	12.8	340
1973-74	37.4	13.7	366
1974-75	37.3	13.8	370
1975-76	38.4	14.6	378
1976-77	40.4	17.3	428
1977-78	36.5	15.8	433
1978-79	36.5	16.9	463
1979-80	35.6	15.5	435
1980-81	34.7	22.3	643
1981-82	33.0	20.8	630
1982-83	31.0	19.5	667
1983-84	31.5	20.7	657
1984-85	28.8	20.2	701
1985-86	35.7	25.8	723
1986-87	28.0	20.1	718
1987-88	27.5	20.0	708
1988-89	25.2	18.8	746
1989-90	24.7	18.5	761
1990-91	24.8	18.2	734
1991-92	23.6	16.7	708
1992-93	22.2	20.0	901
1993-94	23.0	33.3	1448
1994-95	21.0	18.8	895
1995-96	20.3	14.6	719
1996-97	20.4	14.6	716
1997-98	34.6	27.6	798
1998-99	28.8	22.9	795
1999-2K	24.0	19.1	796
2000-01	13.3	10.6	797
2001-02	13.0	10.3	792
2002-03	8.8	6.8	773
2003-04	6.0	5.0	833
2004-05	3.4	2.9	853
2005-06	8.9	6.9	775
2006-07	9.8	8.4	857
2007-08	9.8	8.4	857
2008-09	7.7	6.3	818
2009-10	10.3	10.3	991
2010-11	3.8	5.1	1354
2011-12	2.7	3.6	1319
2012-13	3.1	3.2	1042
2013-14	1.5	4.0	2697
2014-15*	0.6	0.8	1375

Source: http://www.iipr.res.in

Appendix 45: Area, Production & Yield of Total Pulses in Madhya Pradesh

MADHYA PRADESH

Year	A	P	Y
1970-71	4245.8	1991.6	496
1971-72	4328.3	2353.1	544
1972-73	4361.1	2255.6	517
1973-74	4784.3	2074.3	434
1974-75	4439.1	2233.2	503
1975-76	4793.7	2539.6	530
1976-77	4748.5	1910.0	402
1977-78	4640.4	2044.4	441
1978-79	4730.6	2123.0	449
1979-80	4923.3	1606.6	326
1980-81	4576.3	2010.8	439
1981-82	4856.7	2473.1	509
1982-83	5129.0	2608.0	508
1983-84	4972.0	2702.3	544
1984-85	4834.9	2343.6	485
1985-86	5142.2	2610.4	508
1986-87	4769.4	2493.7	523
1987-88	4727.0	2531.9	536
1988-89	4709.0	2488.0	528
1989-90	4535.8	2505.5	552
1990-91	5012.6	3103.9	619
1991-92	4528.4	2792.4	617
1992-93	4745.4	2898.2	611
1993-94	4871.4	3264.6	670
1994-95	5194.8	3653.8	703
1995-96	5176.9	3097.7	598
1996-97	5025.1	3544.0	705
1997-98	5014.8	3282.2	655
1998-99	5023.3	3573.4	720
1999-2K	4938.0	3780.5	766
2000-01	3554.2	2275.4	640
2001-02	4170.2	3224.6	773
2002-03	4137.5	2375.5	574
2003-04	4585.4	3488.0	761
2004-05	4519.7	3429.2	759
2005-06	4284.9	3232..6	754
2006-07	4108.1	3203.1	780
2007-08	4026.2	2453.6	609
2008-09	4559.8	3683.1	808
2009-10	4940.5	4304.6	871
2010-11	5178.0	3391.0	655
2011-12	5179.0	4160.1	803
2012-13	5314.4	5165.9	972
2013-14	5395.8	4644.3	861
2014-15*	5548.9	5224.2	941

Source: http://www.iipr.res.in

Appendix 46: Area, Production & Yield of Total Pulses in Maharashtra

MAHARASHTRA

Year	A	P	Y
1970-71	2490.7	775.9	312
1971-72	2079.7	642.9	309
1972-73	1826.8	451.0	247
1973-74	2750.1	977.4	355
1974-75	2708.0	1041.6	385
1975-76	2911.7	1168.3	401
1976-77	2805.3	977.1	348
1977-78	2842.3	1012.5	356
1978-79	2802.3	1034.1	369
1979-80	2665.5	1037.6	389
1980-81	2804.4	831.2	296
1981-82	2752.2	1039.1	378
1982-83	2574.2	962.9	374
1983-84	2961.1	1257.7	425
1984-85	2833.4	1114.3	393
1985-86	2859.7	1164.1	407
1986-87	2836.1	975.0	344
1987-88	3032.6	1414.0	466
1988-89	3330.4	1729.7	519
1989-90	3289.5	1734.5	527
1990-91	3257.3	1443.9	443
1991-92	3008.4	932.6	310
1992-93	3350.4	1829.1	546
1993-94	3432.8	2205.3	642
1994-95	3595.6	1697.9	472
1995-96	3305.0	1639.2	496
1996-97	3325.0	2036.8	613
1997-98	3260.0	1187.8	364
1998-99	3499.5	2254.9	644
1999-2K	3597.7	2206.1	613
2000-01	3557.3	1637.4	460
2001-02	3388.0	1881.0	555
2002-03	3569.8	2058.1	577
2003-04	3446.1	1960.0	569
2004-05	3384.0	1664.0	492
2005-06	3432.0	2005.0	584
2006-07	3828.0	2304.0	602
2007-08	4056.0	3024.0	746
2008-09	3082.0	1656.0	537
2009-10	3376.0	2370.0	702
2010-11	4070.0	3146.0	773
2011-12	3186.0	2215.0	695
2012-13	3274.0	2306.0	704
2013-14	3953.0	3169.0	802
2014-15*	2977.0	1805.9	607

Source: http://www.iipr.res.in

Appendix 47: Area, Production & Yield of Total Pulses in Manipur

MANIPUR

Year	A	P	Y
2000-01	6.1	3.2	525
2001-02	6.0	3.1	517
2002-03	5.4	2.6	481
2003-04	6.3	3.2	508
2004-05	5.8	3.0	517
2005-06	8.6	4.5	523
2006-07	8.6	4.5	523
2007-08	14.5	7.2	497
2008-09	12.9	6.5	504
2009-10	14.5	7.2	497
2010-11	27.0	24.2	897
2011-12	28.5	26.9	942
2012-13	30.3	28.4	937

Source: http://www.iipr.res.in

Appendix 48: Area, Production & Yield of Total Pulses in Meghalaya

MEGHALAYA

Year	A	P	Y
1976-77	1.5	0.9	600
1977-78	1.6	1.0	625
1978-79	1.6	1.7	1063
1979-80	2.1	1.5	714
1980-81	2.2	1.5	682
1981-82	2.4	1.6	667
1982-83	2.9	2.0	690
1983-84	2.9	2.1	724
1984-85	3.3	2.8	848
1985-86	2.7	2.3	852
1986-87	3.2	2.7	844
1987-88	3.2	2.4	750
1988-89	3.0	2.3	767
1989-90	3.2	2.4	750
1990-91	3.2	2.4	750
1991-92	3.3	2.4	727
1992-93	3.3	2.4	727
1993-94	3.3	2.5	758
1994-95	3.3	2.4	727
1995-96	3.3	2.5	758
1996-97	3.3		
1997-98	3.3	6.7	1634
1998-99	4.6	9.3	1755
1999-2K	6.2	7.3	1780
2000-01	4.7	3.9	1625
2001-02	4.7	3.9	1444
2002-03	4.6	5.0	1087
2003-04	4.7	4.5	865
2004-05	4.7	4.8	1297
2005-06	4.8	7.9	1215
2006-07	3.9	5.8	1160
2007-08	4.0	2.7	529
2008-09	4.5	3.6	900
2009-10	4.0	6.5	1667
2010-11	4.4	6.1	1534
2011-12	4.2	5.3	1389
2012-13	3.6	3.7	1028

Source: http://www.iipr.res.in

Appendix 49: Area, Production & Yield of Total Pulses in Mizoram

MIZORAM

Year	A	P	Y
1976-77	0.4	0.1	250
1977-78	0.4	0.1	250
1978-79	0.3	0.1	333
1979-80	0.3	0.1	333
1980-81	0.3	0.1	333
1981-82	0.3	0	0
1982-83	1.0	0	0
1983-84	0.9	0	0
1984-85	0.9	0	0
1985-86	1.1	0	0
1986-87	0.4	0.5	1250
1987-88	0.5	1.0	2000
1988-89	0.6	1.1	1833
1989-90	0.2	0.7	3500
1990-91	4.0	6.6	1650
1991-92	3.4	6.9	2029
1992-93	4.0	7.9	1975
1993-94	5.4	9.8	1815
1994-95	5.2	9.9	1904
1995-96	4.5	7.0	1556
1996-97	4.3		
1997-98	4.1	6.7	1634
1998-99	5.3	9.3	1755
1999-2K	4.1	7.3	1780
2000-01	2.4	3.9	1625
2001-02	2.7	3.9	1444
2002-03	4.6	5.0	1087
2003-04	5.2	4.5	865
2004-05	3.7	4.8	1297
2005-06	6.5	7.9	1215
2006-07	5.0	5.8	1160
2007-08	5.1	2.7	529
2008-09	4.0	3.6	900
2009-10	3.9	6.5	1667
2010-11	4.0	6.1	1534
2011-12	3.8	5.3	1389
2012-13	3.1	3.3	1065

Source: http://www.iipr.res.in

Appendix 50: Area, Production & Yield of Total Pulses in Nagaland

NAGALAND

Year	A	P	Y
1976-77	2.4	1.6	667
1977-78	4.6	2.3	500
1978-79	4.6	2.3	500
1979-80	1.0	0.5	500
1980-81	1.0	0.5	500
1981-82	0.6	0.3	500
1982-83	0.7	0.4	571
1983-84	0.7	1.0	1429
1984-85	3.9	3.1	795
1985-86	8.8	5.2	591
1986-87	3.6	2.2	611
1987-88	6.7	3.7	552
1988-89	6.1	5.5	902
1989-90	8.9	9.2	1034
1990-91	13.6	12.0	882
1991-92	14.0	15.0	1071
1992-93	9.1	7.5	824
1993-94	11.0	10.0	909
1994-95	10.5	7.9	752
1995-96	10.9	11.8	1083
1996-97	**13.7**		
1997-98	16.7	12.5	749
1998-99	18.8	12.5	665
1999-2K	17.0	13.5	794
2000-01	25.5	21.1	827
2001-02	35.0	29.7	849
2002-03	30.0	28.0	933
2003-04	33.5	33.5	1000
2004-05	31.6	25.2	797
2005-06	31.0	39.7	1281
2006-07	37.5	45.0	1200
2007-08	35.0	41.6	1189
2008-09	33.0	39.7	1203
2009-10	33.5	34.7	1036
2010-11	34.4	36.4	1058
2011-12	31.8	34.7	1091

Source: http://www.iipr.res.in

Appendix 51: Area, Production & Yield of Total Pulses in Odisha

ODISHA

Year	A	P	Y
1970-71	844.8	466.6	552
1971-72	860.9	390.2	453
1972-73	939.8	488.8	520
1973-74	962.6	475.0	493
1974-75	1002.8	428.4	427
1975-76	1133.3	526.7	465
1976-77	967.4	409.6	423
1977-78	1370.2	678.8	495
1978-79	1566.2	791.2	505
1979-80	1651.3	566.8	343
1980-81	1725.5	886.1	514
1981-82	1767.3	944.3	534
1982-83	1642.6	921.6	561
1983-84	1718.3	1054.8	614
1984-85	1649.8	910.8	552
1985-86	1814.9	1031.3	568
1986-87	1968.2	1057.6	537
1987-88	2022.0	1040.0	514
1988-89	1957.5	1066.6	545
1989-90	2013.3	1116.6	555
1990-91	1957.9	1085.8	555
1991-92	2142.9	1132.9	529
1992-93	849.3	362.4	427
1993-94	972.8	498.6	513
1994-95	929.2	410.0	441
1995-96	929.7	431.1	464
1996-97	658.8	225.9	343
1997-98	786.1	286.5	364
1998-99	721.9	249.3	345
1999-2K	684.0	264.8	387
2000-01	604.3	212.7	352
2001-02	713.7	284.2	398
2002-03	548.7	194.4	354
2003-04	714.8	272.7	382
2004-05	642.5	249.6	388
2005-06	809.2	336.3	416
2006-07	791.0	351.8	445
2007-08	859.0	383.5	446
2008-09	804.9	387.3	481
2009-10	867.2	399.4	461
2010-11	852.0	414.0	486
2011-12	757.0	357.2	471
2012-13	827.2	424.4	513
2013-14	780.8	419.3	537
2014-15*	811.0	445.3	549

Source: http://www.iipr.res.in

Appendix 52: Area, Production & Yield of Total Pulses in Puducherry

PUDUCHERRY

Year	A	P	Y
1976-77	4.2	2.2	524
1977-78	5.6	3.2	571
1978-79	5.9	3.2	542
1979-80	6.1	3.4	557
1980-81	6.3	3.1	492
1981-82	3.5	2.0	571
1982-83	2.1	0.6	286
1983-84	1.8	0.7	389
1984-85	3.2	2.4	750
1985-86	3.3	1.7	515
1986-87	6.2	4.0	645
1987-88	4.6	4.4	957
1988-89	5.6	3.8	679
1989-90	10.0	7.2	720
1990-91	9.7	4.9	505
1991-92	9.7	6.1	629
1992-93	12.1	5.7	471
1993-94	9.8	5.3	541
1994-95	11.7	6.7	573
1995-96	10.5	5.6	533
1996-97	6.0		
1997-98	5.7	2.0	351
1998-99	6.0	3.1	517
1999-2K	5.8	3.9	672
2000-01	6.0	3.6	600
2001-02	1.3	0.7	538
2002-03	2.0	0.7	350
2003-04	2.6	0.9	346
2004-05	3.8	1.2	316
2005-06	3.7	1.2	324
2006-07	3.7	1.2	324
2007-08	4.8	0.4	83
2008-09	2.5	0.5	200
2009-10	2.1	0.3	154
2010-11	2.6	1.3	499
2011-12	2.4	1.0	417

Source: http://www.iipr.res.in

Appendix 53: Area, Production & Yield of Total Pulses in Punjab

PUNJAB

Year	A	P	Y
1970-71	413.7	309.1	747
1971-72	384.9	305.2	793
1972-73	380.3	294.0	773
1973-74	429.8	352.3	820
1974-75	327.7	244.6	746
1975-76	438.5	402.3	917
1976-77	392.5	331.1	844
1977-78	400.2	345.5	863
1978-79	409.7	312.6	763
1979-80	305.0	196.6	645
1980-81	337.5	199.9	592
1981-82	320.3	158.9	496
1982-83	208.0	121.6	585
1983-84	197.8	135.6	686
1984-85	203.7	151.4	743
1985-86	224.7	203.7	907
1986-87	224.9	178.9	795
1987-88	157.7	98.3	628
1988-89	170.8	130.4	763
1989-90	136.1	99.1	728
1990-91	146.4	108.4	740
1991-92	102.1	79.6	780
1992-93	102.2	74.6	730
1993-94	101.4	80.7	796
1994-95	103.1	90.5	878
1995-96	102.7	84.0	818
1996-97	97.6	80.1	821
1997-98	87.9	60.0	683
1998-99	77.5	50.7	654
1999-2K	66.0	45.4	688
2000-01	60.0	44.4	740
2001-02	53.7	36.0	670
2002-03	43.3	33.9	783
2003-04	47.9	39.4	823
2004-05	39.6	31.7	801
2005-06	32.6	26.2	804
2006-07	31.9	27.1	850
2007-08	28.6	23.0	804
2008-09	23.9	21.7	908
2009-10	20.1	18.0	895
2010-11	26.0	24.0	923
2011-12	62.8	53.3	849
2012-13	64.4	53.0	823
2013-14	45.4	39.6	872
2014-15*	59.0	54.4	922

Source: http://www.iipr.res.in

Appendix 54: Area, Production & Yield of Total Pulses in Rajasthan

RAJASTHAN

Year	A	P	Y
1970-71	3615.9	1777.3	492
1971-72	3703.7	1317.7	356
1972-73	3140.2	994.9	317
1973-74	3607.9	1279.1	355
1974-75	3571.6	960.9	269
1975-76	4478.4	2083.2	465
1976-77	4242.0	2002.1	472
1977-78	4037.0	2034.4	504
1978-79	3602.2	1890.0	525
1979-80	3030.6	843.7	278
1980-81	3147.2	1169.7	372
1981-82	3786.4	1532.9	405
1982-83	3533.0	1570.0	444
1983-84	3701.7	1679.4	454
1984-85	3376.9	1382.5	409
1985-86	3890.6	1766.9	454
1986-87	3207.0	946.2	295
1987-88	1796.0	472.0	263
1988-89	2935.9	1622.5	553
1989-90	2939.5	1155.6	393
1990-91	3682.8	1718.8	467
1991-92	2830.7	916.9	324
1992-93	3440.7	1475.9	424
1993-94	3328.0	1071.1	322
1994-95	3601.9	1965.5	546
1995-96	3573.9	1455.8	407
1996-97	3760.0	1844.6	491
1997-98	4389.4	2634.9	600
1998-99	4643.8	2444.2	526
1999-2K	2473.1	890.9	360
2000-01	2374.8	731.5	308
2001-02	3357.3	1426.1	425
2002-03	1802.9	484.5	269
2003-04	3860.9	2278.5	590
2004-05	3571.1	1337.4	375
2005-06	3444.6	898.1	261
2006-07	3207.6	1481.3	462
2007-08	3869.9	1552.8	401
2008-09	3672.5	1826.4	497
2009-10	3501.0	713.7	204
2010-11	4710.0	3216.0	683
2011-12	4451.1	2360.0	530
2012-13	3245.6	1956.6	603
2013-14	4197.7	2490.9	593
2014-15*	3574.4	2169.8	607

Source: http://www.iipr.res.in

Appendix 55: Area, Production & Yield of Total Pulses in Sikkim

SIKKIM

Year	A	P	Y
1981-82	4.9	3.9	796
1982-83	7.8	5.5	705
1983-84	8.5	7.1	835
1984-85	9.0	9.3	1033
1985-86	9.5	10.5	1105
1986-87	14.5	14.9	1028
1987-88	15.3	7.9	516
1988-89	17.1	7.2	421
1989-90	27.0	25.5	944
1990-91	24.2	22.8	942
1991-92	14.5	11.0	759
1992-93	12.1	10.3	851
1993-94	10.1	8.6	851
1994-95	10.6	12.1	1142
1995-96	11.3	9.1	805
1996-97	**6.7**		
1997-98	6.7	5.9	881
1998-99	6.7	5.6	836
1999-2K	6.7	6.1	910
2000-01	6.1	5.2	852
2001-02	6.5	5.6	862
2002-03	7.1	6.6	930
2003-04	7.1	6.8	958
2004-05	6.9	6.6	957
2005-06	6.8	6.1	897
2006-07	6.8	6.1	897
2007-08	12.5	11.6	928
2008-09	12.6	11.8	937
2009-10	13.2	12.9	977
2010-11	13.2	11.9	899
2011-12	6.5	5.9	903

Source: http://www.iipr.res.in

Appendix 56: Area, Production & Yield of Total Pulses in Tamil Nadu

TAMIL NADU

Year	A	P	Y
1970-71	481.6	116.5	242
1971-72	525.1	153.7	293
1972-73	621.7	192.0	309
1973-74	663.3	204.8	309
1974-75	464.2	114.6	247
1975-76	470.9	127.4	271
1976-77	545.3	176.5	324
1977-78	582.3	196.4	337
1978-79	616.1	219.0	355
1979-80	605.8	194.9	322
1980-81	532.6	146.7	275
1981-82	583.5	183.3	314
1982-83	600.0	210.0	350
1983-84	707.5	246.4	348
1984-85	617.8	209.9	340
1985-86	840.1	322.3	384
1986-87	689.2	286.9	416
1987-88	930	363.0	390
1988-89	890.6	384.0	431
1989-90	974.2	411.8	423
1990-91	863.4	348.2	403
1991-92	795.5	331.2	416
1992-93	739.0	342.7	464
1993-94	689.9	276.4	401
1994-95	691.3	340.0	492
1995-96	577.3	233.1	404
1996-97	581.6	232.8	400
1997-98	591.5	244.2	413
1998-99	637.1	304.3	478
1999-2K	860.4	370.8	431
2000-01	687.9	312.7	455
2001-02	734.9	313.9	427
2002-03	527.4	182.2	345
2003-04	536.9	200.8	374
2004-05	537.0	245.6	410
2005-06	525.3	177.0	337
2006-07	536.5	290.5	541
2007-08	609.8	185.0	303
2008-09	536.1	164.5	307
2009-10	534.7	204.2	382
2010-11	728.0	296.0	407
2011-12	780.6	399.5	512
2012-13	507.6	209.9	413
2013-14	815.8	613.8	752
2014-15*	886.9	365.3	412

Source: http://www.iipr.res.in

Appendix 57: Area, Production & Yield of Total Pulses in Tripura

TRIPURA

Year	A	P	Y
1976-77	5.6	2.6	464
1977-78	5.6	2.1	375
1978-79	4.7	2.0	426
1979-80	4.6	2.0	435
1980-81	5.1	2.3	451
1981-82	5.4	1.9	352
1982-83	5.3	2.6	491
1983-84	6.3	2.7	429
1984-85	6.1	2.7	443
1985-86	6.0	2.7	450
1986-87	8.8	4.1	466
1987-88	12.6	5.9	468
1988-89	13.1	6.9	427
1989-90	16.7	9.1	545
1990-91	17.1	9.3	544
1991-92	16.8	9.3	554
1992-93	18.2	9.6	527
1993-94	17.4	9.3	534
1994-95	14.9	8.0	537
1995-96	12.7	6.7	528
1996-97	9.8		
1997-98	9.3	5.4	581
1998-99	7.5	4.1	547
1999-2K	7.5	4.4	587
2000-01	10.1	5.9	584
2001-02	8.5	5.5	647
2002-03	8.8	5.5	625
2003-04	8.4	5.2	619
2004-05	8.9	5.5	618
2005-06	8.9	5.6	629
2006-07	8.1	5.3	654
2007-08	6.8	4.7	691
2008-09	6.1	4.4	718
2009-10	6.3	4.5	713
2010-11	7.4	5.2	706
2011-12	8.6	6.0	697

Source: http://www.iipr.res.in

Appendix 58: Area, Production & Yield of Total Pulses in Uttar Pradesh

UTTAR PRADESH

Year	A	P	Y
1970-71	3742.8	3069.2	824
1971-72	3524.7	2919.9	828
1972-73	3509.2	2923.0	833
1973-74	3476.8	1849.6	532
1974-75	3150.3	2185.0	694
1975-76	3155.3	2656.7	842
1976-77	3047.9	2628.8	862
1977-78	2989.0	2421.0	810
1978-79	3016.2	2336.1	775
1979-80	2941.8	1555.2	529
1980-81	2862.3	2523.6	882
1981-82	3043.9	2262.9	743
1982-83	2977.0	2542.0	854
1983-84	2797.3	2487.2	889
1984-85	2853.3	2688.9	942
1985-86	3158.1	2811.6	890
1986-87	3079.6	2627.9	853
1987-88	3090.0	2388.0	793
1988-89	2912.6	2657.5	912
1989-90	3000.0	2413.4	804
1990-91	3040.2	2771.9	912
1991-92	2875.5	2522.1	877
1992-93	2922.0	2526.9	865
1993-94	2874.6	2516.0	875
1994-95	2831.9	2479.2	875
1995-96	2829.9	2189.3	774
1996-97	2832.8	2625.4	927
1997-98	2753.7	2285.2	830
1998-99	2810.0	2323.6	827
1999-2K	2685.9	2598.8	968
2000-01	2691.7	2160.3	803
2001-02	2682.5	2377.0	886
2002-03	2677.7	2202.0	822
2003-04	2698.4	2400.3	890
2004-05	2803.7	2375.0	847
2005-06	2750.7	2231.5	811
2006-07	2724.3	1975.1	725
2007-08	2156.0	1976.9	917
2008-09	2223.3	1998.1	899
2009-10	2540.7	1901.4	748
2010-11	2427.0	2012.0	829
2011-12	2448.0	2426.0	991
2012-13	2367.0	2332.0	985
2013-14	2305.0	1697.4	736
2014-15*	2522.0	2197.2	871

Source: http://www.iipr.res.in

Appendix 59: Area, Production & Yield of Total pulses in Uttarakhand

UTTARAKHAND

Year	A	P	Y
2000-01	28.5	21.8	765
2001-02	29.9	19.0	635
2002-03	46.6	29.1	624
2003-04	42.0	29.0	690
2004-05	45.0	28.0	622
2005-06	61.0	36.0	590
2006-07	53.0	34.0	642
2007-08	63.0	50.0	794
2008-09	64.0	39.0	609
2009-10	64.0	46.0	719
2010-11	77.0	67.0	870
2011-12	55.0	49.0	891
2012-13	61.0	51.3	841
2013-14	65.0	56.5	869
2014-15*	63.1	54.1	857

Source: http://www.iipr.res.in

Appendix 60: Area, Production & Yield of Total Pulses in West Bengal

WEST BENGAL

Year	A	P	Y
1970-71	664.9	375.0	564
1971-72	579.1	317.0	531
1972-73	545.6	284.8	522
1973-74	659.7	349.6	530
1974-75	682.3	375.5	550
1975-76	727.2	410.1	564
1976-77	619.4	351.5	567
1977-78	564.4	334.2	592
1978-79	561.0	267.4	477
1979-80	559.3	304.5	544
1980-81	523.8	237.8	454
1981-82	419.9	224.5	535
1982-83	376.0	198.0	525
1983-84	396.9	258.0	650
1984-85	377.4	221.1	586
1985-86	420.6	264.1	628
1986-87	353.0	199.2	564
1987-88	362.3	226.2	628
1988-89	308.3	208.1	675
1989-90	338.5	211.9	626
1990-91	313.6	193.1	616
1991-92	269.4	174.7	648
1992-93	275.3	198.6	721
1993-94	268.5	170.6	635
1994-95	226.7	134.6	594
1995-96	210.4	141.1	671
1996-97	231.5	171.5	741
1997-98	220.5	151.6	688
1998-99	202.9	126.1	621
1999-2K	275.4	211.5	768
2000-01	274.0	219.3	800
2001-02	248.6	174.9	704
2002-03	241.3	167.6	695
2003-04	251.9	211.7	840
2004-05	225.7	167.1	740
2005-06	222.0	174.3	785
2006-07	219.1	154.0	703
2007-08	186.1	147.6	793
2008-09	182.6	128.5	704
2009-10	181.9	150.3	826
2010-11	192.0	161.0	839
2011-12	206.1	187.8	911
2012-13	201.9	192.3	952
2013-14	286.8	241.7	843
2014-15*	246.7	233.0	945

Source: http://www.iipr.res.in

Appendix 61: All India Area, Production & Yield of Chickpea

CHICKPEA

Year	A	P	Y
1950-51	7.57	3.65	482
1951-52	6.83	3.39	496
1952-53	7.26	4.21	580
1953-54	7.97	4.83	606
1954-55	9.25	5.62	608
1955-56	9.78	5.42	554
1956-57	9.67	6.23	644
1957-58	9.09	4.89	538
1958-59	10.08	7.02	697
1959-60	10.33	5.62	544
1960-61	9.28	6.25	674
1961-62	9.57	5.79	605
1962-63	9.19	5.36	583
1963-64	9.35	4.50	481
1964-65	8.87	5.78	651
1965-66	8.02	4.22	527
1966-67	8.00	3.62	453
1967-68	8.26	5.97	723
1968-69	7.11	4.31	607
1969-70	7.75	5.55	715
1970-71	7.84	5.20	663
1971-72	7.91	5.08	642
1972-73	6.97	4.54	651
1973-74	7.76	4.10	528
1974-75	7.04	4.02	570
1975-76	8.32	5.88	707
1976-77	7.97	5.42	680
1977-78	7.97	5.41	678
1978-79	7.71	5.74	745
1979-80	6.99	3.36	481
1980-81	6.58	4.33	657
1981-82	7.87	4.64	590
1982-83	7.40	5.29	715
1983-84	7.16	4.75	663

1984-85	6.91	4.56	661
1985-86	7.80	5.79	742
1986-87	6.98	4.53	649
1987-88	5.77	3.63	629
1988-89	6.81	5.13	753
1989-90	6.47	4.22	652
1990-91	7.52	5.36	712
1991-92	5.58	4.12	739
1992-93	6.45	4.42	684
1993-94	6.36	4.98	783
1994-95	7.54	6.44	853
1995-96	7.12	4.98	700
1996-97	6.85	5.57	813
1997-98	7.56	6.13	811
1998-99	8.47	6.80	803
1999-2K	6.15	5.12	833
2000-01	5.19	3.86	744
2001-02	6.42	5.47	853
2002-03	5.91	4.24	717
2003-04	7.05	5.72	811
2004-05	6.71	5.47	815
2005-06	6.93	5.60	808
2006-07	7.49	6.33	845
2007-08	7.54	5.75	762
2008-09	7.89	7.06	895
2009-10	8.17	7.48	915
2010-11	9.21	8.22	896
2011-12	8.32	7.70	912
2012-13	8.52	8.83	1036
2013-14	9.93	9.53	960
2014-15		7.59*	

Source: http://www.iipr.res.in

Appendix 62: All India Area, Production & Yield of Pigeonpea

PIGEONPEA

Year	A	P	Y
1950-51	2.18	1.72	788
1951-52	2.45	1.83	748
1952-53	2.40	1.70	709
1953-54	2.40	1.86	775
1954-55	2.40	1.72	715
1955-56	2.29	1.86	814
1956-57	2.29	1.99	867
1957-58	2.36	1.48	625
1958-59	2.47	1.70	688
1959-60	2.43	1.70	700
1960-61	2.43	2.07	849
1961-62	2.45	1.37	559
1962-63	2.45	1.58	645
1963-64	2.52	1.38	549
1964-65	2.58	1.89	733
1965-66	2.56	1.73	678
1966-67	2.52	1.13	448
1967-68	2.67	1.74	653
1968-69	2.53	1.82	718
1969-70	2.67	1.84	690
1970-71	2.66	1.88	709
1971-72	2.35	1.68	718
1972-73	2.42	1.93	795
1973-74	2.65	1.41	532
1974-75	2.53	1.83	725
1975-76	2.67	2.10	786
1976-77	2.57	1.73	672
1977-78	2.63	1.93	735
1978-79	2.64	1.89	716
1979-80	2.73	1.76	643
1980-81	2.84	1.96	689
1981-82	3.00	2.24	745
1982-83	2.93	1.99	680
1983-84	3.22	2.58	801

1984-85	3.16	2.58	819
1985-86	3.18	2.44	767
1986-87	3.15	2.27	722
1987-88	3.33	2.28	685
1988-89	3.49	2.72	779
1989-90	3.60	2.75	763
1990-91	3.59	2.41	673
1991-92	3.63	2.13	588
1992-93	3.58	2.33	652
1993-94	3.53	2.69	762
1994-95	3.31	2.14	644
1995-96	3.45	2.31	670
1996-97	3.51	2.66	756
1997-98	3.36	1.85	551
1998-99	3.44	2.71	787
1999-2K	3.43	2.69	786
2000-01	3.63	2.25	618
2001-02	3.33	2.26	679
2002-03	3.36	2.19	651
2003-04	3.52	2.35	670
2004-05	3.52	2.35	667
2005-06	3.58	2.74	765
2006-07	3.56	2.31	650
2007-08	3.73	3.08	826
2008-09	3.38	2.27	671
2009-10	3.47	2.46	711
2010-11	4.42	2.86	655
2011-12	4.04	2.65	656
2012-13	3.89	3.02	776
2013-14	3.90	3.17	813
2014-15		2.71*	

Source: http://www.iipr.res.in

Appendix 63: All India Area, Production & Yield of Urdbean

URDBEAN

Year	A	P	Y
1965-66	1.84	0.55	300
1966-67	1.86	0.52	280
1967-68	1.69	0.52	305
1968-69	1.84	0.53	289
1969-70	1.93	0.56	290
1970-71	2.07	0.66	318
1971-72	1.87	0.53	286
1972-73	1.96	061	313
1973-74	2.37	0.74	314
1974-75	2.17	0.67	309
1975-76	2.16	0.76	350
1976-77	2.07	0.69	334
1977-78	2.17	0.75	344
1978-79	2.38	0.73	320
1979-80	2.72	0.76	278
1980-81	2.83	0.96	339
1981-82	2.78	1.01	364
1982-83	2.71	0.96	355
1983-84	3.00	1.19	398
1984-85	2.91	1.16	400
1985-86	3.19	1.24	389
1986-87	3.15	1.24	394
1987-88	3.02	1.29	428
1988-89	3.32	1.62	488
1989-90	3.36	1.57	467
1990-91	3.48	1.65	473
1991-92	3.42	1.50	438
1992-93	3.02	1.53	506
1993-94	3.05	1.40	460
1994-95	2.80	1.14	407
1995-96	2.80	1.32	471
1996-97	3.01	1.35	484
1997-98	3.06	1.38	421
1998-99	3.12	1.35	433

1999-2K	2.94	1.33	453
2000-01	3.01	1.29	431
2001-02	3.30	1.50	454
2002-03	3.55	1.47	415
2003-04	3.42	1.47	430
2004-05	3.17	1.33	419
2005-06	2.94	1.23	417
2006-07	3.07	1.44	470
2007-08	3.19	1.46	457
2008-09	2.67	1.17	440
2009-10	2.96	1.24	418
2010-11	3.26	1.76	534
2011-12	3.30	1.77	555
2012-13	3.19	1.90	596
2013-14	3.06	1.70	555
2014-15		1.70*	

Source: http://www.iipr.res.in

Appendix 64: All India Area, Production & Yield of Mungbean

MUNGBEAN

Year	A	P	Y
1965-66	1.83	0.53	288
1966-67	1.87	0.50	269
1967-68	1.98	0.59	297
1968-69	1.98	0.52	264
1969-70	1.67	0.40	237
1970-71	2.07	0.70	339
1971-72	1.84	0.56	306
1972-73	1.96	0.52	267
1973-74	2.38	0.80	334
1974-75	2.29	0.65	284
1975-76	2.52	0.80	317
1976-77	2.40	0.80	331
1977-78	2.44	0.87	357
1978-79	2.55	0.88	344
1979-80	2.59	0.70	269
1980-81	2.84	0.98	344
1981-82	2.85	1.06	372
1982-83	2.63	1.06	403
1983-84	3.06	1.37	447
1984-85	2.84	1.05	371
1985-86	3.00	1.18	392
1986-87	3.06	1.08	352
1987-88	2.95	1.21	411
1988-89	3.29	1.42	432
1989-90	3.35	1.32	393
1990-91	3.36	1.38	413
1991-92	3.49	1.28	368
1992-93	2.99	1.41	471
1993-94	2.94	1.23	419
1994-95	2.90	1.07	369
1995-96	2.71	0.82	303
1996-97	3.06	1.32	430
1997-98	3.02	0.95	316
1998-99	3.05	1.16	380

1999-2K	2.91	1.07	371
2000-01	3.01	1.03	340
2001-02	3.09	1.11	360
2002-03	3.01	0.87	288
2003-04	3.55	1.71	475
2004-05	3.34	1.06	317
2005-06	3.11	0.95	340
2006-07	3.19	1.12	349
2007-08	3.73	1.52	409
2008-09	2.84	1.03	364
2009-10	3.07	0.69	225
2010-11	3.55	1.80	512
2011-12	3.43	1.63	498
2012-13	2.75	1.19	436
2013-14	3.38	1.61	474
2014-15		1.39*	

Source: http://www.iipr.res.in

Appendix 65: All India Area, Production & Yield of Mothbean

MOTHBEAN

Year	A	P	Y
1976-77	2.00	0.49	247
1977-78	1.73	0.43	250
1978-79	1.40	0.21	149
1979-80	1.28	0.08	63
1980-81	1.52	0.21	137
1981-82	1.43	0.21	145
1982-83	1.41	0.22	153
1983-84	1.54	0.42	271
1984-85	1.45	0.27	186
1985-86	1.47	0.10	70
1986-87	1.39	0.09	67
1987-88	0.89	0.05	53
1988-89	1.28	0.44	346
1989-90	1.42	0.32	224
1990-91	1.49	0.47	312
1991-92	1.31	0.15	112
1992-93	1.50	0.40	270
1993-94	1.52	0.21	140
1994-95	1.46	0.40	273
1995-96	1.37	0.20	145
1996-97	1.38	0.37	265
1997-98	1.27	0.36	281
1998-99	1.02	0.16	162
1999-00	0.89	0.06	70
2000-01	1.10	0.18	162
2001-02	1.39	0.34	242
2002-03	0.52	0.04	80
2003-04	1.59	0.83	520
2004-05	1.53	0.22	144
2005-06	1.32	0.18	133
2006-07	1.26	0.23	180
2007-08	1.36	0.44	327
2008-09	1.30	0.35	268
2009-10	1.32	0.05	41
2010-11	1.65	0.80	486
2011-12	1.37	0.47	346
2012-13			
2013-14			

Source: http://www.iipr.res.in

Appendix 66: All India Area, Production & Yield of Lathyrus

LATHYRUS

Year	A	P	Y
1970-71	1.67	0.84	504
1971-72	1.04	0.49	473
1972-73	1.49	0.56	377
1973-74	1.71	0.70	410
1974-75	1.45	0.75	517
1975-76	1.72	0.87	506
1976-77	1.52	0.52	344
1977-78	1.58	0.64	406
1978-79	1.60	0.62	386
1979-80	1.08	0.35	324
1980-81	1.35	0.57	418
1981-82	1.31	0.50	381
1982-83	1.19	0.51	432
1983-84	1.23	0.55	445
1984-85	1.18	0.50	428
1985-86	1.29	0.58	452
1986-87	1.06	0.49	460
1987-88	1.02	0.49	474
1988-89	0.90	0.44	488
1989-90	0.93	0.48	514
1990-91	1.06	0.56	530
1991-92	0.96	0.57	598
1992-93	0.88	0.51	580
1993-94	0.97	0.62	641
1994-95	0.99	0.60	605
1995-96	0.93	0.42	455
1996-97	0.88	0.55	623
1997-98	0.85	0.24	288
1998-99	0.66	0.41	621
1999-00	0.76	0.57	756
2000-01	0.52	0.33	639
2001-02	0.69	0.44	640
2002-03	0.57	0.35	617
2003-04	0.71	0.49	698
2004-05	0.64	0.30	474
2005-06	0.63	0.35	552
2006-07	0.62	0.38	615
2007-08	0.60	0.38	640
2008-09	0.53	0.34	636
2009-10	0.45	0.31	698
2010-11	0.52	0.35	675
2011-12	0.48	0.32	667
2012-13	0.58	0.43	742

Source: http://www.iipr.res.in

Appendix 67: All India Area, Production & Yield of Kulthi (Horsegram)

KULTHI (HORSEGRAM)

Year	A	P	Y
1976-77	1.73	0.48	278
1977-78	2.00	0.70	350
1978-79	2.18	0.74	340
1979-80	2.19	0.66	304
1980-81	2.12	0.62	291
1981-82	2.04	0.72	352
1982-83	1.97	0.64	324
1983-84	1.93	0.76	393
1984-85	1.87	0.63	336
1985-86	1.80	0.61	339
1986-87	1.76	0.66	378
1987-88	1.87	0.69	370
1988-89	1.54	0.55	360
1989-90	1.60	0.65	404
1990-91	1.46	0.59	403
1991-92	1.46	0.60	410
1992-93	1.24	0.49	400
1993-94	1.04	0.43	411
1994-95	1.19	0.44	373
1995-96	1.06	0.40	374
1996-97	0.97	**0.25**	263
1997-98	0.98	**0.24**	248
1998-99	0.92	**0.25**	273
1999-00	0.97	0.39	406
2000-01	0.81	0.35	429
2001-02	0.84	0.35	418
2002-03	0.77	0.24	308
2003-04	0.71	0.23	323
2004-05	0.74	0.28	378
2005-06	0.65	0.26	407
2006-07	0.72	0.27	378
2007-08	0.59	0.27	462
2008-09	0.51	0.21	400
2009-10	0.54	0.25	453
2010-11	0.51	0.26	518
2011-12	0.45	0.21	464

Source: http://www.iipr.res.in

❑❑❑